SOME PREFIXES AND SUFFIXES USED IN MICROBIOLOGY (Continued)

Prefix or Suffix	Meaning	Example
erythro-	red	*erythro*cyte
eu-	good, well, beneficial	*Eu*bacterium (beneficial bacterium)
ex-	out, away from	*ex*cretion (discharge of waste products)
exo-	outside	*exo*toxin (toxin found outside a cell)
extra-	outside of	*extra*cellular
febr-	fever	*febr*ile
fil-	thread	*fil*iform (thread-like)
flav-	yellow	*flav*oprotein (a protein containing a yellow enzyme)
-form	shape	col*iform* (having the shape of *Escherichia coli*)
-fug(e)	flee, avoid	centri*fugal* (receding from the center to the periphery)
-gam-, -gamy	marriage	*gam*ete (a male or female reproductive cell)
-gen-	produce, originate	anti*gen* (a substance that incites production of antibodies)
germ-	bud, a growing thing in its early stages	*germ*inate
glyc-	sweet	*glyc*emia (the presence of sugar in the blood)
haem-, hem-	blood	*hem*oglobin
hapt-	touch, seize	*hapt*en (a reactive portion of an antigen)
hetero-	other	*hetero*graft (transplanted tissue from an individual to another species)
homo-	common, same	*homo*logous (corresponding in structure or origin, as a lock and key)
hydr-	water	de*hydr*ate
hyper-	excessive, above	*hyper*sensitive (abnormally sensitive)
hypo-	under	*hypo*tonic (having low osmotic pressure)
infra-	beneath	*infra*orbital (beneath the eye socket)
iso-	same, equal	*iso*hemagglutinin (an antibody that agglutinates erythrocytes of other individuals of the same species)
-itis	inflammation	aden*itis* (inflammation of a gland)
leuko-	white	*leuko*cyte (white blood cell)
-logy	science, treatise	bio*logy* (the science of life)
ly-, lys-, lyt-	loosen, dissolve	hemo*lysis* (dissolution of erythrocytes)
meso-	middle	*meso*philic (preferring moderate temperatures)
meta-	change	*meta*stasis (transfer of disease from one organ or part to another)
micro-	small, one millionth part	*micro*scope, *micro*meter (10^{-6} meter)
milli-	one thousandth part	*milli*meter (10^{-3} meter)
mito-	thread	*mito*chondrion (small, rod-shaped or granular body in cytoplasm)
mono-	single	*mono*trichous (having a single flagellum)
multi-	many	*multi*nuclear (having many nuclei)
muta-	change	*muta*genic (causing genetic change)
myc-	fungus	*myc*otic (caused by a fungus)
-myel-	marrow	*myel*ocyte (bone marrow cell)

Table continued on back cover.

MICROBIOLOGY

FOURTH EDITION

PHILIP L. CARPENTER

Professor Emeritus of Microbiology
University of Rhode Island

W. B. SAUNDERS COMPANY   Philadelphia   London   Toronto

W. B. Saunders Company: West Washington Square
Philadelphia, Pa. 19105

1 St. Anne's Road
Eastbourne, East Sussex BN21 3UN, England

1 Goldthorne Avenue
Toronto, Ontario M8Z 5T9, Canada

Library of Congress Cataloging in Publication Data

Carpenter, Philip L

Microbiology.

1. Microbiology.   I. Title.

QR41.2.C37 1977   576   76-27056

ISBN   0-7216-2438-3

Listed here is the latest translated edition of this
book together with the language of the translation
and the publisher.

Spanish (*2nd Edition*)—Editorial Interamericana, S.A., de C.V.,
                      Nuéva, Mexico

Cover and title page illustration is a scanning electron micrograph of microcolonies of *Leucothrix mucor* on the carapace of a lobster larva. Courtesy John Sieburth, from *Microbial Seascapes,* Baltimore, University Park Press, 1975.

Microbiology   ISBN 0-7216-2438-3

Print No.:   9   8   7   6   5   4   3   2

PREFACE

The ecologic problems arising from pollution of the environment by wastes of all kinds—detergents, insecticides, hydrocarbons, hot water, and radioactive materials—make it imperative that intelligent citizens have a background in the basic sciences. Inasmuch as biologic forms, from viruses through the most highly evolved plants and animals, are constructed of chemical elements that obey the laws of chemistry and physics, it is necessary to be grounded in these physical sciences before anything more than a superficial, descriptive acquaintance with living organisms can be achieved. The student will therefore profit most from his course in microbiology if he has some knowledge of biology and chemistry, preferably including organic chemistry; otherwise the instructor should review basic principles so that the student will not be "lost" in the areas of microbial physiology and genetics.

Like its predecessors, the fourth edition of *Microbiology* is intended for the student who is being introduced to the field. This may be his only course in microbiology, or he may later take advanced, specialized courses. His primary interest may be in any of the various branches of microbiology; or he may be interested in general biology, molecular biology, medicine, nursing, pharmacy, medical technology, agriculture, home economics, secondary education, or liberal arts.

The organization of the fourth edition is essentially the same as that of the third edition. The chapter on systematic bacteriology has been completely rewritten in accordance with the eighth edition of *Bergey's Manual of Determinative Bacteriology,* and its suggested nomenclature has been followed throughout the text. The chapter on bacterial metabolism has been restructured, and the chapters on bacterial genetics and immunity have been up-dated.

I am indebted to Dr. Norris P. Wood and Dr. Paul S. Cohen of the University of Rhode Island for assistance in revision of the chapters on bacterial metabolism and bacterial genetics. The many suggestions that have been offered, either directly or through the staff of the W. B. Saunders Company, have been very useful and are greatly appreciated. My special thanks go to Mr. Richard Lampert, formerly Biology Editor of the W. B. Saunders Company, for his suggestions and help. Numerous individuals and companies have kindly provided illustrative material, as noted in the figure legends. Special thanks are due to Dr. John McN. Sieburth, Professor of Oceanography and Microbiology at the University of Rhode Island, for the use of several scanning electron micrographs, notably the photograph of *Leucothrix* on the cover and title page.

It is a particular pleasure to acknowledge the important role of my wife, Helen E. Carpenter, who endures calmly the long and antisocial silences of continuous preoccupation and then patiently finds her way through interlineated rough draft and garbled dictation and transcribes it accurately into legible manuscript.

PHILIP L. CARPENTER

WAKEFIELD, RHODE ISLAND

A LETTER
TO STUDENTS

Before you read the first chapter, I would like to welcome you to a new field of study and to wish you pleasure and profit from it. I would also like to give you a bird's-eye view of what is ahead and offer a few suggestions that may help you get the most from your study.

Microbiology is not just a book or a course in college. It is the study of small organisms, which have many of the same attributes as higher forms of life. By observing test tube experiments with microorganisms we can learn many things about how other organisms function.

Glance through the Table of Contents. You will notice that this book is divided into five sections. The first section introduces you to the study of microorganisms—what they are, how they were studied in the early days of microbiology, and how they are studied now.

The second section deals with the biology of the more primitive microorganisms, particularly the bacteria—what they look like and how they are grouped, how they secure energy and building materials, how they make cellular components, how they grow and die, and how their growth and death can be controlled.

In the third section you will become acquainted (or reacquainted) with other microorganisms, the higher protists: protozoa, algae, molds, and yeasts. Most of these are larger than bacteria, and although some consist of multicellular aggregates visible with the naked eye, single cells usually suffice to reproduce the entire organism. Many are handled and cultivated in the same way as bacteria.

The fourth section introduces you to the study of diseases caused by microorganisms. First, it is necessary to know what kinds of microorganisms are normally associated with the body and how other organisms produce disease. Then, since the body is not entirely passive in this situation, you should know how it resists infection. Finally, you will learn about some pathogenic agents that are usually transmitted by the four principal routes: personal contact, water and food, direct inoculation, and air.

By this time you may imagine harmful microbes at every turn, so we close with a section that includes some of the useful aspects of microbiology: the roles of microorganisms in soil and water, in dairy products and other foods, and in industry.

When you study, get an overall picture of the subject first by skimming the subheadings within each chapter. Write a *brief* topic outline of a chapter or subject. Don't memorize a lot of details first; they never make sense by themselves, but if you know the general outline, the details fit into place without much conscious effort. Learning details without knowing how they are connected with one another is like looking at a large portrait only a few inches away; all you see is an eye or a foot, and your impression of the picture is distorted and incomplete until you back away and look at it as a whole.

Many of your fellow students approach microbiology with dread, expecting to be required to memorize long lists of names and other terms that seem to mean nothing. Naturally, there are unfamiliar terms in any new subject, but, as you read, hear, *and use* them, they soon become familiar. Moreover, the words do mean something, as you can learn with only a little trouble. Usually the first Index reference to a new word contains a definition, description, or derivation. If it does not, look the word up in a standard dictionary or in a medical dictionary. Note that the Glossary, immediately following the last chapter, defines many new words and gives Latin and Greek derivations. The end papers inside the front and back covers also contain much useful information: methods of forming Latin plurals with examples from microbiology, metric measures used in science and soon to be in general daily use, and prefixes and suffixes employed in microbiology (with their aid you can often translate an unfamiliar word).

At the end of each chapter is a list of suggestions for supplementary readings, in which further information can be found. Many of these are books, and since no page citations are indicated, you will have to use the index to find the topic you wish. If you are interested in still further information, the results of current research are found in many periodicals. The *Journal of General Microbiology,* for example, is a British publication with excellent papers on the biologic activities of the various microorganisms. Its closest American counterpart is the *Journal of Bacteriology,* published by the American Society for Microbiology. Other publications of the Society include the *Journal of Virology,* the *Journal of Clinical Microbiology,* and *Infection and Immunity,* whose titles are self-explanatory, and *Applied and Environmental Microbiology,* which contains papers on antibiotics, fermentations, enzymes, and the microbiology of manufactured products and of the environment. Papers surveying recent work on a topic are found in *Bacteriological Reviews.* The *Journal of Infectious Diseases* contains articles on the causes, pathogenesis, host response, and laboratory diagnosis of diseases caused by microorganisms.

PHILIP L. CARPENTER

WAKEFIELD, RHODE ISLAND

CONTENTS

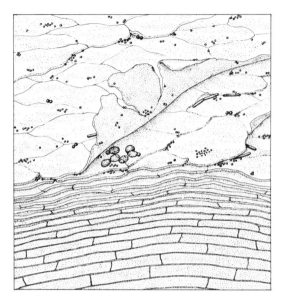

SECTION
ONE

MICROORGANISMS
AND THEIR
STUDY

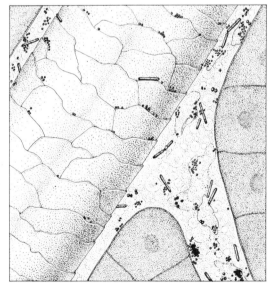

From *Life on the Human Skin* by Mary J. Marples. Copyright © 1969 by Scientific American, Inc. All rights reserved.

THE WORLD OF MICROORGANISMS   1

Bacteria, yeasts, molds, algae, protozoa, and viruses constitute a seemingly heterogeneous group of biologic entities, but they resemble one another in their small size and relative simplicity of structure and organization, and hence are called microorganisms or, as some authorities prefer, *protists* (Greek: *protista,* the very first). Study of them constitutes the science of microbiology.

Many microorganisms are unicellular, some consist of loose aggregates of independent cells, showing little if any specialization of function, and some form long filaments containing several potentially independent vital units or protoplasts within a single cell wall (e.g., *coenocytic* molds and algae). Structural simplicity does not, however, necessarily imply physiologic simplicity. Microorganisms perform the same fundamental activities within their single cells as "higher" organisms do within their many-celled structures: utilization of food and energy, formation of new protoplasm, reproduction. It is important to remember that microorganisms are essentially the same biologically as other organisms.

Their small size makes it difficult to study the anatomy of individual cells, but the homogeneity of a population of microorganisms often permits an experimenter to investigate a particular phenomenon or chemical reaction (e.g., the metabolism of an amino acid) free from some of the complications introduced by the multicellularity of larger organisms. It is for this reason that microorganisms are especially useful tools in the study of genetics, cell physiology, and other aspects of molecular biology. More-

over, the spectrum of forms comprising the microbial world includes those on the border line between obviously living organisms and obviously nonliving matter. Study of these forms provides insight into the fundamental nature of life.

CHARACTERISTICS OF LIVING SYSTEMS

After 1664, when Hooke observed that plants are composed of many smaller individual structures, it became recognized that the cell is the basic unit of life, whether plant or animal. Exactly what this basic unit comprises is largely a matter of definition. At the level of the higher plant or animal there is no disagreement: the cell is a discrete unit with common structural and chemical properties. It is bounded by a wall or membrane, which encloses cytoplasm containing proteins, deoxyribonucleic acid, and ribonucleic acid as necessary constituents. It has certain chemical activities known collectively as metabolism: the synthetic processes by which all its constituents are made from available ingredients and the transformations that convert energy from external sources into energy-rich bonds. Reproduction of the cell takes place by division, following orderly increases in the amount of its chemical components.

At the level of the smallest microorganisms there has been some difference of opinion as to what constitutes a cell or even a living organism. Viruses consist of protein and deoxyribonucleic *or* ribonucleic acid, and

3

in some cases (e.g., the myxoviruses) are bounded by a membrane, but they lack the ability to perform metabolic activities or to replicate outside the proper environment (i.e., a host cell). They are, in fact, replicated only from and in the form of their genetic material; protein may be synthesized simultaneously but separately, and the two components are assembled at a later stage.

Bacteria are approximately the same size as the substructures or *organelles* of the cells of higher plants or animals that carry out various metabolic and reproductive activities, and electron microscopy reveals hardly any organelle in the bacterial cell that is structurally identical with a similar functional unit in the cells of larger organisms. Some authors therefore consider that bacteria are closer to the viruses than they are to "cells." However, bacteria possess both kinds of nucleic acid (DNA and RNA); all their components increase and the cells divide by a process of fission; and they possess many enzymes, some of which are active in converting the energy of foodstuffs into the high-energy chemical bonds essential for biologic syntheses.

The definition of life is essentially a philosophic matter that cannot be settled by argument, although discussion serves the useful purpose of focusing attention on the complexity of the problem. Without entering the realm of controversy we can note that every biologic form contains protein and deoxyribonucleic acid or ribonucleic acid. Protein serves protective and catalytic functions, and in the latter role participates in energy transformation and transfer. Nucleic acids include the genetic material wherein is stored the information necessary to determine the chemical and physical behavior of the system. These ingredients and activities seem to constitute the irreducible minimum consistent with life.

MICROBIAL CELL TYPES: PROCARYOTIC AND EUCARYOTIC

Microbial cells are of two distinct types. The least developed type consists of the bacteria and blue-green algae and is designated *procaryotic* (Greek: *pro,* before, primordial, + *karyon,* nucleus). The more highly evolved cell type, *eucaryotic,* is found in all other biologic forms: higher algae, protozoa, fungi, as well as higher plants and animals.

Procaryotic and eucaryotic cells differ in many important respects. Procaryotic cell walls are rigid and serve to maintain the structural integrity of the cells. They consist of repeating units of mucopeptide in a three-dimensional array (Fig. 1–1). Within the cell wall is a plasma membrane surrounding the protoplasm, which contains the chromatin body or nucleus, ribosomes, chlorophyll-containing membranes or chromatophores (in photosynthetic species), and occasional granules, oil droplets, and vacuoles. One or more flagella are found on some procaryotic cells.

The procaryotic nucleus consists of a single, closed loop of naked DNA that is not bounded by a membrane. Nuclear division is accomplished by replication and splitting of the DNA, and several nuclear divisions may occur before the cell divides, with the result that multinucleate cells are found in rapidly growing cultures.

The cytoplasm of procaryotic cells is packed with ribosomes, which have a characteristic fine structure (Fig. 1–2). They are nearly spherical, hollow, and often arranged in rodlets; their function is to synthesize protein.

Photosynthesis in procaryotic cells takes place in cytoplasmic bodies that differ from the chloroplasts of eucaryotic cells. In blue-green algae, these consist of leaflike lamellar plates or *thylakoids,* distributed throughout the cytoplasm (Fig. 1–3). The chromatophores of photosynthetic bacteria vary from one species to another. They can be viewed as parts of the cytoplasmic membrane, modified in various specific ways. In some species they consist of invaginations of the cytoplasmic membrane, extending inward in the form of connected vesicles and bulged tubules, and comprise 40 to 50 per cent of the cytoplasm (Fig. 1–4). Some species have a system of parallel tubes and others

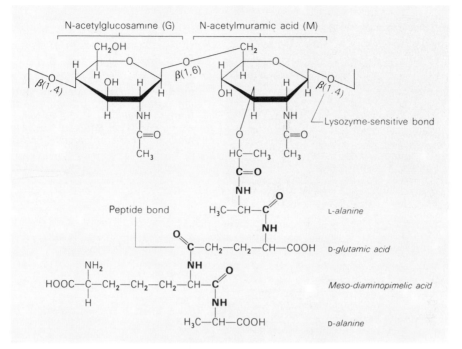

Figure 1–1. *The repeating units that comprise the cell wall mucopeptide of a procaryotic organism,* Escherichia coli. *N-acetylglucosamine (G) and N-acetylmuramic acid (M) alternate in chains, which are cross-linked by peptide bonds (shown in boldface). (From Brock: Biology of Microorganisms. Englewood Cliffs, N.J., Prentice-Hall, Inc., 1970.)*

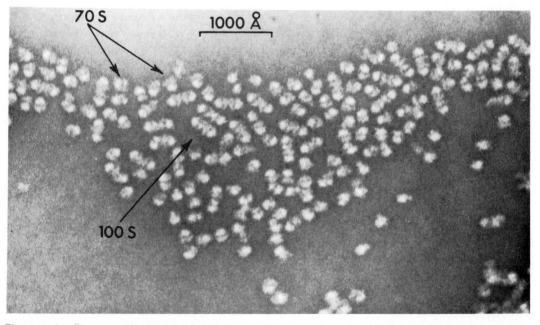

Figure 1–2. *Electron micrograph of microsomal particles (ribosomes) of* E. coli *negatively stained with phosphotungstic acid. This is a mixture of two kinds of particles: (70S) monomers, containing two unequal subunits, and (100S) dimers, composed of two monomers joined at their smaller subunits (200,000X). (From Huxley and Zubay, J. Molec. Biol., 2:14, 1960.)*

Figure 1–3. *Electron micrograph of a cell of the blue-green alga,* Anacystis montana, *showing parallel arrays of photosynthetic membranes (thylakoids). (From Echlin, P., in Gibbs, B. M., and D. A. Shapton (eds.): Identification Methods for Microbiologists, Part B. New York, Academic Press, 1968.)*

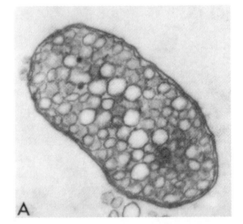

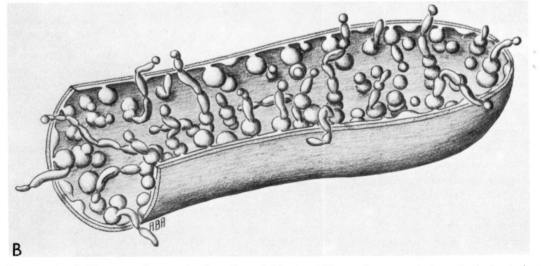

Figure 1–4. A, *Electron micrograph of section of* Rhodospirillum rubrum, *a photosynthetic bacterium, showing vesicular chromatophores (40,000X).* B, *Artist's reconstruction of the three-dimensional appearance of the chromatophore membrane system of* R. rubrum. *(From Holt, S. C., and A. G. Marr, J. Bacteriol., 89:1402–1412, 1965.)*

have an irregular network of branched and bulged tubes. In still others there is a lamellar membrane system, resembling that of the blue-green algae, near the periphery of the cytoplasm and sometimes associated with vesicles. The photosynthetic apparatus in procaryotes is comparatively simple. Since this apparatus is connected with the cytoplasmic membrane, the integrity of the membrane is essential for the photosynthetic process.

Respiratory activities in procaryotes are performed by enzymes associated with the plasma membrane and possibly with the mesosomes, which are bounded by a membrane that seems to be continuous with the plasma membrane.

A flagellum of a procaryotic microorganism consists of a fibril composed of three subfibrils, twisted about each other in a helix like a triple-threaded screw (see Figure 5–19).

Whereas the procaryotic cell is comparatively undifferentiated and possesses few membranous structures except the plasma membrane, photosynthetic thylakoids, and possibly the mesosomes, eucaryotic cells contain many subcellular structures or organelles and are strikingly membranous in nature. The eucaryotic nucleus is enclosed in a membrane and it consists of DNA molecules that comprise the chromosomes, of which there are always more than one per nucleus. The DNA in a eucaryotic nucleus is joined to basic proteins called histones, whereas DNA of procaryotes is not. This is a distinctive difference between the two types of cells. Sexual reproduction is characteristic of most eucaryotes, and in the process of meiosis reassortment of the entire chromosome takes place. Procaryotic cells display only fragmentary evidence of a sexual type of reproduction. Meiosis does not occur, and usually only portions of the genetic information are transferred.

The membranous nature of the eucaryotic cell is illustrated by the electron micrograph of a pancreatic cell shown in Figure 1–5. Most of the cytoplasm is filled with a network of membranes, the *endoplasmic reticulum,* which is the site of much enzy-

matic activity. The granular endoplasmic reticulum is active in protein synthesis.

Respiration in eucaryotic cells is performed in the mitochondria (Fig. 1–6). These are bounded by double membranes; the inner membrane is the origin of a series of thin internal membranes in which are situated the enzymes that participate in the orderly transport of electrons from oxidizable substances to oxygen. Between these membranes are other enzymes involved in the oxidation of carbon compounds to carbon dioxide.

The cells of green plants contain chloroplasts, which are more complicated in structure and function than the chromatophores of procaryotes. They are composed of parallel layers of leaflike membranes, or lamellae, that are not connected with the bounding membranes. They contain the chlorophyll and carotenoid photosynthetic pigments and the enzymes that participate in the fixation of carbon dioxide and formation of carbohydrate. The chromatophores of procaryotes, by contrast, lack the enzymes that fix carbon dioxide. The highly organized structure of a chloroplast is illustrated in Figure 1–7.

The walls of eucaryotic cells, when present, are usually composed of relatively simple organic or inorganic substances, such as the polysaccharide, cellulose, and monosaccharide polymers like mannans and xylans. Most animal cells do not possess a rigid wall.

Membranes of eucaryotic cells contain sterols, whereas those of procaryotes usually lack them. The differences in cell wall and membrane composition are the only consistent chemical differences between procaryotes and eucaryotes.

Flagellated eucaryotic cells possess flagella of microscopic size but of "advanced" design. They are composed of 20 fibrils arranged in a distinctive pattern: nine pairs of fibrils are distributed more or less evenly around two fibrils situated near the center (Fig. 1–8). Each fibril of a eucaryotic cell is approximately the same size as each flagellum of a procaryotic organism.

Differences between procaryotic and eucaryotic cells are summarized in Table 1–1.

Text continued on page 12

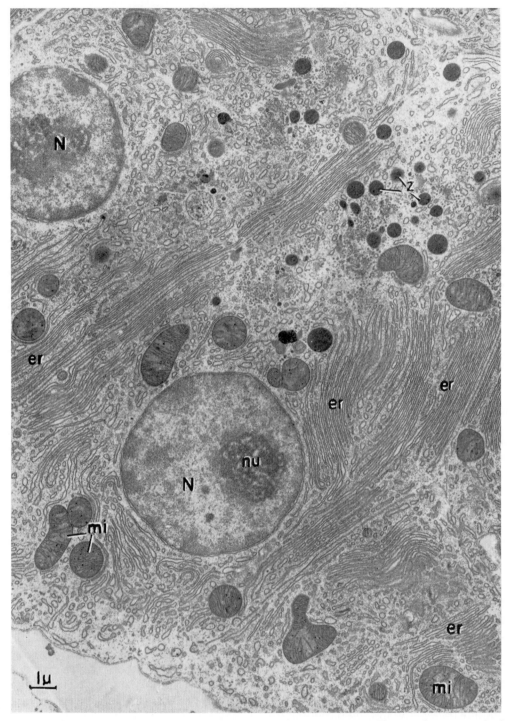

Figure 1–5. *Electron micrograph of a cell of the eucaryotic type (pancreas). The membranous endoplasmic reticulum (er) is conspicuous, as is the nucleus (N), bounded by a double membrane and containing a nucleolus (nu). The mitochondria (mi) are also filled with thin membranes. Magnification: 7000X. (Courtesy of K. R. Porter, from DeRobertis, Saez, and DeRobertis: Cell Biology, 6th ed. Philadelphia, W. B. Saunders Co., 1975.)*

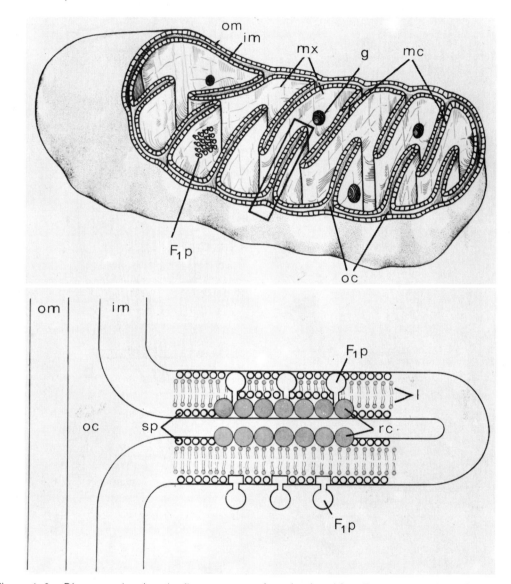

Figure 1–6. *Diagrams showing the fine structure of a mitochondrion.* Top, *a three-dimensional diagram showing the outer membrane* (om), *the inner membrane* (im), *the mitochondrial matrix* (mx), *the mitochondrial crests* (mc), *and granules* (g) *present in the matrix and containing calcium and magnesium. Also shown are the outer chamber* (oc) *and the F_1 particles* (F_1p). Bottom, *the molecular organization of a molecular crest (this corresponds to the inset in the top figure). Shown in detail are the respiratory chains* (rc), *the lipid layer* (l), *and the structural protein* (sp). *(From DeRobertis, Saez, and DeRobertis: Cell Biology, 6th ed. Philadelphia, W. B. Saunders Co., 1975.)*

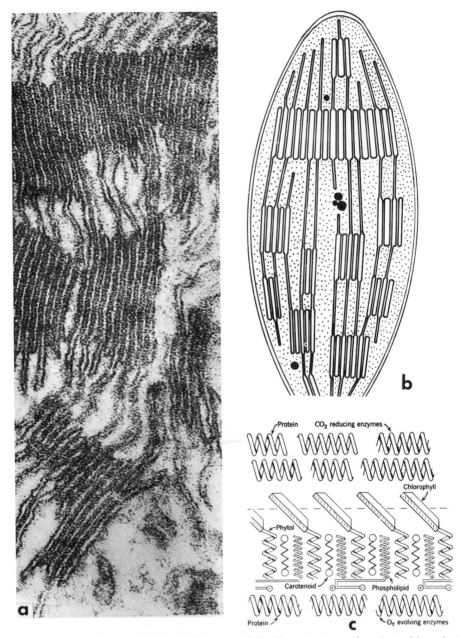

Figure 1–7. *Electron micrograph* (a) *and diagrams of the fine structure of a plant chloroplast. Its membranous nature is obvious, and the internal organization is reconstructed diagrammatically in* (b). *The postulated molecular structure of the leaflike lamellae and adjacent, undifferentiated regions are sketched at* (c). *(Courtesy of P. Weiss, from Allen: The Molecular Control of Cellular Activity. New York, McGraw-Hill, 1962.)*

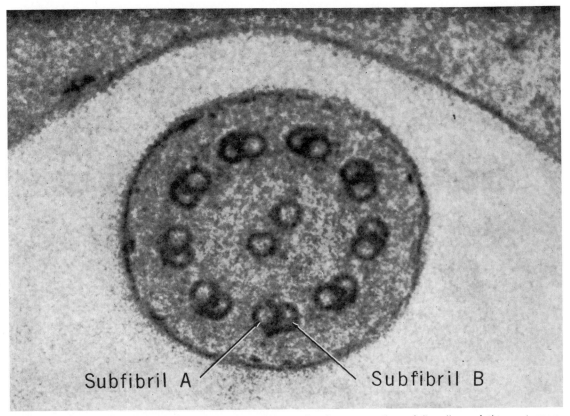

Figure 1–8. *Eucaryotic flagellum. Electron micrograph of cross-section of flagellum of the protozoon, Giardia muris. (300,000X) (From Fawcett, D. W.: An Atlas of Fine Structure. W. B. Saunders Co., Philadelphia, 1966.)*

TABLE 1–1.   **Differences Between Procaryotic and Eucaryotic Cells**

	Procaryotic Cells	**Eucaryotic Cells**
Nucleus	A single DNA molecule, not bounded by a membrane; DNA chains replicate and separate, producing two nuclear bodies; single chromosomes; DNA not complexed with histones.	True nucleus within a membrane; divides by mitosis; always more than one chromosome; DNA complexed with histones.
Respiratory system	Part of the plasma membrane and/or mesosomes.	Membranous organelles, the mitochondria.
Photosynthetic system	Cytoplasmic chromophores or membranous thylakoids.	Membranous organelles, chloroplasts.
Cell walls	Mucopeptides containing muramic acid and diaminopimelic acid or lysine.	If present, consist of simple polysaccharides or inorganic substances.
Membranes	Most species lack sterols.	Contain sterols.
Flagella	Submicroscopic, each equivalent to a single eucaryotic fibril.	Microscopic, contain 20 fibrils (nine pairs surrounding two single fibrils).
Sexual reproduction	Rare and fragmentary, usually only partial genetic reassortment.	Meiosis, with entire chromosome reassortment.

THE ORIGIN OF MICROORGANISMS

It is believed that about four and one-half billion years ago, after the Earth had formed, oceans filled the depths between the mountains, and pools of water filled depressions in the rock. The atmosphere contained water vapor, methane, ammonia, and hydrogen but no oxygen or carbon dioxide.

FORMATION OF ORGANIC COMPOUNDS

The waters of the early Earth were salty with minerals dissolved from the rocks and became the repository of chemicals formed during a period of millions of years. Chance interactions of methane molecules gave rise to short carbon chains and later to longer ones. Further reactions of these hydrocarbons with water yielded sugars, glycerin, and fatty acids. Reaction of carbon compounds with ammonia and water produced amino acids. Some of these reactions occurred in the atmosphere, facilitated by electric discharges or by ultraviolet light that reached the Earth from the sun unhampered by gaseous oxygen, which was absent at that time. The feasibility of forming amino acids by this kind of process was demonstrated in 1953 by Miller. He subjected a mixture of water vapor, methane, ammonia, and hydrogen to the electric discharge from an induction coil in a closed glass system and after a few hours detected glycine, α-alanine, and other amino acids (Fig. 1–9).

Another essential type of compound that formed before life appeared contains nitrogen as well as carbon within a ring structure. Purines and pyrimidines are the principal compounds of this type, and they are present in the nucleic acids of all living cells.

Ultimately these first organic compounds reacted with one another to form molecules of greater size and complexity. These processes were slow, but time was not significant because there were no organisms to utilize and destroy the organic substances as they arose. Some compounds doubtless decomposed, but the steady input of solar energy

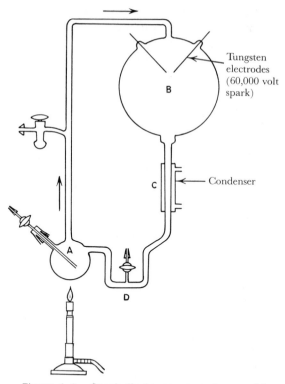

Figure 1–9. *Spark-discharge apparatus used by Miller to produce organic compounds from water vapor, hydrogen, methane, and ammonia gases. The water in flask A is boiled to promote circulation. Products formed by the discharge in B are condensed at C and return to A via U-tube, D, which prevents backflow. (Redrawn from Miller, Ann. New York Acad. Sci., 69:261, 1957.)*

promoted gradual accumulation of substances of increasing complexity, and the various reactions accelerated as the constituents were concentrated by evaporation of water or by adsorption to colloidal clay particles.

The union of amino acids with one another yielded polypeptides, and eventually protein-like macromolecules formed. The development of proteins was important because many proteins, either alone or with the aid of certain metals, possess catalytic properties and can hasten certain chemical reactions. Catalysts of this sort are known as enzymes, and reactions controlled by enzymes are typical of living systems.

The colloidal "proteinoids" may have separated from solution to form microscopic droplets ("coacervates" or "microspheres").

Whereas the original solution consisted of a uniform mixture of chemical compounds, each coacervate droplet was distinctive in composition owing to random inclusion of different compounds when coacervation took place. Some droplets were better balanced and more stable than others and hence could "survive" and develop longer. These could then change in physical and chemical behavior, increase in size, and ultimately become subject to fragmentation by various physical agents in the environment. Fragments containing a representative sampling of the chemicals of the parent coacervate might continue to "grow" and develop and fragment again.

NUCLEIC ACIDS AND THE BEGINNING OF "LIFE"

A somewhat more complicated structure resulted from the union of purines and pyrimidines with sugar and phosphate. These compounds, called nucleotides, are capable of joining each other and forming long chain-like supermolecules in the now familiar helical or coiled form (Fig. 1–10). These nucleic acids possess the remarkable ability to direct the formation of more nucleic acids of the same kind and also the formation of specific proteins. It is at this point that we can *begin* to think of inventing the word "life."

It must be understood that all these chemical reactions occurred very slowly and that many hundred million years elapsed before the first self-reproducing form appeared. At first, chance directed the union of one atom or compound with another atom or compound, and those products persisted that possessed greater stability than the separate components. The process was one of trial and error and displayed in rudimentary form features that later appeared in the evolutionary development of living organisms able to survive under conditions of biologic competition.

THE PRIMORDIAL FORM

The first living form was very simple by present standards. It probably consisted of

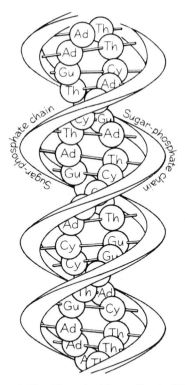

Figure 1–10. *The double coil of DNA. The purines, adenine (Ad) and guanine (Gu), and the pyrimidines, thymine (Th) and cytosine (Cy), are joined in pairs, as shown, by hydrogen bonds, and hold the molecule in a rigid structure. (From Sussman: Animal Growth and Development. Englewood Cliffs, N.J., Prentice-Hall, Inc., 1960.)*

little more than an aggregate of nucleic acid and protein, perhaps surrounded by a layer of protein and nutrient material. This description sounds very much like that of what we call viruses today, and in fact the primordial form, or "ancestral procaryote," has been called a *protovirus* (Fig. 1–11). It should be remarked that this protovirus was

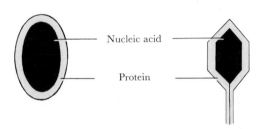

Nucleic acid

Protein

Figure 1–11. *On the left is a sketch of a postulated "protovirus," the first living object. On the right is a sketch of a bacterial virus (T2 coliphage).*

capable of securing its constituents from the organic substances created by chance in prehistoric pools and oceans, whereas modern viruses secure their constituents only from the cells of living hosts: animal, plant, or microbial.

Some authors believe that modern viruses represent degenerate forms of higher microorganisms in which the parasitic state evolved as a consequence of the ready availability of nutrient substances within host cells. It is also postulated that viruses are direct descendants of the primordial protoviruses, or that they consist of nucleic acid and protein fragments originally derived from the cells of higher organisms. Whatever their origin, present day viruses are incapable of replicating in any nonliving nutrient solution so far devised. Conditions two or three billion

years ago were much different from any yet produced in modern laboratories. Moreover, time was not a factor of importance; replication of the protovirus may have required one year, one hundred years, even one million years. Twentieth century investigators decide that their viruses are "dead" if they see no signs of reproduction within a few days; actual replication of the individual virus particle requires only a few minutes.

BIOCHEMICAL EVOLUTION

Postulated evolutionary relationships of microorganisms and higher plants and animals are shown in Figure 1–12. The ancestral procaryote that arose spontaneously almost four billion years ago was a heterotro-

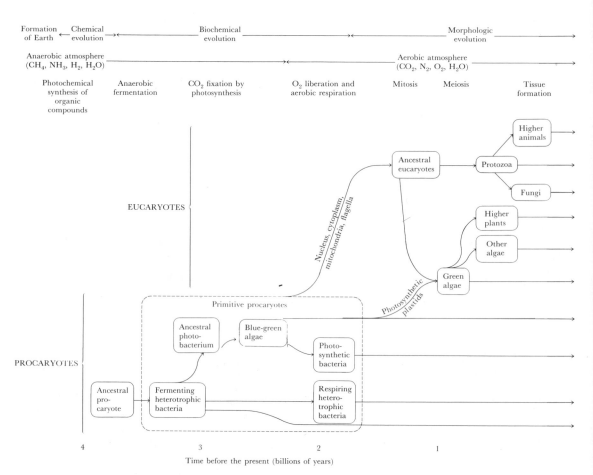

Figure 1–12. *Postulated evolutionary development of procaryotes and eucaryotes from the ancestral procaryote. (Modified from Margulis, 1968.)*

phic organism, that is, it utilized the organic compounds that had been formed by photochemical reactions in the atmosphere and waters of the early Earth. Inasmuch as there was no oxygen in the atmosphere, it secured energy from the chemical bonds of available nutrients by anaerobic processes of rearrangement known as fermentation. This mechanism is not efficient, so the microorganisms grew slowly. Some carbon dioxide was formed as a by-product. When photosynthetic and other pigments necessary for utilization and transformation of light energy into biochemical energy developed, primitive anaerobic photosynthetic bacteria appeared. They were able to fix CO_2 and produce organic compounds, utilizing hydrogen and electrons from such substances as H_2S. The gradual evolution of photosynthetic pigments and related electron transport agents permitted use of reduced nitrogen compounds such as ammonia, hydrazine, and hydroxylamine as sources of electrons and hydrogen, and eventually a system evolved for securing electrons and hydrogen from water. The blue-green algae appeared at this time. Liberation of molecular oxygen was a characteristic of this type of photosynthetic activity; this radically altered the composition of the atmosphere and paved the way for aerobic respiration and the appearance of animal life.

Because eucaryotic organisms are often much larger than procaryotes and contain organelles, such as mitochondria and chloroplasts, that are about the same size as many procaryotes, the hypothesis was advanced that eucaryotes arose through a symbiotic relationship among procaryotic forms or between ancestral eucaryotes and procaryotes. For example, if an ancestral photosynthesizing microbial cell was taken into the cytoplasm of a nonphotosynthesizing organism and continued its photosynthetic activity, the result would be an organism capable of utilizing CO_2 in the presence of light, but able to secure energy in the dark by respiration or fermentation. The photosynthetic partner would then become a chloroplast, losing its identity as a separate organism, and assume a vital role in the metabolic activity of the host cell. Similarly, if a micro-organism capable of carrying out respiration found its way into the cytoplasm of an anaerobic organism, some of the energy it secured by respiration might be utilized by the host organism. The host eventually becomes dependent on the cytoplasmic inclusion, which assumes the character of a respiratory organelle or mitochondrion.

Support for the symbiotic theory of the origin of certain photosynthetic eucaryotes is provided by the observation that the familiar photosynthetic flagellate *Euglena,* when cultivated in the presence of penicillin, no longer produces chloroplasts and becomes indistinguishable from a common protozoon, *Astasia.*

It has been postulated that all eucaryotic organisms in existence today evolved from a common ancestral eucaryote. Green algae arose about a billion years ago, when the progeny of the ancestral eucaryote acquired photosynthetic plastids, and the other modern algae and the higher plants arose from them. Another line of evolution from the ancestral eucaryote led to single-celled animals—the protozoa—and from them to higher animals, and also to modern fungi.

Since evolution is based upon the occurrence of single mutational steps accompanied by preferential multiplication of those mutants best adapted to local environmental conditions, it is not surprising that occasional organisms persist with characteristics of widely divergent forms.

MORPHOLOGIC EVOLUTION

Reproduction by means of particle or cell division restricted the size of the individual to dimensions that doubtless were chemically most efficient. Metabolic activity provided the energy for synthesis of components not immediately available in the environment.

Structural features that improved the survival powers of organisms were gradually added. Some of the most primitive microorganisms known today possess flagella and are independently motile. It is presumed that their extinct direct ancestors were also flagellate, but of course it is not known how many millions of years elapsed between the ancestral procaryote and these ancestral flagellates. Independent locomotion provided a wider choice of food supply and enabled the progeny of a given parent to migrate to a more favorable environment. Each evolutionary form might thus establish itself under those conditions best for its own survival.

Lower Plants. Ancestral green algae were presumably intermediate between the early flagellates and modern algae and other organisms. Modern algae (Fig. 1–13) include green, brown, red, and other species in which the various pigments determine the wavelengths of light that can be utilized by the plant chlorophyll. Some algae are multicellular, and the evolution of a multicellular

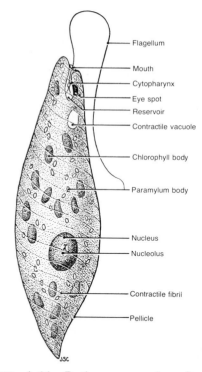

Figure 1–14. Euglena, *a modern flagellate. (From Villee and Dethier: Biological Principles and Processes. Philadelphia, W. B. Saunders Co., 1971.)*

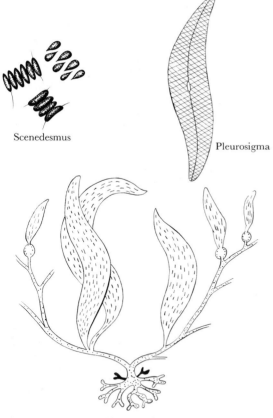

Figure 1–13. *Two small single-celled algae and one large multicellular alga. Each individual of the green alga,* Scenedesmus, *is about* $\frac{1}{2500}$ *inch long. The diatom,* Pleurosigma, *is* $\frac{1}{250}$ *inch in length.* Macrocystis, *a brown alga, may be 150 feet long and anchor in 60 feet of water; it is found on the Pacific coast of the United States.*

habit of growth doubtless led to the development of higher plants.

Animals. The evolution of animal-like forms was characterized by development of improved methods of locomotion and the use of pre-formed food materials. Some ancestral eucaryotes possessed flagella, and evolution from these primitive flagellates must have proceeded in several directions, perhaps simultaneously. One line led by direct descent to the modern flagellates (Fig. 1–14), and others led to ameboid (Fig. 1–15), ciliate, and sporozoan protozoa. There are transitional forms in existence today that possess flagella but may also move in ameboid fashion by pseudopod formation. Ciliates (Fig. 1–16) lack flagella but instead move by means of many short hairlike cilia; these structures may also help to propel food toward the gullet. Sporozoa possess an added feature with survival value, a spore or resistant stage in their life cycle. Most protozoa are single-celled, but there are some colonial types (Fig. 1–17). Higher animals may have evolved from such a multicellular

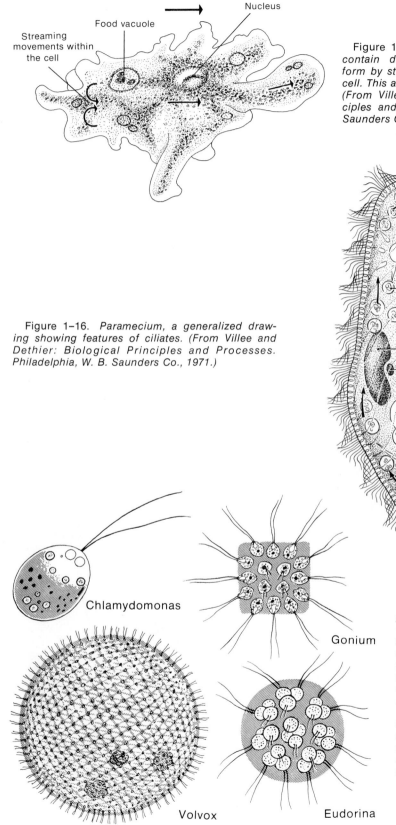

Food vacuole

Nucleus

Streaming
movements within
the cell

Figure 1–15. *An ameba. Food vacuoles contain digesting material. Pseudopodia form by streaming of cytoplasm within the cell. This animal is moving toward the right. (From Villee and Dethier: Biological Principles and Processes. Philadelphia, W. B. Saunders Co., 1971.)*

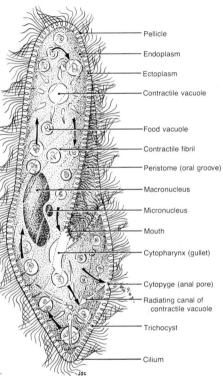

Pellicle

Endoplasm

Ectoplasm

Contractile vacuole

Food vacuole

Contractile fibril

Peristome (oral groove)

Macronucleus

Micronucleus

Mouth

Cytopharynx (gullet)

Cytopyge (anal pore)

Radiating canal of
contractile vacuole

Trichocyst

Cilium

Figure 1–16. *Paramecium, a generalized drawing showing features of ciliates. (From Villee and Dethier: Biological Principles and Processes. Philadelphia, W. B. Saunders Co., 1971.)*

Chlamydomonas

Gonium

Volvox

Eudorina

Figure 1–17. *A series of plantlike flagellate protozoa. Chlamydomonas is single-celled, often called an alga; the others are multi-celled colonial organisms. Each individual cell in the colony feeds itself. The colony as a whole is actively motile.*

Volvox is about $\frac{1}{100}$ inch in diameter; the other organisms are smaller. (From Villee, Walker, and Barnes: General Zoology, 4th ed. Philadelphia, W. B. Saunders Co., 1973.)

form, or perhaps they evolved independently as an offshoot from the line that culminated in modern protozoa.

Fungi. Fungi, including yeasts (Fig. 1–18) and molds (Fig. 1–19), may have arisen from protozoa. Most of them are nonmotile, but some produce flagellated reproductive cells. Like the protozoa, they are heterotrophic, that is, they utilize pre-formed organic nutrients. They lack chlorophyll and hence derive their energy by respiration or fermentation or both.

PROCARYOTES: BACTERIA AND BLUE-GREEN ALGAE

Modern bacteria and the blue-green algae are the only surviving procaryotes. In addition to characteristics already described, most representatives of both groups are relatively small. Their cytoplasm contains no conspicuous vacuoles and is quite immobile, in contrast to that of many larger organisms in which more or less violent activity can be seen. The mucopeptide that characterizes the cell walls and gives them rigidity is unique in composition; it contains muramic

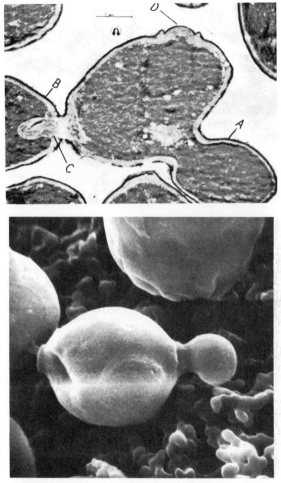

Figure 1–18. Top, *Electron micrograph of a longitudinal section through a budding yeast cell* (Saccharomyces cerevisiae). A, *A young bud with its cytoplasm still continuous with that of the mother cell.* B, *A mature bud with the developing cross-wall between mother and daughter cell.* C, *The extension of newly formed cell wall material into the cytoplasm, a phenomenon which appears to be characteristic of the later stages of the budding process.* D, *A bud scar, the surface of which is always convex. (Degree of magnification indicated by the scale line showing 1 μm at top, center.)*

Bottom, *Scanning electron micrograph of a budding yeast cell. (Original magnification 16,000X; reduced to 55% of original size.) (Courtesy of R. Albrecht and A. MacKenzie. From Frobisher et al.: Fundamentals of Microbiology, 9th ed. Philadelphia, W. B. Saunders Co., 1974.)*

Figure 1–19. *Micrograph of* Rhizopus nigricans, *a common bread mold. Enlarged about 200X. (From Sarles et al.: Microbiology. New York, Harper & Brothers, 1956.)*

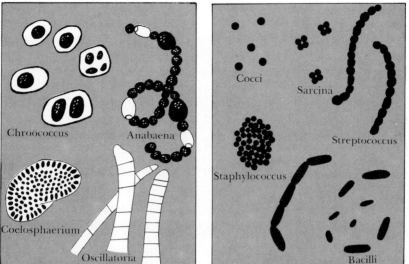

Figure 1-20. *Blue-green algae and bacteria. The algae contain chlorophyll and are photosynthetic organisms; bacteria do not contain chlorophyll.*

acid and either diaminopimelic acid or lysine. It is absent from the cells of higher algae, protozoa, fungi, plants, and animals. Many blue-green algae and bacteria resemble one another morphologically (Fig. 1–20), the principal difference being the presence or absence of chlorophyll.

NUTRITIONAL TYPES OF BACTERIA

It seems evident that the most primitive bacteria were heterotrophic anaerobic organisms and that photosynthetic forms evolved at a later date. Bacteria that secured energy by aerobic respiration obviously did not arise until plant photosynthesis had evolved. After the explosion of photosynthetic activity and growth of plant and animal life, both in the seas and ashore between 250,000,000 and 400,000,000 years ago, there was much organic material available as nutrient for bacterial cells, and presumably saprophytic bacteria multiplied extensively. Some bacteria that associated with living plants or animals acquired parasitic properties, that is, they became adapted to nutrient and other conditions associated with the living hosts, and mutant forms lacking the ability to synthesize certain essential ingredients or growth factors were able to survive.

Parasites. Extension of the parasitic habit led first to partial and ultimately to complete loss of the ability to live independently. Various cell structures were no longer formed. Eventually rickettsiae (Fig. 1–21), pleuropneumonia and pleuropneumonia-like organisms (Fig. 1–22), and the still smaller and less differentiated viruses (Fig. 1–23) appeared. It is argued, as previously mentioned, that modern parasitic viruses evolved directly from the original protovirus or that they represent nucleic acid and protein fragments derived originally from the genetic materials of the cells of higher organisms—that they are, in other words, "wild genes," which have escaped from the control of the host cell, become somewhat altered, and now can enter other similar host cells and pervert their synthetic machinery so that it manufactures virus materials instead of host materials.

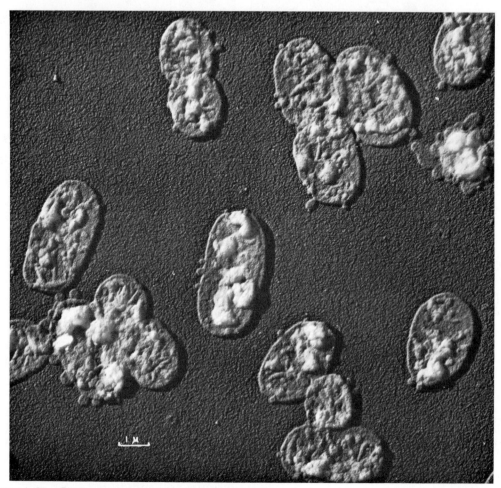

Figure 1–21. *Electron micrograph of* Rickettsia nuseum, *shadowed to show surface structure (8000X). (Courtesy of the U.S. Armed Forces Institute of Pathology.)*

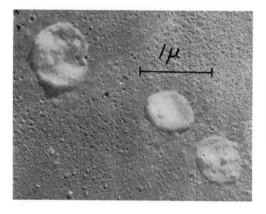

Figure 1–22. *Electron micrograph of a pleuropneumonia-like organism (PPLO) from the human female genital tract. The cells from an agar colony were shadowed with chromium. PPLO are spheroidal to ellipsoidal, lack a rigid cell wall, and are very fragile (19,000X). (From Morton et al., J. Bact., 68:707, 1954.)*

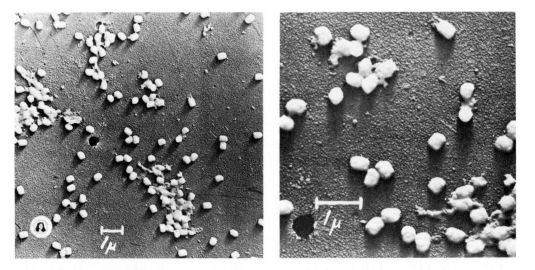

Figure 1-23. *Electron micrographs of the vaccinia virus, shadowed with gold. The picture at the right is an enlargement of the one at the left. The rectangular or brick-shaped appearance of some of the particles is obvious. (Photography by G. G. Sharp; A.S.M. LS-142.)*

MICROORGANISMS: PLANTS, ANIMALS, EITHER, NEITHER?

When Linnaeus proposed his system of classifying living organisms in the mid-eighteenth century, he was unable to decide what to do with certain microorganisms, so he assigned them to the group Chaos! While his problem may have been partly technical —the inadequacy of contemporary microscopic techniques and other methods of study—there were fundamental difficulties that modern techniques have not overcome. Chief among them is a matter of definition. What is a plant and what is an animal?

Gross observations of large animals and plants usually suffice to assign them to the proper kingdom, but classification of smaller forms requires critical analysis and perhaps even arbitrary decision as to what constitute plant and animal characteristics. The problem is compounded, as has been shown, by the borderline organisms that possess properties of both kingdoms.

It has been traditional since the time of Linnaeus to divide all living organisms into two kingdoms, Plantae and Animalia. The most obvious differences between them are their methods of securing nutrients. Plants are primarily photosynthetic organisms,

whereas animals are ingestive organisms. However, distinction should be made between photosynthetic and absorptive modes of nutrition. Green plants, as photosynthetic "producers," differ from most bacteria and all fungi, which are absorptive "reducers." Following the same line of reasoning, animals as ingestive organisms are "consumers."

There are obvious difficulties with a two-kingdom scheme of classification. In addition to the intergrading combinations of plant and animal characters already mentioned, the bacteria and blue-green algae are profoundly different from all other organisms (the contrasting characteristics of procaryotic and eucaryotic cells were discussed earlier in some detail). Moreover, there is little justification for including the fungi in the plant kingdom. They probably evolved from colorless flagellate ancestors: their organization and their reproductive structures and processes differ from those of the plants. They are apparently wholly absorptive, and as far as is known none has ever been photosynthetic.

Over 100 years ago, Haeckel (1866) proposed a third kingdom, Protista, to include the *relatively simple* organisms, such as bacteria, protozoa, fungi, and algae. These are unicellular or coenocytic, or, if multicellular, are not differentiated into separate and distinct tissue regions, as are higher plants or animals.

In 1957, Stanier and his collaborators revived use of the term *protist,* distinguishing two subgroups: the *lower protists,* which included the less developed procaryotic types (bacteria and blue-green algae), and the *higher protists,* which included the more highly evolved eucaryotic organisms (protozoa, fungi, and most algae).

A four-kingdom system of classifying organisms has been in use for many years. In 1969, Whittaker proposed a five-kingdom system based upon the distinction between procaryotic and eucaryotic forms, the type of cellular organization (unicellular or unicellular-colonial, multinucleate, multicellular),

and the mode of nutrition (photosynthetic, absorptive, ingestive). The first kingdom in this system is Monera (Greek: *moneres,* singular or unicellular), which includes procaryotic cells—unicellular or sometimes colonial; they are predominantly absorptive microorganisms, although some are photosynthetic. The Monera includes the blue-green algae and the bacteria. The second kingdom is Protista, which also consists primarily of unicellular or colonial forms, but these are eucaryotic and their modes of nutrition are diverse—photosynthetic, absorptive, ingestive, and combinations of these. Included in this kingdom are the euglenoid microorganisms, diatoms, dinoflagellates, and the various protozoa.

The Plantae are multicellular, eucaryotic,

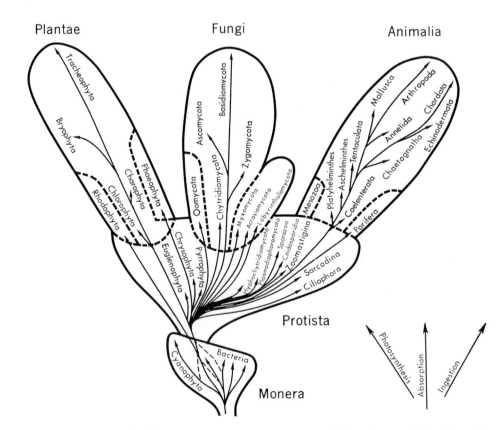

Figure 1–24. *Whittaker's five-kingdom system of classification of living forms, which is based on three levels of organization: procaryotic (Monera), eucaryotic unicellular (Protista), and eucaryotic multicellular and multinucleate. Divergence at each level is based upon the three modes of nutrition: photosynthesis, absorption, and ingestion. The Monera lack an ingestive mode of nutrition, and at the highest level (multicellular-multinucleate) the nutritional modes lead to the different kinds of organization that characterize the three higher kingdoms. (From Whittaker: New concepts of kingdoms of organisms. Science, 163:150–160, 1969.)*

photosynthetic organisms with rigid cell walls. This kingdom includes most algae, the bryophytes (liverworts and mosses), and tracheophytes (vascular plants). The fourth kingdom, Fungi, comprises organisms that are multinucleate, eucaryotic, and that have rigid walls, sometimes divided by cross-walls into a multicellular structure; their mode of nutrition is absorptive. Slime molds are included, together with the other molds, yeasts, mushrooms, toadstools, and bracket fungi. The Animalia are multicellular, eucaryotic organisms lacking rigid cell walls; their mode of nutrition is primarily ingestive, but some forms may also be absorptive. This kingdom includes all forms of animal life from mesozoa through chordates. Whittaker preferred to omit viruses from his sys-

tem, considering that they are not organisms. Some authors have accorded them kingdom status (e.g., Archetista).

The relationships between the five kingdoms proposed by Whittaker are illustrated in Figure 1–24.

According to the eighth edition of *Bergey's Manual of Determinative Bacteriology,* which describes the classification scheme most commonly used in the United States and many other parts of the world, all procaryotic organisms are included in a separate kingdom, *Procaryotae.* This consists of two divisions: Division I, The Cyanobacteria (blue-green algae) and Division II, The Bacteria. All other algae and the fungi are considered to be plants; protozoa are considered to be animals.

SUPPLEMENTARY READING

Blum, H. F.: *Time's Arrow and Evolution,* 2nd ed. Princeton, N.J., Princeton University Press, 1954.

Calvin, M., and Calvin, G. J.: Atom to Adam. *Amer. Scientist, 52*:163–186, 1964.

Keosian, J.: *The Origin of Life,* 2nd ed. New York, Reinhold Book Corporation, 1968.

Margulis, L.: Evolutionary criteria in thallophytes: a radical alternative. *Science, 161*:1020–1022, 1968.

Oparin, A. I.: *The Origin of Life on the Earth,* 3rd ed. New York, Academic Press, 1957.

Stanier, R. Y., and van Niel, C. B.: The Concept of a Bacterium. *Archiv für Mikrobiol., 42*:17–35, 1962.

Wald, G.: The Origin of Life. *Scientific American, 191*(8):44–53, August, 1954.

Whittaker, R. H.: New concepts of kingdoms of organisms. *Science, 163*:150–160, 1969.

2 EARLY DEVELOPMENT OF MICROBIOLOGY

Microbiology, like most other sciences, had its origin in curiosity. As soon as instruments, however crude or imperfect, were devised for producing magnified images of objects too small to be seen with the naked eye, they were used to examine the previously unsuspected minute organisms that populate soil, water, natural foods, body surfaces and secretions, and indeed nearly everything on Earth. At first they appeared to be of little importance and were considered merely objects for speculation: How could such small organisms live, breathe, reproduce? Where did they come from? Much later it was found that they may bring about useful or undesirable chemical changes in their environment and that some produce disease. This knowledge stimulated a tremendous burst of investigation, so the great growth of microbiology was directed along practical lines. More recently, the fact that microorganisms are excellent objects for the study of biologic phenomena in general has come to be appreciated, and microbiology is now recognized as an important branch of biology; in fact, it is often used as the vehicle for presenting general biologic principles in courses in freshman biology.

DISCOVERY OF MICROORGANISMS

The discovery of microorganisms had to await the invention of the microscope. Suitable instruments were developed during the seventeenth century by Janssen, Malpighi, Hooke, Leeuwenhoek, and others (see Figure 2–1). Early microscopes were simple instruments, often just a glass bead in some kind of holder.

As early as 1665, Hooke published a book, *Micrographia,* containing descriptions and illustrations based upon his microscopic examination of higher organisms and of filamentous fungi including molds and rusts.

Anton van Leeuwenhoek (1632–1723) was a versatile and intelligent dry goods merchant of Delft, Holland (see Figure 2–2). He was also "Chamberlain of the Council Chamber of the Worshipful Sheriffs of Delft," city surveyor, and official wine taster. These various positions paid enough to support his large family and still left time for a hobby. His hobby was lens grinding and the production and use of microscopes. His simple microscopes consisted of a single biconvex lens mounted between two pieces of metal (see Figure 2–3). Attachments were provided for focusing and for mounting insects or other large objects or drops of liquid.

Leeuwenhoek's lenses were so perfectly ground that magnifications of approximately 200 times could be obtained. With them, he found protozoa in canal and ditch water and bacteria in rain water that had been allowed to stand in a bowl for several days. Pepper infusions swarmed with bacteria. Scrapings from his teeth contained millions of "animalcules" (Figure 2–4), which horrified him until he found that all people had similar "beasties" in their

Figure 2–1. *An early compound microscope. Christopher Cook of London made these instruments about 1665 after the design of Robert Hooke. The light source is at the left with a lens for focusing the light on the specimen. (From Stanier, Doudoroff, and Adelberg: The Microbial World. Englewood Cliffs, N.J., Prentice-Hall, Inc., 1957.)*

Figure 2–2.   *Anton van Leeuwenhoek at the age of 54. He had discovered bacteria only a few years earlier. (From Dobell: Anton van Leeuwenhoek and His "Little Animals." London, Staples Press, 1932.)*

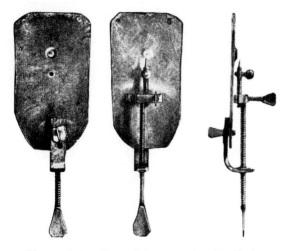

Figure 2–3. *One of Leeuwenhoek's "micro-scopes." This was really a simple magnifying lens, but its quality and method of use permitted objects as small as bacteria to be seen. The lens was mounted between two metal plates. The specimen was placed on the point of the short threaded rod; the various thumbscrews focused and positioned the object properly. (From Frobisher: Fundamentals of Microbiology, 8th ed. Philadelphia, W. B. Saunders Co., 1968.)*

mouths. Some microscopic organisms were obviously alive, as was evident from their active motility, but others merely vibrated in place, exhibiting brownian movement. Leeuwenhoek also discovered that bacteria treated with vinegar or heated to a

sufficiently high temperature "fell dead forthwith."

Leeuwenhoek's observations were recorded in more than 200 letters written to the Royal Society of London from 1673 until the time of his death. He apparently first saw bacteria about 1676.

The existence of viruses was not demonstrated until more than two centuries later, and photomicrography by electron microscopy was not possible for over 250 years.

SPONTANEOUS GENERATION CONTROVERSY

One would expect that the discovery of the first microorganisms would start an immediate, widespread further investigation of their nature and occurrence. Certainly this would happen in the twentieth century. However, communication was much slower in the seventeenth century; Leeuwenhoek's letters were presumably translated into Latin and read at meetings of the Royal Society of London but were not widely distributed, and many years elapsed before much publicity was given them. Consequently, for many decades only a few persons studied bacteria. Then it required a controversy to incite interest and experimental investigation.

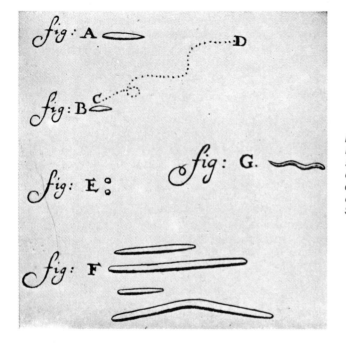

Figure 2–4. *Leeuwenhoek's drawings of bacteria from the human mouth, from his letter of Sept. 17, 1683. Rods, cocci, and a spirochete are shown. The path of motion of a motile short rod form is indicated by the dotted line. (From Frobisher: Fundamentals of Microbiology, 8th ed. Philadelphia, W. B. Saunders Co., 1968.)*

One of the first arguments about microscopic organisms has already been mentioned: the question of their origin. There were, of course, two schools of thought. One maintained that microorganisms arose spontaneously from decomposing organic materials; the other maintained that each organism was the progeny of an identical, pre-existing organism.

Needham in 1748 reported an experiment that seemed to prove that bacteria arose spontaneously where no living organisms had been before. He placed boiled mutton broth in flasks, tightly stoppered them with corks, and at intervals examined the broth with his microscope; bacteria appeared in large numbers. Certainly, he reasoned, boiling should destroy any living cells, and those that appeared later must have arisen spontaneously. Needham postulated the existence of a vegetative force that was necessary to confer life upon the nonliving ingredients of the liquid.

Spallanzani later (in 1765) repeated the experiments of Needham, but with modifications. Instead of closing the flasks with corks, Spallanzani sealed them hermetically in a flame, before heating the contents. He found that if he boiled the broth long enough, no bacteria ever appeared. The broth remained clear when observed with the naked eye and portions examined with the microscope showed no sign of life. It appeared, therefore, that Needham might not have boiled his broth long enough. More importantly, however, the corks may not have been impervious to bacteria.

The fact that bacteria entered nutrient materials such as broth from the air was demonstrated by experiments performed by Schwann (1837). He arranged flasks so that a stream of air could be passed over the broth (Fig. 2–5). The air entering some flasks passed through a tube that was heated to redness, whereas that entering control flasks was unheated. Broth in the control flasks always became cloudy within a short time and showed microscopic evidence of living organisms, but broth that was aerated with heated air remained free from microorganisms.

The proponents of spontaneous generation maintained that such drastic treatment destroyed the life-supporting property of air so that even though spontaneous generation might occur, the organisms generated would be unable to multiply. This argument was countered by Schroeder and von Dusch in 1854 when they introduced the use of cotton into microbiologic practice. They drew air into flasks of broth after passing it through cotton filters that had previously been baked in an oven (Fig. 2–6). The fact that the broth remained clear, while that in flasks that were not protected by cotton promptly became cloudy with a heavy bacterial population, clearly demonstrated that the source of microbial life was the outside air. Schroeder and von Dusch further showed that flasks of broth were protected by a wad or plug of cot-

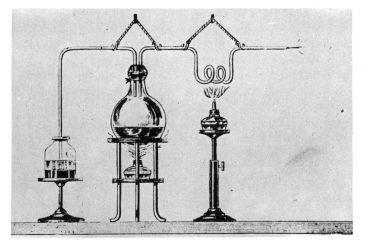

Figure 2–5. *Schwann's experiment to disprove spontaneous generation. The flask of boiled broth, center, received only air that had passed through the heated glass coil. The broth remained sterile. (From Watson-Cheyne's Antiseptic Surgery, 1882.)*

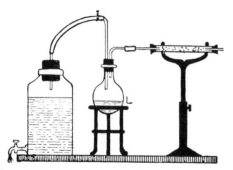

Figure 2–6. *One of the experiments of Schroeder and von Dusch (1854) that showed that microorganisms could be removed from air by a cotton filter. The suction bottle at the left drew air through the sterilized tube of cotton-wool at the right and over the sterile broth in the round flask. The broth remained free from microorganisms. (From Burdon: Textbook of Microbiology. New York, The Macmillan Co., 1958.)*

Figure 2–7. *Louis Pasteur, 1822–1895.*

ton wool in the mouth of the flask. If the flask was then boiled sufficiently, its contents remained sterile indefinitely. This is the origin of the familiar cotton plug of the bacteriology laboratory.

These experiments seemed conclusive, but Pasteur (Fig. 2–7) was starting his work on microbial fermentation at this time, and encountered opposition from the powerful school of chemists who maintained that fermentation and putrefaction were spontaneous chemical processes, and that any microbial activity was an effect rather than the cause of the observed changes. Pasteur realized that he must demonstrate conclusively the source of microorganisms in organic infusions. He became convinced that they floated in the air, perhaps upon dust particles, and that the likelihood of microbial contamination of an organic solution was

greater in a dusty atmosphere than in clean air. He secured air samples in flasks of broth whose drawn-out necks had been sealed while the flasks were boiling so that upon cooling they contained a partial vacuum. A flask was opened wherever a sample was required, air rushed in, and the flask was resealed. Then it was examined after incubation. Flasks opened in the streets of Paris always became turbid with bacteria, whereas similar flasks exposed to the dust-free air near the top of Mont Blanc rarely contained bacteria.

Pasteur also used flasks like those illustrated in Figure 2–8, to show that even filtration through cotton was not necessary to pre-

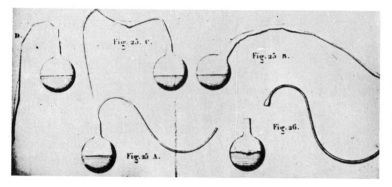

Figure 2–8. *Pasteur's open flasks in which boiled broth remained free from microorganisms because dust particles carrying them were trapped in the bends of the necks. (From Pasteur, 1861.)*

vent contamination by microorganisms. A long, curving neck permitted interchange of air between the inside and outside of the vessel, but dust particles and bacteria were trapped along the moist walls of the neck. This was proved by tilting the flask so that broth washed the inside of the neck. The broth promptly became cloudy.

The final blow to the spontaneous generation doctrine was dealt by the English physicist, Tyndall, in 1877. He had discovered that air containing no dust particles was optically empty; that is, a beam of light passed through it could not be seen, whereas each particle in a dust-laden atmosphere was clearly visible. If Pasteur were correct, optically empty air should be free from microorganisms. Tyndall constructed a box like that illustrated in Figure 2–9. The inside of the box was coated with glycerine, to which dust particles sooner or later adhered. When the box had become optically empty, Tyndall filled the test tubes with broth by means of a thistle tube and sterilized them by raising a pan of boiling brine underneath. The contents of the tubes remained in contact with the air, but as long as

it was free from dust, bacteria did not appear in the broth.

In the course of these investigations, Tyndall discovered a highly resistant bacterial structure, later known as a spore. He observed that infusions made by suspending hay in water were difficult to sterilize. Prolonged boiling was necessary or a process of intermittent sterilization that subsequently became known as tyndallization. The solution was boiled for short periods on each of three successive days and was incubated at a temperature favorable to microbial growth during the intervening time.

The spontaneous generation controversy was important for various reasons. Before the study of microorganisms could be established upon a scientific basis it was necessary to demonstrate that they arose by an orderly process and not haphazardly, at the whim of uncontrollable environmental changes. When it was shown that each microorganism was identical with its predecessor, it became possible to study its properties and compare its activities with those of other organisms. These steps were necessary before the germ theories of fermentation, putrefaction, and disease could be demonstrated.

The spontaneous generation controversy stimulated research and the development of techniques of handling microorganisms. Any argument is likely to arouse the investigative spirit of the protagonists, and this argument was no exception.

GERM THEORY OF FERMENTATION

The souring of milk and the production of alcoholic beverages have been known throughout recorded history, but the fermentation processes concerned have been understood for only about a century. Berzelius, Liebig, Wöhler, and other distinguished and influential organic chemists of the last century interpreted the transformation of sugar into lactic acid or into ethyl alcohol and carbon dioxide as purely chemical phenomena. When it was pointed out that yeasts or other microorganisms are always present, they devised explanations other than the one that

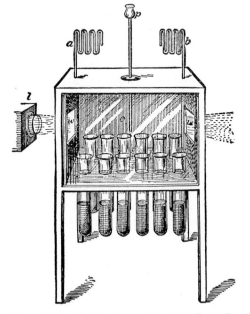

Figure 2–9. *Inside view of the box Tyndall used to demonstrate that "optically empty" air contained no microorganisms. (From Krueger and Johansson: Principles of Microbiology, 2nd ed., Philadelphia, W. B. Saunders Co., 1959.)*

eventually proved to be true. Liebig, for example, noting how readily yeast is destroyed and decomposes, postulated that the process of decomposition was communicated in some way to sugar and other substances in contact with it. Berzelius interpreted fermentation as a contact phenomenon, and Mitscherlich saw in yeast globules a catalyst similar in behavior to spongy platinum in contact with hydrogen peroxide.

None of these investigators accepted the thesis, first proposed by Cagniard de Latour, Schwann, and Kützing between 1835 and 1838, that yeasts actively transform sugar into alcohol and carbon dioxide. They postulated that in other fermentations various specific microorganisms formed characteristic end products during their growth.

Pasteur in 1857 observed the formation of lactic acid from sugar by several kinds of bacteria. He noted that a gray deposit in fermentation vessels consisted of microscopic, very short globules, occurring either singly or in small, irregular masses. These globules were much smaller than those of beer yeast. When they were transferred to a fresh nutrient solution containing sugar, yeast extract, and chalk, lactic acid was produced and the globules increased greatly in number. Pasteur demonstrated that the presence of the globules was a necessary prerequisite to lactic acid formation. He later showed that in alcoholic, acetic, butyric, and other fermentations, the typical end product appeared only when a specific microorganism was present.

A further discovery in connection with butyric fermentation was that the organisms responsible grow only in the absence of air. It was then found that alcoholic fermentation also occurs only in the absence of air, but yeast can grow in the presence of air and, in fact, grows more rapidly and abundantly with than without air. However, oxygen is toxic to the butyric bacterium. This was apparently the first indication that organisms could exist in the complete absence of oxygen—a revolutionary concept.

The germ theory of fermentation, stating that microorganisms bring about specific changes in their substrates, laid the foundation for important industrial developments.

The research necessary to prove the germ theory of fermentation also demonstrated the necessity for strict control of the various factors associated with the fermentation processes: the composition of the fermenting solution, the identity and purity of the microbial population, and incubation conditions such as temperature and aeration.

GERM THEORY OF DISEASE

From the earliest times, disease was regarded as a mysterious or even supernatural phenomenon. Ancient Greek and Roman physicians suspected that invisible, minute, particulate agents caused certain diseases and that they could be transmitted in some way or other from one individual to the next. Fracastoro in 1546 described three modes of transmission of infectious agents: direct contact, fomites (agents contaminated by the diseased individual and later handled by a healthy person), and contagion at a distance (e.g., through the air, as in the case of tuberculosis). Until the nineteenth century, however, there was no direct proof that microbes cause disease. The evidence upon which ancient authors and others through the Middle Ages postulated living, transmissible "seeds of disease" was purely epidemiologic.

The fact that certain bacteria produce disease was first clearly demonstrated by Robert Koch in 1876 (Fig. 2–10). He showed that a spore-forming organism, *Bacillus anthracis* (Fig. 2–11), was the cause of anthrax, which was then epidemic in sheep, cattle, and other domestic animals, and also occurred in man. A few years earlier (1863–1868) Davaine had demonstrated rod-shaped objects in the blood and organs of animals that had died of anthrax, and when material containing these objects was transmitted to healthy animals, the latter promptly died with symptoms characteristic of the disease. Koch passed these rodlike bodies through a long series of microscopic cultures in serum or aqueous humor and observed that they multiplied extensively in each successive culture. He therefore concluded that they were living bacilli. More-

Figure 2–10. *Robert Koch, 1843–1910. His last portrait. (From Bulloch: The History of Bacteriology. London, Oxford University Press, 1938.)*

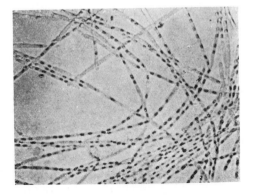

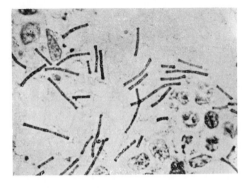

Figure 2–11. *Photographs of stained* Bacillus anthracis *taken by Koch in 1877.*

over, although organisms from the last micro-culture in the series were obviously many generations removed from those in the diseased animal, experimental animals inoculated with them died showing typical symptoms of anthrax, and bacilli with the same characteristics were found in their blood and organs. This was clear-cut demonstration of the causal relationship between the organism and the disease.

Koch had followed four experimental steps, which he subsequently stated in the form of rules that have since been known as *Koch's postulates:*

1. Find the suspected organism in all cases of the disease and demonstrate its absence in healthy individuals.

2. Isolate the organism in pure culture.

3. Reproduce the same disease in suitable experimental animals.

4. Reisolate the same organism from the artificially infected animals.

If carefully obeyed, these simple rules provide a logical basis for concluding that an organism produces a given disease (or any other characteristic change, for that matter).

There are conditions under which all four rules cannot be observed. For example, apparently healthy individuals may be "carriers" of pathogenic microorganisms. Moreover, some infectious agents, such as certain viruses, have in the past been extremely difficult to isolate and cultivate outside the natural host. In these cases indirect evidence is sometimes necessary to establish the cause of an infectious process.

Koch's statement of these rules, together with technical procedures introduced by him and his colleagues, paved the way for an enthusiastic search for the causes of infectious disease, and at least a score were isolated within the last two decades of the nineteenth century.

EARLY TECHNICAL ADVANCES

Improved techniques for handling and studying bacteria were developed rapidly. One important problem was that of isolating pure cultures from the mixed flora usually encountered in natural specimens. Koch was fortunate in his early work because *Bacillus anthracis* grows more rapidly in the animal body than most other bacteria with which it may be mixed. Consequently, serial passage from one animal to another quickly yields a pure culture in the blood and spleen. Serial transfer on laboratory media is effective when the desired organism predominates or multiplies most rapidly under the culture conditions provided. However, a method of separating the bacteria in mixtures, even when the desired organism was in the minority, was obviously needed.

As early as 1872, Schroeder had observed the formation of distinctively colored growths or colonies on the cut surfaces of potatoes. Each colony proved to contain a single kind or organism. Inoculating potato slices with mixtures of microorganisms sometimes made it possible to separate the various components of the mixture, but some organisms did not grow on potato slices.

Brefeld added gelatin to a warm nutrient broth and poured the solution onto sterile glass plates, where it was allowed to cool and solidify. Mixtures of microorganisms were smeared on the surface of the solidified medium or added to the gelatin before it was poured. The plates were incubated in a sterilized box, and bits of the various colonies that developed were transferred to other media. Since gelatin liquefies readily when warmed, the temperature could not be allowed to rise above 20° C., lest the colonies run together. Moreover, many human and animal pathogens grow slowly, if at all, at 20° C. Furthermore, some organisms digest and liquefy gelatin, whereupon their colonies mix.

These disadvantages of gelatin made the introduction of agar by Hesse in 1883 an important contribution. Hesse was a pupil of Koch, and his wife had friends in the Dutch East Indies who were familiar with the use of agar for solidifying jams and jellies. Agar is a polysaccharide extracted from various seaweeds. It is digested by very few microorganisms. It goes into solution only when heated to nearly 100° C. and remains liquid until cooled to about 42° C. It can therefore be inoculated with mixtures of microorganisms at 45° to 50° C., a temperature that most organisms can tolerate for a short time. The addition of 1.0 to 1.5 per cent agar to broth yields a satisfactory liquefiable solid culture medium.

Petri, another pupil of Koch, made what he called a minor modification of Koch's plating technique when he devised the familiar Petri dish, a flat-bottomed dish with a flat cover. The Petri dish is much more convenient to handle and use than the flat glass plates first used by Koch.

These technical advances were of greater importance than their simple nature would indicate, because they made it possible to isolate pure cultures conveniently and thus encouraged the widespread study of microorganisms.

It has been noted that Koch used the method of enrichment to isolate pathogenic bacteria by inoculating animals in which the pathogenic organisms would grow faster than other, contaminating microbes. Winogradsky and Beijerinck used selective enrichment to isolate various interesting soil bacteria, organisms that oxidize ammonia or nitrites, hydrogen sulfide or sulfur, or iron compounds. Winogradsky, for example, studied bacteria that do not grow in laboratory media containing organic compounds but can be cultivated in media containing only inorganic salts. When such a medium, with ammonium chloride as the only source of nitrogen and adjusted to pH 8.5, was inoculated with garden soil and kept in the dark at 25° to 30° C. in the presence of air, the ammonium salt was gradually oxidized to nitrate. Later Winogradsky showed that the oxidation of ammonia occurred in two steps and that two principal kinds of bacteria were responsible: (1) *Nitrosomonas* species, which oxidized ammonia to nitrites, and (2) *Nitrobacter* species, which oxidized nitrites to nitrates. A selectively enriched culture of *Nitrobacter* was obtained by inoculating with soil a salt solution containing $NaNO_2$ at pH 8.5 and incubating it in air in a dark incubator at 25° to 30° C. Cultures of other soil bacteria may be selectively enriched by varying the composition of the medium and the conditions of incubation (Fig. 2–12).

The principle of selective enrichment by control of nutrients and culture conditions (temperature, air supply, light, pH, etc.) is most important. Thoughtfully applied, this method can assist in the isolation of practically any desired organism from natural sources.

Methods for the microscopic study of microorganisms also improved greatly dur-

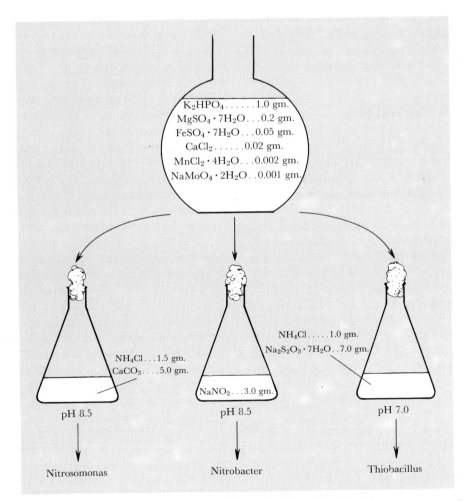

Figure 2–12. *Selective enrichment of soil bacteria in synthetic media of differing composition. To portions of the basal medium from the large flask are added the ingredients indicated in the Erlenmeyer flasks, each of which is then inoculated with a soil sample and incubated aerobically in the dark. The added substances selectively enrich certain kinds of bacteria as indicated. (All amounts shown are in grams per liter.)*

ing the last two decades of the nineteenth century. Koch devised better techniques of using the microscope and was especially skillful in photomicrography. His pictures of *Bacillus anthracis,* showing spores as well as vegetative cells, are excellent. Weigert introduced aniline dyes for staining tissues about 1875, and Koch used many of them to demonstrate the presence of bacteria. Gram devised a staining method in 1884 that distinguished certain bacteria from the tissues in which they were lodged. This technique, when later applied to pure cultures, revealed two great classes of organisms, which shortly became known as gram-positive and gram-negative, according to their ability to retain the dye gentian violet when the background was decolorized with alcohol. The Gram procedure is by far the most commonly used method of staining bacteria, and is almost always employed as a first step in identifying an unknown organism.

DISCOVERY OF VIRUSES

Pasteur suspected that the cause of hydrophobia was a submicroscopic organism that could not be cultivated outside the animal body, but the first disease clearly demonstrated to be produced by a submicroscopic agent was tobacco mosaic. This disease of tobacco plants is characterized by the appearance on the leaves of discolored spots in which the tissue dies and disintegrates, leaving holes. The disease can be transmitted by the sap of sick plants; when this fluid is spread on healthy leaves, they become diseased within two or three weeks. Iwanowski, a Russian, demonstrated in 1892 that the responsible agent passed through unglazed porcelain filters that would retain bacteria. The bacteria-free filtrate transmitted the disease as effectively as unfiltered plant sap, and the number of spots that appeared varied inversely with the extent of dilution of the filtrate. These facts made it appear that the cause of the disease was submicroscopic but particulate.

Beijerinck in 1899, apparently unaware of Iwanowski's work, made the same observations and then proceeded to demonstrate

that the infectious agent could not multiply outside the host plant, although it survived for a considerable time. He showed also that it could withstand drying and precipitation by alcohol, but was inactivated at 90° C.

The first disease of animals attributed to a virus was foot-and-mouth disease of cattle. A German commission consisting of Loeffler and Frosch, appointed to study this disease, reported in 1898 that it was caused by a filterable agent that was too small to be seen under the microscope and could not be cultivated on ordinary media.

Not only plants and animals are subject to virus infections, but many bacteria are also parasitized by viruses. A British investigator, Twort, discovered the first such agent in 1915, but the clearest observations and those that were most thoroughly pursued were made by d'Herelle, in 1917. D'Herelle observed that the feces of a bacillary dysentery patient contained an agent that passed through a porcelain filter and that, when added to a broth culture of the dysentery organism, killed and apparently dissolved the bacteria within a period sometimes as short as a few hours. If a small amount of this lysed bacterial culture was added to a second bacterial culture, killing and lysis occurred again. In his first paper, d'Herelle reported passage of the lytic agent through 50 successive transfers. He further found that when a small amount of bacteria-free lytic filtrate was mixed with bacteria and spread upon an agar medium, the sheet of bacterial growth was punctuated by holes about 1 millimeter in diameter containing no bacteria. These represented colonies of the apparently particulate agent responsible for lysis (see Figure 2–13). D'Herelle referred to the agent as *bacteriophage.* It was soon recognized that bacteriophages are viruses, and they are now commonly referred to as bacterial viruses.

A footnote to the discovery of viruses, one that was important in laying the foundation for the growing field of molecular biology, was the crystallization of tobacco mosaic virus by Stanley in 1935. This naturally stimulated controversy over the question: Is virus material living or dead? Some philosophers again proposed the doctrine of spontaneous

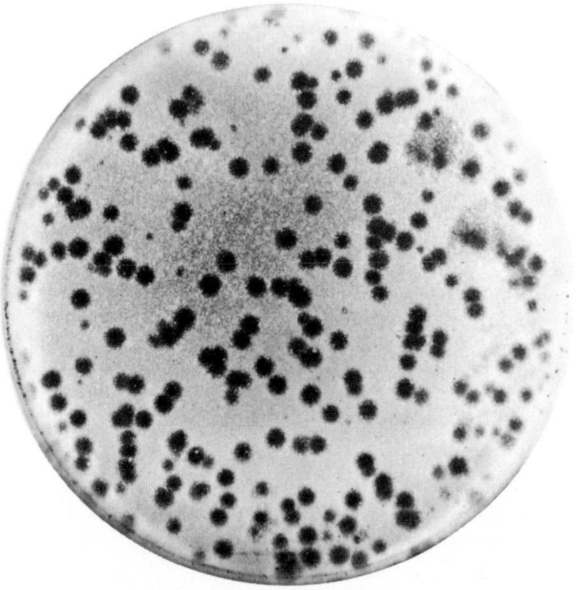

Figure 2–13. *Bacteriophage plaques. The black, fuzzy-edged spots are "colonies" of bacterial virus growing on an agar plate culture of the proper host bacterium. (Photograph by J. Kleczkowska; A.S.M. LS-145.)*

generation, especially when it was found that virus particles consist of nothing more than nucleic acid and protein.

IMMUNITY

The treatment or prevention of infectious disease, that is, disease produced by microorganisms, has always been of interest. It was recognized in ancient times that some diseases occurred only once in an individual, and that thereafter he was immune. The ancient Chinese deliberately inoculated healthy persons with pustular material from the sores of patients with mild smallpox to produce a similarly mild disease that would induce lifelong immunity.

This practice, continued through the ages, spread to other parts of the world, and in the eighteenth century Lady Mary Wortley Montague attempted to introduce it into England on her return from the Near East. She encountered resistance, in part justified, because occasionally severe cases of smallpox resulted or concurrent infection with bacteria caused serious illness. Jenner, a British physician, developed an immunizing procedure known as vaccination. Observing that milkmaids who developed pocklike sores on their hands and arms after milking cows infected with cowpox never afterwards had smallpox, he deduced that there was a relationship between the two diseases. He transferred pustular material from infected cows to the skin of humans, who shortly displayed the typical sore that follows smallpox vaccination. Those who had this reaction became immune to smallpox, as was demonstrated by deliberate inoculation of a few individuals with virulent virus.

Pasteur in 1881 devised procedures for immunization against fowl cholera and anthrax. He used "attenuated vaccines," consisting of bacteria cultivated or treated so they lost most of their disease-producing power or virulence, and immunized animals by one or more injections of the weakened organisms. Fowl cholera bacteria were attenuated by aging cultures for a number of months. Chickens inoculated with them had mild infections from which they recovered, and thereafter they were immune to infection by fully virulent chicken cholera bacteria. The anthrax organism was attenuated by cultivation at an unusually high temperature, 42° to 43° C. Both types of vaccine stimulated satisfactory immunity, and they introduced a new method of controlling disease.

The period from 1880 to 1900 was marked by the discovery of various techniques for

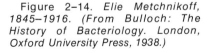

Figure 2–14. *Elie Metchnikoff, 1845–1916. (From Bulloch: The History of Bacteriology. London, Oxford University Press, 1938.)*

studying the so-called immune reactions, both *in vivo* and *in vitro*. It was also marked by controversy over the fundamental nature of immunity. Metchnikoff (Fig. 2–14) in 1884 proposed the cellular theory of immunity, based upon observations made in part on the transparent water flea *Daphnia*. These animals were subject to fatal infection by a yeast, but in some individuals ameboid cells engulfed and digested the infecting yeasts, whereupon the water fleas recovered. Metchnikoff named the ameboid cells phagocytes (Greek: *phagein,* to eat) and proposed that immunity to infectious disease is due to the action of similar cells. Such cells can readily be demonstrated in the blood and tissues of most animals.

The opposing doctrine, the humoral theory, maintained that soluble substances in the body fluids (e.g., the blood serum) are responsible for immunity. It was observed that cell-free serum of certain immune individuals was lethal to the bacteria to which they were immune. Moreover, serum from such immune individuals reacted with the corresponding bacteria in the test tube and caused them to clump together or "aggluti-

nate." Serum from a patient with typhoid fever, for example, agglutinated typhoid bacteria but not cholera bacteria, and vice versa.

Certain pathogenic bacteria produce poisons or toxins, which are responsible for the disease symptoms. Behring and Kitasato demonstrated in 1890 that the blood serum of an animal that had been immunized by the injection of diphtheria toxin contained a substance, *antitoxin,* that was capable of neutralizing the toxin. When administered to an animal before injection of the diphtheria organism or its toxin, antitoxin prevented the disease. If administered shortly after symptoms appeared, it prevented a fatal outcome. Behring and Kitasato simultaneously demonstrated the effectiveness of tetanus antitoxin.

A general term for an immune substance found in serum is *antibody*. Antibody is protein of the globulin type. The controversy between the cellular and humoral theories of immunity raged for several years and was finally resolved when Wright and Douglas showed that, although antibody is not necessary for phagocytosis, it enhances the rate and extent of phagocytosis.

CHEMOTHERAPY

The modern era of chemotherapy began with Ehrlich's search for "magic bullets," beginning in the first decade of the twentieth century (Fig. 2–15). Perhaps his most outstanding success was the compound 606, or arsphenamine, used in the treatment of syphilis. Chemotherapy is an attempt to treat infectious disease by means of chemicals that will kill or interfere with the growth of the pathogenic microorganisms but not damage the host. Inasmuch as both are living agents, and often have similar chemical compositions and metabolic activities, the borderline between toxicity to the parasite and to the host is often very thin. This is, of course, why 605 trials were unsuccessful before Ehrlich found a compound that was effective against syphilis.

Domagk, about 30 years later, tested the antibacterial activity of various chemicals of the sulfonamide type produced by the German dye industry and discovered that Prontosil was active against streptococcal infections in animals but was relatively nontoxic to them. The first human patient treated had a streptococcal septicemia that would otherwise have been fatal. Following other almost miraculous cures, a tremendous amount of research was directed toward the production of other even more effective and less toxic chemotherapeutic agents.

The mechanism of action of the sulfa drugs was deduced by Woods in 1940. He found that the action of sulfonamides was inhibited by p-aminobenzoic acid, a substance essential to the metabolism of certain bacteria. He concluded that sulfonamides inhibited the utilization of p-aminobenzoic acid, and thus the drugs interfered with a normal metabolic process and inhibited growth of the bacteria. Normal antibacterial body defenses then destroyed the organisms, permitting the patient to recover. The discovery that "competitive inhibition" is the mechanism by which these drugs operate started a search for other chemicals that behave in a similar manner—a search that is still in progress.

Another type of antibacterial agent is a natural product of cellular (usually microbial) activity. Antibiotics have been known for nearly a century but have been used only since about 1940. Fleming discovered penicillin in 1929 as the result of a common laboratory mishap: the appearance of a mold contaminant on a Petri dish culture of a staphylococcus. The significant feature of this mishap was that Fleming made note of the fact that staphylococcus colonies in the immediate vicinity of the contaminating penicillium dissolved and eventually disappeared (see Figure 2–16). His keenness in making this observation was rewarded when he found that broth cultures of the mold contained a substance that was nontoxic to animals but inhibited staphylococci and certain other bacteria. Relatively little was done with penicillin until World War II, when its effectiveness as a chemotherapeutic agent was demonstrated and manufacture on a large scale was undertaken in the United States, because the British chemical industry was hampered by wartime difficulties.

At the same time Waksman at Rutgers University and others began to search for additional antibiotics. Streptomycin, one of the earliest and most important, was reported in 1943. Hundreds have since been found, but relatively few are manufactured commercially. Many are toxic; some are effective against only a few kinds of microorganisms;

Figure 2–15. *Paul Ehrlich, 1854–1915. (From Bulloch: The History of Bacteriology. London, Oxford University Press, 1938.)*

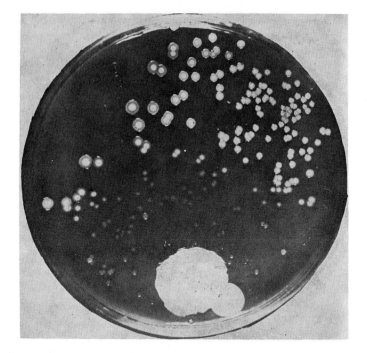

Figure 2–16. *Fleming's original culture of staphylococci contaminated with* Penicillium notatum, *showing lysis of bacterial colonies in the vicinity of the mold. (From Fleming, Brit. J. Exp. Path., 10:228, 1929.)*

many are very expensive to manufacture; and to some antibiotics bacteria easily become resistant.

MICROBIAL PHYSIOLOGY AND GENETICS

Pasteur and his contemporaries studied microbial fermentations and learned that when various microorganisms attack a substrate like sugar, they form a variety of end products such as ethyl alcohol, lactic acid, succinic acid, acetic acid, butyric acid, and carbon dioxide. During this period of discovery much information was amassed by direct observation. Later, early in the twentieth century, came a period of interpretation, when the experimental approach was used to determine the mechanisms by which various fermentation products are formed. Scientists in England, France, the Netherlands, Germany, the United States, and elsewhere undertook to learn the mechanisms of intermediary metabolism. It gradually appeared that the basic metabolic pathways of most organisms are similar. Moreover, cell metabolism in plants and animals markedly resembles that of microorganisms. This was recognized by Kluyver, Biejerinck's successor as chairman of microbiology at the technical university in Delft, Holland.

In a paper published in 1924, he pointed out the biochemical unity displayed by diverse organisms and expressed the belief that this would help to understand metabolism in higher organisms, a view since shown to be correct. Organisms differ with respect to the specific substrates they can utilize and the extent to which they employ oxidative or fermentative methods of attack, but the fundamental mechanisms by which energy is secured are few in number and common to all. They will be discussed in Chapter 8.

As microorganisms were studied more intensively, and physiologic as well as physical properties were determined, it became apparent that mutations occur among them as they do among higher organisms, but more rapidly because of their greater rate of reproduction. Just before the Second World War, Beadle and Tatum reported genetic studies of *Neurospora,* a mold that can be induced to mutate more rapidly than is normal by means of x-irradiation. Some of the mutants displayed metabolic or synthetic capabilities different from those of the parent organism. This means that they possessed different enzymes. Beadle and Tatum correlated the enzymatic differences with genetic differences and concluded that genes control enzyme activity. They introduced the expression "one gene—one enzyme" to indicate that in most cases a single gene controls one enzyme.

Tatum and one of his students, Joshua Lederberg, in 1946 found evidence for a sexual type of conjugation that occasionally occurs in certain bacteria, and other methods of inducing genetic change in bacteria were discovered by various other investigators. A common feature of all these methods was the participation of DNA. In every case, this substance was transferred by some mechanism from one organism to another, whereupon the recipient acquired certain characteristics of the donor.

The speed with which genetic changes can occur and be recognized in microorganisms makes them admirable subjects for genetic studies. Fruit flies, mice, grains, and other higher organisms were formerly used in most genetic studies, but great emphasis is now placed upon microbial genetics. A leading aim of research in molecular biology is to determine the exact relationship between chemical structure and genetic constitution. Much of this work is being done with microorganisms.

SUPPLEMENTARY READING

Brock, T. D.: *Milestones in Microbiology.* Englewood Cliffs, N.J., Prentice-Hall, Inc., 1961.
Bulloch, W.: *The History of Bacteriology.* London, Oxford University Press, 1938.
Clark, P. F.: *Pioneer Microbiologists of America.* Madison, Wis., University of Wisconsin Press, 1961.
DeKruif, P.: *Microbe Hunters.* New York, Harcourt, Brace and Co., Inc., 1926.
Dobell, C.: *Antony van Leeuwenhoek and His "Little Animals."* New York, Staples Press, 1932.
Lechevalier, H. H., and Solotorovsky, M.: *Three Centuries of Microbiology.* New York, McGraw-Hill Book Co., Inc., 1965.
Vallery-Radot, R.: *The Life of Pasteur.* New York, Garden City Publishing Co., Inc., 1926.

3   THE TOOLS OF A MICROBIOLOGIST

Most of the equipment and materials used in microbiology are borrowed from chemistry and physics and the other biologic sciences. Some have been devised by microbiologists for their own purposes. Research often requires considerable ingenuity, and it may be said that a good laboratory investigator should also be handy in the machine shop. As shown in the preceding chapter, microbiology advanced at any one time only as far as the available equipment permitted; further progress had to wait for technical improvements.

THE MICROSCOPE

THE COMPOUND MICROSCOPE

Compound microscopes differ from simple microscopes like those of Leeuwenhoek in possessing two lenses or lens systems: an *objective* and an *ocular* (Fig. 3–1). The objective produces a magnified image of the object, which is further magnified by the ocular (eyepiece). The total magnification is the product of the magnifications produced by the two lenses individually.

It will be recalled that when a light ray passes at an angle from one medium to another of different density (e.g., from air to glass), the ray is bent or refracted. The direction of bending is toward a line perpendicular to the surface between the media when the ray is entering the medium of greater density, and away from the perpendicular when it is entering the medium of lesser density (see Figure 3–2). The density of air is less than

that of all other transparent substances. The intensity of refraction in any other medium is expressed by the *index of refraction,* which is obtained by dividing the angle of incidence in air by the angle of refraction in the second medium (e.g., i_1/r_1 in Figure 3–2). The index of refraction of glass is about 1.52; it varies somewhat according to the composition of the glass.

Light passing from air into glass and out into air again is refracted twice, and if the two glass surfaces are parallel, the ray emerging into the air travels in the same direction as the incident ray did before it entered the glass. However, if the two surfaces are not parallel, as in a biconvex lens, the emerging ray travels in a different direction, which depends in part upon the distance of its point of origin from the center of the lens (see Figure 3–3). For each lens there is a distance from the center of the lens, designated the *focal length,* at which parallel rays entering the lens are brought to a focus; light originating at this point and traversing the lens emerges as parallel rays. Light originating from a source farther from the lens than this point is focused at some finite point when it emerges from the lens.

If, instead of a single point, the source of light is an area (e.g., an object like an illuminated bacterium), each point in that area will serve as the source of a ray that will be refracted by the lens in the same manner and focused to produce an *image* on the opposite side of the lens. The image will be smaller than, the same size as, or larger than the object, depending upon the distance of the object from the lens (see Figure 3–4).

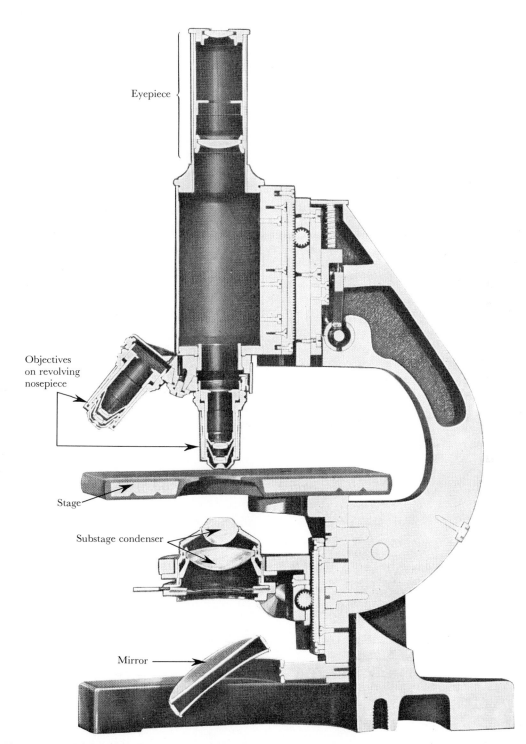

Figure 3–1. *Photograph of a cross section of a compound microscope showing interior construction. Light from the mirror is focused by the substage condenser on the object, placed upon the stage. The image produced by the objective is further magnified by the eyepiece. Some of the lenses of which the objectives are composed are important in reducing various aberrations (e.g., spherical and chromatic aberrations). (Courtesy of Bausch and Lomb.)*

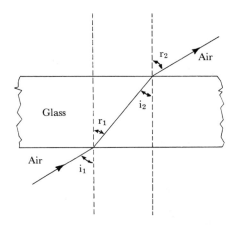

Figure 3–2. *The path of a light ray showing refraction when it passes from air into glass and when it emerges into air again. The ray is refracted toward the perpendicular when entering a medium of greater density (that is, the angle of incidence, i, is greater than the angle of refraction, r) and away from the perpendicular when entering a medium of less density.*

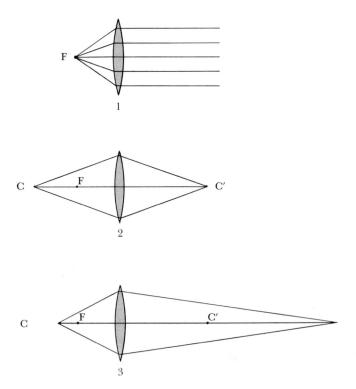

Figure 3–3. *Paths of light through a biconvex lens. 1, Parallel rays from the right are brought to a point (i.e., focused) at* F, *whose distance from the center of the lens is designated the* focal length *of the lens; conversely, a light ray originating at* F *is refracted into parallel rays upon emerging from the lens. 2, A ray originating at a distance twice the focal length from the lens is brought to a focus at the same distance on the other side of the lens;* C *and* C' *are conjugate foci. 3, A ray starting at a point between* F *and* C *is focused at some point between* C' *and infinity. This is the situation when the lens is used to produce a magnified image.*

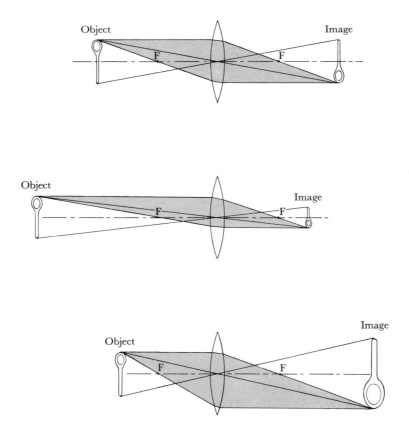

Figure 3–4. *Relationship between the size of the image and the position of the object relative to the lens. Note that the image is inverted.*

Features of a Good Microscope

To be satisfactory, a microscope must provide adequate *magnifying power.* It should magnify the object sufficiently so that the finest details appear far enough separated to be visible to the eye. Magnification by each lens is usually indicated by markings (e.g., 10X, 43X, 97X on the objectives; 5X, 10X, 15X on the eyepieces). A microscope should also produce a visible image of good *definition* and possess high *resolving power.*

Definition. Definition is a matter of contrast and depends upon the quality of the lenses. Ordinary lenses suffer from defects known as spherical and chromatic aberrations. These result from the inability of optical glass to bring all light rays to a focus at a common point. Visible rays of different colors are refracted at different angles, the shorter wavelengths, such as blue, being refracted to a greater extent than the longer wavelengths. Spherical and chromatic aberrations can be partially corrected by the use of various glasses and the mineral fluorite. Ordinary or achromatic objectives are satisfactory for many routine purposes. Apochromatic objectives are more highly corrected and are used extensively for photomicrography because of their better color correction.

Contrast is improved, and hence better visualization is obtained, by various procedures, such as staining (especially with purple or blue-green dyes), or by darkfield or phase contrast optics (see pages 46 and 49).

Resolving Power. Resolving power is the ability to produce separate images of small parts of an object that are only a short distance apart, that is, the ability to distinguish fine detail.

The light rays from an illuminated object on a microscope slide usually form a cone. These rays are brought to a focus by the objective lens, and the more divergent the rays that the lens can admit (i.e., those at the outside of the cone), the greater is the resolving power of the lens. This is determined in part by the lens itself and in part by the index of refraction of the medium (air, water, oil, etc.) between the object and the lens. The mathematical expression of resolving power is *numerical aperture,* NA:

$$\text{N.A.} = \eta \sin \theta$$

in which η (eta) is the refractive index of the medium between the object and the lens, and θ (theta) is the angle between the most divergent rays passing through the lens and the optical axis of the lens. The refractive index of air is 1.0 and that of the glass in lenses and of immersion oil is 1.52. If an oil immersion lens can admit light at an angle of 55°, simple calculation shows that its N.A. is 1.245. Most student bacteriologic microscopes are equipped with an oil immersion objective having a numerical aperture of 1.25. The higher the numerical aperture of a lens, the greater is its resolving power.

The resolution or fineness of detail that can be distinguished depends also upon the wavelength of light used for illumination. Finest detail is observed with light of short wavelength (i.e., blue or blue-green in ordinary microscopy). The relationship between numerical aperture, wavelength of light, and limit of resolution is as follows:

$$\text{Limit of resolution} = \frac{0.61 \times \text{wavelength}}{\text{N.A. of objective}}$$

The unit of measurement most frequently used in describing bacteria is the micrometer (μm), formerly called the micron (μ); this is one millionth (10^{-6}) meter or 0.001 millimeter. Submicroscopic objects, such as viruses, are often measured in nanometers (nm); 1 nm is 10^{-9} meter, and was formerly designated 1 mμ (millimicron).

If blue-green light at a wavelength of 500 nm is used with an oil immersion objective whose N.A. is 1.25, the limit of resolution is

$$\frac{0.61 \times 500}{1.25} = 244 \, \text{nm}$$

In other words, it is just barely possible to distinguish two objects as separate bodies if they are about 0.25 μm apart.

The Substage Condenser. Many improvements were made in the microscope during the three hundred years following its invention. Lens grinding became a science rather than an art, and greater magnification was obtained. Increased magnification necessitated better illumination. A rough comparison may be made with the common experience that newspaper headlines can easily be read by natural light late in the day, but artificial light must be used for reading fine print. Higher magnifications were made possible by a third lens system, which focused and concentrated light from the mirror or lamp upon the object. This lens system was called a *substage condenser* (Fig. 3–5). In the terminology of optics, the substage condenser provides a cone of light of sufficient angle to fill the aperture of the objective. This is not possible with the mirror alone when objectives of high magnifying power are employed. The Abbé condenser, named after its inventor, is found on most student microscopes in bacteriology.

Objective Lenses. Microscopes are usually equipped with a revolving nosepiece upon which are mounted at least three. objectives (Table 3–1). The objective of lowest power ordinarily used provides magnification of ten times, and the intermediate objective magnifies 42 to 45 times. The objective of highest power magnifies about 97 times; this is usually an oil immersion objective, so-called because it must always be immersed in cedar oil or some other medium having approximately the same index of refraction as glass (1.52). Substitutes containing mineral oil as the principal component are often used and offer the advantage that they do not dry on prolonged exposure to air.

Light passing through a glass slide continues in a straight path through the immersion oil in contact with it and the front lens of the objective (see Figure 3–6), whereas light emerging from a slide into air is refracted

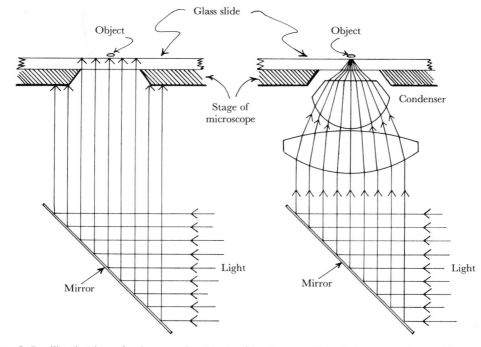

Figure 3–5. *Illumination of microscopic object without and with substage condenser. The condenser focuses all the light from the mirror on the object.*

from the axis of the objective and toward the slide. Details of the observed image become fuzzy, and some light is lost; that is, part of the light passes by the objective and never enters it. Immersion oil enhances the resolving power of the objective so that finer details can be observed, and more of the incident light is utilized.

ULTRAVIOLET MICROSCOPY

It was noted previously that the resolving power of a microscope is determined in part by the wavelength of light employed and that resolution is improved with short wavelengths. Beyond the shortest visible light rays in the ultraviolet region is invisible light, which can be used for photography. A microscope for ultraviolet photography must be equipped with quartz lenses because glass absorbs ultraviolet light. Higher magnification is obtained than with visible light and resolution is approximately twice as great as with visible light; that is, objects about 0.1 μm apart can be distinguished. Focusing is obviously more difficult.

TABLE 3–1. Properties of Commonly Used Microscope Objectives

Objective Magnification	Equivalent Focal Length	Numerical Aperture	Working Distance	Total Magnification with 10X Eyepiece
10X	16 mm.	0.25	8.3 mm.	100X
43X	4 mm.	0.65	0.72 mm.	430X
97X	1.8 mm.	1.25 oil	0.14 mm.	970X

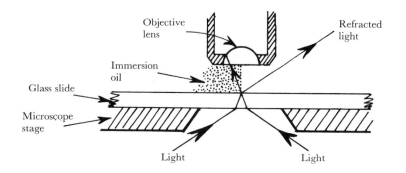

Figure 3-6. *The effect of immersion oil upon the path of light leaving a microscope slide. In the left half of the sketch light passes in a straight line through the slide and oil and into the objective lens. In the right half, drawn without oil, part of the light is refracted away from the lens.*

DARKFIELD MICROSCOPY

The images produced by the darkfield microscope appear as luminous bodies against a black or nearly black background (Fig. 3–7). In the ordinary or brightfield microscope an object is visible if it absorbs or refracts some of the incident light and thereby creates contrast between it and the suspending medium; any object that does not absorb or refract light is difficult to see. In the darkfield microscope, such an object, illuminated by a strong beam of light so directed that none would normally enter the objective, reflects some of the incident light in all directions; part of the reflected light enters the objective and is seen through the eyepiece. The object therefore appears luminous.

The ordinary Abbé condenser fitted with an opaque disk or *dark ground stop* is used with the low power of the microscope (Fig. 3–8). The dark ground stop eliminates all light from the central portion of the condenser and permits only a thin cone of light to reach the object. Objectives of higher magnification require special condensers, which can be substituted for the Abbé condenser. Little or no other modification is required. Intense light is necessary, such as that provided by a carbon arc lamp or various research illuminators. Slides and coverglasses must be perfectly clean and free from scratches because extraneous objects or marks also reflect light and produce a brighter background than desired.

The darkfield microscope is used to study

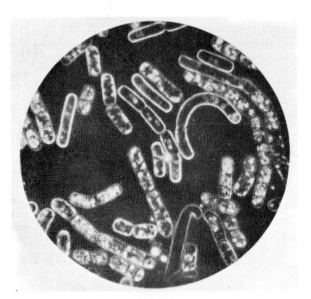

Figure 3-7. *Micrograph of Bacillus megaterium, unstained, darkfield, by ultraviolet illumination (2500X). (From Topley and Wilson: Principles of Bacteriology and Immunology. Baltimore, Williams & Wilkins, 1946.)*

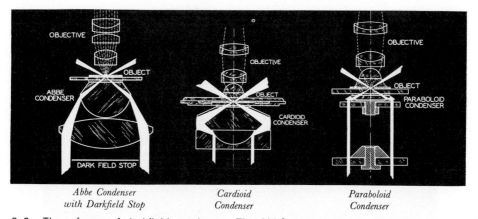

Abbe Condenser   Cardioid   Paraboloid
with Darkfield Stop   Condenser   Condenser

Figure 3–8. *Three forms of darkfield condenser. The Abbé condenser with darkfield stop can be used with low power objectives (10X or 20X); the paraboloid or cardioid condenser is required for higher power objectives. Note that the object is illuminated by peripheral light and only light reflected by the object (dotted lines) enters the objective lens. (Courtesy of Bausch and Lomb.)*

very small or slender bacteria of low refractive index. *Treponema pallidum,* the cause of syphilis (Fig. 3–9), can readily be detected in chancre fluid.

FLUORESCENCE MICROSCOPY

Microorganisms observed with the fluorescence microscope resemble those seen with the darkfield microscope: the cells are bright against a dark background. The principle of the microscope is quite different, however.

A source of intense light is used together with a filter that removes all except the ultraviolet and near ultraviolet rays. The specimen is stained with a yellow dye such as auramine O. A yellow filter is placed in the eyepiece of the microscope. The effect of the blue filter at the source of light and the yellow filter in

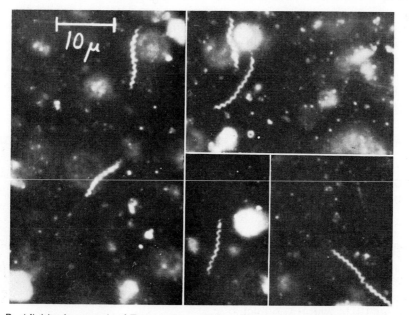

Figure 3–9. *Darkfield micrograph of* Treponema pallidum. *The slender organisms reflect the oblique incident light and therefore appear bright against a dark background (1500X). (Photograph by Dr. Theodor Rosebury; A.S.M. LS-327.)*

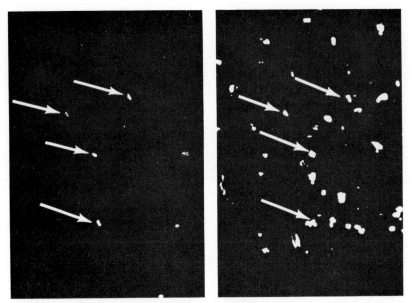

Figure 3–10. *Specific staining of* Pseudomonas pseudomallei *mixed with* Pseudomonas aeruginosa *by means of fluorescein-labeled anti-*P. pseudomallei *globulin. The photograph on the left was taken by fluorescence microscopy; only cells of* P. pseudomallei *fluoresce. The photograph on the right is the same field taken with visible light and a darkfield condenser. Arrows point to* P. pseudomallei *cells; the others are* P. aeruginosa. *(From B. M. Thomason, M. D. Moody, and M. Goldman: J. Bact., 72:362, 1956; A.S.M. LS-361.)*

the eyepiece is to produce a black field. Objects stained with the yellow dye fluoresce in the ultraviolet light and emit yellow light, which passes through the yellow eyepiece filter and readily reveals their presence and shape.

The fluorescence microscope is used to detect *Mycobacterium tuberculosis* in sputum. Lower magnification is required than with conventional methods of examination because the bright yellow organisms stand out clearly against the black background. The field observed is therefore larger, so the smear may be examined more quickly.

IMMUNOFLUORESCENCE MICROSCOPY

Immunofluorescence microscopy is a special extension and refinement of fluorescence microscopy. The stain consists of a particular type of blood serum protein known as an antibody (to be discussed in some detail in Chapter 19) to which fluorescein or rhodamine or some other fluorescent dye has been chemically coupled. An important

property of antibodies is their ability to combine with certain other proteins or protein-polysaccharide complexes which may be part of bacteria or other cells. A fluorescent antibody can therefore be used to stain and hence to detect a specific chemical component of a cell. If this component is found in only a single kind of cell (for example, a hemolytic streptococcus), the fluorescent antibody stain can be used to identify this organism quickly (Figs. 3–10 and 3–11).

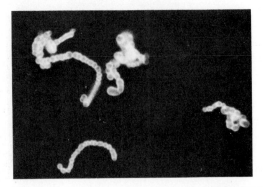

Figure 3–11. *Smear of group A streptococci stained with homologous fluorescent antibody. (Photograph by M. D. Moody.)*

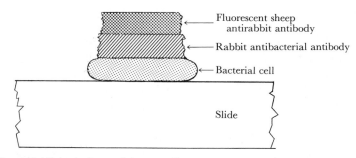

Figure 3–12. *"Sandwich" technique of immunofluorescence microscopy. The rabbit antibacterial anti-body combines with the bacterial cell, and the sheep antirabbit antibody combines with this rabbit antibody. When illuminated with ultraviolet light, the sheep antirabbit antibody fluoresces and can be seen with the microscope.*

Speed is often of importance in the diagnosis of disease caused by microorganisms.

In a modification of this procedure, the indirect method, sometimes called the "sandwich" technique (Fig. 3–12), nonfluorescent antibody produced in a rabbit by immunization with the organism in question is placed on a smear in which the organism is presumed to be present. Excess (uncombined) antibody is washed off, and the preparation is treated with a fluorescent antibody capable of reacting with rabbit serum protein (and hence with the nonfluorescent antibody attached to the organisms). The antibody that reacts with rabbit protein is produced in an animal such as the sheep and then made fluorescent by coupling it with a fluorescent dye. This procedure makes it possible to use the fluorescence technique without preparing each specific fluorescent antibody.

Immunofluorescence microscopy is useful in determining the identity of microorganisms in mixtures (e.g., in a throat smear or fecal specimen), in ascertaining the location of certain organisms within the tissues or cells of an infected host animal or plant, and in studying the chemical structure of cells.

PHASE MICROSCOPY

The refraction that occurs when light passes from one medium into another of different density is utilized in the phase contrast microscope. An ordinary microscope can be adapted for phase microscopy by substitution of appropriate condensers and objectives.

Transparent cells in a transparent medium or transparent structures within the cells are indistinguishable by ordinary microscopy. If, however, the cells or structures differ from their surroundings in refractive index, even slightly, some of the light striking them is deviated. The phase microscope is so designed as to intensify slight differences in contrast produced by this deviation.

The condenser in a phase contrast microscope has a special diaphragm, which permits only a ring of light to strike the object (Fig. 3–13). In the back focal plane of the objective is a transparent disk with a ring upon which light from the annular condenser aperture is focused. This ring alters by a quarter-wavelength the phase of light waves that pass through it, either advancing or retarding them, according to its construction.

If each transparent particle in the object is considered individually, it will be apparent that when a single ray of light strikes a particle, two rays result. One is the direct continuation of the incident ray, which passes through the point and on to the phase-shifting ring, where it is either retarded or advanced. The other ray is refracted or deviated by the particle because of the difference in density between the particle and the material immediately adjacent to it. Since this ray has deviated, it does not strike the phase-shifting ring and hence its phase is unaffected. When both rays are again brought to a focus, the particle appears brighter or darker than the

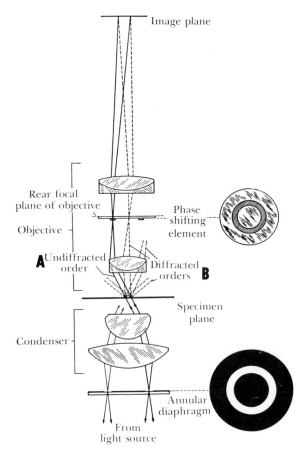

Figure 3–13. *Path of light through the phase contrast microscope. Light enters through the annular diaphragm opening and illuminates the object. Part of the light, A (solid lines), is transmitted directly and focused on the phase-shifting ring at the rear focal plane of the objective, which advances or retards it by $\frac{1}{4}$ wavelength. The remainder of the light, B (broken lines), is diffracted by the object and scattered; most of it does not strike the phase-shifting ring. When undiffracted and diffracted rays (A and B) are finally focused to form an image, additive or subtractive superposition produces differences in brightness that are readily apparent. (Courtesy of Dr. J. R. Benford, Bausch and Lomb Scientific Bureau.)*

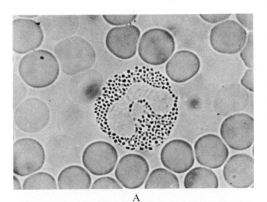

A

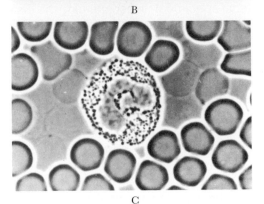

B

surrounding matter, depending on whether superposition of the two rays results in addition or subtraction of their intensities (Fig. 3–14). The object therefore appears either lighter than usual against a dark background (bright contrast) or darker than usual against a bright background (dark contrast).

Phase microscopy is particularly useful in studying the internal structures of microorganisms because structures differing in refractive index from the surrounding protoplasm become visible, and their sizes and locations can be determined.

Figure 3–14. *Appearance of unstained blood cells by A, ordinary, brightfield microscopy; B, darkfield microscopy; C, phase microscopy. A single eosinophilic leukocyte, surrounded by erythrocytes, magnified about 1250X. (From Scope, courtesy of the Upjohn Company.)*

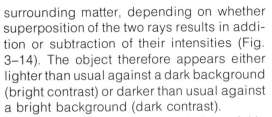

C

PLATE I

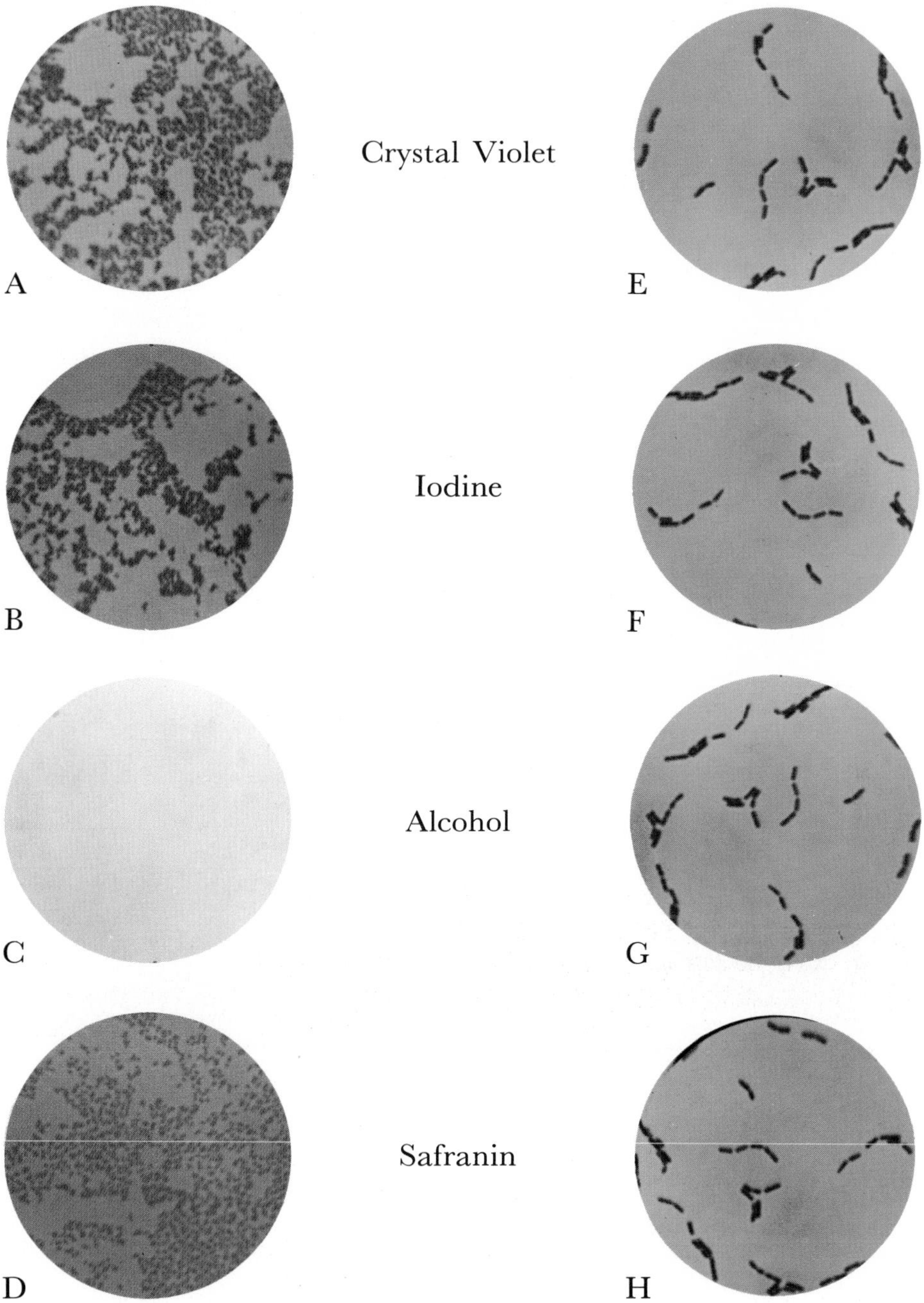

Escherichia coli Crystal Violet *Bacillus subtilis*

Crystal Violet

Iodine

Alcohol

Safranin

Escherichia coli (A–D) *and* Bacillus subtilis (E–H), *stained to show the appearance of a gram-negative and a gram-positive organism after each step of the gram procedure (approx. 930X). (Photographs by Elizabeth S. Watkins.)*

PLATE II

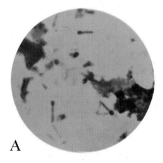

A

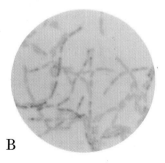

B

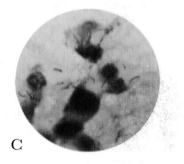

C

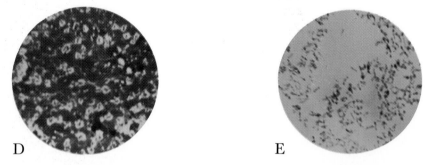

D

E

A, Clostridium tetani *stained with crystal violet to show spherical, terminal (drumstick) endospores (approx. 900X).*

B, Bacillus cereus, *Schaeffer-Fulton stain (approx. 930X). (Photograph by Elizabeth S. Watkins.)*

C, *Tuberculosis sputum stained by Ziehl-Neelsen method. Acidfast bacteria are pink; other bacteria and pus cells are blue (approx. 930X). (Photograph by Elizabeth S. Watkins.)*

D, *Capsule stain of* Enterobacter aerogenes *grown in milk. The bacteria and the milk stain deeply; the capsules are unstained (approx. 930X). (Photograph by Elizabeth S. Watkins.)*

E, Corynebacterium diphtheriae, *Albert's stain, showing metachromatic granules (approx. 900X).*

THE ELECTRON MICROSCOPE

It has been emphasized that a major factor limiting the resolution obtainable with a microscope is the wavelength of light used to illuminate the object. This limitation means that the smallest object that can be seen clearly with the ordinary microscope is about 0.2 μm in diameter and the best magnification obtainable is not much more than 1000X. Ultraviolet light extends the range of useful magnifications to about 2000X.

A new form of microscopy became possible when it was discovered that certain radiations of much shorter wavelength can be focused by suitable electric or magnetic fields in somewhat the same way that glass lenses focus visible light. Radiations used for this purpose consist of electrons emitted by the cathode filament of an electron "gun" at a velocity of a few tens of thousands of volts.

A common type of electron microscope (Fig. 3–15) contains three or more ring-shaped electromagnets, which function as lenses. The first is a condenser which, like its counterpart in the light microscope, focuses the electron stream on the object. The second, the objective, produces a magnified intermediate image of the object. Finally, the projector magnifies a portion of the intermediate image to produce the final image, which is photographed or inspected visually on a fluorescent screen (Fig. 3–16).

Electrons are readily stopped by all forms of matter; therefore, electron microscopy must be carried out in a vacuum. This imposes a limitation on the types of specimens that can be studied. Living organisms cannot be examined because samples must be completely desiccated. Preparations must also be very thin (less than 0.5 μm) to reveal structural differentiation; biologic materials are often mounted on thin films of collodion or cellophane, supported on a metal grid.

Surface structural details are revealed by "shadowing" with a metal such as gold or

Figure 3–15. *R.C.A. electron microscope. A human hair photographed with this instrument would appear as large as the Lincoln Tunnel.*

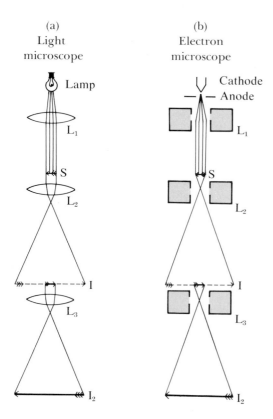

(a)
Light
microscope

(b)
Electron
microscope

Figure 3–16. *The "optics" of the electron and light microscopes. In the light microscope (a) (shown inverted for better direct comparison) light is focused by condenser lens L_1 on the specimen, S; the objective lens, L_2, produces an image, I, which is further magnified by the eyepiece lens, L_3. The source of radiation in the electron microscope (b) is a cathode, which emits electrons accelerated by an electric potential of approximately 50,000 volts. A magnetic coil, L_1, serves as a condenser, deflecting the electrons and focusing them on the specimen, S; a second magnetic coil, L_2, functions like an objective lens, deflecting the electrons and producing an image magnified 100X to 200X, which is further magnified 200X to 250X by the projector coil, L_3, forming an image on a fluorescent screen for visual inspection or on a photographic plate. The total magnification is 20,000X to 50,000X, and the picture can be enlarged to 1,000,000X. (A.S.M. LS-31.)*

chromium. The specimen is placed below and at one side of a tungsten filament, which is charged with the desired heavy metal and heated to vaporize the metal. The operation is carried out in a vacuum, and the metal atoms travel in straight lines in all directions from the filament. Some of the metal deposits on the sample, the thickness of the deposit being greatest on aspects of the surface tissue facing the oncoming atoms and thinnest in regions shaded by the specimen. The heavy metal is opaque to electrons, so an electron micrograph presents a three-dimensional appearance similar to that of a

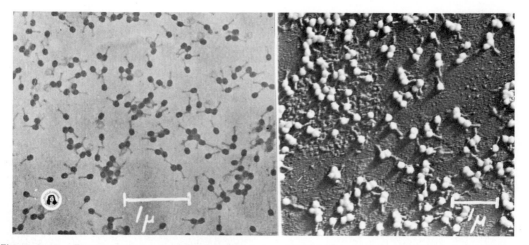

Figure 3–17. *Electron micrographs of a bacterial virus* (Escherichia coli *bacteriophage gamma*) *showing the effect of shadowing with a heavy metal:* left, *unshadowed;* right, *shadowed. (Photographs by D. G. Sharp; A.S.M. LS-139.)*

landscape seen from an airplane by the light of the early morning or late afternoon sun (Fig. 3–17).

Individual cells, even of bacteria, are too thick to show clearly details of internal structure in ordinary, unshadowed preparations. Ultrathin sectioning techniques have been developed, using glass or quartz knives and capable of yielding slices as thin as 20 nm. These reveal that cells possess a complicated fine structure, and they present the opportunity to correlate structure with function.

Another technique that helps to demonstrate the internal structure of cells and even of virus particles is the use of heavy metal salts such as phosphotungstate as "negative" stains. The background and "hollow" regions into which the salt penetrates become opaque to electrons, whereas areas occupied by cell material remain relatively transparent. Much greater structural detail is therefore shown (see Figure 3–18).

The interpretation of electron micrographs is plagued by artifacts due to extraneous materials in the specimen or to faults in the technique of preparing and examining it. The identity of any newly detected structure has to be determined by correlating a variety of methods of examination: light microscopy of stained preparations, phase microscopy, physiologic studies, and any other technique that the experimenter can devise.

The direct magnification theoretically obtainable with the electron microscope is as high as 100,000. Magnifications of 20,000 to 50,000 are common. Pictures obtained at lower magnifications are of sufficient sharpness that photographic enlargements of eight to ten times can be made with a resulting magnification of 1,000,000 or more.

Replica Techniques. The preparation of casts or replicas permits study of the surface structure of objects or materials (e.g., siliceous shells, metals) that are opaque to electrons because of their composition or thickness, or that are volatile in a vacuum or unstable when exposed to the electron beam. The substance to be examined is mounted on a collodion or other film, supported by a grid in the usual way, and then is thinly coated with an electron transparent material such as carbon, collodion, or polyvinyl formaldehyde (Formvar). The specimen is dissolved away with appropriate solvents, leaving only the thin layer of carbon or other coating agent. After this is shadowed with a heavy metal, electron microscopy clearly reveals the surface configuration of the specimen. This method is adaptable to bacteria, pollen grains, mold spores, leaf surfaces, muscle fibers, skin, and many other specimens.

Freeze-etching. Freeze-etching is a technique designed to prevent the formation of artifacts, which sometimes happens during conventional methods of fixing and preparing specimens. The specimen is frozen quickly in a drop of water. The ice is then fractured or cut with a glass or diamond knife so that portions of the specimen are exposed. Some of the ice is then sublimed away, and a replica of the exposed surface is made and shadowed with heavy metal. The result is a preparation in which surface and internal structures can be clearly visualized (see Figure 3–19).

Scanning Electron Microscope. The scanning electron microscope is a recently developed instrument for examining the surface of even fairly large objects. The specimen is coated with a thin layer of a heavy metal such as platinum or gold and palladium, and it is then scanned back and forth by an electron beam. Electrons scattered by the metal produce an image on a viewing screen. Great depth of field and a three-dimensional effect enhance the visual impression (see Figure 3–20).

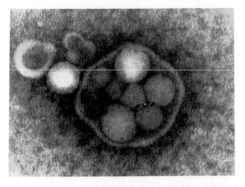

Figure 3–18. *Use of negative staining with phosphotungstic acid and electron microscopy to demonstrate internal structures of cells: spherical bodies within a cell of* Mycoplasma hominis *(50,000X). (From Anderson et al., J. Bact., 90:189, 1965.)*

Figure 3–19. *Freeze-etch preparation of* Escherichia coli. *Intact cell surface is shown at the left, protoplasmic membrane in the center, and cell contents at the right. (From Bayer and Remsen, J. Bact., 101:304–313, 1970.)*

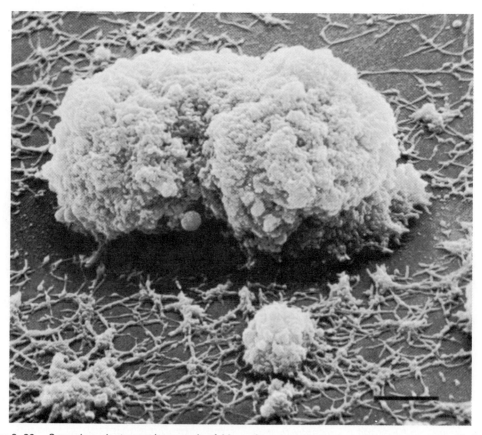

Figure 3–20. *Scanning electron micrograph of* Mycoplasma pneumoniae *grown on a cover slip in liquid medium. Filamentous forms attach to the surface, and colonies contain rounded organisms. The bar represents 10 μm. (From Biberfeld and Biberfeld, J. Bact., 102:855–861, 1970.)*

MICROSCOPIC STUDY OF BACTERIA

OBSERVATION OF UNSTAINED CELLS

Leeuwenhoek and all other microbiologists until the latter half of the nineteenth century had to be content with microscopic observation of unstained organisms. The advantages of direct study of living cells in their natural shape, arrangement, and size were counterbalanced by the fact that the refractive index of microbial protoplasm is so near that of water that the cells and their structures cannot be clearly differentiated from each other and from the mounting fluid.

The *wet mount* was a common type of preparation for observation of bacteria and is still used. A drop of broth culture is placed on a slide and a cover slip is carefully added. The living organisms can be examined by low power and high power dry (i.e., 43X) objectives. The diaphragm of the Abbé condenser must be partly closed to provide a narrow beam of light; otherwise the cells cannot be distinguished from the medium.

A *hanging drop* is preferred if prolonged study of living organisms is to be made (Fig. 3–21). A special slide with a hollow in one side is used. The culture is placed on a *clean* coverglass and inverted over the depression, sealed with water, oil, or petrolatum, and studied with the low power or high power dry objective. This type of mount is especially useful for continuous observation of bacterial motility or for watching the growth of individual cells in a microculture.

BACTERIAL STAINS

Microscopic study of bacteria is greatly facilitated by treating them with dyes or stains. The shapes and relative sizes of stained organisms can be determined more easily, and staining also permits certain cellular structures to be recognized.

Chromophore and Auxochrome Radicals. Dyes are organic compounds containing color-producing *chromophore* radicals and salt-forming *auxochrome* groups. The nitro ($-NO_2$) and azo ($-N=N-$) groups are chromophores; the hydroxyl ($-OH$) and amino ($-NH_2$) radicals are auxochrome groups. Chromophores impart the property of color to a dye, and auxochrome groups permit the dye to unite with fibers or tissues. Most commercial dyes are salts, but are referred to as basic dyes if the colored portion behaves as a base or acidic dyes if it behaves as an acid. Basic dyes are usually available as chlorides and acidic dyes as sodium salts. Acidic dyes are used to stain basic material such as cytoplasm, whereas basic dyes stain nuclei, certain granules, and other acid substances.

Mechanism of Staining. Staining of cells and tissues is probably a combination of physical and chemical processes. The physical phenomena of adsorption, absorption, capillarity, and osmosis play a part. On the other hand, the affinity of basic dyes for acidic cell components and of acidic dyes for basic components indicates that chemical reactions occur and lead to the formation of new compounds.

Preparation of Bacterial Smears for Staining

Before staining, bacteria are usually suspended in water or some other liquid on a *clean* microscope slide and are then spread in a thin, even film. The film is allowed to dry in air, and the organisms are "fixed" to the slide by chemical means or more commonly by gentle heat. The preparation is known as a *fixed smear*.

Figure 3–21. *A "hanging drop" preparation. The drop of culture sealed in the small chamber by water, oil, or petrolatum may be observed for some time without drying.*

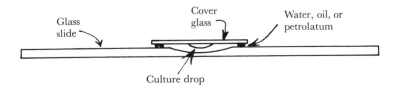

Simple Stains

A *simple stain* is a solution of a single dye, usually in alcohol or water. Most bacteria contain acidic material (e.g., DNA and RNA) distributed more or less uniformly throughout the cell. They therefore stain intensely with basic dyes such as methylene blue, crystal violet, or carbol fuchsin (a solution of basic fuchsin in 5 per cent carbolic acid). The time required for staining varies with the dye solution: one to five minutes is necessary with methylene blue; 15 seconds usually suffices with crystal violet.

Stained smears are rinsed briefly with water to remove excess stain, dried in air, and are then ready for examination.

Simple stains are used to detect the shape, arrangement, and relative size of bacteria and are useful in helping to identify them. They do not, however, reveal details of internal structure.

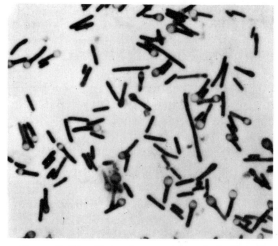

Figure 3–22. *Stained smear of bacteria showing sporangia (cells containing unstained spores) as well as vegetative cells. These spores are spherical, terminal, and greater in diameter than vegetative portions of the sporangia. (Courtesy of Frobisher et al.)*

Differential Stains

Differential staining procedures distinguish structures within a cell or distinguish one type of cell from another. Two dyes are usually employed, and certain structures or cell types appear in one color in the final preparation, with the remaining structures or cell types in the second color. The first dye applied is the *primary* stain. This is usually followed by *differentiation:* application of a solution that removes the primary stain from some cells or structures. The other dye, known as the *secondary* stain or *counterstain,* is then applied. Differential stains are widely employed in identifying bacteria.

Endospore Stain. The Bacillaceae, a large and important family, produce resistant bodies known as endospores at a particular stage in their life cycle. These structures are formed within the cell—hence the name, *endospore*—and contain all the cell components necessary to maintain life. They resist harmful physical and chemical agents, and they also resist staining. Simple stains do not ordinarily appear to penetrate the spore wall, so a cell with its endospore (sporangium; Fig. 3–22) contains a colorless spherical or oval body when stained briefly with crystal violet or methylene blue (Plate IIA).

Endospores may be stained by a drastic procedure. Strong dyes are allowed to remain in contact with the cells for a long time, or staining is facilitated by application of heat. The primary stain in the Schaeffer-Fulton procedure is malachite green; the smear is covered with dye and heated to steaming for 30 to 60 seconds. Differentiation is accomplished by washing with running water for 30 seconds. The green dye, which penetrates the endospores as well as the remainder of the cytoplasm, is thus removed from the vegetative portions of the sporangia but not appreciably from the endospores. A smear examined at this point reveals colorless cells containing green endospores, but the vegetative parts of the cells are difficult to distinguish. A counterstain of contrasting color, safranin, is then applied, and after brief rinsing and drying the preparation is ready to observe. It now consists of pink cells containing green endospores, and the position of the spores and their size can readily be determined (Plate IIB).

Endospore stains are useful in helping to

identify the Bacillaceae, which are practically the only bacteria capable of producing endospores. The endospores of certain species are spherical; others are oval. Spores may be located in the center of the cell, slightly away from the center (excentrically), near the end of the cell (subterminally), or at the very end of the cell (terminally). The diameter of the spores may be less than, equal to, or greater than that of the sporangium. In the latter case, the sporangium bulges. The shape, position, and size of the endospore are fairly characteristic for a given species; hence these properties help to establish the species name of a previously unknown organism. This approach to the classification of bacteria resembles that employed with higher plants and with animals, which are usually identified by morphologic properties.

Acidfast Stain. The mycobacteria (e.g., *M. tuberculosis* and *M. leprae*), like endospores, are difficult to stain, but once stained they retain the dye tenaciously even when washed with dilute mineral acid.

In the Ziehl-Neelsen procedure the primary stain, carbol fuchsin, is heated to steaming for five minutes, after which the smear is decolorized briefly with dilute alcoholic H_2SO_4 or HCl and counterstained with methylene blue to reveal the presence of nonacidfast organisms. Acidfast bacteria retain the pink or red primary stain; all other organisms are decolorized by the acid and stain blue (Plate IIC).

Acidfastness appears to be associated with the presence of large amounts of lipoid material, which may comprise as much as 40 per cent of the cell substance of *M. tuberculosis*. It has been postulated that the phenol-dye complex, carbol fuchsin, is more soluble in the lipoid cellular constituents than it is in the decolorizing agent and hence is retained by mycobacteria but not by organisms that lack a high lipid content.

Gram Stain. The Gram stain is the most widely used staining procedure in bacteriology. It divides bacteria into two great classes, *gram-positive* and *gram-negative;* moreover, the Gram reaction is correlated with certain other properties of an organism (Table 3–2).

The primary stain, crystal violet, is followed by an iodine solution that behaves as a mordant and helps fix the primary dye to gram-positive organisms. The smear is then decolorized, usually with 95 per cent alcohol, and counterstained with a contrasting dye such as safranin. Organisms that retain the purple primary stain are designated gram-positive; gram-negative cells lose the primary stain when decolorized with alcohol and stain with the relatively weak secondary pink dye, safranin (Plate I).

Mechanism of the Gram Stain. Both chemical and cell wall permeability theories of the mechanism of Gram differentiation have held sway in the past. The present view is that the permeability of the cell walls of gram-positive and gram-negative bacteria differs under the conditions of the decolorizing step, and that this is the factor that con-

TABLE 3–2. Some Differences Between Gram-Positive and Gram-Negative Bacteria

Gram-Positive	Gram-Negative
Contain magnesium ribonucleate	Do not contain magnesium ribonucleate
Very sensitive to triphenylmethane dyes (e.g., crystal violet)	Less sensitive to triphenylmethane dyes
Sensitive to penicillin	Sensitive to streptomycin
Not dissolved by 1% KOH	Dissolved by 1% KOH
Apparent isoelectric point pH 2–3	Apparent isoelectric point pH 4–5
Sporeforming rods, many cocci (also *Lactobacillus* and *Corynebacterium* species)	Most nonsporeforming rods, spirals, some cocci
Toxins (if any): exotoxins	Toxins: endotoxins

trols the loss of primary stain from the latter organisms. Protoplasmic constituents of both gram-positive and gram-negative cells combine with crystal violet by an ionic bond between their acidic groups and the basic groups of the dye. Iodine, in aqueous solution, enters the cells and reacts with the dye, either removing it from the cell protein or adding to the dye-protein complex.

It is believed that 95 per cent alcohol dehydrates the mordanted walls of gram-positive cells and forms a barrier, which traps the crystal violet-iodine complex. Such a barrier is not formed in gram-negative cells. It is known that the cell walls of these organisms contain a higher percentage of lipid than those of most gram-positive organisms, and hence the solubility of the surface lipids in alcohol may be a factor in the greater ease of decolorization of gram-negative cells. However, further study is necessary to clarify this point.

There is no sharp line of demarcation between gram-positive and gram-negative organisms; there is a continuous gradation from species that retain the primary stain even after decolorization for several hours, to species that decolorize within a few seconds. Aging cultures of many organisms that are normally considered gram-positive contain increasing numbers of gram-negative cells. Some species and even individual smears contain both gram-positive and gram-negative cells; such cultures are called *gram-variable*. The staining procedure must be carefully standardized to obtain consistent and reliable results, and control organisms of known Gram reaction should also be tested, particularly in critical work.

Nuclear (Chromatinic Body) Stain. The Feulgen reaction and numerous modifications of it are used to demonstrate "nuclear" material in bacteria. This reaction is actually a test for aldehyde, but is considered to indicate the presence of DNA. Smears fixed with osmic vapor are subjected to mild hydrolysis with 1N HCl (7–8 min. at 58° C.) and then treated with Schiff's reagent (fuchsin decolorized with SO_2). A light pink color indicates DNA. RNA is not stained after acid hydrolysis. The color produced by DNA is often not strong, but other stains have been used with success, such as dilute Giemsa stain, azure A, or thionin, following the hydrolysis step. Chromatinic bodies in the cells of *Escherichia coli* and two *Bacillus* species are shown in Figure 3–23.

Flagella Stain. The motility of a culture

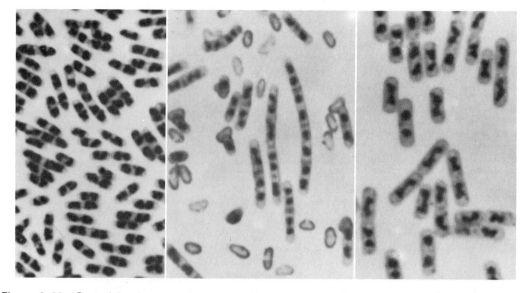

Figure 3–23. *Bacterial cells stained to demonstrate chromatin bodies (nuclei).* Left, Escherichia coli, *4000X;* middle, Bacillus mesentericus, *4400X (note spores in various stages of germination, see page 104);* right, Bacillus cereus, *4400X. (Photograph of* E. coli *by J. Hillier, S. Mudd, and A. G. Smith; A.S.M. LS-239; photographs of* B. mesentericus *and* B. cereus, *Committee on Materials for Visual Instruction; A.S.M. LS-266.)*

to be used for staining of flagella should be determined. Young cultures grown on fresh agar slants are gently washed with 2 to 3 ml. of sterile distilled water and the bacterial suspension is incubated a few minutes. Droplets are transferred by capillary pipette from the top of the suspension, where motile cells are most numerous, to one end of a tilted, *scrupulously clean* microscope slide and allowed to run down the slide and dry in air without mechanical agitation that would break off the flagella. The Leifson stain contains potassium or ammonium aluminum sulfate and tannic acid as a mordant, and the dye basic fuchsin. Slides flooded with the solution are allowed to stand 10 minutes, washed, dried, and examined under oil. Flagella stain pink, their apparent diameter being greatly increased by the deposit of dye. Flagella staining requires very careful technique.

Cell Wall Stain. Cell walls may be stained by alcian blue, by crystal violet following tannic acid as a mordant, or by methyl green. Bisset and Hale's procedure consists of immersing thick smears in 1 per cent phosphomolybdic acid for 3 to 5 minutes and then in 1 per cent methyl green for the same interval. After washing and drying, the smear is examined under oil. Cell walls are dark green or purple; cytoplasm is unstained. Cells of *Bacillus mycoides* stained by this method are shown in Figure 3–24. Complete cross

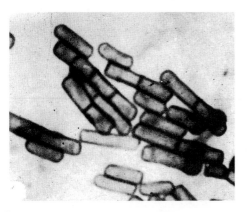

Figure 3–24. Bacillus mycoides *stained with methyl green to demonstrate cell walls. Each bacillus may be divided into as many as four units by cross walls. (From Bisset et al., Exptl. Cell Res., 5:451, 1953.)*

walls are formed and divide the cells into several units.

CULTIVATION OF MICROORGANISMS

Cultivation of microorganisms is accomplished by providing a favorable environment for their growth. This means that the necessary nutrients must be available in suitable form for use as building materials for new cells, a source of energy must be provided, and the pH, temperature, oxygen, and other conditions must be appropriate.

DEFINITIONS:CULTURE, CULTURE MEDIUM

A *culture* is any growth or cultivation of microorganisms. The term is usually employed with reference to deliberate growth of microorganisms in the laboratory. A *pure culture* contains only a single kind of microorganism. Pure cultures are rarely found outside the laboratory because in nature microorganisms are usually associated with one another. In mixed populations one kind of organism may help, harm, or have no effect upon other microorganisms; their social behavior therefore resembles that of higher plants and animals, including man.

A *culture medium* is a substrate or nutrient solution upon which microorganisms are cultivated in the laboratory. Different microorganisms require different nutrient materials, and certain media are used for specific purposes (e.g., to determine the ability of an organism to digest proteins, carbohydrates, and so forth). The number of possible culture media is unlimited.

INFUSIONS

The first solutions for cultivation of bacteria and other organisms consisted of natural materials. Leeuwenhoek employed the liquid obtained by soaking peppercorns in water. This fluid contained organic substances and minerals extracted from the plant tissue. Later investigators soaked other plant materials such as hay in water and obtained liquids in which microorganisms could be grown. Still others extracted animal tissues with water and obtained excellent substrates for the cultivation of bacteria. Tissue extracts are known as *infusions,* and until the latter half of the nineteenth century they were the only culture media for laboratory experimentation with bacteria. Their composition was not known exactly and they could not be reproduced with precision. However, they were usually sufficiently rich in organic materials and growth-promoting substances to favor the multiplication of many kinds of microorganisms and were satisfactory for the types of experiments performed at that time.

CHEMICALLY DEFINED OR SYNTHETIC MEDIA

Pasteur was apparently the first to use culture media of known composition. He demonstrated that various microorganisms could be cultivated in solutions containing only sugar, a source of nitrogen such as ammonium salt or a nitrate, and minerals. A medium of this kind is known as a *chemically defined* or *synthetic* medium. Chemically defined media can be reproduced exactly at any time and by workers in different laboratories because the chemical formulae of all their constituents are known exactly. They are essential for the study of nutritional requirements and are used in the manufacture of microbial products in which a minimum of extraneous organic material is desired.

NONSYNTHETIC MEDIA

Some microorganisms like those studied by Pasteur grow satisfactorily with simple sources of carbon and nitrogen and a few mineral salts. Others require one or more vitamin-like substances or *growth factors.* Certain nutritionally fastidious organisms require a dozen or more of these accessory substances and are more conveniently cultivated on the traditional organic extract or infusion substrates, which are designated *nonsynthetic* media. Meat infusion broth, for example, is commonly employed for certain respiratory pathogens of man. It is prepared by soaking lean meat (e.g., ground beef) overnight in water, straining out the meat particles, boiling, filtering through paper, and adding salt, peptone, and any other desired substances. Commercial peptones are manufactured by partially digesting native proteins (animal or plant) by means of enzymes; they furnish readily available amino acids and other nitrogen compounds.

TABLE 3–3. pH and Hydrogen Ion Concentration

Hydrogen Ion Concentration		pH		Examples
N/1	$= 10^0$ N	0		
N/10	$= 10^{-1}$ N	1		
N/100	$= 10^{-2}$ N	2		Lemon juice: pH 2.3
N/1,000	$= 10^{-3}$ N	3		
N/10,000	$= 10^{-4}$ N	4		
N/100,000	$= 10^{-5}$ N	5		
N/1,000,000	$= 10^{-6}$ N	6	Acid	
N/10,000,000	$= 10^{-7}$ N	7	Neutral	Milk: pH 6.6
N/100,000,000	$= 10^{-8}$ N	8	Alkaline	Blood: pH 7.3
N/1,000,000,000	$= 10^{-9}$ N	9		Sodium bicarbonate (N/10): pH 8.4
N/10,000,000,000	$= 10^{-10}$N	10		Phenolphthalein color change: pH 9.2
N/100,000,000,000	$= 10^{-11}$N	11		
N/1,000,000,000,000	$= 10^{-12}$N	12		
N/10,000,000,000,000	$= 10^{-13}$N	13		Lime water: pH 12.3
N/100,000,000,000,000	$= 10^{-14}$N	14		

ACIDITY OR ALKALINITY OF CULTURE MEDIA

Microbial culture media must possess a "reaction" (acidity or alkalinity) favorable for the organism to be cultivated. Before discovery of the relationship between hydrogen ion concentration and acidity it was customary to adjust culture media by adding alkali until the neutral point of phenolphthalein indicator was reached, and then adding a quantity of acid which previous experience indicated gave satisfactory growth. Discovery of the role of hydrogen ions and formulation of the pH scale revolutionized the adjustment of culture media.

pH. It will be recalled that pH is the negative logarithm of hydrogen ion concentration, and that the pH scale extends from 0 to 14 (Table 3–3). A solution of pH 0 contains hydrogen ions in a concentration of 10^0 or 1 normal. This solution is very acid. A solution of pH 1 is 10^{-1} or 1/10 normal with respect to hydrogen ions. A difference of one pH unit corresponds to a difference of ten times in the concentration of hydrogen ions. The upper limit of the pH scale, 14, represents a hydrogen ion concentration of 10^{-14} or 1/100,000,000,000,000 normal. Aqueous solutions also contain hydroxyl ions, and their concentration varies inversely with that of the hydrogen ions. Neutrality is represented by pH 7 and is the point at which the hydrogen ion concentration of a solution equals its hydroxyl ion concentration; this is the situation in pure water.

Many common bacteria like those found in and on the human body grow best at a reaction near neutrality, or from pH 6.5 to pH 7.5. Numerous pH indicators as well as electrical methods of determining pH are available. Adjustment of culture media by trial and error is relatively simple.

STERILIZATION OF CULTURE MEDIA

Culture media are usually dispensed into test tubes, flasks, or bottles, which are then plugged with nonabsorbent cotton or covered with screw caps or metal or plastic slip-on caps and sterilized. The usual sterilizing

Figure 3–25. *Method of transferring bacteria from one culture tube to another. The two tubes are supported firmly by the index, middle, and ring fingers and are held in position by the thumb. Plugs are grasped by the little and ring fingers of the other hand, which also holds the inoculating needle, as illustrated. Mouths of the test tubes and the needle are flamed, both before and after the transfer.*

agent is steam at a temperature of 120° C. To secure this temperature, the steam must be under pressure (approximately 15 lb. per sq. in.); this is possible in a laboratory autoclave (see Figure 12–3) or a home pressure

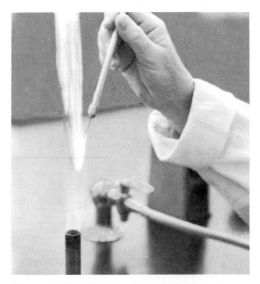

Figure 3–26. *Flaming inoculating needle in the Bunsen flame. The needle is held as nearly vertical as possible so that its entire length reaches red heat rapidly.*

cooker. Exposure for 15 to 30 minutes is sufficient to kill all organisms if the sterilizer is not loaded too heavily.

Sterilization of the mouths of test tubes and bottles to prevent contamination of media during subsequent operations (e.g., pouring Petri dishes, inoculating cultures) and sterilization of inoculating needles are effected by use of the Bunsen flame, as illustrated in Figures 3–25 and 3–26.

SUPPLEMENTARY READING

Committee on Bacteriological Technic, Society of American Bacteriologists: *Manual of Microbiological Methods.* New York, McGraw-Hill Book Co., Inc., 1957.

Conn, H. J.: *The History of Staining,* 2nd ed. Geneva, N. Y., Biotech Publications, 1948.

Kay, D. H. (ed.): *Techniques for Electron Microscopy,* 2nd ed. Philadelphia, F. A. Davis Company, 1965.

Lamanna, C., Mallette, M. F., and Zimmermann, N. L.: *Basic Bacteriology,* 4th ed. Baltimore, The Williams & Wilkins Co., 1973.

Lennette, E. H., Spaulding, E. H., and Truant, J. P.: *Manual of Clinical Microbiology,* 2nd ed. Washington, American Society for Microbiology, 1974.

Salle, A. J.: *Fundamental Principles of Bacteriology,* 5th ed. New York, McGraw-Hill Book Co., Inc., 1960.

Skerman, V. B. D.: *A Guide to the Identification of the Genera of Bacteria.* Baltimore, The Williams & Wilkins Co., 1959.

Wyckoff, R. W. G.: *The World of the Electron Microscope.* New Haven, Yale University Press, 1958.

THE SYSTEMATIC STUDY OF BACTERIA

4

Systematics is the general study of organized nature, including the causes and historical background of observed phenomena. The branch of systematics that deals with the description, classification, and naming of plants, animals, and microorganisms is called *taxonomy* (from Greek: *taxis,* arrangement, + *nomos,* law). The classification of biologic forms into groups is based on their relationships with one another, that is, their greater or lesser similarity. There are various approaches to the problem of determining degrees of similarity and, it must be confessed, systematists do not agree on which is the best. In general, it seems that organisms should be assigned to groups (e.g., species) on the basis of the correlation of many characters, rather than the possession of only a few, so-called *key* characters. Useful characters include morphologic, physiologic, chemical, serologic, ecologic, pathologic, and other properties. The relative importance of various characters is a matter of debate. An experienced biologic taxonomist consciously or unconsciously weights certain characters more heavily than others in his study of particular groups of organisms, whereas a *numerical* taxonomist considers each bit of information of equal value and even utilizes computer analysis to determine relationships. Groups are thus defined by possession of a majority of shared characters.

The ideal goal is a "natural" classification system based upon biologic relationships, and if possible also reflecting evolutionary (i.e., phylogenetic) trends. It should at the same time be practically useful in the identification of unknown specimens.

Natural Classification. Natural or *phylogenetic* classification indicates relationships between organisms on the basis of their probable origin; organisms with a common origin are grouped more closely together than those with diverse origins. The final picture resembles the structure of a tree: the trunk represents the main stream of evolution from its origin at the ground, the branches with their twigs and shoots represent later stages in evolutionary development, and the outermost leaves indicate forms currently in existence.

Phylogenetic classification is possible if enough fossil remains of primitive and intermediate forms can be found to reconstruct the trunk and main branches. This is the reason why there is so much interest in "missing links." A reasonably complete reconstruction of the evolutionary stages in the development of higher animals and plants has been possible, but unfortunately this is not yet the case with most microorganisms, particularly the bacteria.

Electron micrographs of two billion year old flint rock from southern Ontario have shown structures interpreted as bacteria (Fig. 4–1), but otherwise the only bacteria known are those in existence at the present; we see the leaves on the systematic tree but have little basis for connecting them in logical fashion to the main trunk. We may deduce that spherical bacteria are related to each other and came from some common ancestor and that spiral bacteria are derived from another common ancestor—but we find no trace of these ancestors.

Artificial Classification. The alternative to a phylogenetic system of classification is

Figure 4–1. *Electron micrograph of fossil bacteria in Gunflint chert sediments from southern Ontario. These Pre-Cambrian rod-shaped bacteria were alive about 2,000,000,000 years ago. (From Schopf et al., Science 149:1165, 1965.)*

an *artificial* or *phenetic* system based upon easily recognized characteristics of known organisms. Such a scheme provides a practical guide that is useful for identifying unknown organisms; it also shows some relationships between these organisms. The nature of the relationships depends upon the criteria used to establish the classification system; these relationships are often not phylogenetic. Bacterial taxonomy has perforce developed along these lines.

Taxonomy Based on DNA Homology.
A comparatively recent approach to bacterial classification is based on similarities and

differences in DNA. It will be recalled that DNA maintains genetic information in the form of a code whose elements are four nucleotide bases—adenine (A), guanine (G), cytosine (C), and thymine (T). A group of three nucleotide bases comprises a code "word," which directs the synthesis of a single amino acid. A DNA molecule consists of a double helix of nucleotide bases (see Figure 1–10, page 13), and the sequence of base triplets determines the proteins formed and hence the physical and physiologic characteristics of the individual. The structures of the nucleotide bases are such that

G can unite only with C to form a "base pair," and A can unite only with T. In a given species of organism the molar precentages of G + C and of A + T are constant from one individual to another. The base composition of a given DNA could be expressed as the ratio of G + C to A + T, but it has become conventional to express it as the percentage of G + C in the DNA; that is,

$$\frac{G + C}{G + C + A + T} \times 100$$

All individuals of the primordial life form must have possessed DNA with the same base composition, but mutation altered the base composition and selection maintained the alteration, so that ultimately there were marked differences in the GC percentages or in the base sequences (or both) of completely unrelated species. Inasmuch as it is difficult to determine base sequence but relatively easy to determine base percentage, the latter method is widely used in the study of DNA homology. The degree of relationship of organisms is determined in part by the similarity of their GC percentages. A perfect correlation cannot be expected because two organisms with identical percentages may differ in their DNA base sequences. Organisms with different base ratios are obviously unrelated, and those with identical ratios may or may not be related. Within these limitations, DNA base percentages provide helpful information.

The DNA of most members of the gram-negative rod-shaped intestinal bacteria (Enterobacteriaceae) contains 39 to 59 per cent G + C, and that of most *Pseudomonas* species contains 58 to 70 per cent G + C. These groups are fairly homogeneous in their properties. Spore-forming rods of the genera *Bacillus* and *Clostridium* contain DNA with 23 to 62 per cent G + C and appear more heterogeneous. Calculations indicate that if the GC percentages of two organisms differ by more than 10 per cent, they can have few base sequences in common and therefore are not closely related. It is perhaps not surprising that organisms with the survival feature of spore resistance have evolved more broadly than the more easily killed non-spore-formers.

GROUPS OF ORGANISMS

Bacteria at present constitute one of two divisions in the kingdom Procaryotae, which comprises the Monera of Whittaker (see Figure 1–24, page 22). The other division, now called *Cyanobacteria,* consists of the blue-green algae.

The division *Bacteria* is subdivided into 19 Parts, each of which is further divided into subordinate groups. Several Parts contain one or two orders, each consisting of families, which, in turn, include one or more genera. The relationships within many Parts are not clear enough to justify creation of orders; instead, they consist simply of genera, or of genera arranged in families. Genera are composed of "related" species, and families are composed of "related" genera. Decisions regarding the degree of relationship necessary to include a species in a given genus or a genus in a given family are made by experts in bacterial taxonomy. As in any situation influenced by opinion and judgment, there is occasional disagreement. Bacterial classification has so far always been in a state of change.

THE SPECIES CONCEPT

The fundamental group upon which the taxonomic hierarchy rests is the *species.* There is, in general, good agreement regarding what constitutes a species. The definition of species has undergone considerable evolution with reference to higher plants and animals. A species is now often defined as a group of actually or potentially interbreeding forms that do not crossbreed with other groups. This definition presupposes a sexual mode of reproduction.

Bacterial Species. A few strains of *Escherichia coli* reproduce on rare occasions by a sexual method, but it cannot be said that their principal mode of reproduction is sexual. Moreover, sexual reproduction has not been demonstrated among bacteria generally. The definition of a species that applies to higher plants and animals therefore has no meaning with reference to bacteria or, in fact, to any organisms that reproduce principally by asexual means.

A species of bacteria may be defined as a _group of bacteria possessing the same genetic constitution._ For practical purposes, this definition is adequate in view of the necessity for using an artificial rather than a natural system of classification.

The genetic identity of two organisms is presumed to be demonstrated if they are shown to have the same morphology, physiologic behavior, pathogenicity, and other properties. Organisms that differ in one or more characteristics are genetically different. Whether these genetic differences are great enough to warrant classification in different species is left to the judgment of the investigator. This situation differs from that encountered with higher plants and animals, where the criterion is the ability of the two forms to interbreed—a matter in which the judgment of the investigator is not concerned.

The Type Culture. A newly discovered organism is described as completely as possible, and at the same time a typical culture is designated the _type culture._ This culture should be deposited in a central collection, such as the American Type Culture Collection or the British equivalent, the National Collection of Type Cultures. The type culture is then available for any investigators who wish to compare other organisms with it. Indeed, some authorities define a bacterial species as the type culture or specimen and all other cultures or specimens regarded as sufficiently like the type to be grouped with

it. This, it will be noted, is by implication the same as the definition stated earlier.

GROWTH OF BACTERIAL TAXONOMY

Bacteria were discovered by Leeuwenhoek in 1676, but few investigators possessed the interest, facilities, or know-how to study them for almost a century and a half. It is not surprising, therefore, that the famous naturalist Linnaeus, in 1767, included bacteria in his class Chaos, along with various other unrelated forms and ethereal substances. Mueller in 1773 listed two genera containing bacterial species: _Monas_ consisted of spherical or ovoid forms; _Vibrio_ contained longer rods but was composed largely of spiral organisms. Sixty-five years later, Ehrenberg (1838) added four genera: _Bacterium, Spirillum, Spirochaeta,_ and _Spirodiscus._

Serious attempts to classify bacteria began with the work of Cohn (1872), and for a score of years about one new classification scheme was proposed each year. New genera and species were added at an accelerating rate as bacteriologic research expanded into medical, agricultural, and industrial microbiology (Fig. 4–2).

Modern attempts to systematize the classification of bacteria began with the studies of Buchanan (1917), who proposed that the

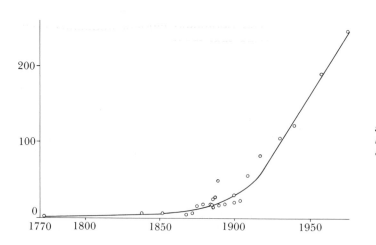

Figure 4–2. "Growth curve" of systematic bacteriology. Numbers of bacterial genera listed by various authors between 1773 and 1975.

class Schizomycetes be divided into six orders. These and many other suggestions were included in the first edition of a *Manual of Determinative Bacteriology,* published in 1923 by a committee of the Society of American Bacteriologists of which the late Dr. D. H. Bergey was chairman. Thirteen families of bacteria were listed and a total of 88 genera.

The *Manual* has been revised at intervals. The eighth edition, published in 1974, represented a departure from traditional systems of classification, recognizing the impossibility of establishing a complete and meaningful hierarchy, as had previously been attempted. The 19 Parts into which the division Bacteria is subdivided include 247 genera. More than 1500 species are listed. Viruses are not included.

The growing number of categories for classifying bacteria reflects the discovery of previously undescribed organisms and changing opinions as to the taxonomic significance of forms already described. The latter consideration also dictates frequent rearrangements in the scheme as different investigators assist in the preparation of succeeding editions. The committee that edits *Bergey's Manual of Determinative Bacteriology* recognizes that it has shortcomings and welcomes suggestions for further revision. Nearly 140 specialists collaborated in writing the eighth edition, almost half of them from outside the United States. Their classification scheme is therefore as nearly international in character as can be achieved. Even so, they realize that it is only temporary and will be greatly changed in succeeding editions as further research dictates.

NAMING OF BACTERIA

SCIENTIFIC NAMES

Bacteria are named according to the *binomial* system proposed by Linnaeus in 1753. The *scientific name* of an organism consists of two words, genus and species. These names are written in a Latinized form and should be italicized (or underscored

when written by hand or typewritten). The generic name is always capitalized, the species need never be capitalized even though derived from a proper noun. Words that are used both as generic names and as morphologic descriptions are capitalized when used to refer to the genus but not when used to indicate the morphologic type; for example, *Streptococcus* designates organisms of the genus, but streptococcus (uncapitalized) refers to spherical cells in a chain form.

A certain amount of abbreviation is permissible if the abbreviations are clearly understandable. Generic names may be abbreviated to the first initial or first few letters; species names are never abbreviated. The first time a scientific name is used in a paper or chapter (or examination) it should be written out in full; thereafter it may be abbreviated.

COLLOQUIAL NAMES

Many bacteria have been studied so much or are discussed so frequently that they have acquired common or *colloquial* names. These names have no scientific status and should not be used in scientific writing; it must be confessed, however, that they frequently appear. *Escherichia coli,* for example, is such a universal inhabitant of the large intestine that it is frequently called the colon bacillus.

DERIVATIONS OF BACTERIAL NAMES

The beginning student frequently complains that there are too many bacterial names to learn and that they do not make sense. Actually, most bacterial names do make sense, but a little search is sometimes required to ascertain their derivation. The effort is rewarding because names are more easily remembered if their associations are known. Scientific names of bacteria usually indicate something distinctive about the organism such as its discoverer (*Escherichia,* by Escherich), its source or habitat (*coli,* the colon), morphology (*Spirillum,* spiral), pigmentation (*aureus,* golden), physiologic peculiarity (*aerogenes,* gas producing), pathogenicity (*typhi,* typhoid fever), or cultural character (*mesentericus,* mesentery or membrane). Derivations are given in *Bergey's Manual.* Names should not be memorized by rote but should be understood and associated with other facts about the organism.

IDENTIFICATION OF BACTERIA

Identification of an organism is the process of determining its species. As many as possible of its characteristics are ascertained by appropriate observations and tests, and the accumulated information is then compared with the published descriptions of the various species. The organism is properly identified when a species description is found that is identical with the observed characteristics.

ISOLATION OF PURE CULTURES

Accurate identification of most bacteria can be made only with pure cultures because many characteristics used in classification

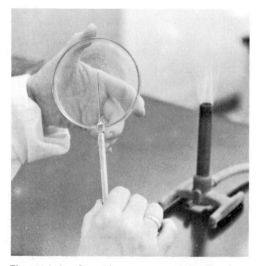

Figure 4–3. *Streaking an agar plate. The inoculum on a straight wire needle is streaked rapidly and lightly back and forth across the medium, starting at the top as the plate is held in the position indicated. When the middle is reached, the plate is rotated 180° and streaking started again at the top. See the diagram in Figure 4–4*(a).

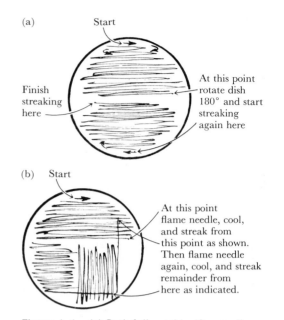

Figure 4–4. (a) *Path followed by the needle on a well streaked agar plate. Fifty to 100 streaks should be made to ensure properly isolated colonies as illustrated in Figure 4–5. This method works well with broth enrichment cultures or liquid suspensions.* (b) *An alternate method of streaking, particularly adapted to isolation of pure cultures from heavily populated specimens (e.g., sputum or feces) or from the mixed growth on solid media.*

depend upon the behavior of populations rather than individual cells. It is true that some bacterial properties change when the organisms are taken from the natural state and cultivated in the laboratory. However, classification schemes are ordinarily based on the properties of bacteria in laboratory cultures, so the net result is usually satisfactory.

Isolation of a pure culture is accomplished by securing the progeny of a single cell (or sometimes a group of identical cells, such as a streptococcus chain). Various techniques are employed: single cell isolation by means of a mechanical micromanipulator, selective enrichment of the desired organism or inhibition of all undesired organisms, or "plating."

"Streak" or "Spread" Plates. Plating is the most widely used method for purifying cultures. "Streak" plates (Fig. 4–3) are prepared by streaking a small amount of the mixed bacterial specimen over the surface of a solid medium in a Petri dish with a platinum or a nichrome wire needle (Fig. 4–4).

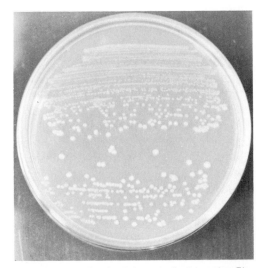

Figure 4–5. *Petri dish streaked as in Figure 4–4(a) showing well isolated colonies in the lower half.*

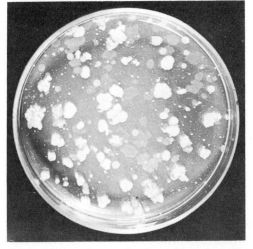

Figure 4–6. *A "poured" plate with at least three different kinds of colonies: large, irregular; medium, smooth; pinpoint. Colonies were "picked" as illustrated in Figure 4–7, and pure cultures were obtained on agar slants (Fig. 4–8).*

Each laboratory has its own favorite streaking procedure. It is essential to cover the agar thoroughly with the material so that the bacteria are well distributed (Fig. 4–5). The inoculum may also be spread with a sterilized, bent glass rod.

"Poured" Plates. "Poured" plates (Fig. 4–6), are prepared by diluting the bacterial mixture serially in tubes of melted and cooled (45° to 50° C.) agar medium which are then poured into sterile Petri dishes and allowed to solidify. Dilutions may be made with the inoculating loop or by pipette.

Poured plates contain subsurface (i.e., within the agar) as well as surface colonies.

Subsurface colonies are smaller than surface colonies of the same species, and their shape is often different. Small, gram-negative rod bacteria, for example, form lens-shaped subsurface colonies, but their surface colonies are smooth, circular, and convex in cross section.

Colony Subculture. After incubation, some well separated colonies should appear on properly inoculated plates. Several such colonies are subcultured by transferring a small amount of each colony to broth or agar medium in test tubes (Figs. 4–7 and 4–8). It is assumed that each colony consists of the

Figure 4–7. *Picking or "fishing" a colony from a Petri dish culture. The plate is held so that the colony can be seen clearly, either by transmitted or by reflected light as circumstances dictate. The inoculating needle is held in such a manner as to give good control, and a small bit of the desired colony is removed.*

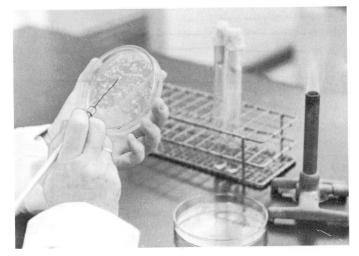

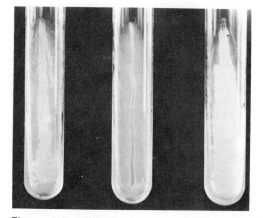

Figure 4–8. *Growth of pure cultures obtained by inoculating agar slants with the three kinds of colonies shown in Figure 4–6.*

progeny of a single bacterium. This is not always true, because occasionally two bacteria stick to the agar so close together that their colonies merge. It is therefore desirable to replate a second or third time.

Special Isolation Methods. Special methods assist in the isolation of certain organisms. If it is desired to purify a spore-forming bacterium, for instance, a suspension containing spores mixed with vegetative cells of the same and other organisms may be heated to 85° C. for five minutes or treated with an appropriate disinfectant to kill all cells except the spores. Plating will then more readily yield a pure culture of the spore-former.

A method related to the foregoing consists of adding inhibitory or germicidal chemicals to culture media to suppress or kill unwanted types of bacteria while permitting the desired species to multiply. Basic dyes such as crystal violet and brilliant green inhibit gram-positive bacteria and aid in the isolation of gram-negative rod bacteria from mixed specimens (e.g., sewage).

Enrichment Cultures. The principle of selective enrichment has already been discussed (page 32). Applications of this method are limited chiefly by the ingenuity of the investigator. Lactose-fermenting bacteria, for example, are selectively enriched or stimulated in a medium containing this sugar, and even though they may occur as a minority in a mixed population (as in drinking water), within 24 to 48 hours they will predominate in a suitable lactose broth and can then easily be isolated.

Serial Dilution. An organism that predominates in a mixture can often be purified by dilution. Inasmuch as a single viable cell is sufficient to initiate growth in an appropriate medium, if the mixture is suitably diluted and aliquots of the dilution are subcultured the laws of probability dictate that some of the subcultures will consist of the desired organism in pure form.

Single Cell Isolation. Single cells can be isolated with a micromanipulator, by which very fine glass pipettes can be manipulated mechanically in the field of the microscope. Droplets containing individual bacterial cells are deposited on the bottom of a coverglass in a special chamber. The droplets are then transferred by sterile micropipettes to suitable nutrient media.

GROWTH REQUIREMENTS

Conditions necessary to cultivate the organism must be determined, both for use in subsequent tests and because these conditions are frequently distinctive. Temperature requirements are ascertained by incubating cultures at a variety of temperatures. Various synthetic and nonsynthetic media are used to determine the nutrient requirements of the organism: whether it requires blood or serum, certain amino acids, or other growth factors. The ability to grow in the absence of atmospheric oxygen is of interest. A simple way to determine this is to inoculate uniformly a deep tube of melted and cooled agar medium, allow the medium to solidify, and then incubate it. Growth at the surface indicates an aerobic organism, growth in the depths of the medium indicates an anaerobe, and growth throughout is obtained with a facultatively anaerobic organism, that is, an aerobe that can grow anaerobically (Fig. 4–9).

Some bacteria grow in a layer a few millimeters below the surface; these have usually been called *microaerophiles,* because they were presumed to be inhibited by oxygen at normal atmospheric tension but not at reduced tension. This may be true in the case

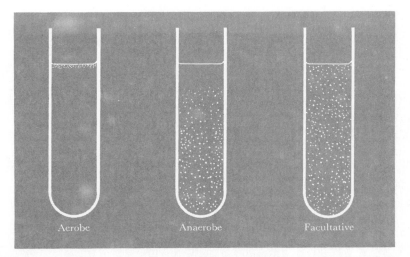

Figure 4–9. *Growth of aerobic, anaerobic, and facultative bacteria in agar "shake" cultures. Deep tubes of a melted and cooled nutrient agar are inoculated, mixed, and allowed to harden. Colony growth occurs as indicated, according to the oxygen requirements of the organism.*

of a few bacteria that possess a vital enzyme that is partially sensitive to oxygen. Other organisms appear rather to be stimulated by carbon dioxide in greater than normal atmospheric concentration—a condition that might obtain within the agar after some metabolic activity has occurred.

MORPHOLOGY

The shape and arrangement of the cells are determined by microscopic examination of wet mounts or hanging drop preparations and stained smears. Examination of the living organisms also reveals motility and gives presumptive evidence of the presence of flagella; nonmotile bacteria frequently display brownian movement, a vibratory motion caused by molecular activity in the fluid. The Gram stain indicates not only the shape and arrangement of the cells but also their Gram reaction; this is useful information. The acid-fast stain, spore stain, capsule stain, and flagella stain are also frequently necessary and provide additional data.

It is important to remember that the morphology of a bacterium varies with the conditions under which it is cultivated. The age of the culture is also significant.

CULTURAL CHARACTERISTICS

The gross appearance of bacterial cultures in various media provides clues to the identity of the organism and should always be noted. Descriptions of colonies on agar plates should include size, shape, color, appearance by reflected and by transmitted light, and appearance when examined with the low power (100X) of the microscope (Fig. 4–10). Similar information can also be secured from agar slant cultures. Growth in broth may consist of uniform turbidity or cloudiness, a sediment at the bottom of the culture, or a membrane or pellicle on the surface. These growth characteristics are frequently typical of certain species, although, as in the case of morphology, temporary variations sometimes occur under different cultural conditions.

BIOCHEMICAL CHARACTERISTICS

The general category of biochemical characteristics includes a variety of physiologic properties.

Carbohydrate Dissimilation. The ability of an organism to attack and break down various carbohydrates can be deter-

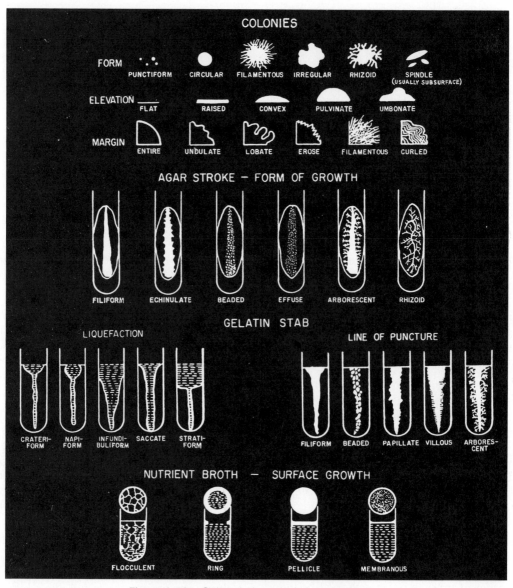

Figure 4–10. *Cultural characteristics of bacteria.*

mined by the use of a suitable nutrient medium containing the carbohydrate and an acid-base indicator, such as bromcresol purple (see Table 4–1). A liquid medium is usually dispensed in Durham tubes, which contain inverted vials to collect some of the gas that may be produced (Fig. 4–11). The formation of acid or of acid and gas is an indication that the carbohydrate is attacked. Chemical analysis of the acids, gases, and other products is not attempted in routine identification (Plate IVG).

The metabolic processes by which the various products are formed in a Durham tube are largely anaerobic because the free oxygen is soon exhausted, especially within the inverted vial. Acid and gas formation under these conditions is called *fermentation*. Some bacteria dissimilate carbohydrates only aerobically. Special methods are required to demonstrate such oxidative formation of acid. Oxidative and fermentative production of acid may often be distinguished in a carbohydrate medium made

TABLE 4–1.   Some Acid-Base Indicators Commonly Used in Microbiology

Indicator	pH Range	Acid Color	Basic Color
Bromcresol green	3.8– 5.4	Yellow	Blue
Methyl red	4.2– 6.3	Red	Yellow
Bromcresol purple	5.4– 7.0	Yellow	Purple
Litmus	4.5– 8.3	Red	Blue
Bromthymol blue	6.1– 7.7	Yellow	Blue
Phenol red	6.9– 8.5	Yellow	Red
Phenolphthalein	8.3–10.0	Colorless	Red

semisolid by the addition of 0.3 per cent agar. Two test tubes of medium are inoculated by stabbing, and the medium in one tube is covered with sterile petrolatum to exclude oxygen. Fermentative organisms produce acid throughout the medium in both tubes, whereas oxidative organisms produce acid in the unsealed tube only, at the surface.

Hydrolytic digestion of starch is demonstrated by the use of an agar medium containing 1 per cent starch. A Petri dish of starch agar is inoculated in a single spot, and after a large colony is formed the medium is flooded with iodine solution, which stains unhydrolyzed starch an intense blue-black. Colonies of bacteria that attack starch are surrounded by a colorless zone.

The digestion of cellulose can be demonstrated in a liquid medium to which strips of filter paper have been added before sterilization. Aerobic cellulolytic organisms visibly decompose the paper at the surface of the liquid, anaerobes at the bottom. Digestion by organisms that grow best at moderate temperatures (20° to 37° C.) usually requires several weeks, whereas thermophilic decomposition (e.g., 55° to 65° C.) is evident within a few days.

Proteolysis Tests.   The ability of an organism to attack proteins can be determined in several ways. Blood serum coagulated in a slanting position in test tubes is digested and liquefied by certain proteolytic bacteria. Deep tubes of nutrient gelatin, which consists of nutrient broth solidified by addition of 10 per cent gelatin, are inoculated by stabbing with a straight wire needle. If these are incubated at 20° C., the undigested gelatin remains solid, and the zones of liquefaction are of characteristic shape (Fig. 4–10). Gelatin may also be incubated at a higher temperature and cooled just before making observations; gelatin that has been digested will no longer solidify below 20° C.

Hydrolysis of gelatin is also demonstrated on Petri dish cultures of Frazier gelatin-agar inoculated in the same manner as starch agar. After colony development, the medium is flooded with $HCl-HgCl_2$ solution, which produces an opaque, white precipitate wherever undigested gelatin remains.

Hydrogen Sulfide and Indole Production.   A further indication of attack on protein breakdown products is provided by tests for hydrogen sulfide and indole.

Some organisms can produce H_2S by reduction of sulfur-containing amino acids such as cysteine. It is usually detected by adding a lead or iron salt to a medium containing protein or peptone and agar. Stab cultures are incubated, and the formation of

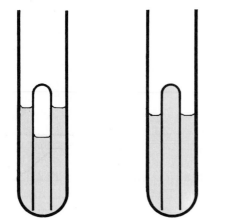

Figure 4–11. *Durham fermentation tubes.* Left, *With gas;* right, *without gas. (How are the inner vials filled with liquid?)*

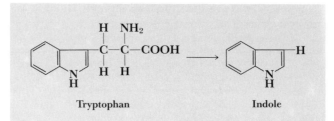

Figure 4–12. *Production of indole from tryptophan by removal of the sidechain.*

black metallic sulfide along the line of inoculation indicates production of H_2S.

The ability of an organism to produce indole from tryptophan (Fig. 4–12) can be tested in almost any medium containing sufficient tryptophan. Nutrient broth is satisfactory, but tryptone or trypticase broth is generally used because it contains more tryptophan. Indole is detected after incubation by adding a solution containing p-dimethyl aminobenzaldehyde. Kovacs' solution contains this reagent together with amyl alcohol, which extracts the indole and brings it to the surface in a thin layer, where a red color appears.

Nitrate Reduction. Many bacteria utilize nitrates as hydrogen acceptors and reduce them to nitrites, free nitrogen, or ammonia. Nitrate reduction tests are usually performed in a liquid medium containing a small amount of peptone and sodium or potassium nitrate. Nitrite is detected by adding sulfanilic acid and α-naphthylamine; a pink to dark red color is the positive test. Nitrogen gas is collected in an inverted vial as in fermentation tests. Ammonia can be detected by adding Nessler's reagent to a few drops of the culture in a spot plate; a yellow to orange color is produced. If the nitrite test is negative, the persistence of unreduced nitrate can be ascertained by adding a knife-point of zinc powder; this reduces residual nitrate to nitrite in the presence of acid, and the nitrite produces the characteristic pink to dark red color when it reacts with the sulfanilic acid and α-naphthylamine.

Fermentation and Proteolysis of Milk. Milk cultures of certain bacteria provide much useful information. Dairy bacteriologists can often identify the organisms with which they are concerned almost entirely on the basis of their morphology and their behavior in milk.

Sterile skim milk containing an indicator such as litmus is a complete medium in which many species grow luxuriantly and in which a variety of physiologic characteristics can be determined. Milk contains lactose, casein and other proteins, and various minerals. Fermentation of the lactose produces an acid reaction; some organisms produce enough acid to curdle (solidify) the milk proteins. If the organism also produces gas, bubbles collect as foam at the top, or break or score the curd as they rise.

Digestion of the casein and other proteins produces a dirty brownish color, and the milk becomes watery; this appearance starts at the top, particularly with an aerobic organism, and gradually progresses downward until the entire contents of the tube are digested. The reaction often becomes alkaline as ammonia is liberated during proteolysis. Protein digestion is known as *peptonization*.

Decolorization or *reduction* of the indicator is characteristic of the behavior of certain bacteria in litmus milk. Litmus shares with some other dyes the property of behaving as an oxidation-reduction indicator as well as an acid-base indicator; that is, it accepts hydrogen from appropriate enzyme systems and is converted into a colorless form known as a leuko-dye. Bacteria that multiply and metabolize rapidly decolorize litmus vigorously; other bacteria decolorize it more slowly or not at all.

Catalase and Oxidase Tests. Two additional tests are useful in identifying many commonly encountered bacteria. The enzyme catalase is easily detected by adding a few drops of 3 per cent H_2O_2 to a colony or to the growth on a slant culture, or by transferring a needlepoint of growth to a drop of the peroxide on a slide or a Petri dish cover. The evolution of gas is the positive test. The reaction is as follows:

$$2H_2O_2 \xrightarrow{\text{Catalase}} 2H_2O + O_2$$

Indophenol oxidase is detected by its reaction with dimethyl or tetramethyl p-phenylene diamine. The reagent, freshly prepared as a 1 per cent solution in water or α-naphthol, may be added dropwise to colonies, which quickly become pink, purple, and then black if the enzyme is present. The test may also be made by rubbing solid growth with a platinum loop or glass rod onto filter paper that has been moistened with the reagent. A purple-black spot is the positive test.

Rapid Biochemical Tests. It will be noted that the various biochemical tests described are simple and are quickly and easily performed after the required period of incubation. Simplicity and speed are virtues if combined with reliability because the investigator does not hesitate to use such tests freely.

The process of identifying bacteria can be shortened or simplified by various rapid tests. One procedure makes use of small amounts of concentrated medium inoculated with large numbers of bacteria; short incubation times are sufficient. Another procedure uses disks of filter paper impregnated with test substances such as carbohydrates. A nutrient agar containing an appropriate indicator is inoculated by streaking with the test organism, and the disks are placed upon the agar surface. The carbohydrates diffuse from the paper, and if acid is produced from a given carbohydrate the corresponding disk will be surrounded by a zone displaying the acid color of the indicator.

Interest in rapid tests is increasing, especially among clinical diagnostic microbiologists, because they decrease the time and expense of ascertaining the cause of infection. Several manufacturers are promoting use of kits or special multitubes with which a large number of tests can be performed with virtually a single inoculation, the results being secured in only a few hours. Before attempting their serious use, however, one should acquire facility with and understanding of the individual tests.

Biochemical Tests, Enzymes, and Genes. It is well to reemphasize that every physiologic test depends on one or more enzymes produced by the bacterium and that these enzymes are under genetic control. Bacterial genetics has not advanced to a point where all genes can be specifically located along chromosomes, but for practical purposes that information is not necessary. The important point is that physiologic studies, like other tests used to characterize bacteria, are fundamentally genetic studies.

SEROLOGIC PROPERTIES

Bacterial cells contain numerous antigenic substances, that is, substances that stimulate animals to produce antibodies capable of reacting specifically with the antigens. Antibodies are found in the sera of the inoculated animals; the sera therefore can be used as reagents to detect the corresponding antigens.

The agglutination test is most frequently used in identifying bacterial antigens. This test is not difficult and is performed routinely in diagnostic laboratories. A saline (e.g., 0.85 per cent NaCl) suspension of the bacteria is mixed with the test antisera; after appropriate incubation, the bacteria that contain antigens corresponding to the antibodies in an antiserum clump together in compact granules or in loose flocculent masses. A suitable set of test antisera provides a means of identifying species within a genus. Usually it is necessary to determine the genus on the basis of morphology and other characteristics. Antisera are available commercially for a few genera; laboratories conducting research in other genera prepare their own sera.

PATHOGENICITY

The value of pathogenicity in bacterial identification is limited because suitable means for testing are not always available, and because some organisms lose their pathogenicity after prolonged laboratory cultivation. It is used as a guide, however, in diagnostic medical bacteriology; for example, specimens from a patient suspected of having typhoid fever are examined for the presence of *Salmonella typhi*. Pathogenicity can sometimes be tested in the laboratory by the inoculation of animals. Certain organisms are accurately identified only in this manner. The most conclusive test for *Corynebacterium diphtheriae* is the production of typical symptoms and death in a guinea pig, especially if the same symptoms are prevented in a second guinea pig by simultaneous administration of diphtheria antitoxin.

IDENTIFICATION BY MEANS OF A KEY

Armed with as complete a description of the unknown organism as possible, the student or investigator will search through a suitable key for an organism with the same characteristics. Most keys are so constructed as to present a series of dichotomies, and it is relatively simple to follow the path through the contrasting characteristics and ultimately arrive at the scientific name of the organism. Figure 4–13 is a diagram of such a key, and its use should be obvious. The keys in *Bergey's Manual* are similar, but alternate choices are indicated by numerals (I vs. II; 1 vs. 2), letters (a vs. b), and by single or repeated letters (a vs. aa). The eighth edition of the *Manual* makes extensive use of tables of characteristics, which may require a longer time to examine carefully but more often lead to a reasonable conclusion.

Occasionally, unknown organisms are encountered that cannot be identified by the keys. There are various reasons why this is so. Many bacteria have not yet been described in sufficient detail to justify their inclusion in a manual; other organisms have not been considered of sufficient importance. The descriptions given in *Bergey's Manual* are based upon the characteristics of the *majority* of strains examined by experts with the group of organisms in question; a given isolate may deviate in one or more particulars from the majority. Finally, and perhaps most important, particularly for beginning investigators, the description of the unknown is sometimes at fault. The culture should be checked for purity and test methods should be reviewed. Positive and negative controls should be examined. Careful technique and a little practice in the use of keys and tables are usually rewarded by increasing success and facility in the identification of unknown bacteria.

NUMERICAL TAXONOMY

The development of computers has made feasible the application of a principle stated by Adanson in 1757. As mentioned earlier

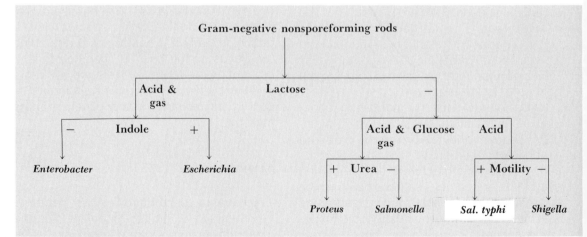

Figure 4–13. *Diagram of a dichotomous key to some members of the Enterobacteriaceae.*

TABLE 4–2. Similarity matrix consisting of similarity percentages calculated from hypothetical data on six bacterial strains and tabulated in random order. The relationships between the strains are not readily apparent from this matrix.

Strain	1	2	3	4	5	6
1	100					
2	10	100				
3	95	0	100			
4	75	15	80	100		
5	10	85	20	30	100	
6	10	70	15	40	90	100

TABLE 4–3. Similarity matrix consisting of percentages from Table 4–2 rearranged to show relationships. The strains in Group A are closely related to one another (similarity 75% or greater), as are those in Group B (similarity 70% or greater). Groups A and B are not closely related to each other (similarity 40% or less).

Strain	Group A			Group B		
	1	3	4	6	5	2
1	100					
3	95	100				
4	75	80	100			
6	10	15	40	100		
5	10	20	30	90	100	
2	10	0	15	70	85	100

(page 63), most taxonomists tend to consider certain characters more significant than others in the study of particular groups of organisms. Classification under these conditions tends to be an art rather than a science. Adanson considered that all characters are equally important in establishing natural groups. As many features as possible should be used, and the relationships between groups is a function of the similarities of the characters compared. Application of these principles results in a purely mathematical method of classification, which is now called *numerical taxonomy.*

The affinity between taxonomic units is evaluated mathematically, and these units are arranged in an order based upon their *similarity coefficients.* The similarity or affinity between strains is defined by the ratio of the number of characters they possess in common to the total number of characters compared and is usually expressed as a percentage. Characters that are negative in both organisms are not included. In practice, a number of strains are studied and as many characters as possible are determined. Similarity percentages (characters in common/total characters tested × 100) are then calculated between each of the different pairs of organisms. These are tabulated in such a manner as to group related organisms together. This can be illustrated by a hypothetical example. Assume six organisms subjected to each of 10 tests and the similarity coefficients calculated as above. The results tabulated in random order in a *similarity matrix,* as shown in Table 4–2, do not indicate clear relationships, but when rearranged as in Table 4–3 it appears that strains 1, 3, and 4 are related to one another, and strains 2, 5, and 6 are interrelated. Each group might be accorded special status (e.g., species rank) in a taxonomic scheme.

Numerical taxonomy provides information that may help to decide how to group organisms. Computers can readily be programmed to perform the necessary calculations and to rearrange the data so that they reveal the relationships indicated. Computers are also ideally suited to assist in identification of bacteria according to the conventional dichotomous key. The observed characters are punched into a card or tape, which is then run through a computer in which are stored the characteristics of all possible organisms.

SUPPLEMENTARY READING

Ainsworth, G. C., and Sneath, P. H. A. (eds.): *Microbial Classification.* Twelfth Symposium of the Society for General Microbiology. New York, Cambridge University Press, 1962.
Buchanan, R. E., and Gibbons, N. E. (eds.): *Bergey's Manual of Determinative Bacteriology,* 8th ed. Baltimore, The Williams & Wilkins Company, 1974.

Gibbs, B. M., and Skinner, F. A.: *Identification Methods for Microbiologists, Part A.* New York, Academic Press, 1966.

Gibbs, B. M., and Shapton, D. A.: *Identification Methods for Microbiologists, Part B.* New York, Academic Press, 1968.

Leone, C. A. (ed.): *Taxonomic Biochemistry and Serology.* New York, The Ronald Press Co., 1964.

Marmur, J., Falkow, S., and Mandel, M.: New approaches to bacterial taxonomy. *Ann. Rev. Microbiol., 17:*329–372, 1963.

Skerman, V. B. D.: *A Guide to the Identification of the Genera of Bacteria.* 2nd ed. Baltimore, The Williams & Wilkins Co., 1959.

Society of American Bacteriologists: *Manual of Microbiological Methods.* New York, McGraw-Hill Book Co., Inc., 1967.

Sokal, R. R.: Numerical taxonomy. *Sci. Amer., 216:*106, 1967.

Sokal, R. R., and Sneath, P. H. A.: *Principles of Numerical Taxonomy.* San Francisco, W. H. Freeman Co., 1963.

SECTION TWO

BIOLOGY OF LOWER PROTISTS

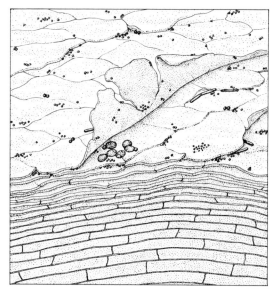

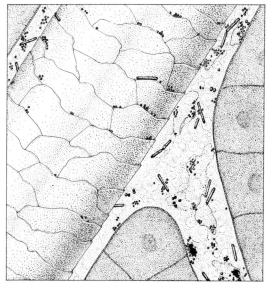

From *Life on the Human Skin* by Mary J. Marples. Copyright © 1969 by Scientific American, Inc. All rights reserved.

MORPHOLOGY AND 5
STRUCTURE OF
BACTERIA

Casual microscopic examination of bacteria, either unstained or stained with the reagents usually found in a diagnostic laboratory, shows little detail beyond the shape, arrangement, and size of the cells. There is no indication of internal structure. Special staining methods, like some of those described in Chapter 3, together with phase microscopy and electron microscopy, are necessary to reveal the internal architecture of the individual cell. These techniques show that bacterial cells are less highly organized than those of higher forms. Indeed, procaryotic cells in general are structurally much simpler than eucaryotic cells.

SHAPES AND ARRANGEMENTS OF BACTERIA

COCCI, RODS, AND SPIRALS

The three principal bacterial shapes are spheres, rods, and spirals (Fig. 5–1). A spherical bacterium is known as a *coccus* (plural, *cocci*). Cocci are not always perfect spheres; they are often flattened like a compressed rubber ball or elongated like a football (Figs. 5–2 and 5–3) and pass through a spherical stage during growth. They appear singly or in pairs (*diplococci*), chains (*streptococci*), irregular, grapelike clusters (*staphylococci*), or boxlike cubical packets (*sarcinae*).

Rodlike bacteria are straight and cylindrical or are long ellipsoids (Fig. 5–4); sometimes they appear slightly curved or wavy.

They occur singly, in chains, or in a parallel (*palisade*) arrangement.

A spiral organism or *spirillum* (plural *spirilla*) is shaped like a corkscrew (Fig. 5–5). Some species consist of only part of a spiral turn and are called *vibrios,* some are composed of one or two loose turns, and others contain many close coils (Fig. 5–6). They are found as single cells or short chains.

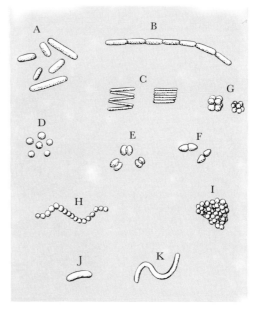

Figure 5–1. *Shapes and arrangements of bacteria commonly encountered. A, Short and long rods occurring singly; B, rods in a chain; C, palisade arrangement of rods; D, single cocci (spheres); E, paired flattened cocci; F, paired elongate cocci; G, cubical packets of cocci (sarcina); H, a chain of cocci (streptococcus); I, an irregular cluster of cocci (staphylococcus); J, comma-shaped or bent rod; K, spiral rod.*

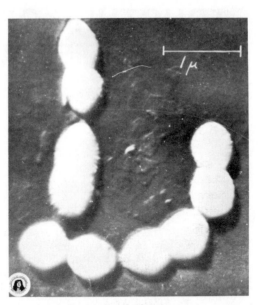

Figure 5–2. *Electron micrograph of* Streptococcus pneumoniae, *type 2, shadowed with chromium. Several cells show the characteristic lancet shape. (Photograph by R. C. Williams; A.S.M LS-162.)*

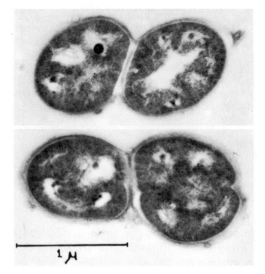

Figure 5–3. *Electron micrograph of ultrathin section of* Branhamella catarrhalis. *(Photograph by G. B. Chapman.)*

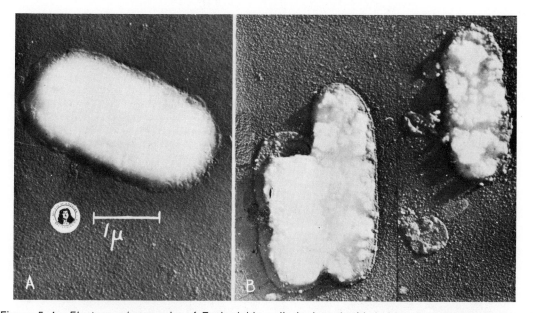

Figure 5–4. *Electron micrographs of* Escherichia coli *shadowed with gold and manganin. These cells were from a one-hour broth culture. The cell at* A *is normal; those at* B *were heated in saline (0.85% NaCl) 10 minutes at 50° C. and show an interesting granulation of the cytoplasm caused by coagulation of some protoplasmic constituents. (Photographs by G. Hedén and R. W. G. Wyckoff; A.S.M. LS-290.)*

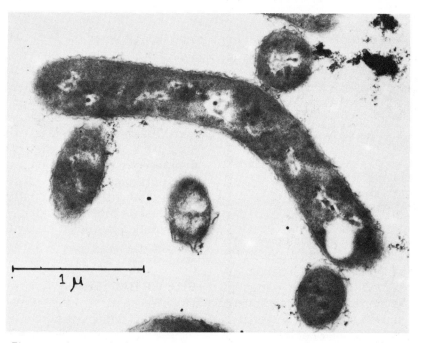

Figure 5–5. *Electron micrograph of an ultrathin section of* Rhodospirillum rubrum. *Part of a coil is shown. (Photograph by G. B. Chapman.)*

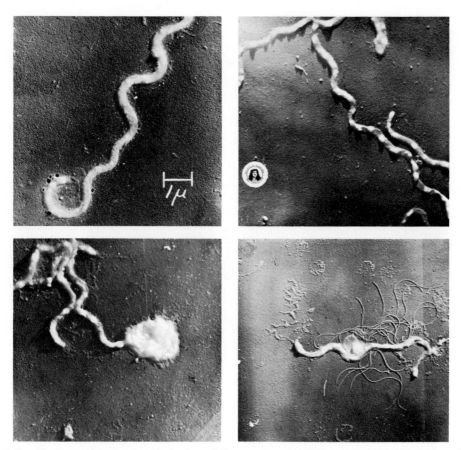

Figure 5–6. *Electron micrographs of shadowed* Treponema vincentii, *an oral spirochete found in cases of trenchmouth associated with a long, tapering rod-shaped bacterium. (Photographs by E. G. Hampp, D. B. Scott, and R. W. G. Wyckoff; A.S.M. LS-248.)*

The distinction between rodlike bacteria and cocci or spirilla cannot be defined accurately and is learned principally by experience. It is impossible to state the exact point at which a cell ceases to be considered a long coccus and becomes a short rod, or at which a rod is sufficiently twisted to be called a spirillum.

MORPHOLOGIC VARIATION; PLEOMORPHIC FORMS

Most bacteria at some time display irregular or variant shapes known as *pleomorphic* forms. These variations may be either permanent or temporary. Permanent variations (mutations) are the result of genetic alteration and are relatively stable and irreversible. Permanent loss of flagella or of the ability to produce endospores illustrates this type of

morphologic variation. Temporary variations, such as changes in cell length or shape, occur during growth and division.

Involution Forms. Involution forms are cells of bizarre shape, filamentous or swollen, formed under abnormal cultural conditions. They are produced as a result of alterations in the osmotic pressure of the medium, the presence of metallic ions, antibiotics, or other chemicals, or the accumulation of waste products. They are particularly likely to appear as angular, budding, or branching forms during the death phase of a culture.

SIZE OF BACTERIA

Bacterial dimensions are difficult to determine with accuracy because considerable shrinkage occurs during the preparation of

TABLE 5–1.   Sizes of Some Bacteria

Organism	Diameter (μm)	Length (μm)
Francisella tularensis	0.2	0.3– 0.7
Brucella melitensis	0.3	0.3– 1.0
Escherichia coli	0.4–0.7	1.0– 3.0
Bacillus anthracis	1.0–1.3	3.0–10.0
Staphylococcus aureus	0.8–1.0	
Streptococcus pyogenes	0.6–1.0	
Sarcina ventriculi	3.5–4.0	
Beggiatoa mirabilis	15.0–21.5	Several cm.
	Trichomes consisting of segments 5.0 to 13.0 μm long	

fixed and stained smears. Stained, dried cells of *Escherichia coli,* for example, may appear less than one-third the size of living cells. Bacteria as commonly observed vary greatly in size; the range among species ordinarily encountered is twenty- to thirtyfold. A few representative and extreme figures are listed in Table 5–1.

It is worth emphasizing that the microscopic appearance of bacteria, including shape, arrangement, and size, depends greatly upon the age of the culture and other factors. Actively growing rod bacteria are usually several times longer than old, dormant, or dying bacteria, which are frequently almost spherical. Rapidly growing cocci may be 50 per cent longer than they are wide. Examination of cells from various growth phases (i.e., young as well as older cultures) is often needed before an unknown organism can be identified as a rod or a coccus.

CHEMICAL COMPOSITION OF BACTERIA

The chemical composition of bacteria is similar to that of other organisms. Although there is considerable variation among species, the approximate composition of a rep-

resentative cell is indicated in Table 5–2.

The DNA in *Escherichia coli* is a single molecule, 1100 to 1400 μm (1.1 to 1.4 mm.) in length. It has what is probably the highest molecular weight of any material in nature and its weight amounts to about 10^{-14} gram per cell.

There are several kinds of RNA: messenger RNA (mRNA), with a molecular weight of 1,000,000; ribosomal RNA (rRNA), associated with protein in cytoplasmic granules, or *ribosomes,* which are the sites of protein synthesis; and transfer RNA (tRNA), sometimes called soluble RNA. For each kind of amino acid incorporated in a protein, a different transfer RNA is necessary. The molecules of tRNA are relatively small, being composed of 80 to 100 nucleotide units and having a molecular weight between 25,000 and 30,000. The amount of RNA in actively growing cells is almost double that in inactive cells.

Nearly all the amino acids have been detected in the proteins. A special amino acid, diaminopimelic acid (DAP), is found in almost all species of bacteria except grampositive cocci and related organisms. It is also present in blue-green algae but is not found in other forms of life. It is part of a polypeptide attached to muramic acid (see Figure 5–7).

Carbohydrates are present in the cell walls and capsules of bacteria. The capsules of some species are composed solely of polysaccharides. Cell walls of gram-positive bacteria contain 35 to 60 per cent carbohydrate; those of gram-negative bacteria

TABLE 5–2.   Approximate Composition of a Bacterial Cell (after Pollard, 1965)

Water   70%
Dry weight   30%, composed of:
  DNA   3% (M.W. = 2,000,000,000)
  RNA   12%
  Protein   70%, found in:
    Ribosomes (10,000) (RNA-protein particles, M.W. = 3,000,000)
    Enzymes
    Surface structures
  Polysaccharides   5%
  Lipid   6%
  Phospholipid   4%

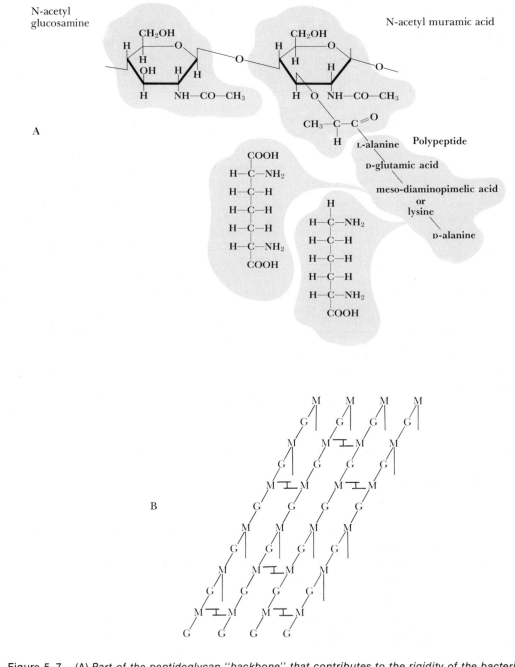

Figure 5–7. (A) *Part of the peptidoglycan "backbone" that contributes to the rigidity of the bacterial cell wall. Diaminopimelic acid is found only in certain bacteria and blue-green algae. (B) Diagrammatic representation of a portion of the peptidoglycan (murein) sheet comprising the cell wall of* E. coli. *G (N-acetyl glucosamine) and M (N-acetyl muramic acid) are joined (diagonal lines) by β-1,4-glycosidic bonds. Polypeptide side-chains are represented by vertical lines and by the symbol* ⊤⌐ *where cross-linking occurs. (From Ghuysen: Bacteriol. Rev. 32:425, 1968.)*

contain 15 to 20 per cent carbohydrate. Glycogen and other polysaccharide granules are present in the cytoplasm.

Lipids are found in the cell wall and in the cytoplasmic membrane, which is a lipoprotein layer. Acidfast bacteria, such as the tuberculosis organism, contain unusually large amounts of lipid—as much as 40 per cent when they are grown in a medium containing a high percentage of glycerin.

In addition to the foregoing constituents there is a "pool" of organic substances of lower molecular weight readily available for use in metabolism, including sugars, organic acids, amino acids, nucleotides, phosphate esters, vitamins, and coenzymes. This reserve of metabolites consists of fewer than 10,000,000 molecules per cell (in *E. coli*), but amino acids and other substances are hundreds of times as concentrated as in the external medium.

BACTERIAL CELL STRUCTURE

Some of the structures of bacteria are indicated diagrammatically in Figure 5–8. The protoplasm is bounded by a discrete, triple-layered *plasma membrane,* with various

intrusions and *mesosomes* (also known as *peripheral bodies* or *chondrioids*) continuous with it (see Figures 5–9 and 5–10). The plasma membrane in turn is enclosed by the *cell wall.* Outside the wall is a *microcapsule,* a *capsule,* or *loose slime.* The protoplasm contains, in addition to thousands of ribosomes, a "nucleus" or *chromatin body* (*nucleoid* or *genophore*). Various species also possess granules of distinctive composition (e.g., iron, sulfur, or polysaccharide). Most motile bacteria possess one or more *flagella,* and sporulating bacteria may contain *endospores.*

CELL WALL

The bacterial cell wall is a strong and rigid structure that protects and supports the weaker and biochemically more active parts of the cell. Its thickness varies from 10 to 25 nm, according to the species of organism. In general, gram-negative bacteria possess thinner walls than gram-positive bacteria (see Fig. 5–9). Simple calculation indicates that the cell wall constitutes about 20 per cent of the total cell volume.

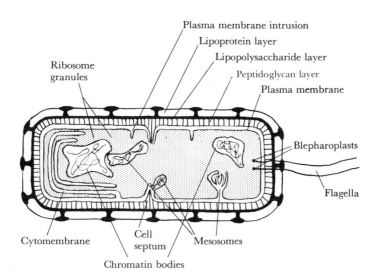

Figure 5–8. *Structures that may be found in various (but not necessarily all) bacterial cells. The layers outside the plasma membrane are considered part of the cell wall. Only a few of the many sites of lipopolysaccharide penetration through the lipoprotein layer are shown.*

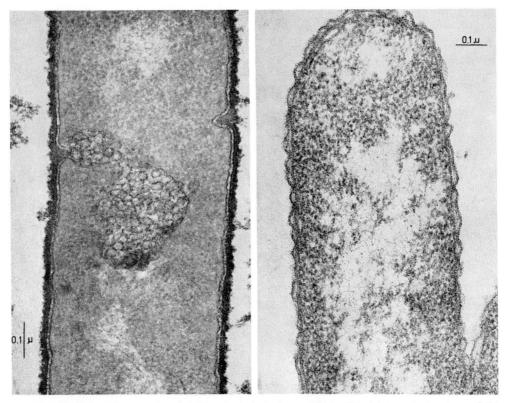

Figure 5–9. *Electron micrographs showing the relatively undifferentiated internal structure of thin sections of* Bacillus subtilis (left) *and* Escherichia coli (right). *The vesiculated organelle ("peripheral body" or "mesosome") in* B. subtilis *seems to participate in formation of the cross wall. Chromatin bodies are the less dense structures near the top and bottom. Chromatin material is scattered widely throughout* E. coli. *Note that the plasma membrane and cell wall are closely associated in* B. subtilis, *but are separate and distinct in* E. coli. *(From van Iterson, Bact. Rev., 29:299, 1965.)*

Cell walls can be removed and isolated by several methods, including mechanical disruption, enzymatic digestion, or sudden immersion in hot water. The inner protoplasm disintegrates readily, and the more resistant walls can be purified for chemical and physical study. Some of the constituents of cell walls are listed in Table 5–3. The cell walls of gram-positive bacteria differ from those of gram-negative species, particularly in their amino acid and lipid composition. The rigidity of bacterial walls is attributed to peptidoglycans (compounds containing amino sugars and amino acids). The basic structure consists of alternating residues of the amino sugars, N-acetyl muramic acid and N-acetyl glucosamine. The polypeptide chain is attached to muramic acid as shown in Figure 5–7. It will be recalled that the rigidity of the walls of blue-green algae, the other group of procaryotic organisms, is also due to their peptidoglycan content. The walls of gram-positive bacteria are relatively amorphous, whereas those of gram-negative bacteria consist of several layers (see Figures 5–8 and 5–9). Some walls are constructed in a regular hexagonal or rectangular pattern of macromolecular units 50 to 140 nm in diameter (Fig. 5–11).

The wall maintains the characteristic shape of the cell and gives physical protection to the cytoplasm, in which most of the vital activities of the cell are performed. It plays little other role in the life of the cell; biochemical activity continues in cells from which the walls have been removed (protoplasts) if they are protected against osmotic lysis.

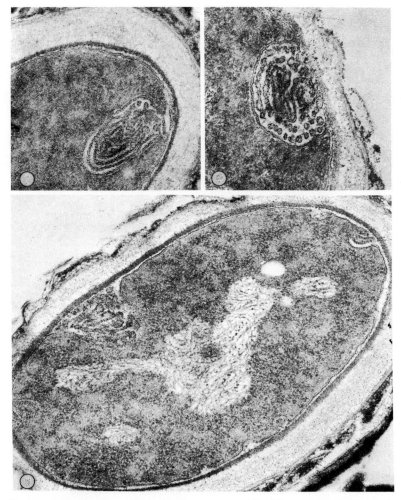

Figure 5–10. *Electron micrographs of germinating spores of* Bacillus megaterium *showing several forms of mesosomes. A simple intrusion is found at the upper end of the bottom cell.* Top left, 60,000X; top right, 80,000X; bottom, 60,000X. (Courtesy of C. F. Robinow and J. F. Marak, from Mazia and Taylor: The General Physiology of Cell Specialization. McGraw-Hill, New York, 1963.)

TABLE 5–3. Chemical Constituents of Cell Walls of Gram-Positive and Gram-Negative Bacteria

	Gram-Positive	Gram-Negative
Amino acids	Three or four principal amino acids, including alanine, glutamic acid, and lysine or diaminopimelic acid. No aromatic amino acids. No sulfur-containing amino acids.	Most amino acids found in ordinary proteins, including diaminopimelic acid.
Muramic acid	Present	Present
Lipids	0–2%	10–20%
Polysaccharides	35–60%	15–20%

Figure 5–11. *Cell walls of bacteria.* Left, *Walls of* Streptococcus faecalis *prepared by grinding the cells; splitting permitted the cell contents to escape. Magnification, 12,000X.* Right, *A portion of the cell wall of* Rhodospirillum rubrum *showing the regular pattern of the spherical bodies of which this wall is composed. Magnification, 42,000X. (Salton and Williams; Salton and Horne.)*

MICROCAPSULE, CAPSULE, AND LOOSE SLIME

Outside the cell wall of most if not all bacteria is a layer of material designated according to its thickness, composition, and solubility as a microcapsule, capsule, or loose slime (Fig. 5–12).

A *microcapsule* is a relatively thin layer composed of protein, polysaccharide, and lipid. Microcapsules are found on gram-negative bacteria and are also known as "somatic antigens" or "endotoxins" (see Chap. 18).

Capsules are thick, viscous, jelly-like structures surrounding the cells of certain species. Some capsules have definite structure; others appear to be amorphous. They stain poorly and are usually demonstrated by a procedure in which the background and cells are colored and the capsules remain colorless (Fig. 5–13 and Plate IID).

Loose slime is similar to capsules but is more soluble in the suspending medium and has less structural integrity.

Capsules and loose slime are accumulated polymers of polysaccharide or polypeptide. The capsules of pneumococci and some streptococci are polysaccharide; each of the several score of pneumococcus "types" possesses a chemically different polysaccharide. Loose slimes of some gram-negative rods are also polysaccharide. Capsules and slime of some of the gram-positive spore-forming rods are polypeptide.

The "extramural" layer is not an integral or

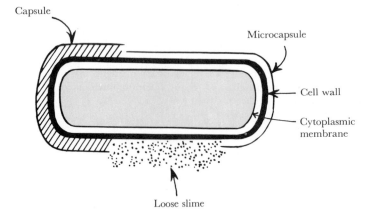

Figure 5–12. *Diagrammatic sketch of surface layers of bacterial cells. (Not all bacteria possess all layers.)*

Capsule

Microcapsule

Cell wall

Cytoplasmic membrane

Loose slime

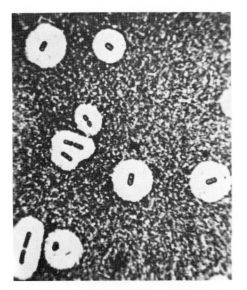

Figure 5–13. *A heavily encapsulated strepto-coccus. The dark background is India ink. Magnification, 1500X. (Courtesy P. M. Borick, Wallace and Tiernan, Inc.)*

become "ropy" when contaminated by certain encapsulated bacteria. Some of the same organisms (e.g., *Leuconostoc* species) are used for commercial production of dextran, which has been used as a plasma "extender" in the treatment of shock caused by loss of blood.

THE PROTOPLAST

The protoplast is that portion of the cell that is within the cell wall. The wall can be removed from cells of some species by treatment with the enzyme lysozyme, derived from egg white, tears, or saliva. Lysozyme digests some of the complex polysaccharides in the cell wall. Certain organisms can also be made to grow without a wall in the presence of penicillin (which interferes with the formation of the glycosaminopeptide layer from its subunits), or by depriving them of diaminopimelic acid. A stabilizing agent such as 0.2 M sucrose must be present to prevent osmotic lysis. The resulting "naked" cells, which lack all traces of cell wall material, are called *protoplasts;* they are globular in shape and relatively stable (Fig. 5–14), although much more sensitive to environmental "discomforts" than intact cells. They are readily lysed by diluting with distilled water the stabilizing solution in which they are suspended.

essential part of the cell. It can be removed without harm and is then replaced by the cell. The presence and amount of capsular and slime material are controlled by the genetic makeup of the organism and by the environment. Mutant forms may possess more or less of such material than normal forms. Capsule and slime formation are often favored by media containing appropriate sugars. Sucrose, for example, promotes the production of capsules or slime by certain organisms that can utilize the fructose portion of the molecule; the unused dextrose portion of the disaccharide polymerizes, and a dextran of high molecular weight accumulates around the cells.

Slime layers and capsules protect bacterial cells against drying and harmful agents. Encapsulated pathogenic bacteria, such as the pneumococcus, resist phagocytosis, a defensive process in which white blood cells or other tissue cells ingest and may digest foreign objects. Noncapsulated variants or organisms from which the capsules have been removed are readily ingested by phagocytic cells.

Bacterial slime and capsules are a cause of economic loss in the dairy and food industries. Milk, syrups, and other sugary solutions

Globular forms possessing partial or modified (e.g., by growth in penicillin or treatment with detergent) cell walls are known as *spheroplasts.*

Protoplasts perform most of the metabolic activities of whole cells, including energy-yielding respiratory processes, synthesis of proteins, enzymes, and nucleic acids. They do not synthesize cell wall material, whereas spheroplasts can do so. This seems to indicate that the wall contains its own synthetic mechanism or that a "starter" or cell wall substance must be present before more can be laid down. Protoplasts can grow and divide, and protoplasts of spore-forming bacteria prepared from cells that have taken the first steps toward sporulation can complete the process of producing spores. Protoplasts of motile organisms may possess flagella but are not motile.

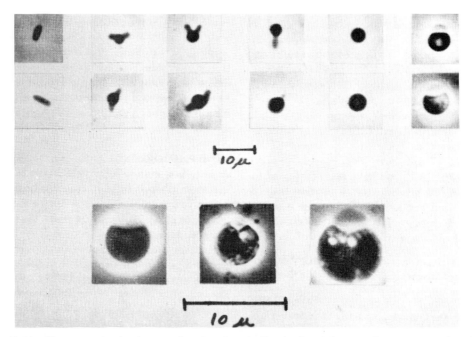

Figure 5–14. *Phase contrast micrographs showing, in the horizontal rows above, successive stages in the formation of spherical protoplasts from typical rod-shaped cells of* Escherichia coli *by cultivation in penicillin sucrose broth for four hours. Below are three late stage protoplasts shown at higher magnification. (J. Lederberg and J. St. Clair, J. Bact., 75:143–160, 1958.)*

PLASMA MEMBRANE

Dilution with water lyses a suspension of protoplasts, releasing the cytoplasm and leaving "ghosts," delicate membranes with some granular material and debris. Protoplast (plasma) membranes are 5 to 8 nm thick and constitute 10 to 20 per cent of the dry weight of the cells. They are composed largely of lipoprotein and contain many enzymes, especially those concerned in biologic oxidations, by which the cell secures energy.

The plasma membrane is a discrete, differentiated outer layer of the cytoplasm just beneath the cell wall (Fig. 5–15). It stains intensely with basic dyes and is said to form the highly reflective layer observed in darkfield preparations. Indirect evidence for the existence of a plasma membrane is derived from the shrinkage of the protoplasm of bacteria suspended in solutions of high osmotic pressure. Water passes outward through the cell wall, and the cytoplasm pulls away from the wall as though bounded by a separate membrane.

The plasma membrane is a membrane of the so-called *unit* type; that is, it is a three-layered structure consisting of a bimolecular "leaflet" of lipid between protein or other hydrophilic layers (Fig. 5–16). It regulates the passage of materials into and out of the cell. Certain substances of low molecular weight, such as urea, glycine, and glycerin, readily enter bacterial cells, whereas the electrolytes NaCl and KCl and larger organic molecules like glucose and sucrose traverse the membrane very slowly. The membrane is essentially impermeable to polar organic substances because of its high lipid content. Enzymes called *permeases* transport particular materials or groups of materials by forming easily dissociable complexes with them. The plasma membrane, together with the membrane intrusions or mesosomes linked with it, comprises essentially a lipoprotein matrix upon which are organized most of the cytochromes (see Chapter 8), succinoxidase and related enzymes, and many other enzymes of the bacterial cell. It is such a vital organelle that Mitchell spoke of it as "not simply . . . an osmotic link between the media on either side of it, but . . . a chemical link."

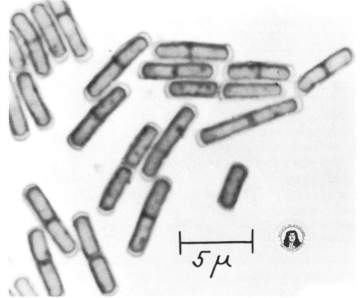

Figure 5–15. *Micrograph of Bacillus cereus, a two-hour culture on agar, showing various stages in growth and cell division. Cell walls and plasma membranes can be distinguished because the cytoplasm has retracted from the walls. (Photograph by C. F. Robinow; A.S.M. LS-235.)*

(a)

(b)

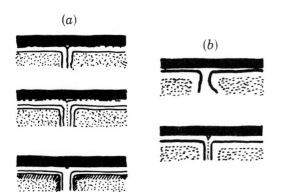

Cytoplasmic Inclusions.   Various granules or inclusions are present in the cells of an aging bacterial culture. Inclusions are nonliving bodies in the cytoplasm. Many seem to be reserve food materials because they accumulate during conditions of good nutrient supply and decrease during starvation. The nature of the inclusions varies with the organism. Volutin granules, sometimes called metachromatic granules (Fig. 5–17),

(c)

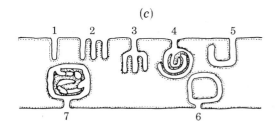

Figure 5–16. *Diagrammatic sketches showing the plasma membrane of bacteria and its intrusions beneath a thick cell wall. (a) Three forms of "unit" membrane; (b) a single, dense layer, either apposed to the cell wall with a low-density layer on the cytoplasmic side (as in the micrococci) or with low-density layers on both sides; (c) various lamellated (1–4) and villous (5–7) mesosomes, originating from the plasma membrane. (Courtesy of R. G. E. Murray, from Mazia and Taylor: The General Physiology of Cell Specialization. McGraw-Hill, New York, 1963.)*

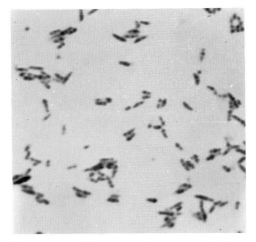

Figure 5–17. *Metachromatic granules in Corynebacterium diphtheriae. The granules stain deeply, and each cell may contain one to five granules. The cells are often club-shaped (1000X). (Courtesy of G. L. Brown.)*

appear in various bacterial species and also in many fungi, algae, and protozoa; they stain intensely with basic dyes and contain polymerized phosphoric acid (Plate IIE).

Polysaccharides may accumulate as glycogen or as a kind of starch. Lipid globules appear in various bacteria, particularly gram-positive organisms. Sulfur and iron are also found in certain species.

Whereas the nucleus in cells of higher plants and animals is an organelle of definite structure bounded by a membrane and containing chromosomes, the nucleoid or chromatin body of bacteria is indeterminate in shape and is not enclosed in a membrane. Special staining techniques are necessary to demonstrate it, as described in Chapter 3 (see page 58). The basic stains customarily used react with any acidic material, RNA as well as DNA. Since the cytoplasm of bacterial cells is filled with RNA-containing granules, the ribosomes, the entire bacterial cell stains deeply unless the RNA is first removed or hydrolyzed by dilute acid. When this is done, the structure of a DNA-containing body or nucleoid can be determined.

Characteristics of Bacterial Chromatin Bodies (Nuclei). *Shape and Arrangement.* Chromatin bodies are more-or-less centrally situated in resting cells and are spherical or oval or rod-shaped (see Fig. 3–23). During active growth they divide along the same axis as the cell, usually a little before cell division; sometimes two to four paired chromatin bodies can be seen in a single rod-shaped cell in the phase of very rapid growth. Eventually cell divisions catch up and the normal ratio of one chromatin body per cell is reestablished.

Lacking a bounding membrane, the shape of a chromatin body is variable. It is, however, a definite structure, readily distinguishable from the remainder of the cell contents by the use of nuclear stains and recognizable by its characteristic relatively low density in electron microscopy. There are two general types: (1) solid structures forming bars or H, V, or butterfly shapes, shown particularly by staining, and (2) small granules enmeshed in fine strands, best seen in electron photomicrographs.

Size. Chromatin bodies vary in dimensions among species and within the same species at different ages. Resting cells of one species of *Staphylococcus* possess chromatin bodies about 0.4 μm in diameter, whereas in growing cells they enlarge to about 0.5 by 0.8 μm. This structure constitutes 5 to 16 per cent of the cell volume. Chromatin bodies of resting *E. coli* occupy 15 to 25 per cent of the protoplasmic space.

Structure and Replication of Chromatin Bodies. Bacterial chromatin bodies appear to be composed of fine fibrils of DNA or desoxyribonucleoprotein 0.3 to 0.4 nm in diameter. In gram-negative bacteria like *E. coli* and *Salmonella typhimurium* these fibrils are arranged in a delicate but compact whorl, whereas in gram-positive bacteria, such as various cocci and bacilli, the dense fibers are aligned in an almost parallel pattern. The DNA constitutes a single chromosome, at least in the bacteria studied so far, and apparently a single two-stranded molecule about 1 mm. long carries all the genetic information of the cell. The chromosome is an endless loop, and during replication it divides, but not by mitosis (Fig. 5–18). New strands of DNA form, complementary in the usual way to each of those in the parent chromosome. The process begins at a certain place in the chromosome and proceeds around the molecule. At the fork in the chain there is presumably a swivel, which permits the parent helix to uncoil as the new strand is formed.

FLAGELLA

Bacterial flagella are slender, spirally coiled appendages found on most freely swimming bacteria. They are generally presumed to be organelles of locomotion.

Physical and Chemical Properties of Flagella. Bacterial flagella are extremely thin but may be very long. Their diameters vary between 0.01 and 0.05 μm, and they may be several times as long as the cell to which they are attached. Flagella more than 70 μm long have been reported. They are composed of an elastic fibrous protein, flagellin,

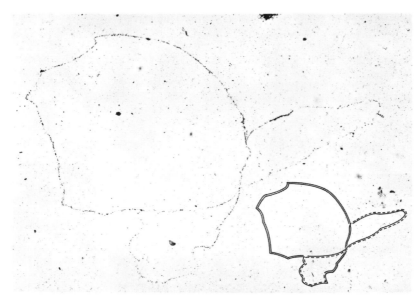

Figure 5–18. *Autoradiograph of the "circular" chromosome of* E. coli *in the process of replicating. The DNA of the chromosome was labeled with radioactive hydrogen (^3H) by growing the organism in tritiated thymidine for two generations. The DNA was then extracted. A photographic plate was exposed to the radiation for two months, with the results shown. The insert is a diagram to explain the replication process. (Courtesy of J. Cairns, from Braun: Bacterial Genetics, 2nd ed. W. B. Saunders Co., Philadelphia, 1965.)*

similar to the actomyosin of skeletal muscle. Electron micrographs and x-ray diffraction patterns indicate that bacterial flagella consist of several fibrils, usually three (Fig. 5–19), surrounding a slender, nonprotein core,

in contrast to the flagella of motile eucaryotic cells, which are composed of nine pairs of fibrils around a core of two fibrils (Fig. 1–8).

Origin of Flagella.   The fact that flagella

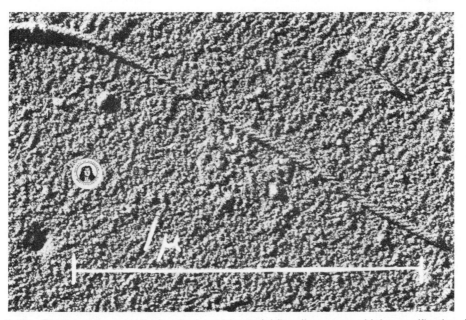

Figure 5–19. *Electron micrograph of a shadowed bacterial flagellum at very high magnification. Its structure is that of a left-handed, triple-threaded screw. (Photograph by M. P. Starr and R. C. Williams; A.S.M. LS-300.)*

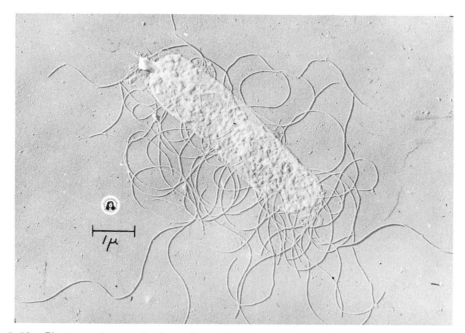

Figure 5–20. *Electron micrograph of shadowed* Proteus vulgaris *demonstrating flagella. A few flagella appear to extend through the cell wall and to originate from a small mass of material within the cell. (Photograph by C. F. Robinow and W. van Iterson; A.S.M. LS-260.)*

arise from the cytoplasm was indicated by the observation that protoplasts of motile bacteria may still possess flagella. Their origin is apparently a granule or *blepharoplast* within the plasma membrane, as has been shown in numerous electron micrographs (Fig. 5–20). Flagella grow rapidly, 0.5 μm per minute, and attain full length in 10 to 20 minutes.

Demonstration of Motility. Motility is usually accepted as presumptive evidence that bacteria possess flagella, although it gives no indication of the number or arrangement of the flagella. Motility is detected directly by microscopic examination of wet mounts or hanging drop preparations (see page 55), usually at a magnification of 400X to 500X. Stab cultures in soft agar (e.g., 0.5 per cent agar instead of the usual 1.5 per cent agar) can also be used (Fig. 5–21); non-motile bacteria grow only along the line of inoculation, whereas motile organisms quickly grow throughout the medium.

Flagellation of Individual Cells. A single cell may possess from one to more than a hundred flagella. Their number and distribution over the bacterial surface are relatively constant for each species. There

are two main groups of organisms: those with polar (terminal) and those with lateral (peritrichous) flagella. Polar flagellation is characteristic of *Pseudomonas, Spirillum,* and *Vibrio* species, whereas peritrichous flagella are found particularly on the Enterobacteriaceae (e.g., *Escherichia, Salmonella, Proteus*), and the spore-forming rods (*Bacil-*

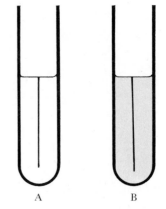

Figure 5–21. *Demonstration of bacterial motility in soft agar. A nonmotile organism grows only along the line of the stab inoculation (A); a motile organism spreads from the line of inoculation throughout the agar to the wall of the test tube, producing turbidity throughout the agar (B).*

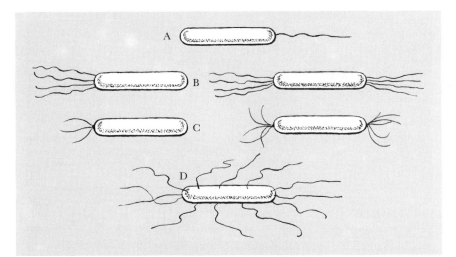

Figure 5–22. *Types of flagellation:* A, *monotrichous;* B, *multitrichous;* C, *lophotrichous;* D, *peritrichous.*

lus species). The length and frequency of the spiral turns vary more or less characteristically.

Leifson (1951) distinguished four types of flagellation (Fig. 5–22):

1. *Monotrichous:* a single flagellum at or near one or both ends of the cell; flagella have more than two curves (Fig. 5–23).

2. *Multitrichous:* more than one flagellum at or near one or both ends of the cell; these flagella also consist of more than two curves (Fig. 5–24).

3. *Lophotrichous:* ordinarily more than one flagellum at one or both ends of the cell; flagella consist of only one or two curves.

4. *Peritrichous:* flagella extending from all sides of the cell and possibly from the ends (Fig. 5–25).

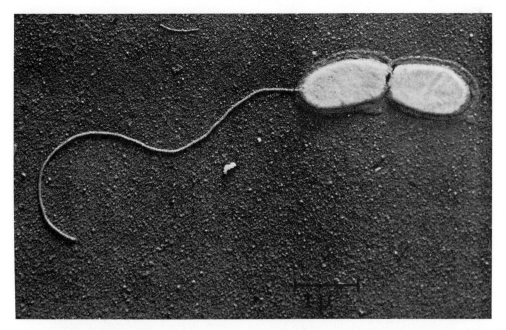

Figure 5–23. *Monotrichous flagellation of* Vibrio cholerae *biotype* proteus. *The single terminal flagellum is present on only one of the dividing pair of cells (15,000X). (Shadowed electron micrograph by W. van Iterson, Biochim. Biophys. Acta, 1:535, 1947.)*

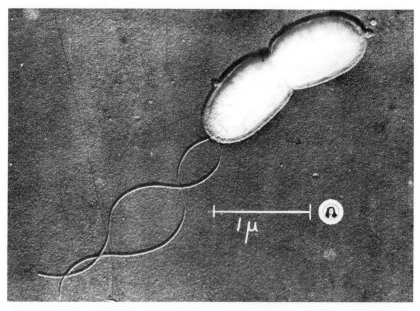

Figure 5–24.   *Multitrichous flagellation of* Pseudomonas fluorescens. *Two terminal flagella seem to arise in the cytoplasm and pass through the cell wall. (Shadowed electron micrograph by A. L. Houwink and W. van Iterson; A.S.M. LS-275.)*

It is probable that bipolar flagellation of a single cell is rare, and a cell that appears to bear flagella at both ends is in reality in the process of dividing. This might be demonstrated if it were possible to stain the developing cell wall in such a bipolar flagellated organism.

Mechanism of Propulsion. The movement of flagella has been variously described as lashing, whiplike, rotary, or corkscrew-like. Lophotrichous flagella appear to rotate in a circular manner; when they are present at both ends of a cell, the forward group curves back toward the body and the

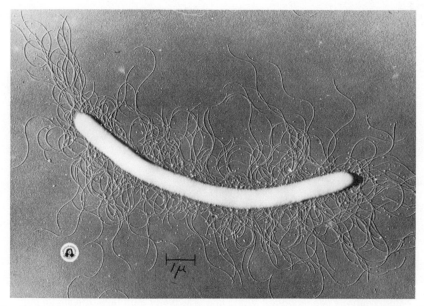

Figure 5–25.   *Peritrichous flagella on* Proteus vulgaris. *(Shadowed electron micrography by C. F. Robinow and J. Hillier; A.S.M. LS-258.)*

posterior group curves out behind. This is observed particularly in certain spirilla; the organism seems to corkscrew itself through the medium. It can reverse its direction without turning around. The flagella of a peritrichous organism trail behind as the cell moves forward.

The driving force seems to be produced by a wave of contraction from the base to the tip of each flagellum; this causes spiral motion of the flagellum and opposite rotation of the bacterial cell. The flagella on a spiral organism rotate at about 40 revolutions per second, but the viscous resistance of the medium slows the rotation of the organism to almost exactly one-third of this rate. The rotary movement propels a spiral organism forward, and in a similar manner rotation of a peritrichate cell applies a torque to the flagellar helix and drives the cell. The speed of forward motion is very great in relation to the size of the cell, but rates of motility vary markedly. Average speeds of 25 μm per second have been recorded for peritrichous intestinal bacteria and a velocity as high as 200 μm per second for the monotrichous, comma-shaped *Vibrio cholerae*.

The energy necessary to maintain flagellar motion is presumably derived from adenosine triphosphate (ATP), a high energy compound formed during photosynthetic and respiratory activity (see Chap. 8). Aerobic bacteria remain motile only as long as they have sufficient oxygen to make ATP by oxidative processes, and the motility of other types of bacteria is correlated with their mechanisms for producing ATP.

Variations in Motility. Flagellar propulsion is the most common mechanism of motility among bacteria. It should be pointed out, however, that bacteria possessing flagella are not necessarily always motile, and loss of motility may occur without loss of flagella. Loss of motility may be caused by environmental changes or by mutation. Certain nonflagellate groups of bacteria are motile, notably some terrestrial forms that creep or glide. Contact with a surface is necessary; the organisms lose their motility when suspended in liquid. Some species secrete a slime, which appears to play some role in gliding motility. The mechanism of this form of locomotion is yet to be learned.

Ecologic Value of Motility. Motility may be looked upon as an adaptation that favors survival by enhancing the chance that an organism will encounter food and other favorable environmental agents, or avoid harmful substances and agents. Movement in response to environmental factors is called *taxis;* it is positive if the movement is toward the factor in question, negative if away from it. Whereas higher organisms can respond to a stimulus directly by turning or moving toward or away from it, bacteria apparently respond only by a "shock reaction," that is, reversal of direction of movement. This phenomenon is called *phobotaxis* (Greek: *phobos,* fear, + *taxis,* influence). When the random motility of a bacterium brings it into a zone containing a stimulating agent that causes it to reverse direction, the response is called *negative* phobotaxis; i.e., the organism tends to leave the vicinity of the stimulus. A *positive* phobotactic response occurs when a bacterium, in the course of its random movement, starts to leave the source of stimulus, but the phobotactic response causes the organism to reverse. The stimulating agent thus serves as a trap, and many cells may accumulate within a small area.

Certain photosynthetic pigmented bacteria exhibit *phototactic* behavior; that is, their movements are influenced by light. If a culture is illuminated by a narrow beam of light, the bacteria will soon congregate in the lighted zone. This is an illustration of positive phobotaxis, and is explained as follows: Those organisms that enter the lighted area in the course of random motility give an immediate shock reaction when they start to swim into the dark portion of the culture; they reverse direction and hence remain in the light, which thus behaves as a trap.

Protozoa and motile aerobic bacteria congregate in a region well supplied with oxygen (e.g., at the edges of a wet mount between a coverglass and a slide). Spirilla prefer a lower oxygen tension and will accumulate some distance from an air bubble or the edge of a wet mount. Anaerobic bacteria will congregate as far from a source of oxygen as possible.

Many other chemical agents exert some sort of *tactic* effect upon microorganisms, causing them to leave the source of the chemical or to be trapped and accumulate near it.

Adaptation is well illustrated by the type of flagellation on aquatic organisms as contrasted with terrestrial species. Terrestrial organisms are adapted to a moist rather than a wet environment and frequently have many peritrichous flagella; aquatic bacteria, on the other hand, usually possess only one or a few polar flagella.

PILI

Bacterial pili (Latin: *pilus,* hair), sometimes called *fimbriae,* are morphologically distinct, nonflagellar appendages, found particularly on gram-negative bacteria freshly isolated from natural sources such as infected urine (see Figure 5–26). Like flagella, they are too slender to be seen by ordinary light microscopy; most pili are 0.003 to 0.007 μm in diameter. They vary from 0.5 to 6 μm in length, and certain pili are as long as 20 μm.

They are straight, and some appear to be rigid. One hundred to 400 are usually distributed over the cell surface. There are several types, which differ in size and structure. Mechanical agitation in a high-speed mixer removes them from the cell, and they can be purified by precipitation and centrifugation to yield a protein, *pilin,* with a minimum molecular weight of about 17,000. A pilus is composed of pilin subunits arranged in a very precise helical structure to form a smooth tube around a longitudinal hole. Physical, chemical, and genetic study has shown that one type of pilus is intimately concerned in the process of sexual mating, and probably provides the channel through which DNA from the donor (male) cell is transferred to the recipient (female) cell. Another type of pilus enables the organism to adhere to erythrocytes and other cells, and another enhances growth when the oxygen supply is limited and the cell population is high, perhaps by mediating the transport of some metabolite. The functions of other pili are not known.

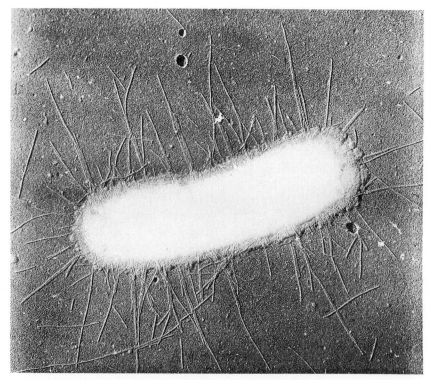

Figure 5–26. *Electron micrograph of* E. coli *showing pili. (From Brinton, Trans. N.Y. Acad. Sci., 27:1005, 1965.)*

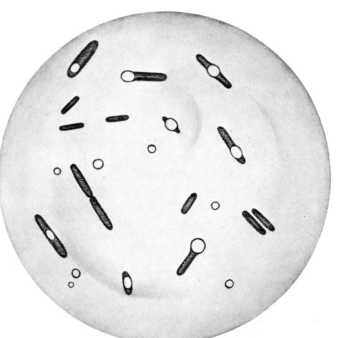

Figure 5–27. *Various types of bacterial spores: terminal, subterminal, central, spherical, oval. Spores are sketched as though unstained, vegetative cells and portions of sporangia as though stained with methylene blue or crystal violet. (From Frobisher: Fundamentals of Microbiology, 8th ed. W. B. Saunders Co., Philadelphia, 1968.)*

ENDOSPORES

Occurrence. Endospores are highly resistant bodies produced within the cells of certain bacteria. They are found in all species of the family Bacillaceae, which is divided into two genera, *Bacillus* (aerobic spore-forming rods) and *Clostridium* (anaerobic spore-forming rods). One bacterial cell normally produces only a single endospore.

Sporulation is therefore not considered to be a method of multiplication of bacteria as it is of yeasts and molds.

Physical and Physiologic Characteristics. Endospores are spherical to elliptical and may be situated anywhere within the parent cell or *sporangium* (Fig. 5–27). Their diameter may be less than, equal to, or greater than that of the rest of the sporangium. A cell with a greatly enlarged central endo-

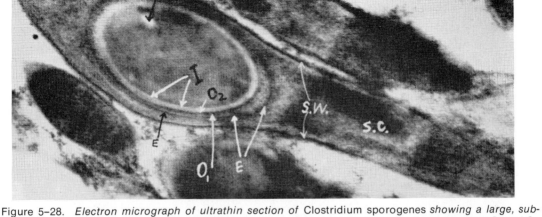

Figure 5–28. *Electron micrograph of ultrathin section of* Clostridium sporogenes *showing a large, subterminal, elliptical endospore. S.C.,* Sporangium cytoplasm; *S.W.,* sporangium wall; *E, exosporium; O_1 and O_2, first and second outer membranes; I, inner membrane; S, spot of unknown nature. (Photograph by T. Hashimoto and H. B. Naylor, J. Bact., 75:647–653, 1958.)*

spore resembles a spindle and is called a *clostridium*. A *plectridium* (Fig. 5–28) is a sporangium containing an enlarged terminal endospore. The sizes of spores differ from one species to another; this property is of some use as a criterion for classification.

Unstained bacterial endospores are highly refractile when observed with the microscope. Ordinary simple staining methods color only the outer layer or spore coat. The inside of a spore can be stained if heat is applied. Apparently this treatment increases the permeability of the spore envelopes and permits strong dyes such as malachite green or carbolfuchsin to penetrate and stain the cytoplasm intensely. Stained endospores resist decolorization and are easily distinguished from vegetative cells or from other portions of sporangia.

The fine structure of endospores differs somewhat from one species to another. In general, however, there is a central core consisting of the spore cytoplasm and nuclear material. This is surrounded by a delicate membrane and the *spore wall*. In many species the spore wall will eventually transform into the cell wall of the future bacillus. Around the wall is a second layer, thicker and of relatively low density, the *cortex*. This is the location of dipicolinic acid (see below), mucopeptide polymers, and calcium, all of which are significant in spore resistance. The cortex, in turn, is enclosed in one or two (depending on the species) *spore coats*. Spore coats may be smooth, grooved, or raised into ridges, sometimes in geometric (e.g., hexagonal) patterns. Lastly, the whole is enclosed in an *exosporium,* which may be somewhat wrinkled and enclose some cytoplasmic material derived from the sporangium. Electron micrographs of carbon replicas of spores of *Bacillus* species show varying degrees and forms of surface sculpturing (see Figure 5–29).

Endospores are the most resistant of all living bodies to heat desiccation, and toxic chemicals, but there is great variation among species. Some endospores are killed within a few minutes at 80° to 90° C., whereas those of other species survive prolonged boiling. Spores of one bacillus resist 100° C. for over 20 hours. Variations occur among the

endospores within a single culture; a few survive exposure to a lethal agent considerably longer than the majority.

The remarkable resistance of endospores implies that their chemical composition or physical structure must differ radically from that of the parent cells. Chemical analysis reveals that endospores contain DNA and RNA, proteins, lipids, carbohydrates, various enzymes, and minerals. Their water content is approximately 25 per cent less than that of vegetative cells. It has long been suggested that the ratio of "bound" to "free" water in spores is greater than in vegetative cells. In addition, spores contain more calcium. Some of the proteins are the same as those in vegetative cells, but numerous proteins peculiar to endospores have been found. Formation of an endospore involves new synthesis of proteins, including enzymes, as well as incorporation of vegetative cell constituents. Some spore enzymes differ qualitatively from their vegetative cell counterparts. For example, catalase from vegetative cells is soluble and heat-sensitive, whereas that from spores is attached to particles and is resistant to heat. Moreover, the two enzymes differ in kinetic properties and in immunologic specificity.

One of the most striking features of endospores is a compound, *dipicolinic acid,*

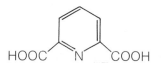

that is present in the cortex of spores and absent from all vegetative cells. It makes up 5 to 15 per cent of the dry weight of the spore. Dipicolinic acid, peptides, and other substances are released from germinating spores, coincident with loss of resistance. It therefore appears that dipicolinic acid is partly responsible for spore resistance.

The metabolic activity of endospores is very low. They contain several active enzymes, but many others are present in a dormant or inactive state.

There is evidence that some of the enzymes and other normally thermolabile substances within spores are bound in the form

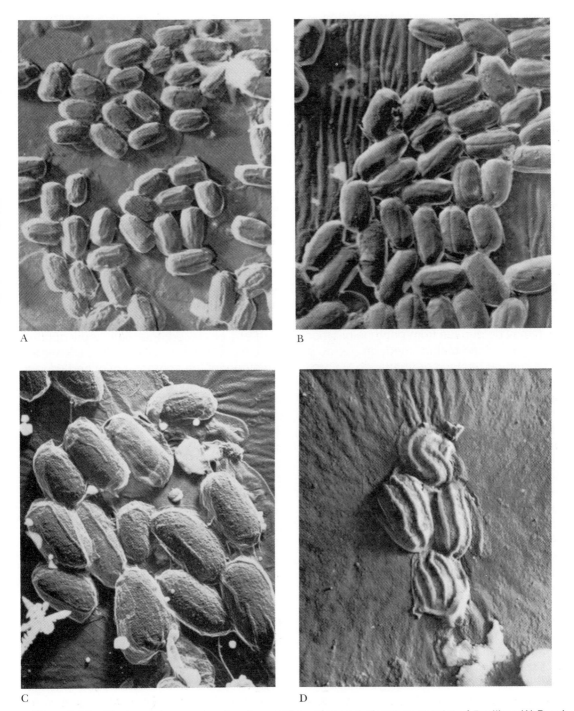

A

B

C

D

Figure 5–29.   *Electron micrographs of carbon replicas of spores of various species of* Bacillus: (A) B. sub-tilis, *ribbed spores, 4000X;* (B) B. licheniformis, *spores slightly ribbed, but with a pronounced longitudinal groove, 5250X;* (C) B. brevis, *ribbed and with a textured surface, 9000X;* (D) B. polymyxa, *marked, curving ribs, 5000X. (From Bradley and Williams, J. Gen. Microbiol., 17:75–79, 1957.)*

of chemical complexes with dipicolinic acid and perhaps also peptide and calcium. Those complexes are highly resistant.

Other factors contribute to the resistance and low metabolic activity of spores. The impermeability of the spore coat undoubtedly prevents the entrance of lethal chemicals. The dehydrated endospore cytoplasm is unfavorable for any kind of chemical activity. This factor has been cited in partial explanation of the observation that intact spores cannot be stained.

Sporulation. Endospores are produced by healthy, well nourished cells growing under favorable conditions, normally just after the period of maximum multiplication rate. Sporulation occurs most frequently at a temperature favorable for vegetative growth and within a narrow range of pH, usually near neutrality but differing from species to species. It is inhibited by certain metabolic byproducts such as straight-chain saturated organic acids of 10 to 14 carbon atoms, which may be derived from peptone or other ingredients of the culture medium. Aerobic spore-forming bacteria ordinarily do not produce endospores in the absence of oxygen nor do anaerobes sporulate in its presence.

It has been suggested that exhaustion of nutrients is a factor in spore formation. Sporulation is favored by cultivation in dilute media; adequate food supplies promote vegetative growth and seem to delay the formation of spores. One of the regulatory mechanisms in sporulation is believed to involve a metabolic signal arising from the deteriorating nutrient situation. This signal releases spore-specific genes that are repressed in the normal vegetative cell. The genes control the activity of a protease (protein-hydrolyzing) enzyme that alters the specificity of an RNA polymerase. The altered polymerase causes major changes in the enzymatic behavior of the cell: thereafter, it produces spore proteins and other constituents instead of vegetative components.

The first sign of spore formation usually detected in ordinary microscope preparations is a faint elliptical envelope or clear patch in the granular cytoplasm at one end of the cell. This area gradually becomes more dense than the rest of the cytoplasm, and stains more intensely with basic dyes until the spore coverings form. Local opacity and viscosity increase, and within a short time the spore wall, cortex, and coat or coats develop.

In a vegetatively multiplying cell (see Figure 5–30), the DNA replicates and divides by processes that will be described later, and each half of the DNA comprises the nuclear material of a new vegetative cell. This cycle of DNA and cellular replication continues as long as growth conditions remain favorable. When conditions change and favor sporulation, half of the DNA (i.e., a complete genome of the longitudinally distributed nuclear body of the stage I cell) becomes segregated into a compartment at one end of the cell, where it is separated from the remaining nuclear and cytoplasmic material by a spore septum (stage II). This compartment develops into a primordial spore or "forespore" (stage III). From this point the two portions of the cell constitute different metabolic systems. Although each contains half the DNA of the original cell, replication of DNA in the forespore ceases. The DNA in the remainder of the sporangium of certain species may continue to replicate until lysis occurs, and vegetative components continue to be produced. Biosynthetic activity in the forespore yields only spore-specific components, which differ from vegetative components. New envelopes then form around the forespore: the cortex, immediately outside the spore wall (stage IV), and one or two spore coats outside the cortex (stage V). The nucleus becomes somewhat less dense as it is distributed throughout the spore cytoplasm. The structure of the cortex changes as calcium dipicolinate accumulates, the outer portion of the cortex becomes less dense, and the thermoresistance of the spore increases (stage VI). These various stages have each occupied 30 to 90 minutes in some species, and a total of six to seven hours has elapsed since rapid vegetative multiplication ceased. Through the next several hours the vegetative portion of the sporangium undergoes autolysis, liberating the free, refractile, resistant spore. *Autolysis* is the process whereby the enzymes of a cell that has ceased to metabolize digest the cell's own protoplasm. The stages in sporula-

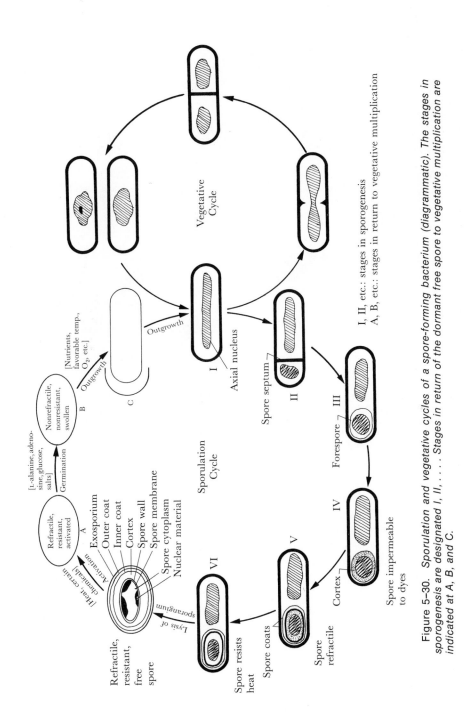

Figure 5-30. *Sporulation and vegetative cycles of a spore-forming bacterium (diagrammatic). The stages in sporogenesis are designated I, II, ... Stages in return of the dormant free spore to vegetative multiplication are indicated at A, B, and C.*

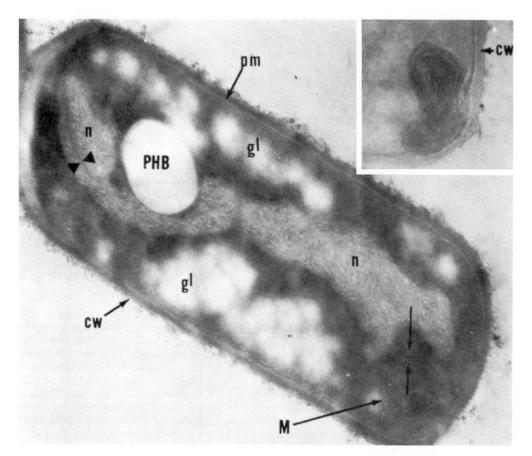

Figure 5–31. *Electron micrograph of* B. cereus *in stage I of sporulation. The axial filament of nuclear material (n) is associated at one end with a mesosome (M). The cell wall (cw) and protoplasmic membrane (pm) are shown, and a poly-β-hydroxybutyrate (PHB) inclusion and bodies believed to contain glycogen (gl). The inset is from a Stage I cell showing the concentrically laminated mesosome in greater detail. (From Ellar and Lundgren, J. Bact., 92:1748–1764, 1966.)*

tion are illustrated in the electron micrographs in Figures 5–31 through 5–36.

Transformation of Endospores into Vegetative Cells. The transformation of endospores into vegetative cells occurs within a very few hours after transfer to a favorable environment. This is true whether the spores have only recently been formed or have been dormant for a long time. Spores can survive for many years; bacteriology is so young that no one knows just how long.

Favorable conditions include the presence of water and nutrients, suitable temperature and oxygen tension, and the presence of certain "trigger" substances and conditions. Three processes occur in sequence during the transformation of the spore into the vegetative cell: (a) *activation* conditions the spore

to germinate in a suitable environment; (b) *germination* is a process in which the typical characteristics of a dormant spore are lost; and (c) *outgrowth*, characterized by the formation of new proteins and structures, converts the spore into a vegetative cell.

Activation. Activation is a reversible process that prepares a dormant spore for germination if appropriate conditions are available. It is most commonly accomplished by "heat shocking," that is, by heating at 80° to 85° C. for a few minutes. Reducing agents and low pH are also activating agents. An activated spore is still dormant and retains the characteristics of spores. The nature of activation is not well understood, but it may consist of reversible denaturation of a macromolecule such as protein.

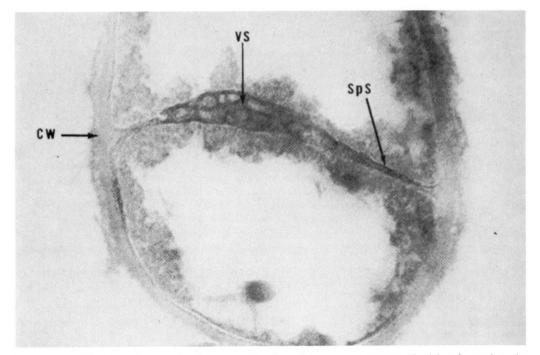

Figure 5–32.   *Electron micrograph of* B. cereus *in stage II. The spore septum (Sps) has formed, and some vesicles (vs) can be seen between the layers of the septum. In this preparation the nucleus and cytoplasm appear as structureless areas. (From Ellar and Lundgren, J. Bact., 92:1748–1764, 1966.)*

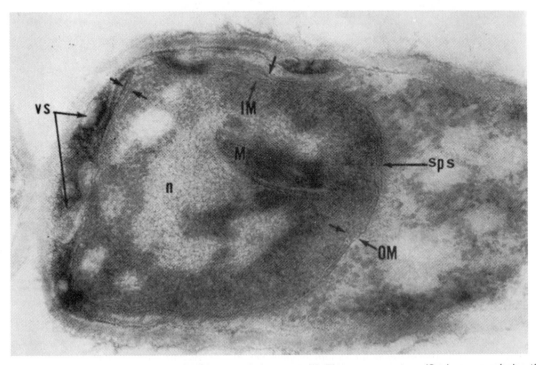

Figure 5–33.   *Electron micrograph of* B. cereus *in stage III. The spore septum (Sps) now encircles the forespore; the outer membrane (OM) and inner membrane (IM) components of the septum are clearly distinguished. (From Ellar and Lundgren, J. Bact., 92:1748–1764, 1966.)*

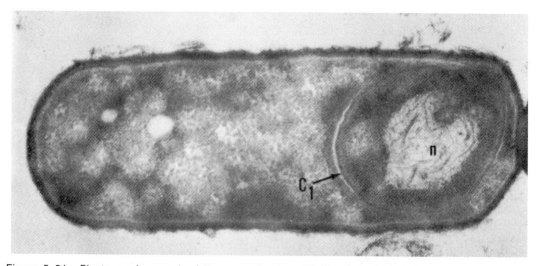

Figure 5–34. *Electron micrograph of* B. cereus *in stage IV. The first step of cortex development* (C_1) *is shown. This layer persists throughout maturation of the spore and may form the new vegetative wall after germination of the spore. (From Ellar and Lundgren, J. Bact., 92:1748–1764, 1966.)*

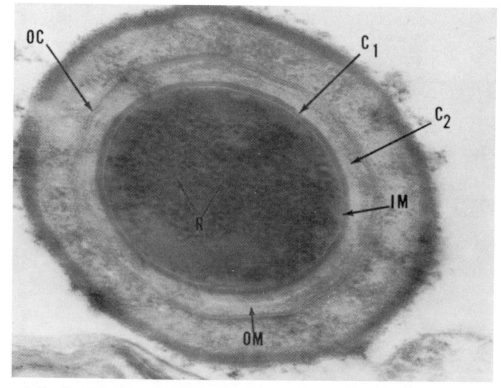

Figure 5–35. *Electron micrograph of* B. cereus *in stage V. This is a transverse section through the spore. The inner membrane (IM) of the spore septum immediately bounds the core, and between it and remnants of the outer membrane (OM) are the denser* (C_1) *and less dense* (C_2) *layers of the cortex. An outer coat layer (OC) surrounds the cortex. Ribosomal aggregates (R) are present in the cytoplasm. (From Ellar and Lundgren, J. Bact., 92:1748–1764, 1966.)*

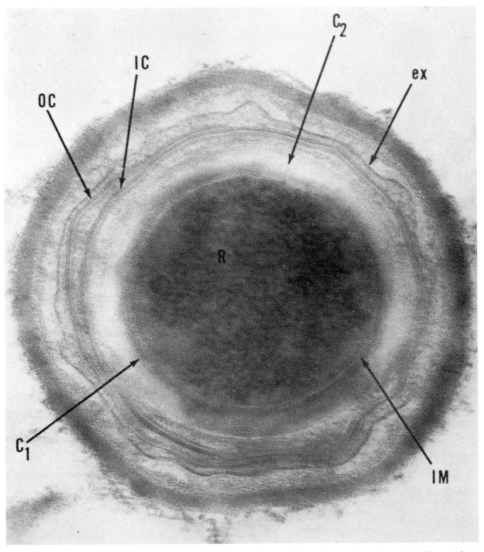

Figure 5–36. *Electron micrograph of transverse section of* B. cereus *in stage VI. The various layers, previously identified, are well defined, and in addition the inner* (IC) *and outer* (OC) *coats and exosporium* (ex) *can be clearly distinguished. (From Ellar and Lundgren, J. Bact., 92:1748–1764, 1966.)*

Germination. Germination is triggered by chemical agents such as L-alanine, adenosine, glucose, or salts; most of these are normal metabolites and are utilized in the process of germination. During germination, the spore swells, the coat may rupture or be absorbed; the properties and substances that distinguish a spore are lost—resistance, refractility, impermeability, dipicolinic acid, calcium, mucopeptides—and metabolic activity increases. Since most of these processes require enzymatic activity, one of the first steps is activation or derepression of spore enzymes.

Outgrowth. Outgrowth occurs during the period immediately following germination. Proteins and structures characteristic of vegetative cells are synthesized during this interval, which covers the period up to the time of cell division and return to vegetative growth. Conditions necessary for outgrowth are usually different from those that support germination. These processes have different temperature optima, and outgrowth often requires nutrients not needed for germination. Germination of an activated spore may require only five minutes, but following this, synthesis of RNA and proteins, including en-

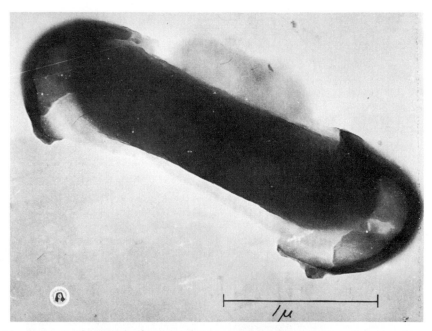

Figure 5–37. *Electron micrograph of outgrowing spore of* Bacillus cereus *variety* mycoides. *The cell still carries the two halves of the completely ruptured spore coats at its ends. The culture was grown on nutrient agar at 35° C. for 105 minutes. (Photograph by G. Knaysi, R. F. Baker, and J. Hillier; A.S.M. LS-204.)*

zymes, requires 10 to 40 minutes, and cell wall synthesis may not occur before two hours. Another hour may elapse before cell division gets under way. Figure 5–37 shows an outgrowing spore nearly ready for vegetative division.

Germination is characterized by loss of heat resistance and refractility, and by increased stainability. A rapid breakdown of the structures that protect vital functions is followed by the release of the dipicolinic acid-peptide complex, activation of dormant enzymes, and increased respiratory activity. Nuclear material is readily detected. The chromatin body divides twice and the resulting four bodies eventually appear in four vegetative cells.

Significance of Endospores. The biologic significance of endospores is not known. Their great resistance to heat, drying, and chemicals tempts us to argue teleologically that spores are produced to permit survival under these unfavorable conditions. However, the imminence of such conditions does not seem to be the trigger that initiates sporulation. Rather, spore formation occurs only in cultures that have been adequately supplied with food and energy, under fairly restricted conditions of pH and temperature, and in the absence of "antisporulation factors," such as certain organic acids. A change in the metabolic activity of the cell apparently derepresses genes that control the formation of spore proteins.

It has been mentioned that spore formation is not a method of multiplication in bacteria. Neither does it seem to be a means of rejuvenation by nuclear fusion or a sexual process. An endospore is clearly a resting stage in the life cycle. Its metabolic activity is low; it contains only small amounts of enzymes, less water than vegetative cells, and the unusual stabilizing agent dipicolinic acid. These properties evidently provide the remarkable resistance typical of spores. They are therefore adapted to survival, although it must be admitted that they can withstand conditions more rigorous than any likely to be encountered today except in rare circumstances (e.g., in hot springs). Resistance to desiccation may be significant in the survival of spore-forming bacteria. The predominant bacteria found in a study of the Egyptian desert during periods of drought were three species of *Bacillus: B. subtilis, B. licheni-formis,* and *B. megaterium;* very few non-spore-formers were observed.

It has long been believed that the resistance of spores is due to their marked dehydration, which can be expected to diminish greatly all chemical activity within the cytoplasm. Thimann pointed out that the volume of a spore is one-quarter to one-tenth that of the parent vegetative cell and that a large fraction of this is wall structure, so that the water content of the cytoplasm must be much less than that of vegetative cell cytoplasm. Therefore, the cytoplasm of a spore might contain protein in as high a concentration as 90 per cent, whereas in a vegetative cell its concentration is approximately 10 per cent. Obviously, it would be difficult to dehydrate spore protein further, either by desiccation at

physiologic temperatures or by the use of heat. Moreover, the low moisture content greatly retards the action of chemical disinfectants. It appears, therefore, that the resistance of spores to desiccation, chemicals, and heat has a common basis, namely the lack of "free" water.

VEGETATIVE REPRODUCTION OF BACTERIA

Vegetative reproduction, as followed in a series of stained preparations studied at ordinary magnification, begins with the formation of new protoplasm and cell growth. The process of cell division is initiated in some way when the ratio of cell mass to DNA (i.e., nuclear mass) exceeds a critical con-

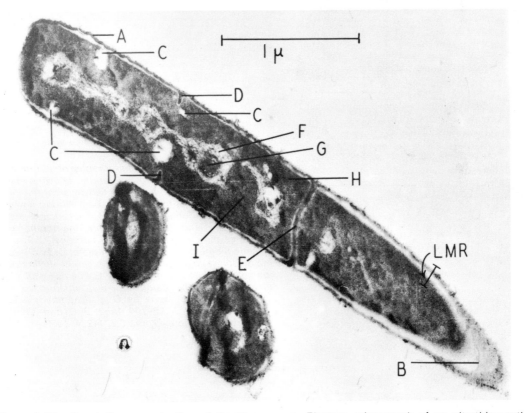

Figure 5–38. *Vegetative reproduction of* Bacillus cereus. *Electron micrograph of an ultrathin section close to the median plane of one cell and passing obliquely through the next cell; cross sections of two other cells. A, Cell wall showing evidence of the shrinking of the cytoplasm. B, Very oblique section of the cell wall showing dense particles and the dense inner layer. C, Four peripheral bodies cut at different levels. D, Beginning of the centripetally growing transverse cell wall. E, Completed transverse cell wall before thickening. F, Low density fibrous component of nuclear apparatus. G, Dense body in nuclear apparatus that may be inclusion of cytoplasmic material. H, Small dense particles that appear to be main constituent of cytoplasm. I, Unidentified cytoplasmic inclusions. LMR, Scale indicating the limit of resolution of a light microscope using visible light. (Photograph by G. B. Chapman; A.S.M. LS-325.)*

Figure 5–39. *A plasmodesm (at right) connecting adjacent cells of* Bacillus cereus. *(Electron photomicrograph by F. H. Johnson; A.S.M. LS-58.)*

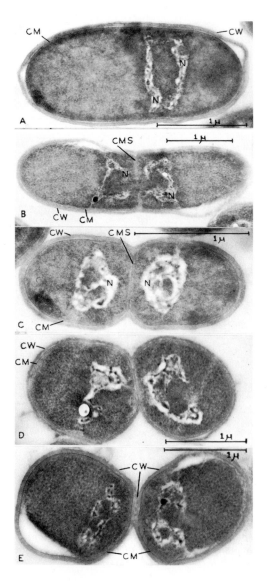

Figure 5–40. *Electron micrographs of ultrathin sections of an unidentified bacterium showing various stages in cell division. A, Before cell division; the nuclear material is in the form of two bars through which a threadlike component extends. B, The nuclear material has divided and the cytoplasmic membrane nearly separates the cytoplasm into two portions. C, The cytoplasmic membrane septum is complete. D, Two layers of the cytoplasmic membrane septum can be distinguished; the cells are becoming rounded. E, The cell wall is complete between the daughter cells, which have nearly separated. (Photographs by G. B. Chapman; A and C are from J. Bact., 78:96–104, 1959; B, D, and E are from J. Biophys. Biochem. Cytol., 6:221–224, 1959.)*

stant value. Cells elongate, often to many times their original length, and nuclear division occurs. Cell division seems to be initiated by inward growth of the cytoplasmic membrane producing a cross-plate and forming two independent sister cells. A cross wall develops and splits this plate into two layers, and the sister cells may then separate.

Chapman and Hillier in 1953 examined ultrathin sections of *Bacillus cereus* with the electron microscope. Their reconstruction of the events in vegetative reproduction (Fig. 5–38) indicated that cell growth is accompanied by one or two nuclear divisions yielding two or four nuclei. Rings of six or eight "peripheral bodies" about .02 μm in diameter appear within the cytoplasm near the edge of the cell and approximately midway between the nuclei. The peripheral bodies are probably the structures later designated mesosomes (see Figures 5–8, 5–9, 5–10, and 5–16), which seem always to originate as plasma membrane intrusions. They also appear in other situations involving protoplast division, such as the formation of endospores. They gradually move toward the axis of the cell, and the cell wall grows inward like a slowly closing iris diaphragm. The inward-growing wall follows closely behind the peripheral bodies, and it is assumed that these bodies synthesize and secrete cell wall material. Cross wall formation is sometimes initiated before the adjacent nuclei have completely separated, and several cross walls may be in various stages of growth within a single cell. Occasionally a transverse wall is not completed and a small central hole remains, through which the cytoplasm of one cell is connected with that of the sister cell. The connecting link is called a *plasmodesm* (Fig. 5–39).

The completed transverse wall becomes thicker; a less dense layer differentiates in the middle; and eventually the wall splits into two layers, one for each sister cell. These end walls are continuous with the lateral walls of the cells. Indentation occurs at the surface of the cell at the transverse wall, and the sister cells may separate. As the sister cells grow they set up turgor pressure, which pulls on their walls at regions of contact with adjacent cells so that separation begins at the outside and progresses toward the axis of the cells (Fig. 5–40). This is true of bacteria that characteristically appear as single cells. Chain-forming bacteria apparently possess tougher walls, which resist the tension produced by the growth of sister cells.

SUPPLEMENTARY READING

Bisset, K. A.: *The Cytology and Life-History of Bacteria,* 2d ed. Baltimore, The Williams & Wilkins Co., 1955.

Brieger, E. M.: *Structure and Ultrastructure of Microorganisms.* New York, Academic Press, 1963.

Brinton, C. C.: The Structure, Function, Synthesis and Genetic Control of Bacterial Pili and a Molecular Model for DNA and RNA Transport in Gram-negative Bacteria. *Trans. N.Y. Acad. Sci.* II, 27:1003–1054, 1965.

DeRobertis, E. D. P., Saez, F. A., and DeRobertis, E. M. F., Jr.: *Cell Biology,* 6th ed. Philadelphia, W. B. Saunders Company, 1975.

Dubos, R. J.: *The Bacterial Cell.* Cambridge, Mass., Harvard University Press, 1945.

Ellar, D. J., and Lundgren, D. G.: Fine structure of sporulation in *Bacillus cereus* in a chemically defined medium. *J. Bact., 92:*1748–1764, 1966.

Gillies, R. R., and Dodds, T. C.: *Bacteriology Illustrated,* 3rd ed. Baltimore, The Williams & Wilkins Co., 1973.

Gunsalus, I. C., and Stanier, R. Y. (eds.): *The Bacteria, A Treatise on Structure and Function. Vol. I: Structure.* New York, Academic Press, Inc., 1960.

Halvorson, H. O., Vary, J. C., and Steinberg, W. In Clifton, C. E. (ed.): *Annual Review of Microbiology. 20:*169–188, 1966.

Knaysi, G.: *Elements of Bacterial Cytology,* 2nd ed. Ithaca, N.Y., Comstock Publishing Co., Inc., 1951.

Murray, R. G. E.: The Organelles of Bacteria. In Mazia, D., and Tyler, A. (eds.): *The General Physiology of Cell Specialization.* New York, McGraw-Hill Book Company, Inc., 1963.

Salton, M. R. J.: *The Bacterial Cell Wall.* Amsterdam, Elsevier Publishing Company, 1964.

Sieburth, J. M.: *Microbial Seascapes.* Baltimore, University Park Press, 1975.

Spooner, E. T. C., and Stocker, B. A. D. (eds.): *Bacterial Anatomy,* Sixth Symposium of Society for General Microbiology. New York, Cambridge University Press, 1956.

6   GROUPS OF BACTERIA

It was noted in Chapter 1 that some authorities classify bacteria in a separate kingdom, Protista. According to this system, bacteria and blue-green algae comprise the *lower protists,* and the more highly developed algae, together with fungi and protozoa, are *higher protists.* At present, microbiologists assign the lower protists to the kingdom Procaryotae. Plant-like higher protists (eucaryotic algae and fungi) remain in the plant kingdom; the protozoa remain in the animal kingdom. The systematic status of viruses is not established inasmuch as they lack a cellular organization comparable to that of either procaryotes or eucaryotes.

The kingdom Procaryotae consists of two divisions. Division I, The Cyanobacteria, formerly known as Cyanophyceae or Schizophyceae, is the blue-green algae. They have a procaryotic structure and are phototrophic; that is, they utilize energy from light and perform photosynthesis, using water as electron donor and producing oxygen. Their cell walls are rigid owing to an inner layer of peptidoglycan, which is enclosed within several additional layers of cell wall substance (see Figures 5–7 and 5–8). Most cyanobacteria are motile at some stage of development by gliding in contact with a surface, rather than by flagella or other organs of locomotion. Photosynthesis is performed in paired lamellae or thylakoids distributed throughout the cytoplasm (see page 4 and Figure 1–3).

Unicellular cyanobacteria reproduce by simple binary fission, by multiple fission, or by pinching off cells (exospores). Filamentous forms, which may be simple or branched, grow by repeated divisions within the chain of cells. The filaments may fragment randomly, or short motile chains of cells (hormogonia) may be released from the rods. Some filamentous species produce specialized reproductive or resting cells called heterocysts.

Photosynthetic activity is associated with the presence of chlorophyll *a.* In addition, other pigments known as phycobiliproteins give the organisms their characteristic color. These include phycocyanin, allophycocyanin, and phycoerythrin.

Division II, The Bacteria, comprises procaryotes that are unicellular or that are associated in chains or masses owing to failure to separate after division. Multiplication usually consists of growth and simple fission; in some forms unequal division or budding occurs. Motility is principally by flagella, but some groups glide, twitch, snap, or dart on solid surfaces. Some species produce endospores, others form arthrospores or cysts, and many produce no such structures.

Most bacteria possess rigid or semirigid cell walls containing peptidoglycan. The Mollicutes are exceptions that lack a rigid wall.

The few bacteria that are photosynthetic utilize special pigments—bacteriochlorophylls—under anaerobic conditions. They require an electron donor other than water and, hence, do not produce oxygen. They lack chlorophyll *a* and phycobiliproteins.

The 19 Parts of which Division II, The Bacteria, is composed are distinguished by physiologic and morphologic characteristics that appear to be fundamental and are relatively easy to determine. Primary subdivision is based upon the source from which the organisms secure energy. *Phototrophic* bacteria utilize the energy of light, and *chemotrophic* bacteria use chemical bond energy. The latter group contains *chemolithotrophic* organisms, which use CO_2 as the principal source of carbon for growth and obtain energy by oxidizing nitrogen, sulfur, or iron compounds. These groups are further divided by morphologic, staining and physiologic characters as indicated by the key in Table 6–1. The remainder of this

TABLE 6–1.   Key to the 19 Parts of The Bacteria*

I. Phototrophic   *Part 1, Phototrophic Bacteria*
II. Chemotrophic
  A. Chemolithotrophic
    1. Derive energy from the oxidation of nitrogen, sulfur, or iron compounds; do not produce
      methane from carbon dioxide
      a. Cells glide   *Part 2, Gliding Bacteria*
      aa. Cells do not glide
        b. Cells ensheathed   *Part 3, Sheathed Bacteria*
        bb. Cells not ensheathed   *Part 12, Gram-negative Chemolithotrophic Bacteria*
    2. Do not oxidize nitrogen, sulfur, or iron compounds; produce methane from carbon
      dioxide   *Part 13, Methane-producing Bacteria*
  B. Chemoorganotrophic
    1. Cells glide   *Part 2, Gliding Bacteria*
    2. Cells do not glide (exceptions in Part 19)
      a. Cells filamentous and ensheathed   *Part 3, Sheathed Bacteria*
      aa. Cells not filamentous and ensheathed
        b. Products of binary fission not equivalent (have appendages other than flagella and
          pili or reproduce by budding)   *Part 4, Budding and/or Appendaged Bacteria*
        bb. Not as above
          c. Cells not rigidly bound
            d. Cells spiral-shaped, have cell wall   *Part 5, Spirochetes*
            dd. Cells not spiral-shaped, no cell wall   *Part 19, Mycoplasmas*
          cc. Cells rigidly bound
            d. Gram-negative
              e. Obligate intracellular parasites   *Part 18, Rickettsias*
              ee. Not as above
                f. Curved rods   *Part 6, Spiral and Curved Bacteria*
                ff. Not curved rods
                  g. Rods
                    h. Aerobic   *Part 7, Gram-negative Aerobic Rods and Cocci*
                    hh. Facultatively anaerobic   *Part 8, Gram-negative*
                        Facultatively Anaerobic Rods
                    hhh. Anaerobic   *Part 9, Gram-negative Anaerobic Bacteria*
                  gg. Cocci or coccobacilli
                    h. Aerobic   *Part 10, Gram-negative Cocci and Coccobacilli*
                        Part 7, Gram-negative Aerobic Rods and Cocci
                    hh. Anaerobic   *Part 11, Gram-negative Anaerobic Cocci*
            dd. Gram-positive
              e. Cocci
                f. Endospores produced   *Part 15, Sporeforming Rods and Cocci*
                ff. Endospores not produced   *Part 14, Gram-positive Cocci*
              ee. Rods or filaments
                f. Endospores produced   *Part 15, Sporeforming Rods and Cocci*
                ff. Endospores not produced
                  g. Straight rods   *Part 16, Gram-positive Nonsporeforming Rods*
                      Part 17, Actinomycetes and Related Organisms
                  gg. Irregular rods (coryneform) or tend to form filaments or
                    filamentous
                    Part 17, Actinomycetes and Related Organisms

---

*From Buchanan, R. E., and Gibbons, N. E. (eds.): Bergey's Manual of Determinative Bacteriology, 8th ed. Baltimore, The Williams & Wilkins Co., 1974, pp. 18–19.

chapter will consist of a brief description of each Part, with mention of a few groups or species of particular interest.

PART 1, THE PHOTOTROPHIC BACTERIA

These organisms make up a single order of predominantly aquatic bacteria, Rhodospirillales. They are morphologically diverse, including spherical, straight or bent rod, and spiral forms. All members possess bacteriochlorophyll and carotenoid pigments, and their colors range from red and green to purple and orange-brown. Photosynthesis occurs only under anaerobic conditions and requires oxidizable electron donors such as reduced sulfur compounds, organic compounds, and molecular hydrogen. Carbon dioxide is assimilated in the presence of light and an oxidized form of the electron donor is released or may accumulate within the cell (e.g., sulfur globules).

Family I, Rhodospirillaceae. These are called the purple nonsulfur bacteria because they are red or purple but do not utilize elemental sulfur as an electron donor in their photosynthetic activity. Instead, molecular hydrogen, sulfide, thiosulfate, or simple organic substances serve as electron donors for the reduction and photoassimilation of CO_2 or simple organic compounds (see Figures 6–1 and 6–2).

Family II, Chromatiaceae. These organisms are called purple sulfur bacteria, because most species use elemental sulfur or sulfide as the electron donor when they assimilate CO_2 anaerobically. Sulfur may accumulate temporarily within the cells (Figs. 6–3 and 6–4) or, in one genus, outside the cells and the final product of oxidation is sulfate.

Family III, Chlorobiaceae. The so called green sulfur bacteria contain green or brown carotenoid pigments in addition to bacteriochlorophyll c or d. Carbon dioxide is photoassimilated anaerobically when sulfide or sulfur is the electron donor. Many strains utilize molecular hydrogen as the electron donor if H_2S is the source of sulfur. Sulfur globules may accumulate outside the cells as an intermediate product in the oxidation of sulfide; sulfate is the final oxidation product. The green sulfur bacteria occur in anaerobic sulfide-containing aquatic environments such as ditches, ponds, lakes, rivers, sulfur springs, and various marine habitats.

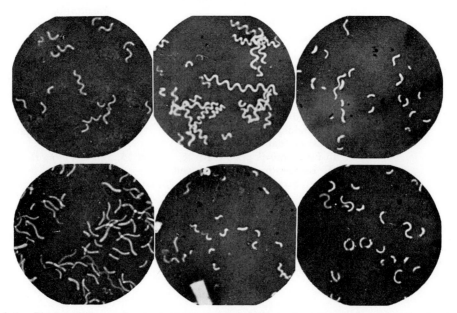

Figure 6–1. Rhodospirillum rubrum, *a species of the Athiorhodaceae (purple non-sulfur bacteria), 7 day anaerobic cultures on various media (800X). (From van Niel, Bact. Rev., 8:24, 1944.)*

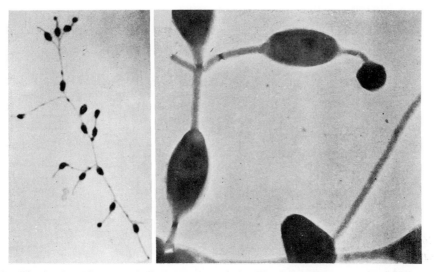

Figure 6-2. Rhodomicrobium vannielii, *a member of the Rhodospirillaceae. Left, 1800X; right, electron micrograph, 10,000X. The cells are approximately 1.2 × 2.8 μm and are connected by filaments 0.3 μm in diameter. A bud forms at the end of the filament, and this eventually becomes a new bacterial cell. (From Duchow et al., J. Bact. 58:411, 1949.)*

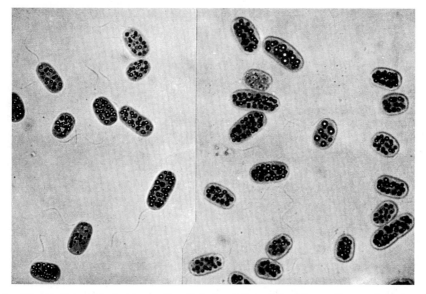

Figure 6-3. Chromatium okenii, *a large, flagellated member of the Chromatiaceae (purple sulfur bacteria) containing many sulfur granules (960X). (From Schlegel and Pfennig, Arch. Mikrobiol. 38:4, 1961.)*

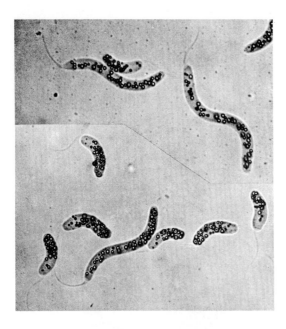

Figure 6–4. Thiospirillum jenense, *a spiral member of the Chromatiaceae. Numerous granules of sulfur are within the cells (960X). (From Schlegel and Pfennig, Arch. Mikrobiol. 38:5, 1961.)*

PART 2, THE GLIDING BACTERIA

Some gliding bacteria are chemolithotrophic and some are chemoorganotrophic. The gliding bacteria comprise two orders, which are separated on the basis of their production of fruiting bodies.

Order I, Myxobacterales

These are unicellular rods less than 1.5 μm in diameter that can aggregate and form fruiting bodies. The normal single-celled rod is gram-negative and is a uniform cylinder with tapered or blunt, rounded ends. It multiplies by transverse fission. Migration and aggregation of cells to form a fruiting body occurs under special conditions, such as when nutrient supplies become depleted. The fruiting body, which is macroscopic in size, contains slime and resting cells called myxospores (see Figure 6–5); it is sometimes situated at the top of an acellular stalk (see Figure 6–6). The life cycle of *Myxococcus xanthus* is shown in Figure 6–7.

The myxobacteria are strictly aerobic and their energy-yielding mechanisms are respiratory rather than fermentative. They are chemoorganotrophic and can utilize substrates consisting of macromolecules such as proteins, nucleic acids, and polysaccharides. Two principal groups can be discerned based upon nutritional habits. The bacteriolytic myxobacteria utilize as nutrients living or dead bacteria or yeast which may be incorporated in an agar medium. Cellulolytic species grow in an inorganic medium supplemented with a single carbohydrate such as cellulose or a simple sugar. Bacteriolytic types are often found in rich organic deposits such as animal dung.

Motility or swarming occurs only on a solid substrate and is not associated with any detectable organ of locomotion. The cells bend and flex actively in the vegetative stage and when swarming they advance as groups, forming tongue-like clumps or streams.

Order II, Cytophagales

These are gram-negative rods or filaments that do not form fruiting bodies and are motile by gliding in at least one morphologic stage. They may produce hormogonia, gonidia, or resting cells. *Hormogonia* are groups of cells that detach from parent filaments, become motile, and develop into new filaments.

Family I, Cytophagaceae. Cytophaga-

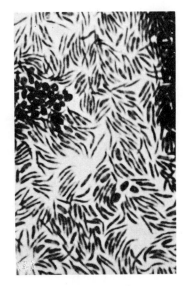

Figure 6–5. *Myxococcus, showing oval or spherical microcysts and long, thin, tapering, rod-shaped vegetative cells (1000X). (From Stanier et al.)*

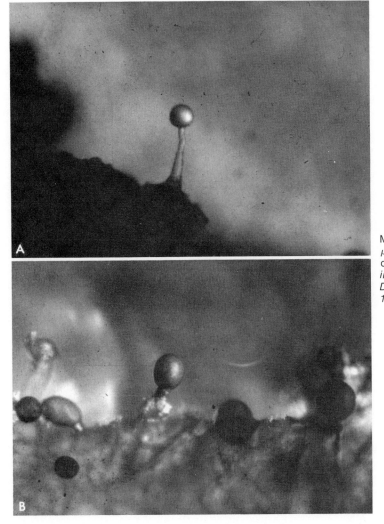

Figure 6–6. A, *Fruiting body of* Myxococcus stipitatus; *it is about 200 μm tall.* B, *Fruiting bodies of* Myxococcus fulvus; *they are 75 to 150 μm in diameter. (From Wireman and Dworkin, Science 189:516–523, 1975.)*

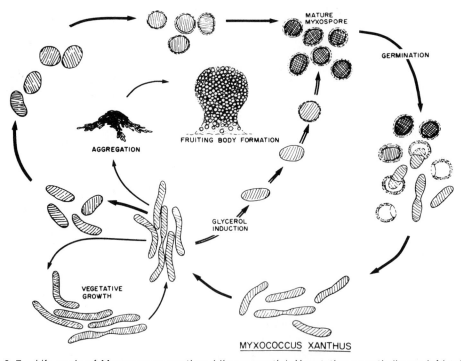

Figure 6–7.   *Life cycle of* Myxococcus xanthus *(diagrammatic). Vegetative growth (lower left) takes place when nutrients are abundant. Deprivation of nutrients induces aggregation (center) and formation of a fruiting body. Formation of the myxospores of which it is composed is shown clockwise at the left. Myxospore formation may also be induced by adding glycerol to a liquid medium. Germination of myxospores occurs when nutrients are again available. (From Wireman and Dworkin, Science 189:516–523, 1975.)*

ceae contain carotenoid pigments and are colored yellow, orange, or red. Some genera can attack agar, cellulose, or chitin.

Family II, Beggiatoaceae. Their cells occur in cylindrical filaments (see Figure 6–8). Motile forms move over a solid surface by gliding or by a slow, rolling, jerky motion. They multiply by transverse fission of single cells or of the individual cells in a filament. Sulfur globules may form on or within the cells. Species of Beggiatoa closely resemble blue-green algae of the genus Oscillatoria but are not photosynthetic. They are found in fresh water and marine habitats and in soil, where they can easily be mistaken for colorless members of the cyanobacteria.

Family IV, Leucotrichaceae. These long, unbranched, colorless filaments attach to solid substrates. They produce gonidia, which may glide jerkily and aggregate in a rosette pattern (see Figure 6–9).

PART 3, THE SHEATHED BACTERIA

Most sheathed bacteria are found in slowly running natural waters or in contaminated waters, and many species attach to submerged surfaces. Some are chemolithotrophic and some are chemoorganotrophic. Two genera, Sphaerotilus and Leptothrix, are motile by means of polar or subpolar flagella, the other five genera are nonmotile.

Leptothrix species are rods, either swimming freely as single cells or short chains or enclosed within a sheath. The sheaths are often covered or impregnated with ferric or manganic oxide. Leptothrix is chemoorganotrophic. It is strictly aerobic; its metabolism is respiratory, and molecular oxygen serves as its electron acceptor. It is found in running, unpolluted, iron-containing water or in polluted water; various species have been iso-

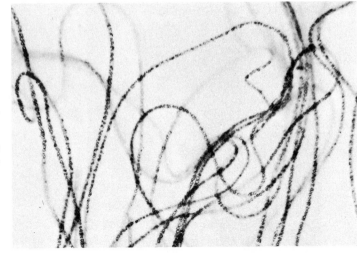

Figure 6–8. *Filaments of Beggiatoa, which structurally resemble blue-green algae but are non-photosynthetic. They obtain energy by oxidizing H_2S, and sulfur droplets accumulate within the filaments (1000X). (From Stanier et al.: The Microbial World, 2d ed. Englewood Cliffs, New Jersey, Prentice-Hall, Inc., 1963.)*

lated from activated sludge in sewage treatment plants.

There is a single species of Sphaerotilus, *S. natans* (see Figure 6–10). The rod-shaped cells are arranged in chains surrounded by a thin sheath, which may attach by means of a holdfast to submerged stones or plants. These sheaths are not encrusted with ferric or manganic oxide. This organism may be found growing as long tassels of sheaths filled with cells attached to submerged surfaces in slowly running water contaminated with sewage.

PART 4, BUDDING AND/OR APPENDAGED BACTERIA

Part 4 consists of more than a dozen genera of bacteria characterized by budding as a mechanism of reproduction and/or by the presence of some sort of nonlocomotor appendages such as slender hyphae. In addition, some of the organisms are motile by means of flagella at one stage in their life cycle.

These organisms are soil or water forms. Most species are chemoorganotrophic, and

Figure 6–9. *Leucothrix mucor on the shell of a 3-day old lobster. Scanning electron micrograph shows long filaments, gonidia within the filaments, motile gonidia that have separated from the filament, and rosettes formed from motile gonidia, some of which have begun to grow into filaments. (1200X) (From Sieburth, J. M.:Microbial Seascapes. Baltimore, University Park Press, 1975.)*

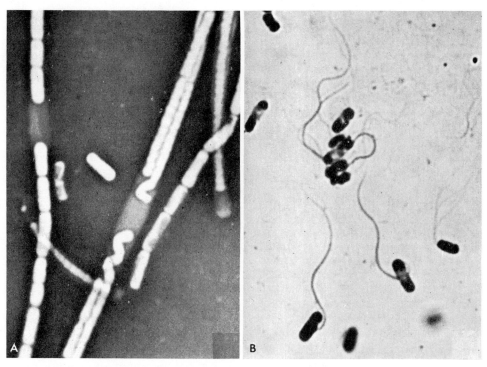

Figure 6–10.   Sphaerotilus natans, *a sheathed iron bacterium of the Chlamydobacteriales.* A, *Negative stain with nigrosin showing two strands of cells within a single sheath (2500X); B, tannic acid-fuchsin stain to show flagellated swarm cells that have emerged from their sheath (2700X). (From Stokes, J. Bact. 67:278, 1954.)*

many are respiratory, requiring a highly aerobic atmosphere. A few genera are characterized by iron or manganese deposition when growing in water containing high concentrations of these elements.

Hyphomicrobium (see Figure 6–11) cells are rod-shaped or oval, and produce one or two slender, filamentous hyphal outgrowths. Buds form at the tips of these hyphae. The mature buds are motile and may separate from the hyphae and attach to a surface.

Caulobacter (see Figure 6–12) cells are rod-shaped or fusiform. They bear a slender stalk on one end, and the other end possesses a single flagellum. Cell division occurs by transverse, asymmetrical fission, and after division the stalk end develops a flagellum at the opposite pole, and the flagellate end develops a stalk. These organisms may attach to one another by means of the stalks in the form of a rosette.

The genus Gallionella consists of kidney-shaped or rounded cells borne transversely on the ends of long, twisted stalks (see Figure 6–13). They may be covered by deposits of iron hydroxide. These organisms appear to be chemolithotrophic, since they oxidize ferrous iron to ferric iron and grow in inorganic media and in natural waters containing minimal amounts of nutrient material. Iron hydroxide may constitute as much as 90 per cent of the cell dry weight.

PART 5, THE SPIROCHETES

The spirochetes are a single order, Spirochaetales, with one family, Spirochaetaceae. They are slender, flexuous, unicellular, and coiled in a spiral consisting of at least one complete turn. All spirochetes are motile, but the mechanism of motility differs from that of most other bacteria. The cells consist of a protoplasmic cylinder and one or more axial fibrils, which wind about each other (see Figures 6–14 and 6–15). Approximately equal numbers of fibrils originate near each end of the protoplasmic cylinder. An outer envelope

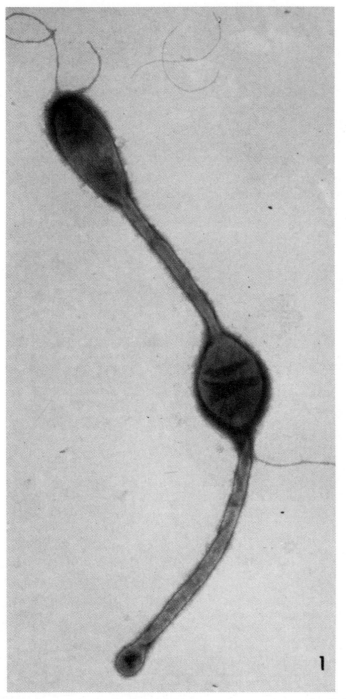

Figure 6-11. *Electron micrograph of a Hyphomicrobium species, a mature cell with hyphae and terminal buds. (20,500X). (From Bergey's Manual of Determinative Bacteriology, 8th ed. Baltimore, The Williams & Wilkins Company, 1974.)*

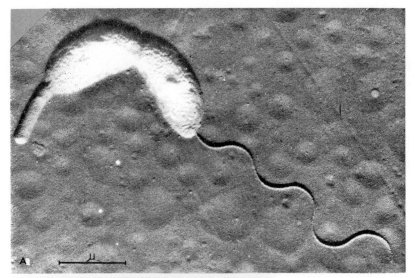

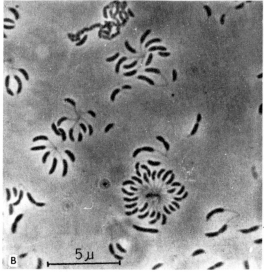

Figure 6–12. *Stalked bacteria of the genus Caulobacter. A, Electron micrograph of a dividing cell, with flagellum at one pole and stalk at the other (17,500X). B, Phase contrast micrograph showing rosettes formed when stalks of several cells adhere to each other's holdfasts (3800X). (From Poindexter, Bact. Rev. 28:231, 1964.)*

Figure 6–13. *Gallionella; line drawing from a micrograph by Cholodny (1950X). Curved rods attach to the substrate by means of twisted stalks, which may be impregnated with iron. (From Dorff:* Die Eisenorganismen, Pflanzenforschung, *1934. Courtesy of Gustav Fischer, Stuttgart, 1950.)*

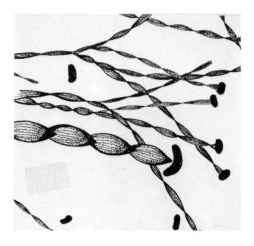

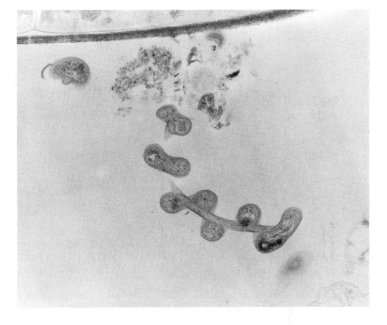

Figure 6–14. *Transmission electron micrograph of fresh water spirochete showing a longitudinal section of the spiral protoplasmic cylinder wound around the axial fibrils. In the upper left is a cross-section in which five axial fibrils appear in a bundle located at the center of the bottom edge of the section (62,400X). (Photograph by J. M. Sieburth and P. Johnson.)*

encloses both the cylinder and the fibrils. The fibrils appear to be contractile and seem to induce flexion of the cell, rapid rotation, and corkscrew locomotion.

Spirochetes require complex organic compounds as a carbon source. Some species are free-living, but many are commensal or parasitic and several are pathogenic. Some spirochetes are aerobic, some are facultative, and some are anaerobic.

There are five genera of spirochetes. They vary from 3 to 500 μm in length, and the filament of which the coil is composed varies from 0.09 to 3.0 μm in width. The amplitude

Figure 6–15. *Transmission electron micrograph of* Cristispira *from the crystalline style of an oyster. A longitudinal section is shown with an oblique cross-section at the upper right. A bundle consisting of more than 40 axial fibrils is clearly shown. Each fibril attaches to one side of the cell at an insertion point that appears as a dark disc (27,500X). (Photograph by P. Johnson, P. Willis, and J. M. Sieburth.)*

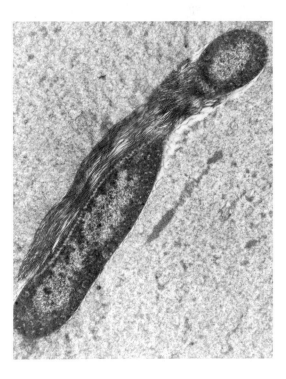

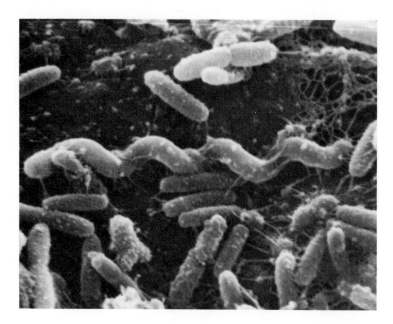

Figure 6–16. *Scanning electron micrograph of decaying salmon tail, showing a spirochete among a large population of bacterial rods (11,100X). (Photograph by J. M. Sieburth and P. Johnson.)*

of the coil is often no greater than 1 μm. Microscopic observation is difficult, and silver impregnation or other procedures that increase the diameter of the filament, or negative staining, darkfield, phase, or electron microscopy must be employed (see Figure 6–16).

Spirochaeta cells are long, free-living, and common in sewage and foul waters, but also present in fresh and sea water, especially in the presence of H_2S. *S. plicatilis* was the first species and one of the earliest bacteria to be described (Ehrenberg, 1838). It has never been cultivated in ordinary laboratory media.

Cells of Treponema are 0.09 to 0.5 by 5 to

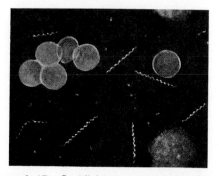

Figure 6–17. *Darkfield preparation showing* Treponema pallidum *in serous exudate from a primary syphilitic chancre. Several red blood cells and parts of two leukocytes are also present (1000X). (From Gillies and Dodd: Bacteriology Illustrated. Baltimore, The Williams & Wilkins Co., 1965.)*

10 μm. They consist of tight, regular or irregular spirals. The best known species is *T. pallidum* (see Figure 6–17), which causes syphilis and will be discussed in Chapter 20. This organism is highly parasitic and cannot be cultivated *in vitro;* it can be kept alive for several days in specially enriched media under anaerobic conditions and at a constant temperature near that of the human body. It is easily killed by chemical disinfectants and by slight variations in temperature.

Several species of Treponema are part of the normal flora of the mouth. When *T. vincentii* is associated with a gram-negative, long tapering rod or "fusiform bacillus," it causes trenchmouth or Vincent's angina.

PART 6, SPIRAL AND CURVED BACTERIA

There is one family, Spirillaceae, containing two genera, and four additional genera, for which the systematic position is unclear.

Cells of Spirillaceae are rigid, in contrast to the flexuous cells of spirochetes. They vary from less than one complete turn to many turns and are motile by means of a single polar flagellum or a bundle of flagella situated at one or both ends of the cell. They need organic nutrients and they range in oxygen requirements from strictly aerobic to anaerobic.

Members of the genus Spirillum possess bundles of flagella, usually at both poles. They also usually contain cytoplasmic granules of polyhydroxybutyrate. All but one of the 19 species are normally found in fresh or salt water. *S. minor* is parasitic; it occurs naturally in the blood of rats and mice and causes one type of rat bite fever in man.

The genus Bdellovibrio is of interest because its small, curved, motile cells may be parasitic in other bacteria. When first isolated they require a living host. These parasitic strains attach to and penetrate into the cells of specific host bacteria, where they multiply intracellularly. Facultatively parasitic and host-independent strains are derived from parasitic strains in the laboratory by mutation. They can grow in complex media. Host strains include both gram-positive and gram-negative bacteria. A schematic diagram of the life cycle of parasitic, facultative, and host-independent strains of *B. bacteriovorus* is shown in Figure 6–18.

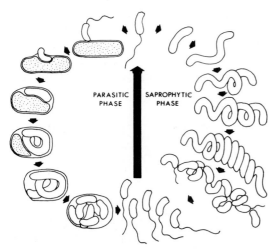

PARASITIC PHASE   SAPROPHYTIC PHASE

Figure 6–18. *Stages in the growth cycle of* Bdellovibrio bacteriovorus *(diagrammatic). In the parasitic phase, the small, motile, comma-shaped rod invades the host bacterium (e.g.,* E. coli*), grows, and divides by a constrictive process to produce many comma-shaped cells, which are released. In the saprophytic phase, growth occurs with formation of a long spiral, which fragments. (From Burnham, J. C., T. Hashimoto, and S. F. Conti, J. Bacteriol.,* 101*:1004, 1970.)*

PART 7, GRAM-NEGATIVE AEROBIC RODS AND COCCI

This group is physiologically heterogeneous, ranging from species that can utilize a wide variety of organic compounds as sources of energy to some that derive energy by oxidation of molecular hydrogen. It includes those that fix atmospheric nitrogen as well as those that secure nitrogen only from substances found in the human or animal body or from other complex sources.

Family I, Pseudomonadaceae. Pseudomonadaceae consists of straight or curved rods that are motile by means of polar flagella. They are chemoorganotrophic and respiratory rather than fermentative. They do not fix nitrogen and can use compounds containing more than one carbon atom as the sole source of carbon.

Pseudomonas is the best known genus. Twenty-nine species are recognized at present, but more than 200 species have been assigned to this genus at one time or another. They are common soil bacteria and are frequently found in fresh and marine water environments, where they contribute significantly to the mineralization of organic matter. Some species are facultative chemolithotrophs, able to use molecular hydrogen or carbon monoxide as a source of energy. The universal electron acceptor is molecular oxygen, but some species can use nitrate as an electron acceptor and bring about denitrification.

Pseudomonas is especially noted for its nutritional versatility; species can be found that are able to degrade a wide variety of organic compounds such as starch, cellulose, agar, chitin, phenols, naphthalene, hydrocarbons, and resins. A few common species produce water-soluble blue-green pigments, and some produce diffusible fluorescent pigments that can be detected by their behavior in ultraviolet light.

A considerable number of species of Pseudomonas are parasitic or occasionally pathogenic on plants; two species are pathogenic to animals. The most significant human pathogen is *P. aeruginosa*. This organism can be isolated from soil and water but its principal importance derives from its presence in wounds, burns, and urinary tract infections. These infections are particularly difficult to treat because the organism resists most common antibiotics. Moreover, it produces toxic substances that aggravate the clinical symptoms.

Family II, Azotobacteraceae. Members of this family are large and are normally oval or bluntly rod-shaped; they are often found in pairs, end-to-end (see Figure 6–19). These organisms fix molecular nitrogen when growing in a nitrogen-free medium in the presence of organic carbon.

Family III, Rhizobiaceae. Rhizobiaceae cells are normally rod-shaped. They incite the formation of cortical hypertrophies on plants. Species of Rhizobium cause the production of nodules on the roots of legumes, and species of Agrobacterium incite the formation of galls on the roots and stems of various plant species. The bacteria can be isolated from nodules or galls. They are best identified by appropriate plant inoculation tests.

Rhizobium species are noted for their symbiotic relationship with legume plants. The bacteria within the root nodules fix free molecular nitrogen and contribute dramatically to the nitrogen economy of the plant; properly nodulated plants grow luxuriantly in nitrogen-free sand if provided minerals and other nutrients. This will be discussed later (see Chapter 24).

Family V, Halobacteriaceae. These are rods and cocci that require approximately 2 molar (12 per cent) NaCl for growth. They differ from other gram-negative organisms with respect to the composition of their cell walls, which are lipoprotein and lack diaminopimelic and muramic acids. The halobacteria are found wherever salt occurs in adequate concentration, as in salt lakes, the Dead Sea, salterns (where sea water is evaporated to prepare solar salt), and in fish and other products preserved with solar salt.

Parvobacteria. Part 7 contains six genera that do not fit into any recognized family. Three of these genera were formerly known as parvobacteria, since they are extremely small. They are also described as coccobacillary, because the rods are so short that they appear coccoid.

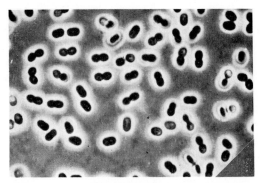

Figure 6–19.   Azotobacter vinelandii *showing typical paired coccoid cells; phase contrast (1000X). (From Jensen, Bact. Rev. 18:214, 1954.)*

Brucella consists of nonmotile, gram-negative cells, only 0.5 to 0.7 by 0.6 to 1.5 μm. They require complex organic media, including various vitamins. They are strictly aerobic, and some need added CO_2, especially when first isolated from human or animal sources. Brucella species occur in cattle, goats, and swine, from which they are transmitted to man. In the animal hosts they produce a generalized infection, sometimes with abortion as a symptom. The disease in man is usually chronic, prolonged, and debilitating. Undulant fever is one manifestation.

Bordetella also consists of coccobacillary, gram-negative cells, which are 0.2 to 0.3 by 0.5 to 1.0 μm. They require growth factors such as nicotinic acid, cysteine, and methio-

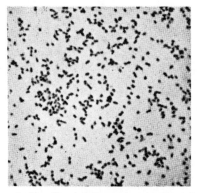

Figure 6–20.   Bordetella pertussis, *the whooping cough organism, showing minute, coccobacillary cells typical of the parvobacteria. (From D. T. Smith et al.: Zinsser Microbiology, 14th ed. New York, Appleton-Century-Crofts, 1968.)*

nine. The principal human pathogen is *B. pertussis,* which causes whooping cough (see Figure 6–20).

Francisella contains very small, coccoid or ellipsoidal, pleomorphic, gram-negative rods that require especially enriched media. *F. tularensis,* which causes tularemia, occurs in a great variety of animals, including rabbits, ground squirrels, various rodents, and wild animals throughout North America, continental Europe, and Asia. It readily infects humans who have contact with animals, and it appears to penetrate unbroken skin and mucous membranes. Like brucellosis, tularemia is prolonged and debilitating, but not highly fatal.

PART 8, GRAM-NEGATIVE RODS

Two families, comprising 17 genera, and an additional 10 genera or other groups of uncertain affiliation are included.

Family I, Enterobacteriaceae. This family consists of small, gram-negative rods, which possess peritrichous flagella if they are motile. They are chemoorganotrophic, and their metabolism is both fermentative and respiratory. Nearly all representatives are catalase positive and oxidase negative. Most strains reduce nitrates to nitrites.

Tribe I, Escherichieae. The 12 genera of which the family is composed are sometimes arranged in five groups, loosely referred to as tribes. Escherichieae includes five genera. The first genus, Escherichia, consists of the single species *E. coli,* one of the most widely studied bacteria. It is almost universally present in the lower intestine of warm blooded animals. When it gains access to other locations, such as the urinary tract, it can produce serious infection that resists antibiotic treatment. *E. coli* is characterized by the fermentative production of acid and gas from glucose and lactose, among other carbohydrates, and failure to produce acetylmethyl carbinol (negative Voges-Proskauer test).

There are two genera of intestinal pathogens in Escherichieae: Salmonella and Shigella. *Salmonella typhi* (see Figure 6–21) causes typhoid fever. There are approximately 2000 other species or types of Salmonella distinguished from one another principally by their antigenic properties. The nontyphoid salmonellae produce enteric fevers and gastroenteritis. The genus Shigella includes only four species, subdivided into about 35 serologic types. These organisms cause bacillary dysentery. Salmonella and Shigella will be discussed further in Chapter 21.

Tribe II, Klebsielleae. Klebsiella and Enterobacter are the most commonly en-

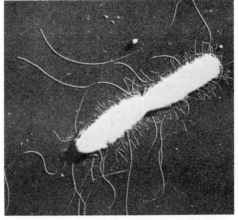

A   **B**

Figure 6–21.   Salmonella typhi, *the cause of typhoid fever. A, A preparation stained to demonstrate flagella. B, Electron micrograph of shadowed preparation of dividing rods showing many pili and a very few flagella (12,500X). (From Frobisher et al.: Fundamentals of Microbiology, 9th ed. Philadelphia, W. B. Saunders Company, 1974.)*

countered genera in this tribe. *K. pneumo-niae* is widely distributed in soil, water, on grain and other vegetable matter, and in the intestinal tract of man and animals. Encapsulated strains cause one form of pneumonia in man. Species of Enterobacter are present in the feces of man and other animals, in sewage, soil, water, and dairy products. Klebsiella is nonmotile, whereas Enterobacter is motile.

Tribe III, Proteae. Proteus species are found in the feces of many animals, including man, and hence in sewage and soil. They may contribute significantly to the pathogenesis of local infections such as cysts.

Tribe IV, Yersinieae. An important member of this tribe is *Yersinia pestis*. It is the cause of bubonic plague in man and of sylvatic plague in ground squirrels, rats, and other rodents. Sylvatic plague is endemic in the southwestern United States, where a few cases of bubonic plague have recently occurred.

Tribe V, Erwinieae. The genus Erwinia is associated with plants as part of the normal flora or as pathogens. The genus is currently divided into three clusters of organisms: (a) the *amylovora* group, which causes necrosis and produces plant wilt diseases; (b) the *herbicola* group, which is found on plant surfaces and occurs secondarily in lesions caused by plant pathogens; and (c) the *carotovora* group, which produces soft rot.

Family II, Vibrionaceae. The Vibrionaceae are rigid, gram-negative rods, either straight or curved. They are usually motile by means of polar flagella. These organisms are chemoorganotrophic and are both fermentative and respiratory. Their habitat is fresh or sea water, but they are sometimes found in fish or humans.

Two species of Vibrio are significant in human disease. *V. cholerae* causes Asiatic cholera, which is still a major problem in some southern Asiatic countries such as Bangladesh. *V. parahaemolyticus* is found in water and is usually isolated from diseased fish. It causes an acute enteritis in man.

Haemophilus. Among the genera of uncertain affiliation is Haemophilus. These organisms are minute to medium in size, often coccobacillary, and frequently pleomorphic. They are strictly parasitic, requiring growth factors present in blood, particularly hemin or NAD (nicotinamide adenine dinucleotide) or both. *H. influenzae* was originally thought to cause influenza, because it was first isolated from cases of this disease. Later it was found normally in the human nasopharynx and in other locations in the body. It is now known to be significant principally as the cause of a highly fatal form of meningitis in infants.

PART 9, GRAM-NEGATIVE ANAEROBIC BACTERIA

These are nonsporeforming rods, straight, bent, or pleomorphic, and obligately anaerobic. There is one family, Bacteroidaceae, containing three genera; an additional half dozen genera of uncertain affiliation are also included. The three genera of Bacteroidaceae are distinguished primarily according to the products formed in a peptone-glucose medium. Species of Bacteroides produce a mixture of acids including succinic, acetic, formic, lactic, and propionic, but butyric acid is usually not a major product. Fusobacterium species produce butyric acid as a major product. Leptotrichia produces only lactic acid as a major fermentation product.

Bacteroides. As indicated in Figure 6–22, these organisms vary from minute, coccoid or coccobacillary forms to straight rods and, under certain conditions of cultivation, extremely pleomorphic forms. They are found in the body cavities of man and other animals, in the oral cavity and gingival crevice area of man, and in the rumens of various animals. They are the most abundant bacteria in the human intestine, amounting to as many as 10^9 per gram of feces. Some species are pathogenic. Diseases in which they have been found include appendicitis, peritonitis, heart valve infections, rectal abscesses, pilonidal cysts, and urogenital tract lesions.

Fusobacterium. These species tend to be pointed on one or both ends, but they are also quite pleomorphic (see Figure 6–23). They do not grow on the surface of agar plates incubated aerobically. They are found in the body cavities of man and other ani-

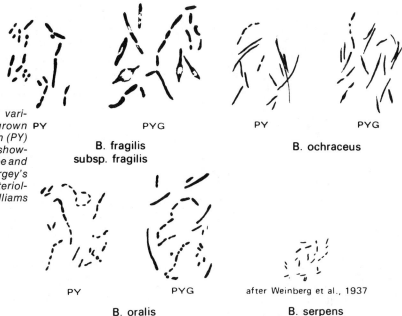

Figure 6–22. *Sketches of various species of Bacteroides grown in peptone-yeast extract broth (PY) or PY-glucose broth (PYG), showing the great variability in shape and size of the cells. (From Bergey's Manual of Determinative Bacteriology, 8th ed. Baltimore, The Williams & Wilkins Company, 1974.)*

PY   PYG   PY   PYG

B. fragilis
subsp. fragilis

B. ochraceus

PY   PYG   after Weinberg et al., 1937

B. oralis   B. serpens

mals, and some species occur in necrotic abscesses and local infections.

Genera of Uncertain Affiliation. *Desulfovibrio.* Species of this genus are principally water forms, often present in polluted or brackish waters or in water-logged soils rich in organic materials. Their metabolic peculiarity is the reduction of sulfate or other sulfur compounds to H_2S by anaerobic respiration. The hydrogen and electrons needed for the reduction are derived from lactate, pyruvate, and malate, which are oxidized to acetate and carbon dioxide.

PART 10, GRAM-NEGATIVE COCCI AND COCCOBACILLI

There is one family composed of four genera and two additional genera of uncertain affiliation. They include spherical organisms arranged in pairs or masses, with the adjacent sides flattened or even slightly concave, and rod-shaped organisms in pairs or short chains. Some species have complex nutrient and environmental requirements for primary isolation but grow better after an interval of laboratory cultivation. Most species produce catalase and oxidase.

The best known genus in the family Neisseriaceae is Neisseria. The cells are often shaped like coffee beans with their flattened sides opposing each other. All species are parasitic, and some are also pathogenic. *N. gonorrhoeae* (Fig. 6–24) is the cause of gonorrhea, and *N. meningitidis* produces cerebrospinal meningitis. These organisms are nutritionally fastidious and require specially enriched media. *N. gonorrhoeae* may be isolated from clinical specimens on chocolate (heated blood) agar; it will not ordinarily grow on blood agar. A moist atmosphere containing 5 to 10 per cent CO_2 and a temperature very close to 37° C. are also required. *N. meningitidis* will grow on blood agar, but it also requires a tempera-

Figure 6–23. *Sketches of two species of Fusobacterium grown in peptone-yeast extract broth (PY) or PY-glucose broth (PYG). (From Bergey's Manual of Determinative Bacteriology, 8th ed. Baltimore, The Williams & Wilkins Company, 1974.)*

PY   PYG   PY   PYG

F. nucleatum   F. necrophorum

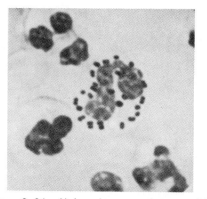

Figure 6–24. Neisseria gonorrhoeae *within a leukocyte in urethral pus. (From Joklik, W. K., and D. T. Smith: Zinsser Microbiology, 15th ed. New York, Appleton-Century-Crofts, 1972.)*

ture of 37° C. for primary isolation and grows better in a moist atmosphere containing extra CO_2. Gonorrhea and meningitis are discussed further in Chapter 20.

PART 11, GRAM-NEGATIVE ANAEROBIC COCCI

The gram-negative anaerobic cocci include a single family, Veillonellaceae. There are three genera, differing in cell size, sources of energy, and products formed in growth media. The best known is Veillonella. Members of this genus are very small (0.3 to 0.5 μm). They use lactate, pyruvate, or succinate as sources of carbon and energy and produce CO_2 and H_2 and two- and three-carbon volatile fatty acids. These organisms are strictly anaerobic and they require complex media and an atmosphere containing CO_2. They are parasitic in the mouth and respiratory tract and intestine of man and other animals.

PART 12, GRAM-NEGATIVE CHEMOLITHOTROPHIC BACTERIA

These bacteria use CO_2 as their principal source of carbon for growth and they obtain energy by oxidation of inorganic compounds.

Family I, Nitrobacteraceae. Bacteria included in this family are nonsporeforming and they are rod-shaped, ellipsoidal, spherical, or spiral. They are soil or water forms, some of which play a vital role in the nitrogen cycle by oxidizing ammonia or nitrites to a form that can be utilized by plants. There are two physiologic groups of these *nitrifying* bacteria. Both groups secure carbon by fixation of CO_2. Nitrosomonas, Nitrosococcus, and two other genera derive energy from the oxidation of ammonia to nitrite; Nitrobacter and two other genera secure energy by oxidizing nitrite to nitrate. These bacteria are aerobic and nearly all are obligately chemolithotrophic. Their importance in the soil economy will be discussed in Chapter 24.

Sulfur-metabolizing Bacteria. Some of these organisms secure energy by oxidizing reduced or partially reduced sulfur compounds, including elemental sulfur, sulfides, thiosulfate, and sulfite. The final oxidation product is sulfate, but sulfur and other products may accumulate under certain conditions.

The genus Thiobacillus (Fig. 6–25) has been studied intensively and is of interest because of the high acidity it may produce under natural conditions. Certain species yield sufficient sulfuric acid to lower the reaction to pH 1 or less. These bacteria cause great damage by corroding concrete and steel in mines. Other species are found in seawater, fresh water, sewage, sulfur springs, and soil. They are especially common in locations where H_2S is produced or sulfur is deposited.

Family II, Siderocapsaceae. These spherical, ellipsoidal, or rod-shaped organisms deposit iron or manganese oxides on or in capsules or other extracellular material.

PART 13, METHANE-PRODUCING BACTERIA

The Methanobacteriaceae are gram-positive or gram-negative, nonsporeforming rods or cocci. They are strict anaerobes and secure energy by oxidizing hydrogen and formate or fermenting acetate and methanol. The electrons and hydrogen generated in

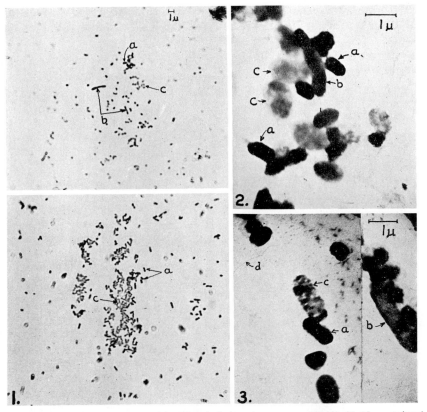

Figure 6–25. Thiobacillus thiooxidans, (1) *stained, light microscopy, 1200X;* (2, 3) *unstained, electron microscopy, 8000X.* a, b, c, *and* d *indicate cells of different size, staining, and electron opacity. (From Umbreit and Anderson, J. Bact. 44:318, 1942.)*

these processes are used to reduce CO_2 or (possibly) CO to methane, CH_4. The organisms do not utilize carbohydrate, protein, or other organic compounds than those already listed as sources of energy. They are therefore a very highly specialized physiologic group.

Methane-producing bacteria are widely distributed in anaerobic habitats such as lake sediments, soil, anaerobic sewage digestors, and the gastrointestinal tracts of animals. Most species grow best at or slightly above 37° C.; one species grows best at 65 to 70° C.

PART 14, GRAM-POSITIVE COCCI

Aerobic or facultative gram-positive cocci are included in Micrococcaceae and Strep-

tococcaceae; anaerobic cocci comprise Peptococcaceae.

Family I, Micrococcaceae. These spherical bacteria divide in more than one plane to form regular or irregular clusters or packets. Species of the genus Micrococcus are strictly aerobic, and they are oxidative rather than fermentative. Their optimum growth temperature is 25 to 30° C. Most strains are pigmented, yellow or red.

Staphylococcus species are facultatively anaerobic; their metabolism is both respiratory and fermentative. Their optimum growth temperature is 35 to 40° C. Most strains can grow in the presence of 15 per cent NaCl. *S. aureus*, the best known species, is part of the normal body flora of many individuals and it is also pathogenic, producing a variety of diseases in any part of the body, as will be discussed in Chapter 20. *S. epidermidis* is almost universally present in great numbers on the skin and upper respiratory mucous membranes of normal individuals. It is usually not pathogenic.

Family II, Streptococcaceae. Members of this family are spherical or ovoid, and the cells are arranged in pairs or chains of varying length or in tetrads. Their metabolism is fermentative, and their nutritional requirements are complex.

Streptococcus species ferment glucose by the hexose diphosphate pathway (see Chapter 8) and yield chiefly dextrorotatory lactic acid. Their optimum growth temperature is generally about 37° C.

Many of the 21 species are highly pathogenic. For diagnostic purposes they are often classified primarily by their action on blood agar. *Hemolytic* streptococci dissolve erythrocytes; the reaction is called beta hemolysis. *S. pyogenes* is the principal human pathogen in this group. *Viridans* streptococci produce a greenish or brownish discoloration on blood agar known as the alpha reaction. They are normally present on the oral and upper respiratory membranes. If they gain access to the bloodstream or the heart valves, they may produce serious disease.

S. pneumoniae is an alpha streptococcus that causes pneumonia. It can be distinguished from most other alpha streptococci by its solubility in bile, its ability to ferment inulin, and its sensitivity to optochin.

Leuconostoc species are found principally on vegetable matter (e.g., sauerkraut, pickles) and in dairy products. Their optimum growth temperature is between 20 and 30° C. They ferment glucose via the hexose monophosphate pathway and produce levorotatory lactic acid, together with CO_2 and ethanol or acetic acid.

Family III, Peptococcaceae. These anaerobic cocci occur singly, in pairs, in tetrads and irregular masses, and occasionally in cubical packets. Their nutritional requirements are complex. They are found in the mouth, respiratory tract, and intestinal tract of man and lower animals, in the human female urogenital tract, in the soil, and on the surface of cereal grains. Peptostreptococcus occurs principally in chains and Sarcina in cubical packets.

PART 15, ENDOSPORE-FORMING RODS AND COCCI

All 114 species of endospore-forming bacteria except five are included in the genera Bacillus and Clostridium.

Bacillus species are aerobic or facultative rods that produce catalase. Their metabolism is either respiratory, fermentative, or both. Most species occur in such natural habitats as soil and water and on plant materials. A few (e.g., *B. anthracis*) are animal and human pathogens.

The cells of Clostridium species are anaerobic rods that do not reduce sulfate to sulfide. Catalase is usually not produced, and the anaerobic nature of these organisms has been attributed in part to lack of this enzyme, because otherwise the toxic substance H_2O_2 might accumulate. Endospores are ovoid to spherical and their diameter is usually greater than that of the rest of the cell, so the rods are distended. A cell with a swollen terminal spore is a *plectridium,* one with a swollen central spore (i.e., spindle-shaped) is a *clostridium*. Clostridium species are common in soil, marine and fresh water sediments, and in human and animal intestines. A few are pathogenic, such as *C. tetani* and *C. botulinum*. These organisms release very potent toxins, which produce the diseases tetanus and botulism, respectively.

PART 16, GRAM-POSITIVE, ASPOROGENOUS ROD-SHAPED BACTERIA

This part consists of one family and some additional genera of uncertain systematic position. It should be pointed out that Part 17 also contains gram-negative nonsporeforming rods, but they are irregular or filamentous in shape.

Family I, Lactobacillaceae. These organisms are straight or occasionally curved rods occurring singly or sometimes in chains. Most strains are nonmotile, facultative or anaerobic, and catalase negative. They require complex organic media, are highly saccharolytic, and they produce considerable lactic acid when they ferment

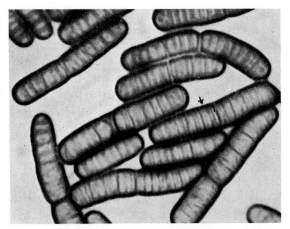

Figure 6–26. Caryophanon latum, *a large bacterium of unusual structural complexity; stained with tannic acid and crystal violet (2700X). Different stages in transverse fission are shown. Closely spaced surface striations are in focus in the cell indicated by the arrow, whereas unstained nuclear structures are in focus within other cells. (From Pringsheim et al., J. Gen. Microbiol. 1:267, 1947.)*

other animals. It causes meningitis, encephalitis, endocarditis, septicemia, and abscesses and other local infections.

Erysipelothrix. A single species, *E. rhusiopathiae,* is the cause of swine erysipelas and is transmissible to man and other animals. The organisms are rod-shaped and tend to form long filaments. They are nonmotile and catalase negative.

Caryophanon. *C. latum* is interesting because it consists of rods or filaments that may be 3 μm in diameter and 6 to 30 μm long. These large bodies are divided by crosswalls into numerous disc-shaped cells (see Figure 6–26). They are gram-positive, motile, and strictly aerobic. They were originally isolated from cow dung.

PART 17, ACTINOMYCETES AND RELATED ORGANISMS

This large and heterogeneous assortment of bacteria is related by the fact that its members are gram-positive, nonsporeforming rods, many being irregular or filamentous in shape.

CORYNEFORM GROUP

These are club-shaped bacteria. The genus Corynebacterium consists of straight or slightly bent rods that stain irregularly, sometimes showing definite granules, especially when treated with alkaline methylene blue. Their characteristic snapping division produces angular and palisade arrangements. They are aerobic or facultative and generally grow best as a surface pellicle on liquid media. Pathogenic strains of *C. diphtheriae,* the cause of diphtheria, produce a potent exotoxin when harboring a prophage carrying a specific genetic determinant called *tox*[+]. This will be discussed in Chapter 23. *C. pseudotuberculosis* resembles the diphtheria organism but produces ulcerative lymphangitis, abscesses, and various chronic infections in sheep, goats, horses, and other warmblooded animals. It also secretes an exotoxin. There are a number of bacteria like these species in morphology,

sugars. Species of Lactobacillus, the only genus, are found in decomposing animal and plant materials that contain carbohydrates. They are also present in the mouth and intestinal tract and vagina of warmblooded animals, including man. In general, their growth temperature requirements are high: many strains grow best at 37 to 45° C. Lactobacilli are active in natural fermentations of food substances and are also added in "starters" to promote industrial fermentations, as in the manufacture of fermented milk products. There are two major physiologic types: *homofermentative* strains produce principally lactic acid from glucose; *heterofermentative* lactobacilli produce considerable amounts of CO_2, acetic acid, and ethanol, in addition to lactic acid.

Listeria. This genus consists of small, coccoid, gram-positive rods that tend to produce chains of three to five cells. Diphtheroid, palisade forms (see Part 17) are also seen on occasion. They are motile when grown at 20 to 25° C., but may appear nonmotile at 37° C. Listeria is usually catalase positive. *L. monocytogenes* has been isolated from healthy animals, sewage, vegetable matter and lesions of various organs and the blood and spinal fluid of man and but part of the normal skin and mucous membrane flora of man and animals. They are known as diphtheroid bacteria.

Plant pathogenic Corynebacteria are morphologically similar to human and animal pathogens, but they are not as pleomorphic and show less evidence of the palisade arrangement. They produce a variety of disease conditions, including stem hypertrophy, gumming, and wilts or leaf spots.

Cells of Arthrobacter species undergo a marked change in form during growth on complex media. Older cultures (2 to 7 days) contain largely or entirely coccoid cells. When these coccoid forms are transferred to fresh medium, one or two parts of each cell swell and elongate to rod forms that are usually narrower than the coccoid cell. Subsequent growth and division yield irregular rods that vary in size and shape and include straight, curved, wedge- and club-shaped forms (Fig. 6–27). Some show a V arrangement or a more complex angular grouping. Arthrobacter species are considered gram-positive, but under some conditions they are gram-variable. They are common soil organisms.

Family I, Propionibacteriaceae. Members of the family are pleomorphic, branching or regular rods or filaments. They are anaerobic to aerotolerant. Saccharolytic strains produce CO_2, propionic and acetic acids, or mixtures of organic acids. Propionibacterium species are found in dairy products, on the skin of man, and in the intestinal tracts of man and animals. A few species have been isolated from blood, pus, infected wounds, and abscesses.

Order I, Actinomycetales

Actinomycetales are branching, filamentous bacteria; in some families they develop into a mycelium. Filaments may be short, as in the Actinomycetaceae and Mycobacteriaceae, or they may be well developed, as in the Streptomycetaceae. When filaments break into small fragments, coccoid or diphtheroid forms are produced. Some families produce spores on aerial hyphae similar to the spores of certain molds. These are not the same as endospores of the Bacillaceae.

Among the best known families are Mycobacteriaceae and Streptomycetaceae. The former are acid alcohol-fast, at least in some stages of growth. *Mycobacterium tuberculosis,* the cause of tuberculosis, will be discussed further in Chapter 23. Several other species of Mycobacterium are interesting, either as cause of disease or because of their resemblance to the tuberculosis organism. Many species are obligate parasites, but there are also saprophytic strains capable of growing on simple substrates. The rate of growth is slower than that of many other bacteria—several days or weeks are required for visible growth of *M. tuberculosis*.

Streptomycetaceae includes gram-positive organisms with a mycelial form of growth that does not fragment readily. Reproduction occurs by formation and germination of aerial spores or conidia (see Figures 6–28 to 6–30) and sometimes by growth of bits of the vegetative hyphae. They are primarily soil forms, and members of the genus Streptomyces are noted for production of antibiotic substances. They are aerobic, highly oxidative, and heterotrophic. They can utilize glucose for growth and generally hydrolyze complex substances such as gelatin, casein, and starch. Colonies of many species are hard and brightly colored or chalk-white. Some species produce the

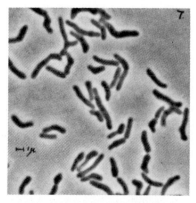

Figure 6–27. Arthrobacter globiformis *from a 24 hour yeast-soil extract agar culture. (From Bergey's Manual of Determinative Bacteriology, 8th ed. Baltimore, Williams & Wilkins Company, 1974.)*

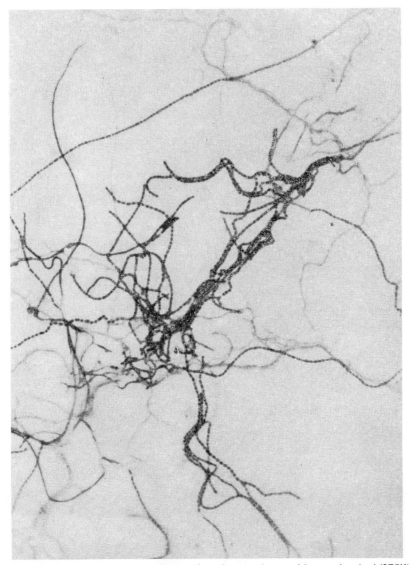

Figure 6–28. Streptomyces venezuelae, *the organism that produces chloramphenicol (975X). Gram stain of branching mycelium and chains of oval spores. (Prepared by Littman, Armed Forces Institute of Pathology.)*

odor characteristic of decaying leaves, both in soil and in laboratory cultures. This is a very large genus: 416 species and 47 subspecies are listed.

PART 18, THE RICKETTSIAS

The Rickettsias are small, gram-negative, parasitic bacteria, most of which multiply only within living host cells. Many are transmitted by arthropod vectors.

Order I, Rickettsiales

These organisms are rod-shaped or coccoid and often pleomorphic (see Figure 6–31). They have typical bacterial cell walls and are nonmotile. Most species multiply only inside host cells; they may be cultivated in chick embryos (see Figure 6–32), living tissues, or vertebrate cell cultures. They are parasitic or mutualistic. The parasitic species are often found within endothelial cells of the vascular system or are as-

STRAIGHT   FLEXUOUS   FASCICLED

OPEN - LOOPS
PRIMITIVE SPIRALS   OPEN   CLOSED
HOOKS   SPIRALS   SPIRALS

Figure 6–29. *Various forms of conidial chains pro-*
duced by different species of Streptomyces. (From
Frobisher: Fundamentals of Microbiology, 8th ed.
Philadelphia, W. B. Saunders Co., 1968. Adapted from
Pridham, Hasseltine, and Benedict, Appl. Microbiol., vol.
6.)

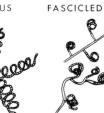

MONOVERTICILLATE   MONOVERTICILLATE
NO SPIRALS   WITH SPIRALS

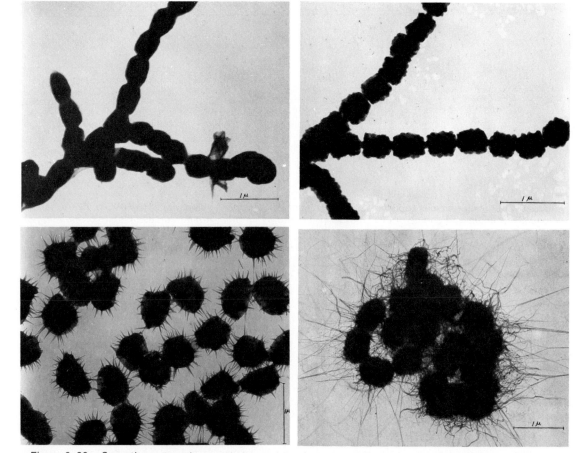

Figure 6–30. *Smooth, warty, spiny, and hairy conidia of species of Streptomyces. (From Tresner, Davies, and*
Backus, J. Bact., 81:70–80, 1961.)

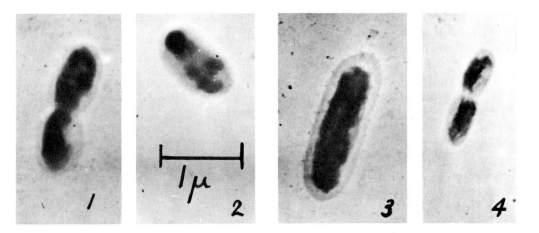

Figure 6–31.   *Electron micrographs of rickettsiae: 1, epidemic typhus fever; 2, endemic typhus fever; 3, Rocky Mountain spotted fever; 4, American Q fever. (Photographs by Plotz, Smadel, Anderson, and Chambers; A.S.M. LS-25.)*

sociated with erythrocytes of vertebrates. They may also be present in arthropods, which serve as primary hosts or as vectors. Mutualistic species are also found in insects.

Among rickettsial diseases are typhus fever, caused by *Rickettsia prowazekii,* and Rocky Mountain spotted fever, caused by *R. rickettsii.* The former is transmitted by the human louse and the body louse. The bacteria multiply within the digestive tract of the arthropod, which succumbs to infection within 1 to 3 weeks. Human infection occurs when Rickettsiae in the louse feces are introduced into the skin in the process of scratching a louse bite. The organisms invade the blood vessels and are distributed to the entire body. The disease in man is characterized by general symptoms, including chills, fever, headache, other aches and pains, and a rash that appears on the trunk and spreads to the extremities.

Order II, Chlamydiales

These organisms are coccoid, and, when multiplying intracellularly, they change from a small, rigid walled form (elementary body) into a larger, thin walled noninfectious form (initial body) that divides by fission. The daughter cells then reorganize and condense, becoming elementary bodies, which can survive extracellularly and infect other host cells.

Chlamydia trachomatis is parasitic in man, in whom it causes a variety of diseases of the eyes and the urogenital tract: trachoma, a form of conjunctivitis, lymphogranuloma venereum, urethritis, and proctitis. Transmission occurs by contact.

PART 19, THE MYCOPLASMAS

Most of the Mycoplasmas are included within a single class, Mollicutes (Latin, *mollis,* soft, pliable + *cutis,* skin). These are procaryotic organisms that are surrounded by a triple layered membrane; they lack a true cell wall and are unable to synthesize cell wall precursors such as muramic acid and diaminopimelic acid. The cells are small,

Figure 6–32.   *Section through a developing (10 to 12 day) chick embryo showing how inoculation can be made into the head of the embryo, the allantoic cavity, and the yolk sac. (From Burrows: Textbook of Microbiology, 19th ed., Philadelphia, W. B. Saunders Co., 1968.)*

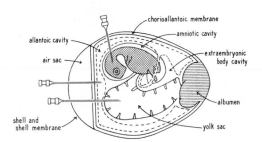

often submicroscopic, pleomorphic, and coccoid (see Figure 6–33) or filamentous. Filamentous forms tend to produce branched, myceloid structures (see Figure 6–34). Reproduction appears to take place by the formation of coccoid "elementary bodies" within the filaments and their subsequent release when the filaments disintegrate; binary fission and budding may also occur.

Species recognized to date will grow in artificial, cell-free media of varying complexity. Colonies are minute and grow into the surface of a solid medium with a "fried egg" appearance (see Figure 6–35). Inasmuch as these organisms lack a cell wall, they are resistant to penicillin and other antibiotics for which the site of action is the cell wall. Mycoplasmas may be saprophytic or parasitic, and many species are pathogenic, causing diseases of animals and possibly of plants. Some appear to be transmitted by arthropods.

Various species of Mycoplasma are human or animal pathogens or parasites. *Mycoplasma pneumoniae* has been recognized in recent years as the cause of primary atypical pneumonia in man. The infection is

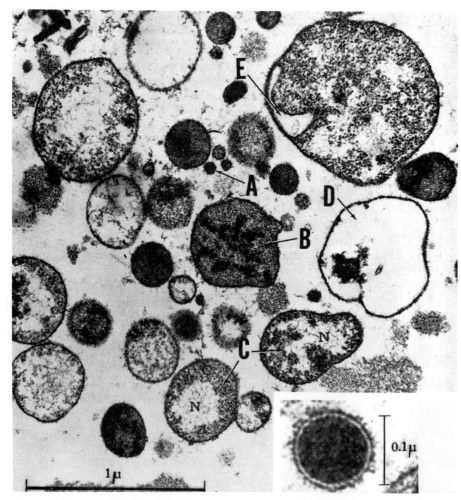

Figure 6–33. *Electron micrograph of* Mycoplasma hominis, *a pleuropneumonia-like organism, grown in a broth of brain and heart infusion and horse serum and thin-sectioned to show internal structures. A variety of forms are apparent, ranging in size from about 0.1 μm (A and inset) to about 0.9 μm (E). Cell B has finely granular cytoplasm, cells C contain "nuclear" areas (N) with netlike strands and cytoplasm with ribosomelike granules, and D consists of an empty plasma membrane. (From Anderson et al., J. Bact., 90:181, 1965.)*

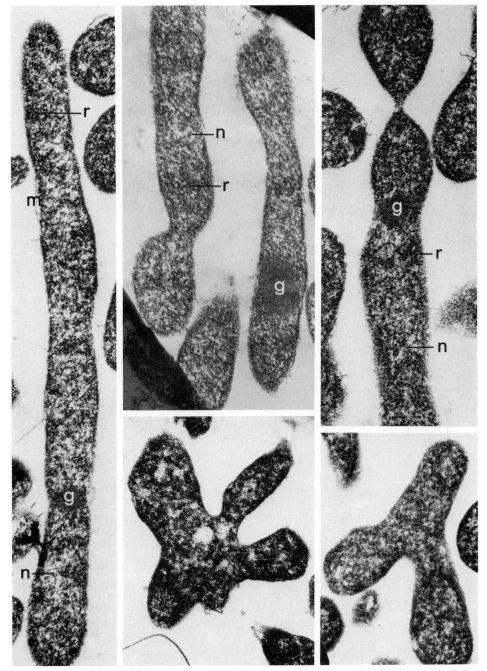

Figure 6–34. *Electron micrographs of cells of* Mycoplasma laidlawii, *showing a variety of filamentous, asteroid, and other irregular forms. Ultrastructural details that are indicated include the membrane* (m), *nuclear material* (n), *ribosomes* (r), *and granular region* (g). *(From Maniloff, J. Bact., 102:561–572, 1970.)*

Figure 6–35. *Longitudinal section through an agar colony of* Mycoplasma salivarium. *Subsurface growth in the agar and thinner, spreading surface growth are responsible for the typical "fried egg" appearance of a colony viewed from above. (From Knudson and MacLeod, J. Bact., 101:609–617, 1970.)*

accompanied by development of a cold hemagglutinin; that is, an antibody that agglutinates (aggregates) erythrocytes at low temperatures (e.g., 5° C.) but not at normal body temperature (37° C.). This reaction has been known to occur *in vivo*: patients who received a transfusion of compatible blood that had not been warmed to body temperature had a severe reaction similar to that which occurs when incompatible blood is administered, due to intravascular clumping of the blood cells.

SUPPLEMENTARY READING

Buchanan, R. E., and Gibbons, N. E.: *Bergey's Manual of Determinative Bacteriology,* 8th ed. Baltimore, The Williams & Wilkins Company, 1974.

VIRUSES 7

VIRUSES AND PARASITISM

It was pointed out in Chapter 1 that the ancestral procaryote or protovirus probably consisted of nucleic acid and protein, and these turn out to be the essential and in many cases the only components of modern viruses. Whereas the first organism secured its chemical components from the supply of organic compounds that had formed by chance during the preceding billions of years in the pools and oceans of the early Earth, twentieth century viruses are made in the living cells of animals, plants, or bacteria, where their proteins and nucleic acids are assembled. Thousands or even millions of years were probably required to replicate a protovirus; a few minutes suffice for a modern virus.

The origin of viruses is not known. It has long been considered that they are the culmination of a series of evolutionary transformations beginning with the primordial organism—first, through forms of increasing efficiency in the activities associated with independent life, and later through forms of increasing dependence as the parasitic habit developed. The path from the postulated intermediate organism to modern viruses is a trail along which the capacity to live independently was lost by disuse, because the intracellular habitat provided preformed protoplasmic constituents, energy, a stable environment, and other favorable conditions. Structural and chemical components disappeared as they were no longer used; the final form is hardly larger than a macromolecule, retaining only the ability to direct its own reproduction or replication.

However, some authorities point out that viruses are essentially only bits of genetic material and might be nothing more than misplaced cell components—that is, organelles that have been transferred in some manner from the cells of which they were originally a part to other cells. When they encounter appropriate conditions in the recipients, they participate in the normal functions of the latter and in so doing are replicated. Students who pursue the subject of virology further will find many arguments consistent with the tenets of modern molecular biology to support this hypothesis.

CULTIVATION OF VIRUSES

HOST SPECIFICITY

Routine cultivation of viruses requires the continuous availability of suitable living cells. The word *suitable* is used deliberately, because host specificity is a significant characteristic of viruses; that is, a given virus ordinarily grows in only one or a very few host species. For plant viruses, this usually means the appropriate specific host plant or certain closely related plants. Early workers with animal viruses formerly needed a constant supply of susceptible animals. A laboratory had to be equipped with several kinds of animals in order to work with a variety of viruses.

CHICK EMBRYOS

The discovery that many viruses could be cultivated in the developing chick embryo (see Figure 6–32, page 139) was an important advance. The chick embryo provides a suitable culture medium in its own "test tube"

143

and is easily secured in large numbers. In addition to permitting rapid diagnosis of numerous virus infections, the chick embryo is also used for mass cultivation of viruses in the production of immunizing vaccines and for large scale research. Some viruses, when inoculated upon the chorioallantoic membrane of the chick embryo, produce discrete, visible lesions or pocks on the membrane, each lesion representing an area of destruction caused by the progeny of a single virus particle. An estimate of the concentration of the original virus suspension can be made from the number of pocks produced.

CELL CULTURES

Another method of propagating viruses arose out of the earlier development of techniques for cultivating human and animal tissue cells *in vitro* (i.e., in test tubes or flasks). Susceptible tissues are grown in a suitable nutrient medium and inoculated with virus; upon continued incubation the virus infects and multiplies in the cultivated cells. Monkey kidney cells are used to propagate the poliomyelitis virus in the manufacture of Salk vaccine; the virus is harvested when a satisfactory concentration has been attained and is then inactivated with formalin.

Tissue cells can be made to grow on agar in a layer only one cell thick; in such a culture many human and animal viruses produce characteristic plaques or areas of cell destruction readily visible to the naked eye (Fig. 7–1). Theoretically, each plaque is formed as the consequence of replication of a single virus particle originally planted on the tissue cell monolayer, so the number of plaques provides an indication of the number of virus particles in the inoculum.

BACTERIAL VIRUSES

Bacterial viruses also exhibit host specificity; a given virus lyses and multiplies within a certain species of bacterium, or perhaps a few closely related species. Some viruses attack only particular strains of a species. Bacterial viruses are cultivated

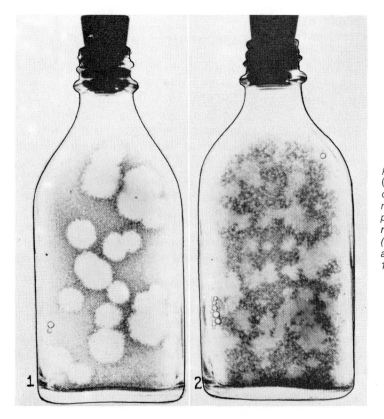

Figure 7–1. *Plaques of type III poliovirus (1) and type 6 ECHO virus (2) in a thin layer of monkey epithelial cells growing on a special nutrient medium solidified with agar. The plaques resemble those of the bacterial virus shown in Figure 2–13. (Photos courtesy of Drs. G. D. Hsuing and J. L. Melnick; from Virology, 1955, Vol. 1.)*

upon the appropriate bacteria growing on agar or in broth.

A young broth culture of an organism such as *Escherichia coli* inoculated with coliphage becomes increasingly turbid for two to four hours and then clears completely within 10 to 15 minutes. The culture may contain as many as 10 billion virus particles per milliliter. Turbidity sometimes gradually reappears during continued incubation, and the secondary bacterial growth is resistant to the virus. A normally susceptible bacterial culture contains about one resistant cell for every 1,000,000 to 100,000,000 susceptible cells. The susceptible cells are lysed by the virus, but the resistant cells multiply freely and their progeny are thereafter resistant.

The particulate nature of bacteriophage is shown by the formation of plaques in the growth on agar media. The solid sheet of bacterial growth on a heavily inoculated nutrient agar plate, upon which a small amount of bacterial virus lytic filtrate has also been spread, becomes riddled with holes where no bacteria can be found. The number of these plaques is proportional to the amount of lytic filtrate spread upon the agar. A bit of material from one of these plaques produces typical lysis when transferred to a young broth culture of the same bacterium. It therefore represents a colony of the bacteriophage.

PURIFICATION OF VIRUSES

The usual sources of viruses are infected plant or animal tissues, body fluids, secretions, feces, sewage, and so forth. All are crude mixtures containing bacteria and other microorganisms as well as the virus, tissue cells, and debris. Moreover, more than one virus may be present.

FILTRATION

Partial purification of viruses in tissue extracts, fecal suspensions, sewage, etc., can be effected by the method used by Iwanowski and other early investigators, that is, filtration to remove bacteria and other large particles. Filters are made of unglazed

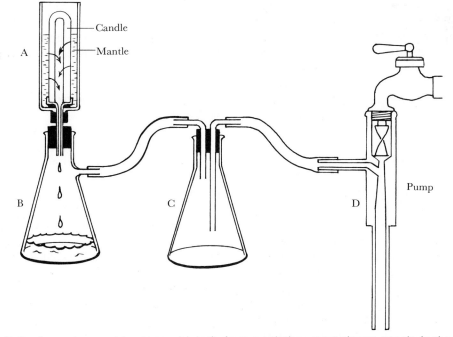

Figure 7–2. *Apparatus used to remove bacteria from a solution; most viruses remain in the liquid after filtration. The solution in the mantle at A passes through the diatomaceous earth or unglazed porcelain candle and collects in the filter flask, B; C is a flask to trap water that might accidentally be sucked over from pump, D. Two other filters are shown in Figure 7–3.*

Figure 7–3.   *Various filters.* Left, *A sintered glass filter. The filter disk is made of finely powdered glass that has been heated just enough to fuse together. Filters of different porosities can be made.* Center, *A diatomaceous earth filter.* Right, *A Seitz filter. The filtering agent is an asbestos pad clamped tightly between the upper and lower portions of the unit.*

porcelain, diatomaceous earth, asbestos, paper, cellulose acetate, sintered glass, or collodion (Figs. 7–2 and 7–3). The effective pore size of most filters can be determined indirectly with a fair degree of accuracy, and collodion filters of predetermined pore size can be made. In addition to removing bacteria and other large particles, filters have been used to estimate the dimensions of viruses. Early rough approximations have been confirmed in many instances by electron microscopy and by ultracentrifugation.

ULTRACENTRIFUGATION

The ultracentrifuge is a machine that whirls test tubes or other containers at very high speed (e.g., 70,000 revolutions per minute). Particles suspended in a liquid are thrown outward from the axis of rotation at a velocity that varies with the size and density of the particles relative to the density of the suspending medium, the speed of rotation, and other factors. Large (i.e., heavy) particles travel most rapidly, so after a short period of centrifugation the sediment consists mainly of large particles, and the supernatant liquid

contains the small particles. Longer centrifugation deposits smaller and smaller particles. The ultracentrifuge therefore separates viruses according to their sizes. Optical devices can be employed to measure the rate of sedimentation during operation of the centrifuge and from such data the sizes and relative purity of viruses can be calculated.

DILUTION AND ENRICHMENT; ANTIBIOTICS

Purification of viruses is often accomplished by inoculating appropriate hosts, chick embryos, or tissue cultures with a series of dilutions (e.g., 1:10, 1:100, 1:1000, 1:10,000, . . .) of the crude mixture. The mixture may be treated with streptomycin or other antibiotics to inhibit bacteria. If the desired virus predominates in the mixture, it should theoretically appear in pure form in one or more of the higher dilutions. Serial passage of virus-containing material from one individual host or chick embryo to another is sometimes used to enrich the virus; after several transfers it can be isolated in a pure state.

PLAQUE FORMATION

Isolation of bacterial viruses from plaques on solid media has been mentioned. The same method can be used with animal viruses that produce plaques on tissue cultures. Some plant viruses form leaf spots or other localized lesions and can be purified by a similar procedure (Fig. 7–4).

PHYSICAL AND CHEMICAL PROPERTIES OF VIRUSES

MORPHOLOGY OF VIRUSES

The electron microscope not only made possible accurate measurements of virus particles but also permitted investigators to determine their shapes. It was found that viruses are at least as heterogeneous in gross appearance as bacteria (Fig. 7–5). They range from 10 nm to more than 300 nm in diameter. Many are spherical or oval; some are long, narrow cylinders; a few are brick-shaped or consist of other regular geometric forms; and some are shaped like tadpoles, possessing a spherical or oval head with a slender tail.

RESISTANCE OF VIRUSES

Viruses withstand freezing very well. Many laboratories preserve their virus collections by storage in dry ice or mechanically refrigerated cabinets at approximately −75° C. Smallpox virus has been kept alive as long as 15 years frozen in chick embryo membranes.

Viruses are about as resistant to heat as many nonsporulating bacteria; most species are inactivated or killed within 30 minutes at 53° to 56° C.

Certain viruses (e.g., poliomyelitis) possess considerable resistance to phenol, cresol, and ether, withstand a wide range of pH, and survive indefinitely in 50 per cent glycerin. The infectious hepatitis virus survives about three times the concentration of chlorine necessary to kill pathogenic intestinal bacteria; a dosage of one part per million is required to ensure its destruction in water.

COMPOSITION OF VIRUSES

Viruses vary in chemical composition from those of small particle size, which contain only nucleic acid and protein, to the large vaccinia (cowpox) virus, with several distinct proteins (including nucleoprotein), lipid, carbohydrate, copper, and one or two vitamin-like substances. Even this complex form does not carry on independent metabolism in the absence of living susceptible host cells.

Nucleic Acids of Viruses. The nucleic acids that carry genetic information are the most vital constituents of any organism. In

Figure 7–4. *Tobacco leaf with local lesions produced by cabbage black ringspot virus. (A. R. C. Virus Research Unit, Cambridge.)*

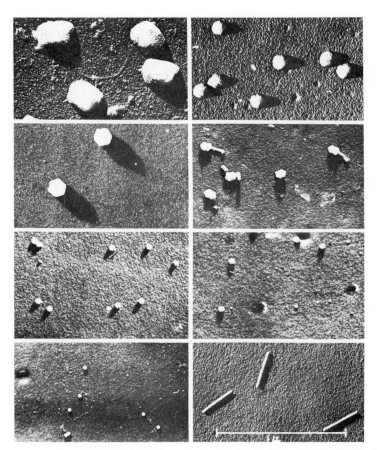

Figure 7–5. *Electron micrographs showing the variation in size and shape of animal, plant, and bacterial viruses. From top to bottom:*

Left	Right
Vaccinia	*Influenza*
Tipula iridescens	*T2 bacteriophage*
T3 bacteriophage	*Rabbit papilloma*
Poliomyelitis	*Tobacco mosaic*

(From Fraenkel-Conrat: Design and Function at the Threshold of Life: The Viruses. New York, Academic Press, 1962.)

nearly all plants, animals, and microbes, as previously noted, these are of the DNA type and are found principally in cell nuclei. RNA may also be present in nuclei, but is consistently found in cytoplasm, usually associated with granules where protein synthesis is most active. RNA therefore translates the information from nuclear DNA into proper structural terms as proteins (which include enzymes) are made.

The nucleic acid of most bacterial viruses is DNA. Most animal viruses also contain DNA, but some of the smaller viruses (e.g., poliomyelitis and influenza) contain RNA and no DNA. RNA is the only nucleic acid in plant viruses.

Microstructure of Viruses. A striking feature of the microstructure of viruses is their symmetry. A complete infectious virus particle, or *virion,* consists of a nucleic acid *core* surrounded by a protein coat, or shell, known as a *capsid.* The core with its enveloping capsid is called a *nucleocapsid.* The nucleic acid is a single- or double-stranded chain of nucleotides arranged in the usual helical pattern, and the capsid is composed of protein subunits (*capsomeres*) assembled in a precise array according to comparatively simple geometric principles. The subunits are protein molecules synthesized to specifications supplied by the virus nucleic acid, and they are arranged in one of the two

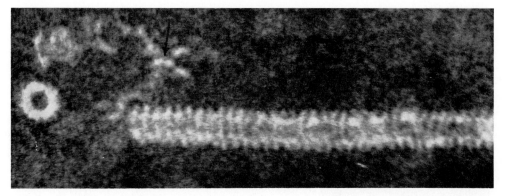

Figure 7–6. *Electron micrograph of negatively stained tobacco mosaic virus showing helical symmetry (700,000X). The cross-section at the left shows the central hollow. The long, rod-shaped virion is composed of 2130 capsomeres helically arranged around the central hole, as shown by the model in Figure 7–7. (From Horne et al., Virology, 15:348, 1961.)*

ways in which a regular, symmetrical structure can be constructed from asymmetrical building stones: either a helix or a closed shell. There are therefore two types of capsid symmetry: *helical* and *isometric* (or *cubic*).

Some viruses are surrounded by a lipid or lipoprotein envelope acquired during the later stages of maturation. Although the exact composition of envelopes is not known, when the lipids of some viruses are extracted by ether, the virus becomes noninfectious. Many contain constituents that are identical with components of the host cell, in addition to components peculiar to the virus. In some cases the envelope is formed within the cytoplasm of the host cell after assembly of the virion from the nucleic acid and protein capsid components; in others it is acquired as the virus is extruded from the cell, and herpes virus acquires its envelope as the virus passes from the nucleus to the cytoplasm.

The most thoroughly studied example of helical symmetry, also the first to be discovered, is that of the tobacco mosaic virus (TMV). This virus (Fig. 7–6) consists of a spiral protein capsid, 300 nm long, composed of 2130 identical capsomeres. Between the coils of the protein capsomeres is the helix of RNA, with a diameter of 8 nm, and in the center of the entire structure is an open hole 4 nm in diameter (Fig. 7–7). The molecular weight of each protein subunit is 17,400, and that of the RNA molecule is 2.06×10^6, so the entire virion has a molecular weight of 39×10^6 daltons.

The TMV capsid is rigid, which indicates that the successive coils are held together by relatively strong bonds. The capsids of various other plant viruses and of many animal viruses are less rigid. Some can bend irregularly, and some are wound or folded within an envelope, which surrounds them like a cocoon (Fig. 7–8).

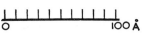

Figure 7–7. *Model of tobacco mosaic virus showing a portion of the helical array of protein capsomeres that comprise the capsid of the virion. The RNA helix near the central hole of the virion is also shown. (From Klug et al., Adv. Virus Research, 7:225, 1960.)*

Figure 7–8. *Electron micrograph of negatively stained Orf (contagious pustular dermatitis) virus. A complex structure is formed when a helically symmetrical capsid is wound or folded within an envelope (143,000X). (From Nagington and Horne, Virology, 16:248, 1962.)*

Isometric symmetry is characteristic of many viruses, both large and small. The capsids are nearly spherical, but x-ray diffraction and electron microscopy reveal that the subunits of which they are constructed are actually arranged in definite geometric patterns—triangles, pentagons, and hexagons—so that the entire capsid is an icosahedron; that is, a regular polyhedron with 12 vertices, 20 faces, and 30 edges (Fig. 7–9). The number of capsomeres (N) in a capsid is given by the equation:

$$N = 10x(n - 1)^2 + 2$$

in which x is either 1 or 3, and n is the number of capsomeres along one edge of a triangular face of the capsid. The value of x is 1 in all

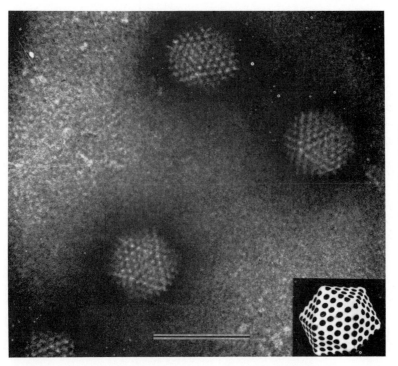

Figure 7–9. *Electron micrograph of negatively stained particles of adenovirus type 2 showing isometric (cubic) symmetry. The icosahedral form of the virions is clearly illustrated by the model in the lower right corner. This virus has 252 capsomeres. (Bar = 100 nm.) (From Frobisher et al.: Fundamentals of Microbiology, 9th ed. Philadelphia, W. B. Saunders Company, 1974.)*

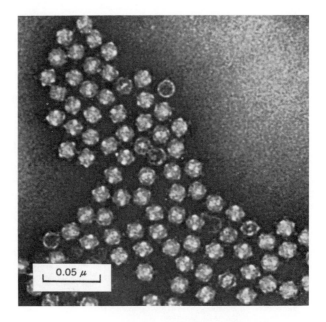

Figure 7–10. *Electron micrograph of negatively stained* E. coli *phage* ΦX174. *(From Luria and Darnell: General Virology, 2nd ed. New York, John Wiley & Sons, 1967.)*

icosahedral viruses studied so far with the exception of turnip yellow mosaic virus, in which x is 3. From the equation it can readily be calculated that as the number of capsomeres in each edge increases in the series 2, 3, 4, 5, 6 . . . , the total number of capsomeres comprising the capsid shell increases as follows: 12, 42, 92, 162, 252 . . . .

Bacterial viruses vary in structural complexity from the small and comparatively simple ϕX174, which contains only 12 capsomeres arranged with cubic symmetry (Fig. 7–10), to the tadpole-like *E. coli* phage λ (Fig. 7–11), and the so-called T-even *E. coli* phages (Fig. 7–12). The latter consist of a cubically symmetrical protein head contain-

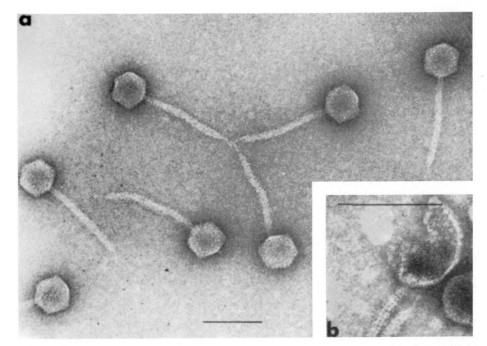

Figure 7–11. *Electron micrograph of negatively stained* E. coli *phage* λ; *inset (b), disrupted phage head showing capsomeres. (After Eiserling and Boy de la Tour.)*

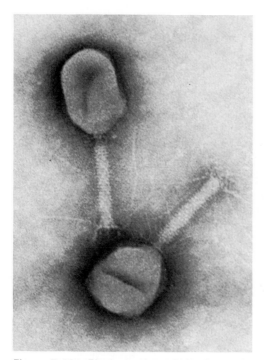

Figure 7–12. *Electron micrograph of negatively stained* E. coli *phage T4, showing base plate, tail fibers, and subunits of tail sheath. (From Rhodes and van Rooyen: Textbook of Virology, 5th ed. Baltimore, Williams & Wilkins Co., 1968.)*

has been most thoroughly investigated in bacterial viruses such as coliphage T2, and it is their reproduction that will be described; it should be understood that the multiplication of animal and plant viruses differs in various details from that described here.

A virus alone is unable to replicate *in vitro;* it must first enter a susceptible host cell. When a virus encounters an appropriate receptor site on such a cell, union and adsorption occur *immediately* (Fig. 7–14). In the case of T2 bacteriophage, the tail fibers attach to cell wall receptors, and a viral en-

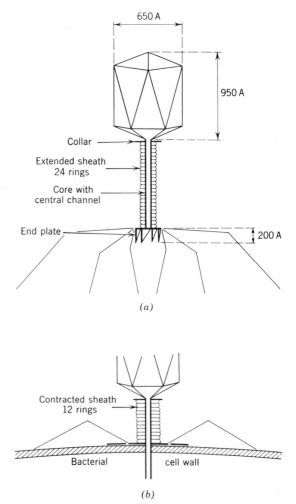

ing the nucleic acid and a complex tail with a central channel. The tail sheath is formed by a number of rings, which contract when the end plate and fibers of the tail make proper contact with the host bacterial cell wall and expel viral nucleic acid through the channel into the bacterium. Some details of the structure of the T4 virion are shown diagrammatically in Figure 7–13.

REPRODUCTION OF VIRUSES

BACTERIAL VIRUS REPRODUCTION

Virus multiplication is a complex process, divisible for descriptive purposes into five phases: (1) infection: adsorption and entry of virus into the host cell; (2) synthesis of enzymes needed for replication of virus nucleic acids; (3) intracellular synthesis of virus components; (4) assembly of new virus particles; (5) liberation of virus. Reproduction

Figure 7–13. *Diagrammatic sketch showing component parts of* E. coli *phage T4. (A) Complete virion with tail fibers extended and tail sheath in the normal form. (B) Lower part of the virion with the tail fibers in contact with a bacterial cell wall, tail spikes flattened, and tail sheath in the contracted condition. (From Luria and Darnell.)*

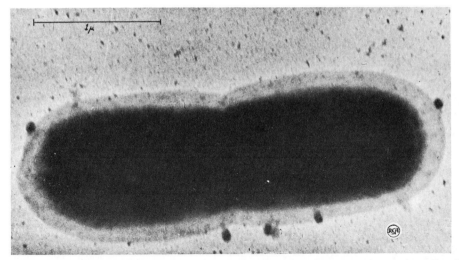

Figure 7–14. *Electron micrograph showing five particles of bacteriophage attached to the cell wall of* E. coli. *(Photograph by S. E. Luria, M. Delbrück, and T. F. Anderson; A.S.M. LS-37.)*

zyme apparently starts digesting the muramic acid-peptide complex that confers rigidity on the bacterial cell wall. The elastic protein tail sheath contracts, and the viral DNA is injected into the bacterium. The protein coat remains outside, having performed its function of protecting the virus during its extracellular existence (Fig. 7–15). Plant and animal viruses gain entrance to their respective host cells by different and less dramatic mechanisms. Plant viruses are usually injected by insect vectors, but occasionally enter through areas of mechanical injury. Animal viruses are adsorbed to the host cell membrane and are ingested by phagocytosis.

A period of "eclipse" then follows, during which the virus disappears as an independent infective agent. This is the interval when virus components are being formed. The metabolism of the host bacterium suddenly changes and production of bacterial DNA, RNA, and protein ceases. In fact, microscopic evidence indicates that the chromatin body of the bacterium, which contains the cellular DNA, disintegrates and its contents are dispersed within the first minute after infection (Fig. 7–16B). Pools of virus-type DNA precursors and protein precursors accumulate. Part of these materials is derived from the breakdown of host components and part from the external medium.

After the eclipse period, virus DNA is removed from the pool for incorporation into progeny phage. The rate of formation of phage is approximately the same as the rate of synthesis of phage DNA, and the precursor DNA pool contains the equivalent of 50 to 100 phage units throughout the remainder of the latent period.

The protein pool is derived almost entirely from the external medium instead of from the host cell. Synthesis of the internal protein components of the virus begins within two or three minutes, but the polypeptides of the head and tail proteins are not formed until several minutes later.

Assembly of an infective phage unit—that is, enclosure of the nucleic acid in a protein coat and addition of the protein tail—begins about 10 minutes after infection with *condensation* of a single viral DNA macromolecule from the vegetative precursor pool into a polyhedral body that can be detected by electron microscopy (Fig. 7–16D–E). This body is stabilized by its internal protein, and the protein head membrane forms around it by aggregation of a thousand or more identical units from the protein pool. The tail is added and finally the virus particle attains maturity when its various components "set" sufficiently to remain intact after being liberated by lysis of the host cell. Lysis may be the result of continuation of

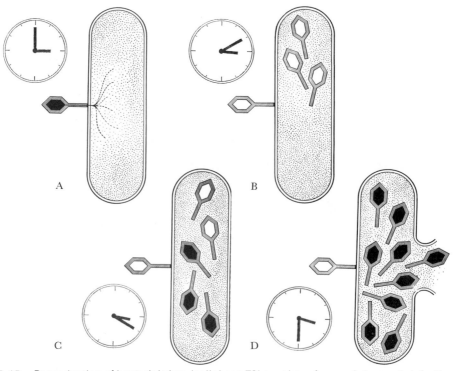

Figure 7–15. *Reproduction of bacterial virus (coliphage T2) consists of several stages:* A, *Infection, in which the virus attaches to the host bacterium by the "tail," and the nucleic acid core is emptied into the bacterial cell;* B, *the "dark" period during which synthetic enzymes form, virus nucleic acid is produced and induces formation of new protein "coats";* C, *the "rise" period when nucleic acid appears within the protein coats; and* D, *the burst, when some 200 new virus particles are released. The bacterium lyses and disappears. (From The Multiplication of Bacterial Viruses by Gunther S. Stent. Copyright © 1953 by Scientific American, Inc. All rights reserved.)*

the enzymatic process which enabled the infecting phage to inject its DNA into the host. The various stages in the development of a coliphage are diagramed in Figure 7–17.

Normally, 100 to 300 virus particles form within an infected bacterium. These are suddenly released when lysis occurs. This is known as the "burst" and is shown graphically in Figure 7–18, which presents the results of a *one-step growth* experiment. In this experiment, the infective phage units in broth cultures of appropriate bacteria were counted at short intervals. All the units were situated within bacterial cells during the first 24 minutes, which is called the *latent period*. During the *rise period* of 12 minutes which followed, 100 phage particles were released for every one that had initially infected a bacterial cell.

Virus particles are released from animal cells by extrusion through the cell membrane; the cell does not necessarily lyse.

It is possible to infect host cells with free nucleic acid secured by extraction from virus particles. The infected cells then produce viral nucleic acid and protein, and complete virions are assembled and released. The efficiency of infection is only 1/1000 to 1/1,000,000 as great as with intact particles, so it appears that the protein capsid provides a mechanism for adsorption of the virion to the host cell surface and thus facilitates penetration. Host specificity is determined by the physicochemical nature of adsorption sites on the virus and on the host cell.

LYSOGENY

The events just described, which occur after infection of a bacterium by a bacterial virus, can be considered the consequence of partial replacement of one system of genetic information by another. The new genetic

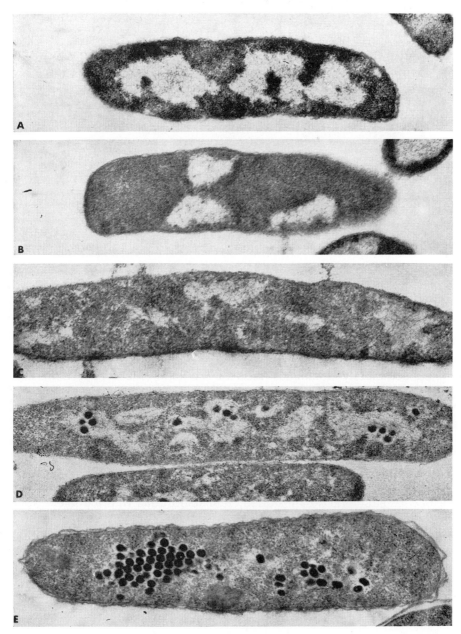

Figure 7–16.   *Electron micrographs of ultrathin sections of E. coli at various stages in the growth of infecting T2 bacteriophage. A, A normal cell at the moment of infection showing typical electron-transparent chromatin bodies. B, A cell two to four minutes after infection; the chromatin bodies have changed in shape and migrated toward the cell wall. C, Ten minutes after infection the chromatin bodies have disappeared and have been replaced by vacuoles filled with fibrillar phage DNA. D, The first polyhedral condensates of phage DNA appear 14 minutes after infection. E, Forty minutes after infection many condensates and phage heads are present. (From Viruses and Genes by Francois Jacob and Eli Wollman. Copyright © 1961 by Scientific American, Inc. All rights reserved.)*

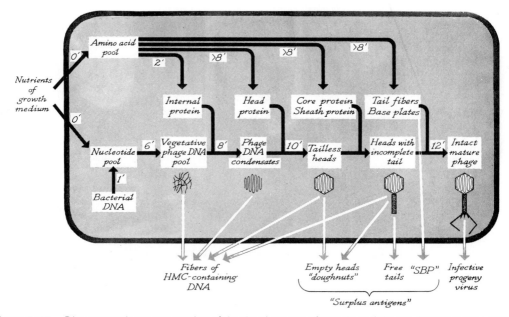

Figure 7–17. *Diagrammatic representation of the development of certain coliphages, indicating the timing of the successive intracellular events. The origin of various phage constituents is indicated in the lower part of the diagram. HMC-containing DNA is phage DNA, which possesses the unusual pyrimidine, 5-hydroxymethylcytosine, whereas bacterial DNA contains cytosine. "Surplus antigens" are incompletely assembled phage proteins. (From Stent, Molecular Biology of Bacterial Viruses. San Francisco, W. H. Freeman & Co., 1963.)*

mechanism so completely dominates the physiologic activity of the host cell that the latter is destroyed in fulfilling the mission of the former. However, many bacterial strains and phage strains can exist compatibly together. The DNA of such phages replicates at the same rate as the remainder of the host DNA and the cell itself.

Synthesis of virus structural proteins does not occur because most of the viral genes have been repressed by *repressor* proteins formed early following infection of the host

bacterium. This inhibits both replication of virus and expression of the virus genes. Spontaneous derepression may take place in a few cells, however, whereupon the viral DNA triggers a cycle of normal development with consequent lysis and liberation of complete virus particles.

This phenomenon is called *lysogeny,* and infected cells with the latent ability to produce mature virions are said to be *lysogenic*. A bacteriophage that can enter into this essentially commensal relationship, replicat-

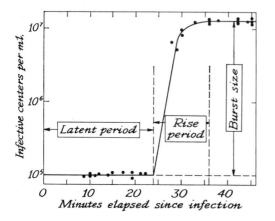

Figure 7–18. *One-step growth curve of a coliphage. Each infective center represents one or more phage particles assayed by formation of a single plaque. (From Doermann, J. Gen. Physiol., 35:645, 1952.)*

ing at the same rate as the host cell DNA, is described as *temperate,* and when present within the cells it is called *prophage.* In certain cases (e.g., *E. coli* K12 and phage λ), prophage has been shown to be actually incorporated into the host DNA as part of the chromosome; in others it seems to attach to specific sites in the cell membrane.

The mechanism of phage repression in the case of the *E. coli* K12–phage λ system involves a gene (C_1) that causes production of the repressor protein. The protein can bind specifically to a receptor site on the phage, and this inhibits phage replication and expression of all genes except C_1. When the receptor site is unoccupied, phage DNA replicates normally and its genes are expressed by the formation of virus proteins.

Prophage per se is not infective, but it endows the cell with the ability to form infective phage without the entrance of phage from outside. Some phages, such as phage λ, are inducible, that is, treatment of the host cell with agents that interfere with DNA replication (e.g., ultraviolet light, alkylating agents, or mitomycin C) is followed by inactivation of the phage repressor, whereupon complete virulent phages develop and all host bacteria are lysed. Cells in which synthesis of host DNA is blocked by the inducing agent apparently accumulate a metabolite that combines with and inactivates the repressor.

VIRAL INFECTIONS

PLANT VIRUSES

More than 400 plant viruses have been described and named since 1898; the number of diseases they cause has not been determined. A given virus may produce disease in a variety of plants, and any one plant species may be susceptible to many different viruses. At least 25 viruses are known to infect potatoes, and a dozen or more infect tomato plants.

As already noted, the nucleic acid of plant viruses is RNA. The free RNA of many viruses is infectious, but its infectiousness is only about 1/1000 that of the whole virus.

Plant viruses often multiply within host cells to tremendous numbers. A single cell of a tobacco plant may contain 60,000,000 tobacco mosaic virus particles, and the virus constitutes as much as 10 per cent of the dry weight of the plant (see Figure 7–19).

Viruses are transmitted to plants largely by arthropods, such as aphids and leaf hoppers; other vectors include mites, white flies, mealy bugs, and thrips. Insect transmission is sometimes a mechanical process in which the virus on infected mouth parts is inoculated into the plant tissue. However, leaf hoppers may become infected with certain plant viruses, which replicate and produce disease, sometimes even killing the insect host. This raises interesting questions regarding host specificity and the cellular requirements for virus replication.

Plant viruses are also transmitted by nematodes, by soil fungi and their spores, and to a limited extent by mechanical means (e.g., via grafts).

Virus inoculated mechanically, as by insect bites, spreads slowly from the initial site. Multiplication probably precedes distribution from cell to cell via the plasmodesmata or intercellular bridges, and the rate of progress is no more than 1 mm per day. If the vascular tissue is infected, virus travels rapidly—25 mm per minute—and it may spread to all parts of the plant. Actively growing tissues are generally attacked first.

The effects of viruses on plants are varied but are probably indirect results of interference with various metabolic processes: photosynthesis, respiration, growth regulation, and transport of water and nutrients. External symptoms include mosaic, other color changes, and tissue distortion or malformation. The mosaic condition is shown by mottling of the leaves and stems in characteristic colored patterns, often raised and blister-like; the pattern may consist of single or concentric rings around a central spot, sometimes chlorotic or necrotic. Aside from mottling, color changes include yellowing or development of a deep green or bluish-green leaf color. Alterations in flower color are illustrated by the appearance of flecks or streaks of yellow in tulips—the so-called *tulip break.* Tissue distortion takes the form of

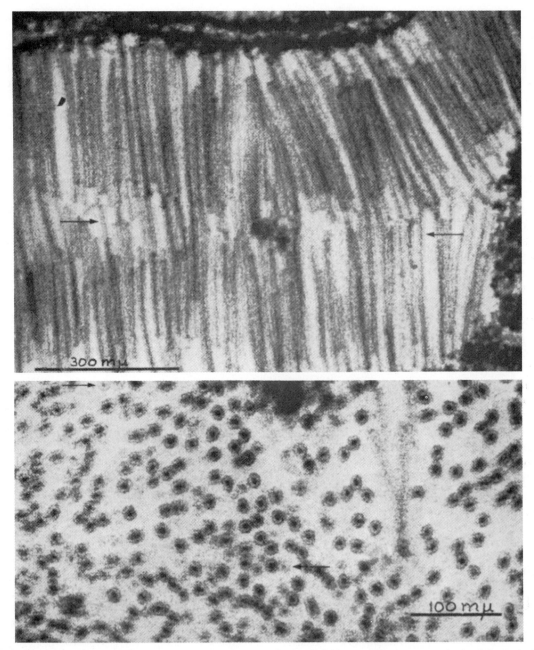

Figure 7–19. *Electron micrograph of part of a cell of a leaf of* Chenopodium amaranticolor *infected with tobacco mosaic virus; particles in parallel array are viewed in profile* (above) *and in cross section* (below). *(From Milne, Virology, 28:520, 1966.)*

thickened, crumpled, or curly leaves, tattered or perforated leaves, misshapen fruits, and formation of root tumors.

Internal evidences of virus infection are detectable by microscopic examination of tissue sections. Histologic changes occur,

especially in distorted or necrotic tissues, and in some cases crystalline or amorphous intracellular inclusions appear; some of these contain virus (see Figure 7–20).

A few groups of plant viruses and some of their characteristics are listed in Table 7–1.

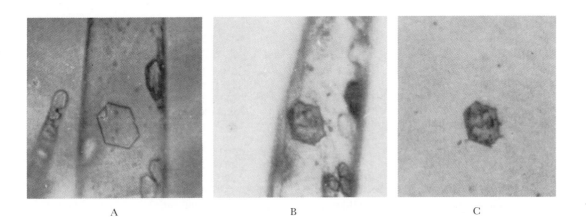

A   B   C

D

Figure 7–20.   *Crystalline inclusion of tobacco mosaic virus in a hair cell of tobacco:* (A) *before freeze-drying,* (B) *after freeze-drying,* (C) *removed by micromanipulation while frozen-dried,* (D) *characteristic rod-shaped virions released by dissolving such crystals in water. (From Steere and Williams, Am. J. Bot., 40:81, 1953.)*

TABLE 7–1.   Some Groups of Plant Viruses*

| Group | Shape | Properties of Virions | |
		Size	Presence of Envelope
Tobacco mosaic	Rods, rigid	18 × 300 nm	–
Tobacco rattle	Rods, rigid	20 × 175 nm	
Potato X	Rods, flexible	13 × 500 nm	–
Potato Y	Rods, flexible	ca. 700 nm	
Tobacco etch	Rods, flexible		
Henbane mosaic	Rods, flexible		
Soybean mosaic	Rods, flexible		
Sugar beet yellows	Rods, flexible	10 × 1250 nm	
Alfalfa mosaic	Prolate ellipsoid	18 × 36–60 nm	
Potato yellow dwarf	Rodlike	75 × 380 nm	+
Gomphrena	Rodlike	100 × 260 nm	+
Tobacco ringspot	Spherical (icosahedral)	26 nm	
Tobacco necrosis	Spherical (icosahedral)	30 nm	
Tobacco bushy stunt	Spherical (icosahedral)	28 nm	–
Southern bean mosaic	Spherical (icosahedral)	28 nm	–
Turnip yellow mosaic	Spherical (icosahedral)	28 nm	–

*Modified from Luria and Darnell: General Virology, 2nd ed. New York, John Wiley & Sons, Inc., 1967.

ANIMAL VIRUSES

At least 500 animal viruses have been isolated. As mentioned previously, they display considerable host specificity and, in fact, many viruses invade only certain types of tissue. *Neurotropic* viruses, for example, multiply in the brain, spinal cord, and peripheral nerves; *respiratory* viruses infect cells of the respiratory tract; *enteric* viruses are associated with the gastrointestinal tract; and *viscerotropic* viruses attack the abdominal viscera, especially the liver, or produce generalized infection.

Animal viruses are classified according to the nucleic acid (RNA or DNA) they contain, their size, sensitivity to ether, presence or absence of an envelope, type of symmetry displayed by their capsids, and the number of capsomeres in each capsid. Size can be determined by use of the electron microscope, but for practical purposes it is often sufficient to estimate it roughly by attempting to pass the virus through filters of known pore size. Ether, a lipid solvent, greatly reduces the infectivity of certain viruses but has little effect on others, which either lack lipid or contain lipid that is not essential for infectivity or is not extractable by ether. The *nucleocapsids* of certain viruses are "naked," those of others are enclosed within a membrane or envelope, partly derived from the nuclear or cytoplasmic membrane of the host cell and partly composed of virus protein.

It is obvious from Table 7–2 that animal viruses vary greatly in physical characteristics. The range of size is nearly 20-fold, and the actual structures of the smallest have not yet been determined with certainty. There are 14 principal groups, divided according to the characteristics just listed.

The *picornaviruses* (Spanish: *pico,* small quantity + RNA) are the smallest animal viruses known. They exhibit cubic symmetry, but their small size makes it difficult to determine the number of capsomeres in each virion. There are more than 140 viruses in this group, including enteroviruses of man and lower animals (polio, Coxsackie, ECHO, foot and mouth disease), over 60 rhinoviruses (common cold), and several encephalomyelitis and encephalomyocarditis viruses of lower animals. The ECHO viruses (**E**nteric, **C**ytopathogenic, **H**uman, **O**rphan) were originally isolated from the feces of supposedly normal persons, and were detected by their destructive effect (cytopathogenicity) on tissue cultures; they were called orphan viruses because they had not been shown to produce disease.

Togaviruses are part of the *arbovirus* (**ar**thropod-**bo**rne) group, which consists of viruses that multiply in a blood sucking insect and are transmitted by bite to a vertebrate. The togaviruses are spherical and contain single-stranded RNA in a core of icosahedral symmetry. They are enclosed within a lipoprotein envelope. Many togaviruses produce nonapparent viremic infections of birds, reptiles, or mammals, being transmitted by the bite of an arthropod, such as a mosquito. When man is bitten, mild infection is the usual result; however, the fatal diseases, yellow fever and encephalitis, are among those caused by mosquito-borne togaviruses.

Reoviruses are found in the **r**espiratory and **e**nteric tracts of animals and man, and are **o**rphan viruses. Unlike that of other RNA viruses, their RNA is double-stranded and is present in very high amount. So far, these viruses are of primarily academic interest as biologic oddities.

Rhabdoviruses are large and bullet-shaped. They are enclosed within an envelope, and they contain single-stranded RNA. The best studied representative is an arbovirus that causes vesicular stomatitis in horses. The principal human disease associated with them is rabies, which is transmitted by the bite of an infected animal.

The *coronaviruses* are intermediate in size, somewhat pleomorphic, and they contain single-stranded RNA and are surrounded by a lipoprotein envelope. The chief human disease caused by coronaviruses is the common cold; other members of the group produce mouse hepatitis and infectious bronchitis in chickens.

The *leukoviruses* also contain single-stranded RNA in a lipid envelope. They are animal pathogens, causing abnormal tumors in birds and rodents. None has yet been found in man.

The *myxovirus* and *paramyxovirus* groups (Greek: *myxa*, mucus) are so named because of the affinity of their members for mucoproteins. Both spherical and filamentous forms are found: characteristic features include a helical internal ribonucleoprotein

TABLE 7–2. Distinguishing Characteristics of Major Groups of Animal Viruses*

Group	Shape	Size (nm)	Presence of Envelope	Capsid Symmetry	Number of Capsomeres
Riboviruses					
Picornavirus	Spherical	20–30	−	Icosahedral	?60
Togavirus	Spherical	50–60	+	Icosahedral	?
Reovirus	Spherical	70–80	−	Icosahedral	180 or 270
Rhabdovirus	Bullet-shaped	70×180	+	Helical	
Coronavirus	Spherical	80–120	+	?	
Leukovirus	Spherical	100–120	+	Helical	
Myxovirus	Spherical or filamentous	80–120	+	Helical	
Paramyxovirus	Spherical or filamentous	100–300	+	Helical	
Deoxyriboviruses					
Parvovirus	Spherical	20	−	Icosahedral	32
Adeno-associated virus	Spherical	20	−	Icosahedral	?
Papovavirus	Spherical	45–55	−	Icosahedral	72
Adenovirus	Spherical	70–80	−	Icosahedral	252
Herpesvirus	Spherical	150–200	+	Icosahedral	162
Poxvirus	Brick-shaped	100×200×300	−	Helical	

*Modified from Fenner, F. J., and White, D. O.: Medical Virology. New York, Academic Press, Inc., 1970.

and a lipoprotein envelope with radial projections. Many members of these groups react with erythrocytes of one or more species and may cause them to clump together (agglutinate). The myxoviruses and paramyxoviruses differ in size and in various other physical and biologic properties. Myxoviruses include the etiologic agents of influenza. Among the paramyxoviruses are those that cause parainfluenza, mumps, measles, distemper, and a number of other animal diseases.

The *parvoviruses* are very small, icosahedral, single-stranded DNA viruses. They appear to infect mice and rats, but little is known about them.

The *papovaviruses* comprise a group of small, DNA viruses that induce tumor formation. The name was coined from the first two letters of the names or three viruses: **pa**pilloma, **po**lyoma, and **va**cuolating agent. A virus of interest to man is the human papilloma (wart) virus.

There are many types of *adenovirus* (Greek: *adenos*, gland). They are medium-sized, ether resistant, of uniform icosahedral structure, immunologically (i.e., chemically) related, and found in man, monkeys, and certain other animals. Some are associated with epidemic or sporadic cases of respiratory disease, others have been isolated from tonsils and adenoids and from fecal specimens. Pathogenicity of some types has not yet been demonstrated.

Adeno-associated viruses are found as contaminants in laboratory stocks of adenoviruses. They are defective viruses; that is, they replicate only in the presence of a "helper" virus—any adenovirus—within the nucleus of a host cell. They probably do not cause disease in man, but their presence in man is indicated by the fact that antibodies to them have been found in human serum.

Members of the *herpesvirus* group are moderately large DNA viruses enclosed in an envelope consisting of host cell material. The 10 viruses include herpes simplex, found in cold sores, and the varicella-zoster virus, which occurs in chickenpox and in herpes zoster or shingles. There are also several animal pathogens.

Herpesviruses consist of two related groups distinguished by their antigenic and biologic properties. Type 1 viruses usually cause oral lesions, and type 2 viruses produce genital lesions. Recovery from infection does not lead to complete immunity; instead, the viruses tend to remain latent, apparently within nerve tissue, and can be reactivated by stress such as fever, menstruation, or exposure to sunlight. Reinfection, especially by type 2, may also occur.

The viruses are found naturally only in humans. They are highly labile and hence spread only in secretions: type 1 by the oral-respiratory route, type 2 by venereal transmission. Virus can be recovered from the secretions of asymptomatic carriers as well as from patients having the disease. The normal incubation period of both types is three to nine days, and lesions occur in a variety of tissues: skin, the oral cavity, conjunctiva, vagina, and the nervous system. The typical primary lesion is a painful, superficial vesicle, which ruptures and becomes a shallow ulcer. Initial infection with type 1 virus is usually an acute inflammation of the gums and mouth; recurrent manifestations include cold sores and fever blisters. Type 2 lesions in the female commonly occur on the vulva, vagina, cervix, and perineum; there is evidence that cervical cancer may later develop. Symptoms in the male include a painful vesicular eruption on the glans, prepuce, or shaft of the penis, urethritis, and prostatitis. Generalized or localized disease, often fatal, occurs in newborn infants infected from the birth canal of the mother.

The *poxviruses* are large, brick-shaped DNA viruses that infect a variety of species and produce typical skin lesions. Among the twenty-odd poxviruses are variola (smallpox), vaccinia (used to immunize humans against smallpox), and poxes of a dozen or more animals. Several tumor viruses of lower animals are also included in this group. The poxviruses are about the size of rickettsiae and other small bacteria, and approach them in chemical organization. However, they possess only a single type of nucleic acid and their manner of replication is virus-like rather than bacteria-like. They are therefore considered to be the most complex viruses.

Modification of Viruses. Tissue specificities are not absolute and unchangeable. Initial invasion of the customary susceptible tissue may be followed by spread to other tissues. Moreover, the tissue affinity of certain viruses can be altered deliberately. Viscerotropic strains of yellow fever virus become neurotropic after repeated inoculation into the brains of mice, and will produce fatal encephalitis in suitable experimental animals.

The modification of viruses by passage through an unnatural tissue or host is important to an understanding of the natural history of disease and the evolution of microorganisms. It illustrates processes that doubtless occur constantly in nature. In a normal infection with a given virus the customary tissues are affected, and ordinarily the virus in its usual form passes to the next host. An unusually large infecting dose or overwhelming multiplication in a highly susceptible individual may be followed by invasion of tissues not ordinarily affected. This provides opportunity for a transformation in the virus (more properly, selection of a mutant virus) and the appearance of a modified parasite capable of producing a new disease, as is illustrated by the yellow fever virus. Host specificity may also be altered in a similar manner.

SUPPLEMENTARY READING

Burnet, F. M.: *Principles of Animal Virology*, 2nd ed. New York, Academic Press, 1960.

Burnet, F. M., and Stanley, W. M. (eds.): *The Viruses*. New York, Academic Press, 1959.

Cohen, A.: *Textbook of Medical Virology*. Oxford, Blackwell Scientific Publications, 1969.

Fenner, F. J., and White, D. O.: *Medical Virology*, 2nd ed. New York, Academic Press, 1976.

Goodheart, C. R.: *An Introduction to Virology*. Philadelphia, W. B. Saunders Company, 1969.

Horsfall, F. L., Jr., and Tamm, I. (eds.): *Viral and Rickettsial Infections of Man*, 4th ed. Philadelphia, J. B. Lippincott Co., 1965.

Kaplan, A. S. (ed.): *The Herpesviruses*. New York, Academic Press, 1973.

Knight, C. A., and Fraser, D.: The Mutation of Viruses. *Scientific American*, *193*(1):74–78, July, 1955.

Luria, S. E.: The T2 Mystery. *Scientific American*, *192*(4):92–98, April, 1955.

Luria, S. E., and Darnell, J. E., Jr.: *General Virology*, 2nd ed. New York, John Wiley & Sons, Inc., 1967.

Pollard, E. C.: The Physics of Viruses. *Scientific American*, *191*(6):62–70, December, 1954.

Rhodes, A. J., and Van Rooyen, C. E.: *Textbook of Virology*, 5th ed. Baltimore, The Williams & Wilkins Co., 1968.

Stanley, W. M., and Valens, E. G.: *Viruses and the Nature of Life*. New York, E. P. Dutton and Co., Inc., 1961.

Stent, G. S.: *Molecular Biology of Bacterial Viruses*. San Francisco, W. H. Freeman and Co., 1963.

8 BACTERIAL METABOLISM

Metabolism is the sum of the physical and chemical processes by which living organized substances are produced and maintained; important among these processes are the transformations that make energy available for the uses of the organism.

Living cells consist of proteins, carbohydrates, lipids, and nucleic acids. Most of these are constructed internally by the organism as needed from ingredients available in the environment. Nonmotile microorganisms are unable to search out the required nutrients actively, so their growth is limited by the availability of these materials. Moreover, the growth of any organism, motile or nonmotile, depends upon the presence of needed substances, and the rate of growth varies, within limits, with the concentrations of these substances.

Microorganisms differ greatly with respect to the nature of the substances they can utilize. Photosynthetic organisms may be content with CO_2 as a source of carbon, nitrate salts as sources of nitrogen, and various mineral salts. From these they manufacture the components of living cells. Highly parasitic bacteria, such as *Treponema pallidum,* the cause of syphilis, require special organic sources of carbon and nitrogen, often including amino acids, sugars, purines, and pyrimidines, together with vitamin-like substances that assist the metabolic processes.

Energy is required for all activities of an organism. In the same way that synthetic reactions require energy, it is needed for growth, reproduction, and repair; physical movement of the organism or any of its parts also requires energy.

SOURCES OF BUILDING MATERIALS AND ENERGY

Organisms are commonly classified according to their ability to use CO_2 as a sole source of carbon. *Autotrophs* use CO_2 or carbonates as the sole source of carbon, whereas *heterotrophs* do not. These distinctions are not as clear-cut as was first thought; organisms have been found that normally appear to be heterotrophic, yet are able to utilize CO_2; some soil bacteria can live either autotrophically or heterotrophically.

There are two principal sources from which living organisms secure energy: light and chemical bonds. *Photosynthetic* organisms utilize the energy of light and *chemosynthetic* organisms secure energy from chemical bonds. One common way in which chemical bond energy is made available is by oxidation; therefore, any oxidizible substance may serve as a source of energy for an organism that possesses the proper equipment for performing the oxidation and for using the energy in a profitable manner, such as synthesizing protoplasm or moving from one place to another.

NUTRITION

NUTRIENTS

Nutrients (foods) are extracellular substances that, on entering a cell after passing across the cell membrane, can be used by the cell for building material or for obtaining energy.

Practically any material on Earth can nourish one microbe or another. An astounding list of materials can be compiled, ranging from the usual growth substances—proteins, sugars, purines, and pyrimidines—to the unusual, such as rubber, paper, leather, oil, turpentine, carbon monoxide, iron, and elemental sulfur. No one organism is capable of utilizing all nutrients, and some nutrients can be used by only a small number of species. Many cannot attack native proteins but grow luxuriantly when supplied with products resulting from partial digestion of proteins. Most organisms use several carbohydrates, but a few do not use any.

Differences in the ability to utilize nutrients constitute a part of the basis for the identification of microorganisms. Nearly all species, and in some instances genera, are classified according to the compounds they utilize and the products formed from them.

Except for viruses and holozoic protozoa (see p. 22), microorganisms are holophytic in their nutrition. These latter organisms can transport molecules of low molecular weight across the cell membrane but not the large molecules of proteins, fats, and starch. In most natural environments there are bacteria that can secrete enzymes (discussed in the next section, page 172) that hydrolyze large particles (predigestion) to substances of lower molecular weight. These extracellular enzymes diffuse away from the organisms and into the surrounding environment. Their activity furnishes nutrients not only for their own organisms, but for neighboring organisms as well.

Hydrolysis of Complex Foods. Hydrolysis is a process by which proteins, polysaccharides, fats, and other large structural components are converted into their constituent parts by the introduction of water at points of cleavage of the large molecules, liberating the constituent molecules (Fig. 8–1). The removal of water during the synthetic process by which the complex food was made is usually accomplished by a mechanism that differs from the hydrolytic reaction. The hydrolysis of proteins yields the consecutively smaller proteoses, peptones, peptides, and amino acids. Proteoses are large fragments that can be precipitated by ammonium sulfate. Peptides are smaller fragments of two or more amino acids united through the carboxyl group of one and the amino group of the next. These fragments differ according to the protein from which they are derived and the method of digestion that produced them. The hydrolysis of polysaccharides such as starch or cellulose yields disaccharides and monosaccharides; glycerin and fatty acids are the products of fat digestion.

Penetration of Nutrients. The selection and transport of nutrients is the function of the cell membrane. The bacterial wall appears to play a minor part, that of excluding large molecules. Some compounds enter the cell by diffusion, but in most instances the nutrients are transported across the cell membrane by a process called *active transport*. Active transport refers to the ability of an organism to accumulate substances within the cell in high concentration from an external environment in which the substances are

Figure 8–1. *Hydrolysis of a disaccharide.*

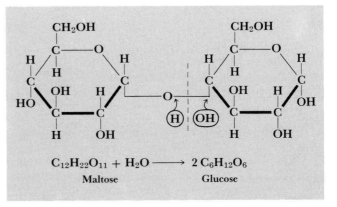

$$C_{12}H_{22}O_{11} + H_2O \longrightarrow 2\ C_6H_{12}O_6$$

Maltose   Glucose

in low concentration (movement against the concentration gradient). The transport process requires energy and is catalyzed by enzymes called *permeases*. Some permeases are produced only when the substrate (the substance upon which the enzyme acts) is present. Such is the case in certain strains of *Escherichia coli* that do not readily metabolize lactose. The organism cannot transport lactose until the cell has produced the permease (enzyme induction). After the lactose crosses the cell membrane, the organism may have to synthesize the enzymes to metabolize it. Thus the inability of an organism to metabolize a nutrient may reflect its inability to transport the nutrient; on the other hand, the organism may have a transport mechanism but lack the necessary metabolic enzymes. *E. coli* has a mechanism for metabolizing citrate, but it is unable to transport it into the cell.

NUTRITIONAL TYPES OF BACTERIA

Bacteria can be divided into four categories according to their carbon source and manner of securing energy (Table 8–1).

Autotrophic bacteria utilize CO_2 as the source of carbon with the aid of energy from sunlight (photosynthetic) or from the oxidation of inorganic compounds (chemosynthetic). The ability to use organic carbon varies among autotrophs. *Obligatory* autotrophs use only inorganic carbon and growth is usually inhibited by organic compounds. *Facultative* autotrophs can use organic or inorganic carbon, and *assimilatory* autotrophs use carbon dioxide as the main carbon source but can incorporate certain organic compounds. All autotrophs incorporate carbon dioxide into cell material by the carboxydismutase reaction (Fig. 8–2). Since this process is peculiar to autotrophs, some microbiologists have defined autotrophic bacteria as those that incorporate carbon dioxide by the carboxydismutase reaction.

Among the *chemosynthetic* microorganisms, thiobacilli can obtain energy through the oxidation of thiosulfate (Table 8–2). *Thiobacillus thiooxidans* can oxidize elemental sulfur to sulfuric acid and grow in an environment of 5 per cent sulfuric acid. *Nitrosomonas* species oxidize ammonia to

TABLE 8–1. Nutritional Types

Type	Oxidizable Substrate (Hydrogen Donor)	Energy Source	Carbon Source	Organisms
Photosynthetic autotroph[1]	Inorganic	Light	CO_2	Chlorobiaceae (green sulfur bacteria) Chromatiaceae (purple sulfur bacteria) [Also green plants]
Chemosynthetic autotroph[2]	Inorganic	Oxidation–reduction reactions	CO_2	*Pseudomonas spp.* *Thiobacillus* *Beggiatoa* *Nitrosomonas* *Nitrobacter*
Photosynthetic heterotroph[3]	Organic	Light	CO_2 and organic	Rhodospirillaceae (purple nonsulfur bacteria)
Chemosynthetic heterotroph[4]	Organic	Oxidation–reduction reactions	Organic	Most microorganisms except algae [Also animals]

Note: Also referred to by the following names:

1. Photolithotroph
2. Chemolithotroph
3. Photoorganotroph
4. Chemoorganotroph

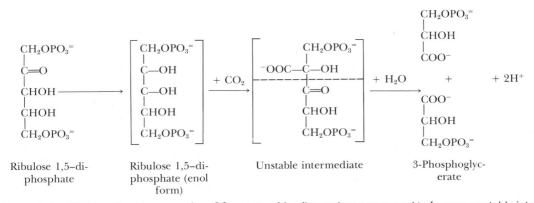

Figure 8–2. *Carboxydismutase reaction; CO_2 reacts with a five carbon compound to form an unstable intermediate, which splits into two molecules of phosphoglycerate.*

nitrite; *Nitrobacter* species oxidize nitrite to nitrate; five species of *Pseudomonas* oxidize hydrogen to water, and *Thiobacillus ferrooxidans* oxidizes the ferrous ion to the ferric form.

The photosynthetic autotrophs are members of Part 1 of The Bacteria (see Chapter 6). Cells of the family Chromatiaceae, the purple sulfur bacteria, can grow with sulfide or sulfur as the only photosynthetic electron donor; sulfide may be oxidized to elemental sulfur and further oxidized to sulfate. Chlorobiaceae, the green sulfur bacteria, may also oxidize sulfide to elemental sulfur.

Our knowledge and techniques of cultivating autotrophs stem from the pioneer work of Beijerinck, Winogradsky, Kluyver, and van Niel. The classic reports by Serge Winogradsky on the sulfur bacteria (1889) and the nitrifying organisms (1890) should be read by all students.

Heterotrophic bacteria require complex organic compounds for their main carbon source, although they may use carbon dioxide to a lesser extent. The energy source may be photosynthetic or chemosynthetic. *Photosynthetic heterotrophs* consist of one family—the Rhodospirillaceae (the purple nonsulfur bacteria). *Rhodopseudomonas* and *Rhodospirillum* are representative genera of the Athiorhodaceae. These organisms can use organic compounds as their carbon source and hydrogen donor. They require vitamins (e.g., *p*-aminobenzoic acid), and some will grow aerobically in the dark by oxidizing organic compounds *(facultative phototrophs).*

Chemosynthetic heterotrophs constitute the vast majority of bacteria commonly studied or dealt with in general or applied microbiology. A type organism is *E. coli*. This species has the ability to grow on inorganic salts and glucose; the ammonium ion serves as the nitrogen source. *E. coli* can also use preformed materials such as amino acids, vitamins, purines, and pyrimidines. A second type organism is *Streptococcus faecalis* (various *Lactobacillus* species are of the same type). This organism does not have the ability to grow on a glucose-salt medium but requires vitamins, amino acids, purines, pyrimidines, and peptides as well. Sometimes the amino acids are supplied as peptides to facilitate transport across the cell membrane and to prevent decarboxylation of

TABLE 8–2.   Medium for Cultivation of Thiobacilli

Constituent	Amount per Liter
Sodium thiosulfate	5.0 gm.
Ammonium sulfate	0.4 gm.
Monopotassium phosphate	4.0 gm.
Calcium chloride	0.25 gm.
Magnesium sulfate	0.5 gm.
Ferrous sulfate	0.01 gm.
Sodium bicarbonate	1.0 gm.

essential amino acids (e.g., tyrosine). *S. faecalis* will grow well on the relatively complicated medium illustrated in Table 8–3. This is a complete medium for the growth of *S. faecalis,* but it is not complete enough to elicit all the capabilities of the organism. In this medium *S. faecalis* will not oxidize pyruvate (an intermediate in sugar metabolism), because the vitamin (coenzyme) lipoic acid is omitted (see next section). *S. faecalis* does, however, produce the protein (apoenzyme) to which lipoic acid can attach, and thus when lipoic acid is added to the cells the complete active enzyme is formed. In this manner cells containing the apoenzyme can be used to assay food for lipoic acid content.

Microorganisms require of a growth medium a source of carbon, nitrogen, energy, and inorganic ions. The inorganic needs of most bacteria can be met by the addition of K^+, Mg^{++}, Mn^{++}, Fe^{++}, $PO_4^\equiv$, and $SO_4^=$ to a synthetic medium. A requirement can be demonstrated for these ions, but others (Zn^{++}, Cu^{++}, molybdate) are needed in such small amounts that a requirement is difficult to establish, mainly because these ions are contaminants of glassware and the chemicals used in the medium. For example, many bacteria synthesize vitamin B_{12}, which contains cobalt, but no quantitative requirement for cobalt has been demonstrated.

Potassium is needed for attachment of amino acyl RNA to ribosomes in protein synthesis and for activation of some enzymes (e.g., pyruvic kinase). Iron participates in electron transport, bound to proteins such as ferredoxin or cytochromes. Divalent ions (e.g., Mg^{++} or Mn^{++}) may complex with an enzyme-substrate complex and also bind to membranes and cell particles, including ribosomes, to stabilize the structure.

Most bacteria can utilize ammonia for the synthesis of nitrogen compounds, but some require additional nitrogenous growth factors. Certain species require vitamin B_6 before ammonia can be utilized. Some bacteria can convert nitrate or nitrite to ammonia or degrade organic nitrogenous compounds to ammonia. A few *(Azotobacter, Rhizobium,* and *Clostridium pasteurianum)* are able to utilize molecular nitrogen for growth (see p. 438).

TABLE 8–3.   Semisynthetic Medium for Lipoic Acid Assay*

Constituent	Amount per Liter	
Acid-hydrolyzed casein (H_2SO_4)	10	gm.
Enzymatic casein hydrolyzate	7.5	gm.
Glucose	3	gm.
Dipotassium phosphate	5	gm.
Sodium thioglycolate	100	mg.
DL-tryptophan	200	mg.
L-cystine	200	mg.
Adenine, guanine, uracil (each)	25	mg.
Nicotinic acid	5	mg.
Riboflavin	1	mg.
Pyridoxine (HCl)	1	mg.
Thiamine (HCl)	1	mg.
Calcium pantothenate	1	mg.
Folic acid	10	μg.
Biotin	1	μg.
Salts B†	5	ml.

*Final pH 7.0 to 7.3; autoclave for 15 minutes at 121° C.

†Salts B, per 250 ml.: 10 gm. of $MgSO_4 \cdot 7H_2O$; 0.5 gm. of NaCl; 0.5 gm. of $FeSO_4 \cdot 7H_2O$; 0.5 gm. of $MnSO_4 \cdot 4H_2O$; and 0.5 gm. of ascorbic acid.

From I. C. Gunsalus, and W. E. Razell: *In* S. P. Colowick and N. O. Kaplan (eds.): *Methods in Enzymology,* 3rd ed. New York, Academic Press, Inc., 1957, p. 941.

The requirement for sulfur is met for most microorganisms by supplying them with the sulfate ion. These organisms reduce sulfur to the sulfhydryl (—SH) form and attach it to amino acids to form cystine, methionine, glutathione, etc. Some bacteria have lost their ability to reduce sulfate and require sulfur in the reduced form—hydrogen sulfide, cysteine, methionine.

The requirement for preformed amino acids, vitamins, purines, pyrimidines, and other organic compounds represents a loss in the ability of a microorganism to synthesize or transport these compounds. This in turn reflects the genetic capabilities of the organism, as will be discussed in Chapter 9.

ENZYMES

The activities of living cells, whether bacterial, plant, or animal, are performed with the help of enzymes. Enzymes are proteins, and are produced only by living cells. Each cell possesses many enzymes, the kind and number depending on the genetic capabilities of the organism. They can be extracted from cells and studied *in vitro*.

THE NATURE OF ENZYMES

An enzymic protein behaves as a catalyst, altering the rate of a chemical reaction, yet remaining unchanged at the end of the reaction. The protein has a special site that can react specifically in some way with a certain chemical grouping. It performs the function of bringing together and orienting substrates so that they can interact in close proximity. In forming a complex with the substrate(s), the enzyme brings about an energy redistribution within the substrate(s), and thus lesser amounts of energy are required to initiate the reaction.

It is a principle of thermodynamics that only chemical reactions in which there is a decrease in free energy (i.e., exothermic reactions) can proceed spontaneously. A reaction that involves an increase in free energy must be coupled to a source of free energy. A thermodynamically possible reaction does not necessarily start spontaneously —"frictional" intermolecular forces tend to oppose it. The molecules must be in the reactive state, and only those molecules with high energy content are likely to react and form a product. In ordinary test tube reactions, the activation energy barrier is often overcome by heating the mixture. In biologic systems, this is obviously not done. Instead, enzymes decrease the activation energy required by permitting a large proportion of the molecular population to react at any given instant. Each enzyme molecule is structurally designed to react with substrate molecules and bring them into close apposition, whereupon they react spontaneously. The enzyme-substrate complex quickly decomposes, releasing unaltered enzyme and product. It may be said that the enzyme provides a pathway through the activation energy barrier.

In addition to protein, some enzymes contain an organic component of low molecular weight, either tightly bound (prosthetic group) or loosely bound (coenzyme). A metallic ion, such as magnesium, zinc, copper, iron, or manganese (cofactor), may be necessary for enzyme activity. Both cofactors and coenzymes may modify the structure or surface forces of the substrate or enzyme so a better "fit" or a lower activation energy is obtained. Coenzymes usually serve as donors or temporary acceptors of atoms or groups that are added to or removed from the substrate (see Table 8–4).

TABLE 8–4. Coenzymes

Coenzyme	Abbreviation	Group Transferred
Hydrogen-transferring coenzymes:		
Nicotinamide adenine dinucleotide	NAD	Proton
Nicotinamide adenine dinucleotide phosphate	NADP	Proton
Flavin mononucleotide	FMN	Proton
Flavin adenine dinucleotide	FAD	Proton
Lipoic acid	Lip (S_2)	Proton and acyl
Cytochromes	cyt.	Electrons
Group-transferring coenzymes:		
Adenosine triphosphate	ATP	Phosphate
Coenzyme A	CoA	Acyl
Biotin	—	Carboxyl
Thiamin pyrophosphate	TPP	C_2-aldehyde
Coenzymes of isomerases and lyases:		
Uridine diphosphate	UDP	Sugar (isomerization; e.g., glucose to galactose)
Pyridoxal phosphate	PALP	Carboxyl (decarboxylation of amino acids)

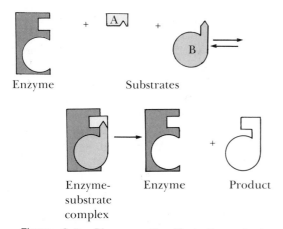

Enzyme   Substrates

Enzyme-   Enzyme   Product
substrate
complex

Figure 8–3. *Diagrammatic illustration of the reaction between an enzyme and two substrate molecules. Reaction between the substrates A and B is possible in the absence of enzyme, but it occurs more rapidly when they are brought into close apposition in the enzyme-substrate complex, which quickly dissociates and releases the enzyme and product.*

A reaction between an enzyme and molecules of two different substrates to form a product is illustrated diagrammatically in Figure 8–3. The enzyme active sites conform stereochemically to regions of the substrates, which are enabled to react with each other more readily than they would in the absence of the enzyme.

Reversibility of Enzyme Action

Enzyme reactions, like most chemical reactions, are reversible under proper conditions. The long arrow in the following equation:

$$A + B \rightleftharpoons C + D$$

indicates that the forward reaction (i.e., from left to right) proceeds until most of A and B are used up in the formation of C and D. A point is reached eventually at which some uncombined A and B remain mixed with the products C and D. The addition of more C and D at this time reverses the reaction and yields some A and B. The equilibrium in this example is said to lie "to the right." It is apparent that the relative concentrations of

the reactants (A and B) and the products (C and D) determine the direction in which the reaction is to proceed.

Many enzyme reactions in living cells *appear* to be irreversible, but apparent irreversibility is attributed to the removal of the products of reaction as fast as they form. The products may be used in subsequent synthetic reactions, which yield insoluble substances; they may be passed to another enzyme that catalyzes a further step in degradation; or they may be eliminated in a form (e.g., gaseous carbon dioxide) that escapes from the cell and even from the medium. Some reactions, however, are practically irreversible, because a large amount of energy is needed to reverse them.

Factors Affecting Enzyme Action

Enzyme action is affected by physical or chemical agents that modify the properties of proteins.

Temperature. The rates of chemical reactions, including those in which enzymes participate, approximately double for each 10 degree (C.) rise in temperature. However, as the temperature increases, bonds within the enzyme protein become unstable and rearrange to form an inactive enzyme (denaturation). It follows that for each enzyme an optimum temperature can be determined at which the apparent activity (i.e., chemical change per unit of time) is greatest (Fig. 8–4). Most enzymes are inactivated very quickly above 70° C.

Enzymes differ in their optimal temperature characteristics. Bacteria that occur in

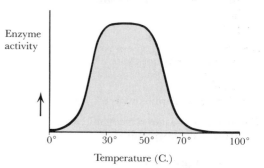

Enzyme
activity

0°   30°   50°   70°   100°
Temperature (C.)

Figure 8–4. *Effect of temperature on the activity of common enzymes.*

moderate climates have enzymes with temperature optima between 20° and 45° C. Those that customarily reside in hot springs, manure piles, or other very warm habitats possess enzymes having greater stability above 45° C. The enzymes of organisms found in cold, deep water and in pickling brines are active below 10° C., although their temperature optima may be only slightly lower than those of bacteria from more moderate habitats.

pH. The activity of each enzyme is greatest in a certain pH range and is less in the more acid and alkaline solutions (Fig. 8–5). The optima of many bacterial enzymes are between pH 6 and 8. Some yeast and mold enzymes are most active between pH 3 and 5. The influence of pH on enzyme action has been attributed to its effect on ionization of the enzyme and of the substrate, and the consequent alteration of their stereochemical structures.

Inhibition by Chemicals. *Competitive Inhibition.* A chemical that closely resembles the normal substrate of an enzyme may compete with the substrate for attachment to a reactive site of the enzyme. Adsorption of the usual substrate is decreased or completely prevented, and hence the normal reaction is inhibited. This phenomenon, competitive inhibition, is illustrated by the effect of sulfanilamide on bacteria that require *p*-aminobenzoic acid (see Figure 8–6). The two chemicals differ only in the nature of the group attached to the para position: $-SO_2NH_2$ vs. $-COOH$. *p*-Aminobenzoic acid is an *essential metabolite* for these organisms, but its utilization is

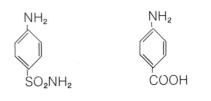

sulfanilamide   *p*-aminobenzoic acid

Figure 8–6. *Sulfanilamide, an antimetabolite, interferes with the utilization of p-aminobenzoic acid, which is an essential metabolite for many bacteria. Growth of these organisms is therefore inhibited.*

blocked by the presence of sulfanilamide, which is therefore called an *antimetabolite.*

Noncompetitive Inhibition. Heavy metals or their salts and various other chemicals that denature proteins may inactivate or "poison" enzymes. Some of these combine with $-SH$ groups rather than with the specific combining sites of the enzymes, but in so doing, they alter the stereochemical configuration or the distribution of forces in the enzyme, so that it can no longer function catalytically.

Concentration of Reagents. The concentration of an enzyme or of any required cofactors or coenzymes determines the amount of chemical reaction brought about, provided sufficient substrate is present. Likewise, the velocity of the reaction depends upon the presence of sufficient substrate to saturate the active sites on the enzyme.

Enzymes and Living Cells

Enzymes are not considered to be living, although they are vital components of all living cells. Enzyme inactivation may lead to the death of a cell containing the enzyme by depriving the cell of certain functions necessary to life. It will be shown in Chapter 12 that the same agents that inactivate enzymes also adversely affect the growth of bacteria and may even kill them.

Nomenclature, Classification, and Activities of Enzymes

Enzymes are named and classified according to the recommendations of the Inter-

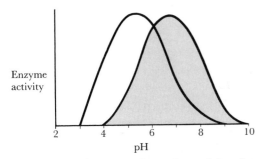

Figure 8–5. *Effect of pH on the activity of enzymes. Different enzymes have different optima.*

national Union of Biochemistry. The over-all reaction and the name of the substrate are used as the base for naming an enzyme; the suffix "ase" designates that the substance is an enzyme.

There are six principal types of enzymes: (1) *oxidoreductases* are involved in oxidation-reduction reactions; (2) *transferases* transfer a group such as acetyl (CH_3CO-) from one compound to another; (3) *hydrolases* split compounds by hydrolysis; (4) *lyases* remove groups from their substrates by some method other than hydrolysis, leaving double bonds, or they add groups to double bonds; (5) *isomerases* convert compounds into isomeric forms; (6) *ligases* (*synthetases*) couple groups or molecules to form larger compounds, complexes, polymers, or macromolecules.

Extracellular Enzymes. Exoenzymes or extracellular enzymes are secreted from a cell and catalyze reactions outside the cell (Table 8–5). They are usually hydrolases, and their function is the digestion of complex food materials.

Intracelluar Enzymes. Enzymes whose activity is confined to the cell are called endoenzymes or intracellular enzymes. The following are particularly significant:

Permeases. Permeases, or *translocases*, transport certain nutrients from the medium through the cell membrane and permit the accumulation of these substances within the cell in much greater concentration than that in which they exist outside the cell. Their role in active transport was discussed on page 165.

Hydrolases. Intracellular hydrolases perform certain hydrolytic reactions within the cell. For example, organic phosphates like acetyl phosphate and ATP are hydrolyzed and release inorganic phosphate and energy.

Group Transfer Enzymes: Transferases. Enzymes that transfer phosphate groups are essential in carbohydrate metabolism and also in the formation of high-energy phosphate bonds (see page 175). Other groups that may be added or removed by specific enzymes include the

TABLE 8–5.   **Typical Extracellular Enzymes**

Name	Equation or Type of Reaction
Proteinases	Hydrolyze proteins to proteoses, peptones, peptides.
Gelatinase	Hydrolyzes gelatin and destroys its ability to solidify as a gel; produced by many bacteria.
Caseinase	Hydrolyzes casein of milk; produced by many bacteria.
Pepsin	Hydrolyzes proteins in the animal intestine.
Trypsin	Hydrolyzes proteins in the animal intestine.
Carbohydrases	Hydrolyze polysaccharides, disaccharides, etc.: $(C_6H_{10}O_5)_n + nH_2O \rightarrow nC_6H_{12}O_6$
Cellulase	Cellulose $\rightarrow \underset{\text{cellobiose}}{C_{12}H_{22}O_{11}}$
Amylase	Starch $\rightarrow \underset{\text{maltose}}{C_{12}H_{22}O_{11}}$
Maltase	$\underset{\text{maltose}}{C_{12}H_{22}O_{11}} + H_2O \rightarrow \underset{\text{glucose}}{2C_6H_{12}O_6}$
Lactase (β-galactosidase)	$\underset{\text{lactose}}{C_{12}H_{22}O_{11}} + H_2O \rightarrow \underset{\text{glucose}}{C_6H_{12}O_6} + \underset{\text{galactose}}{C_6H_{12}O_6}$
Sucrase (Invertase)	$\underset{\text{sucrose}}{C_{12}H_{22}O_{11}} + H_2O \rightarrow \underset{\text{glucose}}{C_6H_{12}O_6} + \underset{\text{fructose}}{C_6H_{12}O_6}$
Lipases	Hydrolyze fats to glycerin and fatty acids: $C_3H_5(O \cdot CO \cdot R)_3 + 3H_2O \rightarrow C_3H_5(OH)_3 + 3RCOOH$

methyl, acetyl, amino, and carboxyl groups. These reactions are important in the inter-conversion of amino acids and other organic compounds and make possible the manufacture of protoplasm from whatever materials are at hand.

Oxidoreductases. Oxidation-reduction enzymes catalyze the electron (and hydrogen) transfers by which energy is abstracted from or built into chemical bonds, and hence they participate in many of the vital activities of a cell. Enzymes that remove hydrogen from a hydrogen donor are commonly called *dehydrogenases*. They are usually further identified according to the substrate. The enzyme that dehydrogenates lactic acid, for example, is known as lactic dehydrogenase.

A dehydrogenase enzyme is usually specific for a particular substrate; that is, it will combine only with that substrate (lactic acid or succinic acid, for example). Coenzymes are less specific and may function with several different dehydrogenases.

The type reaction for dehydrogenation is

$$AH_2 + B \longrightarrow A + BH_2$$

AH_2 represents the hydrogen donor and B represents the hydrogen acceptor. A few dehydrogenases utilize oxygen as the hydrogen acceptor and produce hydrogen peroxide. One of the hydrogens is removed from the substrate as a hydride ion (a hydrogen atom with its pair of electrons), and the other is removed as a hydrogen ion (proton). The hydride ion attaches to the coenzyme NAD, but the hydrogen ion remains in the medium:

$$AH_2 + NAD^+ \longrightarrow A + NADH + H^+$$

The reduced NAD can then pass electrons and hydrogen to a flavoprotein containing FMN or FAD:

$$NADH + H^+ + FAD \longrightarrow NAD^+ + FADH_2$$

If the organism does not contain cytochromes, the reduced flavoprotein forms hydrogen peroxide by transferring the hydrogen to oxygen (Fig. 8–7). In most bacteria, the final carriers are iron-containing proteins known as *cytochromes*. Cytochromes are proteins that have an iron molecule in a ring structure called *heme* (as in *hemoglobin*). Heme, which is firmly bound to the protein, accepts electrons from reduced flavoproteins:

$$FADH_2 + 2Fe^{+++} \longrightarrow FAD + 2H^+ + 2Fe^{++}$$

In most bacteria more than one cytochrome is involved (cyt c and cyt b). The reduced cytochrome transfers electrons via the enzyme cytochrome oxidase to oxygen:

$$4Fe^{++} + O_2 \longrightarrow 4Fe^{+++} + 2O^=$$

Finally, hydrogen ions from the medium react with activated oxygen to form water:

$$2H^+ + O^= \longrightarrow H_2O$$

The ultimate fate of the hydrogen depends upon the enzymes or carriers produced by the cell in question. Aerobic bacteria, for example, usually produce cytochromes and can oxidize a substrate completely. Strict anaerobes, on the other hand, do not produce cytochromes and are therefore unable to transfer hydrogen to oxygen. Instead they utilize other final hydrogen acceptors and yield products that are only partially oxidized.

Two other related enzymes should be mentioned. *Catalase* splits hydrogen peroxide into water and gaseous oxygen:

$$2H_2O_2 \longrightarrow 2H_2O + O_2$$

This reaction provides a mechanism for the disposal of hydrogen peroxide, which might otherwise accumulate in a culture and kill the organisms that produce it. Anaerobic bacteria (which cannot grow in the presence of air) do not produce catalase. *Peroxidase* decomposes hydrogen peroxide to form water and active oxygen, which is then reduced to form a second molecule of water.

Specificity of Enzymes

A given enzyme is capable of adsorbing and acting upon only a given substrate or

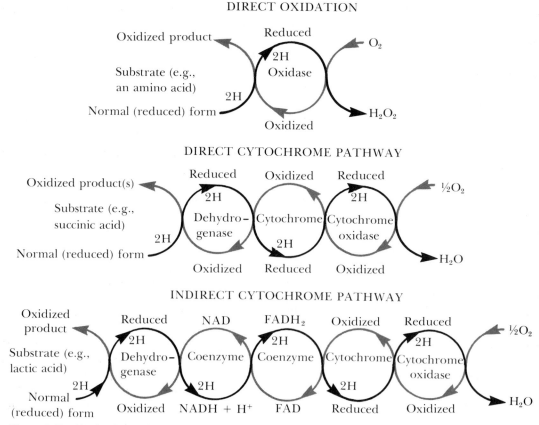

Figure 8–7. *Mechanisms of respiration; some of the pathways to free oxygen (diagrammatic). Hydrogen and electrons follow the heavy lines.*

perhaps certain closely related substrates. Maltase, for example, hydrolyzes maltose but not sucrose or lactose. The fact that enzymes are so specific indicates that an organism that attacks many different substrates or causes a great variety of reactions must produce a correspondingly large number of enzymes. The total might well be in the hundreds.

Inductive and Constitutive Enzymes

Certain enzymes are produced by bacterial cells only when their specific substrates are present in the medium. These are known as *inducible enzymes*. Enzymes that are constantly present are called *constitutive enzymes*.

An example of enzyme induction is provided by experiments with *Leuconostoc*

mesenteroides (Table 8–6). This organism was grown in media containing either glucose, lactose, arabinose, or no sugar, and then transferred to test solutions containing one of the three sugars. Glucose was fermented in any case, so the responsible enzymes were constitutive. Enzymes necessary for fermentation of lactose and arabinose

TABLE 8–6.　Inductive Enzyme Formation by *Leuconostoc mesenteroides*

After Growth in Medium Containing:	Leuconostoc Can Ferment:		
	Glucose	Lactose	Arabinose
Glucose	+	–	–
Lactose	+	+	–
Arabinose	+	–	+
No sugar	+	–	–

were inductive: they were produced only by organisms cultivated in media containing the same sugar.

It has been suggested that trace amounts of inductive enzymes are present in all cells, but the formation of useful amounts depends upon a stimulus provided by an "inducer," which is often the substrate but may be a related substance. The ability of bacterial cells to produce inducible enzymes helps to solve the problem of providing hundreds of enzymes only when needed.

METHODS OF SECURING ENERGY

It was mentioned earlier that all the physical activities of microbial (and other) cells as well as their synthetic activities require energy. Energy exists in many forms, but living organisms can utilize only one or two: the energy of chemical bonds and, in some cases, the energy of light (i.e., the sun). Light energy is incorporated into chemical bonds by the processes of photosynthesis.

The second law of thermodynamics states that in any given system the available energy tends to decrease. The lost energy is often given off as heat. This happens in spontaneously occurring chemical reactions, commonly known as *exothermic* or *exergonic* reactions. An *endothermic* or *endergonic* reaction will proceed only if energy is supplied from an external source. In a nonbiologic system the source of energy is often

heat; in biologic systems it is usually a simultaneously occurring exergonic reaction. Phosphates often participate in this type of coupled reaction. A particularly good source of energy is obtained by the hydrolysis of adenosine triphosphate (ATP) (see Figure 8–8) to adenosine diphosphate (ADP) and inorganic phosphate (P_i). A common method of indicating its participation in an endergonic reaction, $A \longrightarrow B$, is:

Chemical bond energy is made available through the displacement of electrons and, inasmuch as different kinds of bonds involve different numbers of electrons, they provide different amounts of energy. The bonds by which phosphate is attached to organic compounds vary in energy content from 2300 calories per mole to more than 12,000 calories per mole. Hydrolysis of ATP to ADP releases 7000 calories per mole. Compounds containing these high-energy phosphate bonds, symbolized as $\approx P$, are especially important in biologic systems as means of storing and transferring energy.

The fundamental operation in biologic energy release is electron transfer. In biologic oxidation, electrons are removed from a substrate. Electrons do not remain free in the living cell but must immediately be transferred to something else. The substrate,

Figure 8–8. *Adenosine triphosphate (ATP). The phosphate attached to ribose is low-energy phosphate. The high-energy phosphates may be removed one at a time to form adenosine diphosphate (ADP) and adenosine monophosphate (AMP), or two may be removed together to form AMP and pyrophosphate (P—P).*

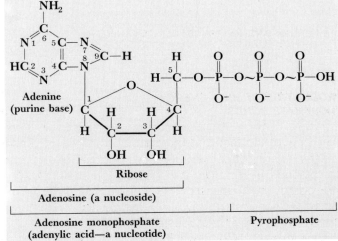

which may be called an *electron donor,* is oxidized as a result of losing electrons, and the recipient of these electrons *(electron acceptor)* becomes reduced. Whenever an oxidation occurs, there is an accompanying reduction. Oxidation and reduction are therefore always coupled reactions.

Oxidation may perhaps be more readily visualized when oxygen is added directly to a substrate, as in the following examples:

$$H_2 + \tfrac{1}{2} O_2 \longrightarrow H_2O + energy$$
$$NaNO_2 + \tfrac{1}{2} O_2 \longrightarrow NaNO_3 + energy$$

However, in these exergonic reactions there is electron transfer from the substrate (H_2, $NaNO_2$) to oxygen; the substrate becomes oxidized and O_2 is reduced. Biologic oxidation in the absence of atmospheric oxygen frequently involves the loss of hydrogen, which is transferred to a suitable hydrogen acceptor along with electrons. For example, in the oxidation of ethyl alcohol to acetaldehyde in the presence of NAD, hydrogen and electrons are transferred to the coenzyme:

$$CH_3 \cdot CH_2OH + NAD^+ \rightleftarrows CH_3CHO + NADH + H^+$$

    ethyl alcohol   acetaldehyde

Biologic oxidation in the absence of oxygen also occurs when the positive valence of an element such as iron is increased:

$$Fe^{++} - (e^-) + electron\ acceptor \rightarrow Fe^{+++} + reduced\ acceptor$$

In all of these varied reactions, the common feature is electron transfer.

PRODUCTION OF HIGH-ENERGY PHOSPHATE BONDS

In view of the fact that biologic systems transfer energy through high-energy phosphate bonds, it is appropriate to review some methods by which these bonds can be produced. As suggested previously, hydrolysis of the ordinary ester or glucoside phosphate linkage yields only 2300 to 3300 calories per mole, but high-energy bonds yield 7000 to nearly 13,000 calories per mole. Low-energy compounds may be brought to the high-energy level by adding more inorganic phosphate and removing hydrogen and electrons to effect an oxidation; the consequent rearrangement of electrons produces a compound containing high-energy phosphate. Removal of a molecule of water from an organic phosphate leaves a double bond ($-C{=}C-$), and the accompanying electronic rearrangement also increases the energy level of the phosphate.

High-energy phosphates can be transferred to AMP or ADP with formation of ADP or ATP, respectively. The latter are then available as sources of energy to activate molecules in the synthesis of cellular constituents.

Substrate Phosphorylation. The addition of phosphate to an organic compound is known as phosphorylation. There are several methods by which this is accomplished, with the generation of high-energy bonds. In substrate phosphorylation, inorganic phosphate attaches to an organic compound and yields an organic phosphate whose electrons are appropriately arranged to produce the high-energy state.

Glucose is an important source of energy for many organisms. Its intracellular breakdown *(dissimilation)* is initiated by the widely used glycolytic pathway known as the Embden-Meyerhof scheme. This sequence of steps, diagrammed in Figure 8–9, yields a three-carbon compound. In the first step, high-energy phosphate of ATP is added to glucose, which is then sufficiently reactive to participate in subsequent steps. Rearrangement of the glucose-6-phosphate to fructose-6-phosphate and addition of a second high-energy phosphate from another molecule of ATP modifies the electronic balance within the molecule, which is then cleaved to two three-carbon molecules. One of these, glyceraldehyde-3-phosphate, is phosphorylated by addition of inorganic phosphate and oxidized by removal of hydrogen and electrons. The resulting diphosphate then contributes high-energy phosphate to ADP. The remaining 3-phosphoglycerate rearranges and loses water,

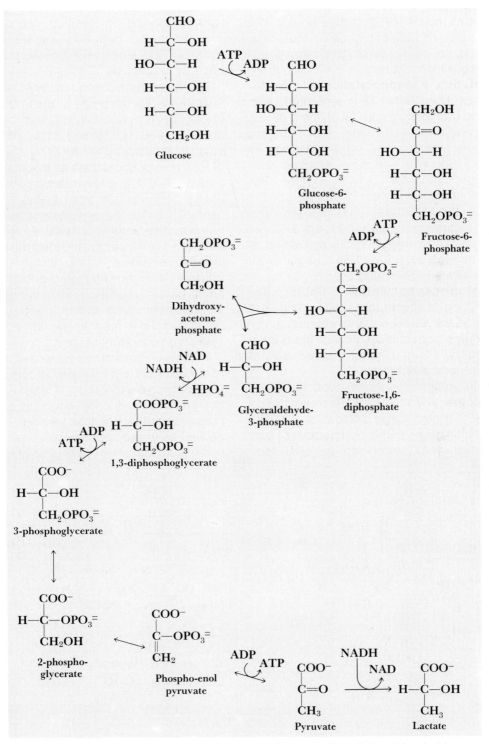

Figure 8–9. *The Embden-Meyerhof pathway of glucose dissimilation (glycolysis).*

forming energy-rich phospho-enol pyruvate. Transfer of high-energy phosphate from this compound yields another molecule of ATP. Since two molecules of pyruvate were derived from the original glucose molecule, there is a net gain of two high-energy phosphates per mole of glucose in this scheme.

Oxidative Phosphorylation. In oxidative phosphorylation, ATP is generated by an electron transport system involving a series of enzymes and coenzymes that transport electrons from an oxidizable substrate or electron donor to a terminal electron (and hydrogen) acceptor such as oxygen. The electron transport system in most aerobes is illustrated by the direct and indirect cytochrome pathways shown in Figure 8–7. For each pair of hydrogen atoms converted to water through such a pathway, as many as three ATPs are generated.

Photophosphorylation. Photophosphorylation is the process by which light energy is transformed into chemical energy in living cells. In photosynthesis, the chemical energy is then used for the synthesis of cellular constituents from CO_2.

Photosynthesis depends upon the presence of the green pigment, chlorophyll. The pigment in bacteria is somewhat different from that in green plants (chlorophyll *a*) and is designated bacteriochlorophyll (see Figure 8–10) in the purple bacteria (Rhodospirillaceae and Chromatiaceae) and chlorobium chlorophyll in the green bacteria (Chlorobiaceae). In addition, red and yellow carotenoid pigments are reported to transfer light, albeit inefficiently, to chlorophyll by resonation.

Two fairly distinct sequences of reactions are required for the photosynthetic reduction of CO_2. In the *light reactions,* light energy is absorbed and transformed into chemical energy (ATP). Chlorophyll absorbs a quantum of light energy and expels an electron, which then goes to a "primary electron acceptor," presumably the iron atom in the "reaction center," located in the photosynthetic membrane system or its equivalent. From the primary electron acceptor, the electron is transferred to ferredoxin and then through an electron transfer system including cytochromes *b* and *c* before returning to chlorophyll. In the step between cytochromes *b* and *c*, ATP is generated in a manner similar to that in oxidative phosphorylation.

Ferredoxin is a low molecular weight (5000 to 12,000) protein containing three to 10 atoms of iron per molecule. It is active at very low oxidation-reduction potential and participates in other electron transport systems as well as in photosynthesis.

In the *dark reactions* of photosynthesis, the

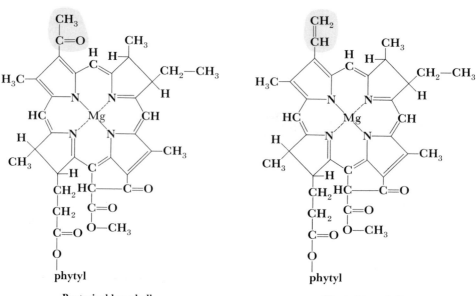

Bacteriochlorophyll   Plant chlorophyll

Figure 8–10. *Structural formulas of bacterial and plant chlorophylls.*

TABLE 8–7. Energy-Yielding Reactions

Type	Example
Respiration	$C_6H_{12}O_6 + 6O_2 \rightarrow 6CO_2 + 6H_2O + 688{,}000$ cal.
Anaerobic respiration	$C_6H_{12}O_6 + 12KNO_3 \rightarrow 6CO_2 + 6H_2O + 12KNO_2 + 429{,}000$ cal.
Fermentation	$C_6H_{12}O_6 \rightarrow 2CO_2 + 2C_2H_5OH + 54{,}000$ cal.

chemical energy of ATP assists in reducing CO_2 to organic compounds. A reducing substance is necessary and is often obtained from the environment. It may be H_2S or some other sulfur compound or an organic substance (e.g., in the purple nonsulfur bacteria). Green plants and algae do not use H_2S or organic compounds, but instead they obtain reducing power from H_2O and then liberate the oxygen in the gaseous form.

The overall reactions in photosynthesis may be summarized as follows:

$$CO_2 + 2H_2A \longrightarrow (CH_2O) + H_2O + 2A$$
$$(H_2A \text{ may be } H_2S)$$
Bacterial photosynthesis

$$CO_2 + H_2O \longrightarrow (CH_2O) + O_2$$
Plant photosynthesis

The CO_2 absorbed during photosynthesis is coupled to the 5-carbon sugar, ribulose diphosphate, by the carboxydismutase reaction (see Figure 8–2), and the resulting 6-carbon compound splits into two molecules of 3-phosphoglyceric acid.

BIOLOGIC OXIDATIONS

Bacteria also secure energy by oxidation reactions that do not necessarily involve phosphorylation. As mentioned earlier, when a substance becomes oxidized it loses electrons, and another substance receives the electrons and becomes reduced. Every oxidation is accompanied by an equivalent reduction of some other substance, and no oxidation occurs in the absence of reduction.

ELECTRON TRANSFER

In the majority of biologic oxidations, a hydrogen atom (a proton and its electrons) is transferred from a substrate molecule, designated the *hydrogen donor,* and eventually passed to another substance, the *hydrogen acceptor.* Between the donor and ultimate acceptor there may be several *carriers,* substances that readily accept and release both electrons and protons (FAD, FMN, NAD, NADP).

Fermentation. In fermentation, organic substances serve as both electron and proton donor and acceptor; indeed different parts of the same molecule often fill these roles (Table 8–7). Since the products contain considerable amounts of energy, the yield in available energy during fermentation is low.

Glycolysis. The best known process by which energy is obtained from glucose anaerobically is glycolysis, otherwise known as the Embden-Meyerhof pathway (Fig. 8–9). In homolactic bacteria (bacteria producing mainly lactic acid), glucose is metabolized to lactic acid, although traces of other products may form. These may increase in amount with a change of pH. With yeast, pyruvate is not converted to lactate, but instead to ethanol and carbon dioxide (see Figure 8–11). In either case a net gain of two ATP is obtained.

Hexosemonophosphate Pathways. In heterolactic fermentation, only parts of the Embden-Meyerhof pathway may function because of the absence of certain enzymes.

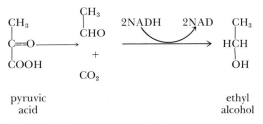

pyruvic acid   ethyl alcohol

Figure 8–11. *Formation of ethyl alcohol and carbon dioxide by reduction of pyruvic acid, which may be produced by glycolysis of glucose (see Figure 8–9).*

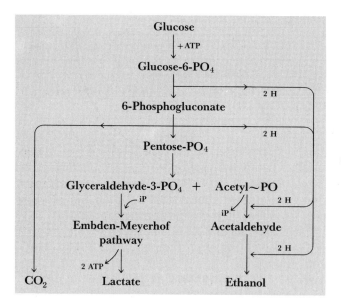

Figure 8–12. *Heterolactic fermentation (hexose monophosphate pathway) of* Leuconostoc mesenteroides.

The organisms use a different pathway—the hexose monophosphate pathway, which has a number of variations and is found in a variety of microorganisms: *Escherichia coli, Pseudomonas fluorescens, Azotobacter vinelandii, Bacillus subtilis, Streptococcus faecalis.* In some bacteria both the glycolytic and monophosphate pathways are present but function under different conditions.

Leuconostoc mesenteroides oxidizes glucose-6-phosphate to 6-phosphogluconate, which is then decarboxylated. The resulting pentose-phosphate splits into a three-carbon and a two-carbon unit. The final products are lactate and ethanol (Fig. 8–12). A net gain of one ATP is obtained from this reaction.

Respiration. Biologic oxidation in which molecular oxygen is the ultimate hydrogen acceptor is called respiration—sometimes more specifically designated *aerobic respiration* (Table 8–7). The product, carbon dioxide, is the most highly oxidized form of carbon, and thus the greatest possible yield of energy is obtained. Chemosynthetic autotrophs secure energy by oxidizing inorganic substances: hydrogen gas, hydrogen sulfide, elemental sulfur, iron, thiosulfate, ammonia, and nitrite (Table 8–8). Energy is obtained by oxidative phos-

TABLE 8–8. Respiration by Oxidation of Inorganic Substrates

Organism	Reaction
Pseudomonas spp. (formerly *Hydrogenomonas*)	$6H_2 + 2O_2 + CO_2 \rightarrow 5H_2O + (CH_2O)$ cell material
Beggiatoa	$2H_2S + CO_2 \rightarrow 2S + (CH_2O) + H_2O$ (S stored cell material in cells) or, in absence of H_2S: $2S + 5H_2O + 3CO_2 \rightarrow 2SO_2^= + 3(CH_2O) + 4H^+$ cell material
Thiobacillus	$S_2O_3^= + H_2O + 2O_2 \rightarrow 2SO_4^= + 2H^+$
Nitrosomonas	$2NH_3 + 3O_2 \rightarrow 2NO_2^- + 2H^+ + 2H_2O$
Nitrobacter	$2NO_2^- + O_2 \rightarrow 2NO_3^-$

phorylation. For some heterotrophs, glucose is converted to pyruvic acid by glycolysis and then to carbon dioxide and water by processes outlined in the tricarboxylic acid cycle (Krebs citric acid cycle). At present there are only two well established mechanisms by which microorganisms oxidize food-stuffs to carbon dioxide and water. These are the tricarboxylic acid cycle and the oxidative pentose phosphate cycle (a hexose monophosphate pathway).

Krebs Citric Acid Cycle. The most thoroughly studied and best known of the respiration processes is the oxidation of acetylcoenzyme A (acetyl-CoA) by the Krebs tricarboxylic acid cycle (Fig. 8–13). In respiration, pyruvate formed from glucose is not converted to lactate as in glycolysis but is converted to acetylcoenzyme A (active acetate) by decarboxylation and dehydrogenation. A small amount of pyruvate also reacts with carbon dioxide to form oxalacetate, which condenses with acetyl-CoA to produce the tricarboxylate, citrate. A series of reactions follows, some of which release hydrogen atoms and carbon dioxide, until eventually oxalacetate is regenerated.

Four pairs of hydrogen atoms are removed from the intermediate substrates by NAD or FAD and appropriate enzymes and are ultimately transported to oxygen by the cytochrome pathway to form water. For each acetyl-CoA fed into the cycle, 12 ATP are generated. During the conversion of glucose to pyruvate by glycolysis and then to carbon dioxide and water, 38 ATP are generated.

Oxidative Pentose Phosphate Cycle. The oxidation of glucose by successive dehydrogenation, decarboxylation, and group transfer produces carbon dioxide and water. Energy is generated by oxidative phosphorylation. This cycle can be demonstrated in many microorganisms (Fig. 8–14).

Anaerobic Respiration. Anaerobic respiration is an oxidative process in which inorganic substances other than oxygen serve as the terminal hydrogen and electron acceptor. Some organisms oxidize glucose completely in the absence of atmospheric oxygen when an oxidizing agent such as potassium nitrate (KNO_3) is present (Table 8–7). Other electron and hydrogen acceptors used by certain bacteria include sulfate and carbon dioxide (Table 8–9). Sulfate reduction may yield hydrogen sulfide (H_2S), and various bacteria produce methane (CH_4) from carbon dioxide. The energy formed from these oxidations is considered to be generated by an electron transport system. The mechanisms by which these reactions take place, although not completely understood, are presently receiving intensive investigation.

DISSIMILATION

Dissimilation is the intracellular breakdown of food materials; it yields compounds that can be incorporated into new protoplasm and also the energy that makes this possible.

CARBOHYDRATE DISSIMILATION

Carbohydrate is the most generally and easily utilized source of energy. The dissimilation of glucose has been extensively studied, largely because it is so widely distributed and readily used by animals and by microorganisms.

Methods of Study. Glucose dissimilation by microorganisms is studied by cultivating an organism in a suitable broth containing the sugar. The nature and amounts of the various metabolic by-products are then determined by chemical analysis. They might include, for example, succinic, lactic, and acetic acids; ethyl alcohol; carbon dioxide; and hydrogen. This experiment provides only limited information—the final result of dissimilation—it does not tell how dissimilation occurred.

The intermediate steps are ascertained by trial and error. A mechanism to account for the various products is postulated and tested. It might be suggested, for example, that a gram-negative rod of the Enterobacteriaceae produces CO_2 and H_2 by the decarboxylation of formic acid:

$$HCOOH \longrightarrow H_2 \uparrow + CO_2 \uparrow$$

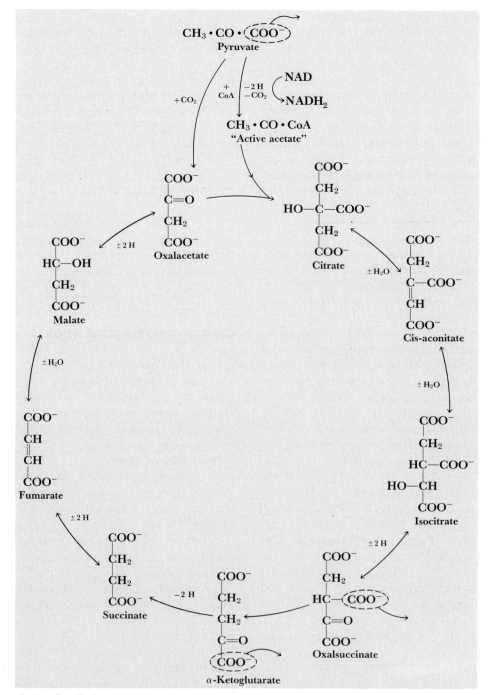

Figure 8–13. *The Krebs cycle (abridged); aerobic oxidation of pyruvate. Oxalacetate formed by carboxylation of pyruvate is joined with acetyl-coenzyme A; two carbon atoms are lost by oxidation to CO_2, and oxalacetate is regenerated. Succinyl-CoA is an intermediate between α-ketoglutarate and succinate, and during its formation two hydrogen atoms are transferred to NAD.*

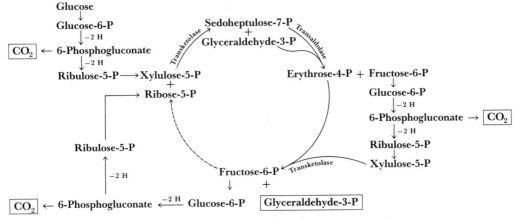

Figure 8–14. *The oxidative phosphate pentose cycle.*

The organism is therefore grown in a medium containing a formate salt, and the production of these gases is presumptive evidence that this hypothesis is correct. If it is then proposed that formate is derived from pyruvate, a pyruvate medium is similarly tested.

A radioactive or other isotopic element can be employed as a tracer. Pyruvic acid containing ^{14}C in the carboxyl group is used in the culture medium, and the reaction

$$CH_3 \cdot CO \cdot {}^{14}COOH + H_2O \longrightarrow$$
pyruvic acid

$$CH_3 \cdot COOH + H{}^{14}COOH$$
acetic acid   formic acid

not only shows that formic acid is produced from pyruvic acid, but proves that the pyruvate carboxyl group is the source of carbon in formic acid.

Products of Carbohydrate Dissimilation. Microorganisms produce a variety of

TABLE 8–9. Anaerobic Respirations

Organism	Reaction
Many species	$NO_3^- \xrightarrow{H} NO_2^- \nearrow^{H} N_2\uparrow$ $\searrow_{H} NH_4^+$
	"Sulfate reduction"
Desulfovibrio	$CH_3COOH + SO_4^= \rightarrow 2CO_2 + H_2S + 2OH^-$
	"Methane fermentation"
Various species	Organic cmpds. $\xrightarrow{\text{(anaerobic)}} CO_2 + CH_4 + (CH_2O)$ cell material
Methanobacterium omelianskii	$2C_2H_5OH + CO_2 \rightarrow 2CH_3COOH + CH_4$
Various species	$4H_2 + CO_2 \rightarrow CH_4 + 2H_2O$
Various species	$CH_3COOH \rightarrow CH_4 + CO_2$
Various species	$4CH_3OH \rightarrow 3CH_4 + CO_2 + 2H_2O$
Clostridium sp.	$4H_2 + 2CO_2 \rightarrow CH_3COOH + 2H_2O$

compounds in their dissimilation of carbohydrates: acids, alcohols, gases (see Table 8–10). Some organisms produce only one or two compounds (e.g., lactic acid); others yield a mixture of many. The products are determined by the nature of the organism, that is, the enzymes it possesses, and the environmental conditions. Respirative (i.e., aerobic) dissimilation by a facultative microorganism differs from fermentative (anaerobic) dissimilation by the same organism. Yeast, for example, produces carbon dioxide and water from glucose when growing in air, but yields carbon dioxide and ethyl alcohol in the absence of air.

Mechanism of Dissimilation. Most dissimilation processes are complex, requiring many intermediate steps before the final products are formed. Each step is a *more or less* independent reaction, but successive steps follow one another so rapidly that the individual stages cannot ordinarily be detected. Once the chain of reactions is initiated, it usually goes to completion, just as a tennis ball bounces down an entire flight of stairs when given a slight push at the top.

Fermentation of Pyruvic Acid. Some of the products formed by microbial fermentation of pyruvic acid are shown in Figure 8–15. Many of the overall reactions indicated consist of several steps, and each step is catalyzed by a specific enzyme. It is obvious that the products of dissimilation by a given organism depend upon the enzymes it produces, and that a lack of a single enzyme may interrupt a whole series of reactions. Most products of fermentation are incompletely oxidized and hence contain available energy. Fermentation is therefore a poor method of securing energy.

Some of the products of anaerobic dissimilation are utilized in assimilation, the process whereby protoplasmic constituents are manufactured. One assimilatory reaction is indicated in Figure 8–15: the formation of the amino acid alanine from pyruvic acid and ammonia. Amino acids, it will be recalled, are the building blocks of proteins.

The Cell and Dissimilation of Carbohydrate. The numerous intermediate compounds in fermentation and respiration are rarely present within a metabolizing cell in appreciable quantities. The essential mechanism in oxidation is electron transfer, and it seems likely that nature uses the simplest possible means of passing electrons from substrate to acceptor. In the attempt to visualize this process we postulate a sequence of chemical reactions as though they occur one at a time. Intracellular conditions doubtless permit rapid passage of electrons from one atom to another in succession without the accumulation of the assumed intermediate compounds. The result is the gradual liberation and transfer of energy to synthetic reactions.

TABLE 8–10. **Partial List of Products of Microbial Dissimilation of Carbohydrates**

Acids	Alcohols, Etc.	Gases
$H \cdot COOH$ formic	C_2H_5OH ethyl	CO_2
$CH_3 \cdot COOH$ acetic	C_3H_7OH propyl	H_2
$C_2H_5 \cdot COOH$ propionic	$CH_3 \cdot CHOH \cdot CH_3$ isopropyl	
$C_3H_7 \cdot COOH$ butyric	C_4H_9OH butyl	
$CH_3 \cdot CHOH \cdot COOH$ lactic	$CH_3 \cdot CO \cdot CHOH \cdot CH_3$ acetylmethyl carbinol	
$COOH \cdot CH_2 \cdot CH_2 \cdot COOH$ succinic	$CH_3 \cdot CHOH \cdot CHOH \cdot CH_3$ 2,3-butylene glycol	

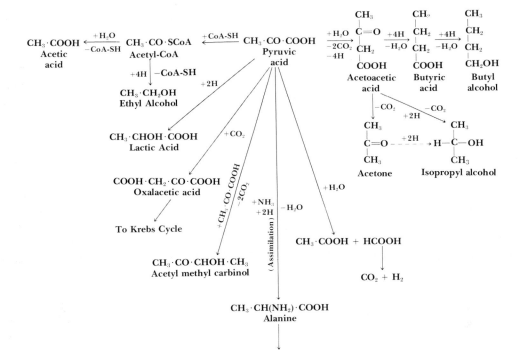

Figure 8–15. *A few products of anaerobic dissimilation of pyruvic acid by various bacteria. No one organism forms all the products shown. Phosphate and various coenzymes participate in some of these reactions but are not shown. Two molecules of pyruvic acid are required to make acetoacetic acid.*

PROTEIN DISSIMILATION

Dissimilation of Amino Acids. Amino acid dissimilation within the cell usually consists of decarboxylation or deamination or both, sometimes followed by further degradation (see Figure 8–16). *Decarboxylation* yields amines, many of which are foul smelling compounds. Putrescine is derived from the amino acid ornithine, and cadaverine from lysine. These and other amines were formerly known as ptomaines and were thought to cause food poisoning following the ingestion of spoiled meat or fish. True food poisoning by bacteria is now known to be caused by toxins; other digestive upsets are attributed to the actual invasion of the intestine by microorganisms in the food.

Deamination of an amino acid yields an organic acid and ammonia. The nature of the acid depends upon the conditions (aerobic, anaerobic, etc.) under which deamination occurs.

Transamination is a special type of deamination in which the amino group is transferred to an α-keto acid (see Figure 8–17). This is an important source of new amino acids in protoplasm.

The manner in which an organism dissimilates amino acids is partly controlled by the pH of the culture medium. An acid medium promotes the formation of decarboxylases, whereas an alkaline medium stimulates the production of deaminases. An organism growing in an acid medium may therefore decarboxylate amino acids and produce amines, which are more alkaline than the parent amino acids, and thus raise the pH of the medium (Fig. 8–18). In an alkaline medium the same organism deaminates amino acids and produces organic acids, which lower the

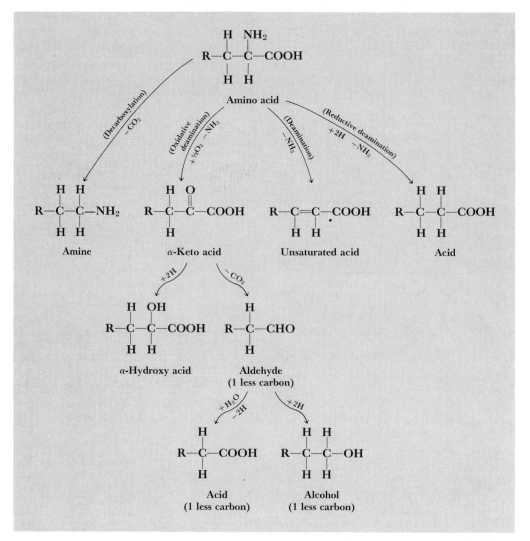

Figure 8–16. *Dissimilation of amino acids. A variety of compounds may be formed, depending on the enzymes and conditions available.*

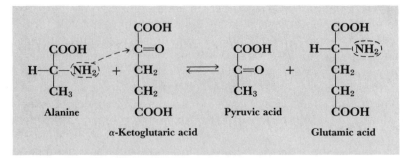

Figure 8–17. *An example of transamination.*

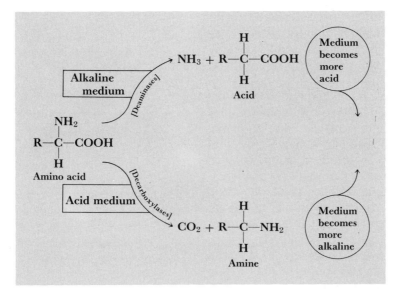

Figure 8–18. *Effect of the reaction of the medium on the production of enzymes that dissimilate amino acids.*

pH. This neat, built-in mechanism for controlling the reaction of the environment is the result of an evolutionary process of great survival value.

ASSIMILATION

Assimilation is the constructive activity by which food materials are transformed into cell constituents. Proteins, carbohydrates, fats, and other substances are produced from the simpler compounds that result from dissimilation or diffuse directly from the culture medium.

Growth Factors. Vitamins, coenzymes, and other organic substances essential in trace amounts for energy transformations are synthesized by many organisms from simpler inorganic or organic materials. The syntheses can be represented as series of reactions. Four of the postulated steps in the formation of coenzyme I (NAD), for example, are shown in Figure 8–19. Many organisms, like *Escherichia coli,* perform all four steps and produce coenzyme I from the substrate *x. Proteus vulgaris* lacks the enzymes that convert *x* into nicotinic acid (step 1), but can perform steps 2, 3, and 4. Since it can produce coenzyme I if supplied with nicotinic acid, this substance is said to be a growth factor for *P. vulgaris.* Other more fastidious bacteria can perform steps 3 and 4 or 4 alone

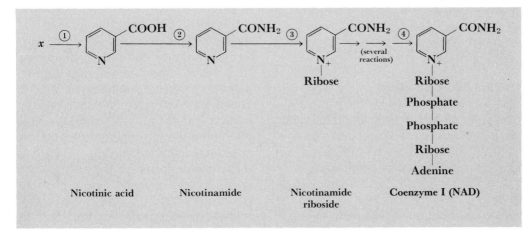

Figure 8–19. *Postulated steps in the biosynthesis of coenzyme I (NAD).*

and hence must be supplied with nicotinamide or nicotinamide riboside, respectively. Some organisms require not only growth factors but also one or more amino acids.

It will probably be no surprise to learn that bacteria with specialized nutrient requirements like the foregoing are often parasitic. Continued residence in a living host that supplies certain required substances seems to invoke some natural principle of conservation, and the parasite loses some of its synthetic ability. Not all fastidious organisms are parasitic, however. Specific growth factor requirements are also characteristic of saprophytes that have become adapted to natural environments like milk, where certain growth factors are found.

SYNTHESIS OF POLYSACCHARIDES

Many microorganisms synthesize starch or glycogen or some other polysaccharide from glucose. Glucose phosphate forms first and then polymerizes, with elimination of the phosphate groups. Starch (α-1,4-polyglucoside) is a linear chain of glucose molecules; glycogen is a branched chain (Fig. 8–20).

Some polysaccharides are produced from disaccharides. A dextran (α-1,6-polyglucoside) can be formed from sucrose by union of the glucose residues; the fructose units are set free. Similarly, an organism that couples the fructose moieties and releases glucose produces a levan (β-2,6-polyfructoside). Dextrans and levans are found in the capsules of certain bacterial species.

SYNTHESIS OF FATTY ACIDS

The substrate for building long chain fatty acids in bacteria is malonyl-CoA. To produce this compound from acetate, 3 ATP are required; the activation of acetate and a subsequent carboxylation form malonyl-CoA (Fig. 8–21). Malonyl-CoA then increases in chain length through a stepwise series of reactions to yield a long chain fatty acid such as palmitic acid ($C_{15}H_{31}COOH$).

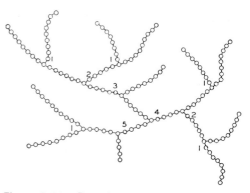

Figure 8–20. *Branched chain structure of glycogen. Each circle represents a glucose molecule. Starch (amylose) is a linear chain.*

SYNTHESIS OF PROTEINS

The synthesis of proteins will be discussed in Chapter 9. A necessary prerequisite to protein formation is an adequate supply of the proper amino acids. They may be assimilated from the surrounding medium. Bacteria accumulate and concentrate certain amino acids within their cells in a "metabolic pool."

Amino acids that are not taken into the cell intact must be synthesized. Ammonia appears to be the only inorganic form of nitrogen converted into amino groups; two pathways known to accomplish this are shown in Figure 8–22. Glutamic acid, produced from ammonia and α-ketoglutaric acid, is the key intermediate in the formation of other amino acids by transamination, as previously described (p. 185).

Some amino acids are formed in step-wise fashion from smaller molecules (see Figure 8–23).

Metabolic pathways that function for both biosynthesis and degradation are referred to as amphibolic pathways. These include glycolysis, the hexosemonophosphate pathway, and the tricarboxylic acid cycle. These pathways furnish both energy and compounds for biosynthesis. The degradative pathways are regulated by the product of energy metabolism (e.g., ATP), whereas biosynthetic pathways are controlled by the specific end product (e.g., phosphoenol pyruvate inhibits the conversion of fructose-6-phosphate to fructose diphosphate). These inhibitors combine with the enzyme at a site different from the substrate

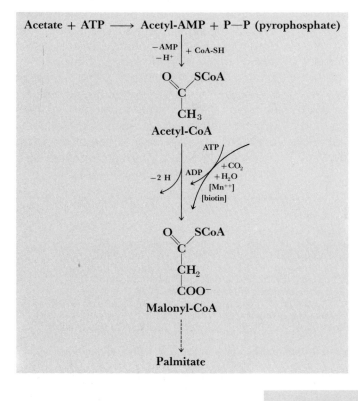

Figure 8–21. *Steps in the formation of malonyl-CoA, preliminary to the synthesis of long chain fatty acids.*

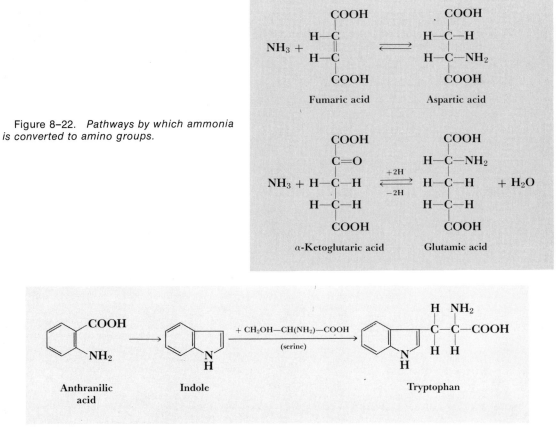

Figure 8–22. *Pathways by which ammonia is converted to amino groups.*

Figure 8–23. *Stepwise biosynthesis of tryptophan.*

site, distorting the enzyme so that it no longer provides a good fit for the substrate. An enzyme that can be controlled by this method is referred to as an *allosteric* enzyme. The controlling process is called *feedback inhibition.* Allosteric enzymes are involved in metabolic processes other than amphibolic pathways. For example, amino acid synthesis can be shut off by amino acids in the growth medium by feedback inhibition.

Animal, plant, and microbial cells perform many of the same chemical processes, both dissimilative and assimilative. It is of interest that so many different forms of life perform similar chemical reactions. Either the primitive organism from which all arose had developed a successful metabolic pattern, which persisted in the various evolutionary branches, or else these different branches ultimately arrived at the same pattern after separate trial and error processes.

SUPPLEMENTARY READING

Brock, T.: *Biology of Microorganisms.* Englewood Cliffs, N.J., Prentice-Hall, Inc., 1970.

Brock, T.: *Milestones in Microbiology.* Englewood Cliffs, N.J., Prentice-Hall, Inc., 1961.

Davis, B. D., Dulbecco, R., Eisen, H. N., Ginsberg, H. S., and Wood, W. B., Jr.: *Principles of Microbiology and Immunology,* 2nd ed. New York, Harper and Row, Publishers, 1973.

Doelle, H. W.: *Bacterial Metabolism,* 2nd ed. New York, Academic Press, 1975.

Gunsalus, I. C., and Stanier, R. Y. (eds.): *The Bacteria: A Treatise on Structure and Function,* Vols. II and III. New York, Academic Press, Inc., 1961, 1962.

Mandelstam, J., and McQuillen, K.: *Biochemistry of Bacterial Growth.* New York, John Wiley & Sons, Inc., 1968.

Peck, H. D.: Energy-coupling mechanisms in chemolithotrophic bacteria. Ann. Rev. of Microbiol., *22*:489–518, 1968.

Sanwal, B. D.: Allosteric controls of amphibolic pathways in bacteria. Bacteriol. Rev., *34*:20–39, 1970.

Sokatch, N. R.: *Bacterial Physiology and Metabolism.* New York, Academic Press, 1969.

Stanier, R. Y., Adelberg, E. A., and Ingraham, J. L.: *The Microbial World,* 4th ed. Englewood Cliffs, N.J., Prentice-Hall, Inc., 1976.

Vernon, L. P.: Photochemical and electron transport reactions of bacterial photosynthesis. Bacteriol. Rev., *32*:243–261, 1968.

VARIATION AND 9
GENETICS OF
BACTERIA

It has already been shown that bacteria possess most of the properties of other living organisms and function according to the same general principles. Each bacterial cell is a complete and independent individual in which all vital activities are performed; a pure culture therefore consists of a relatively homogeneous population of cells carrying on the same processes, whereas a multicellular organism contains a variety of cells performing somewhat different activities.

The fact that bacteria multiply more rapidly than higher forms presents a unique opportunity to study genetic phenomena on a greatly foreshortened time scale: 10 generations in the life of higher animal or plant species require years, whereas a few hours suffice in many bacterial species. Bacteria are therefore excellent subjects for the study of metabolic and genetic problems.

It is well established that bacterial chromatin bodies contain most of the genetic material of the cells. Preceding and during cell division, the chromatin bodies enlarge, undergo characteristic changes in shape, and divide (Fig. 9–1). The process is not identical with that observed in cells of higher plants and animals, but the net result is the same, i.e., replication of genetic units.

DNA is the principal reservoir of genetic information in bacteria as in other organisms. The DNA molecule (Fig. 9–2) consists of a double strand of nucleotides, which are held in a coiled configuration by chains of alternating sugar and phosphate molecules (Fig.

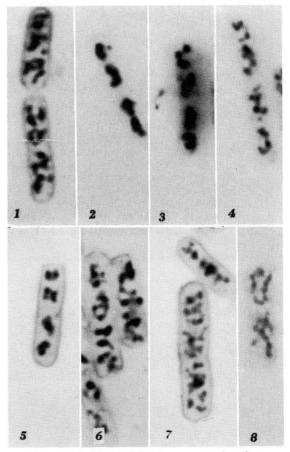

Figure 9–1. *Cells of* Bacillus megaterium *in successive stages of chromosome division. Linear chromosomes condense (1 and 2), duplicate (3), separate (4 and 5), elongate (6 and 7), and form long, beaded threads (8). Three chromosomes are shown in the lower nucleus of the upper cell in (1). Magnification, 4450X. (Photographs and interpretations by E. D. DeLamater; from W. Braun: Bacterial Genetics. Philadelphia, W. B. Saunders Co., 1953.)*

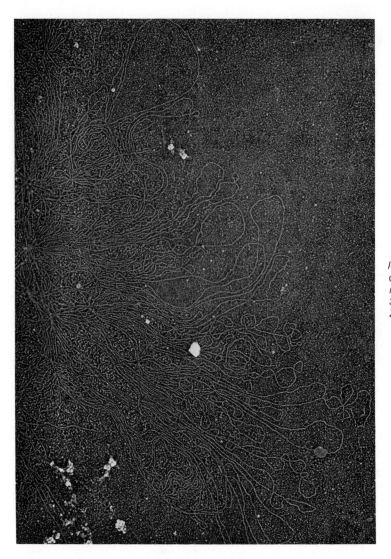

Figure 9–2. *Shadowed DNA released from lysed cells of* Micrococcus lysodeikticus. *Electron micrograph at approximately 31,000X. (From Kleinschmidt et al., Z. Naturforsch., 16b:730, 1961.)*

1–10). There are only four different nucleotides in DNA, and because of their structural affinities the bases in these nucleotides are always associated in pairs: adenine of one strand with thymine of the other, and guanine with cytosine.

The replication of DNA consists of the formation of a second double helix that is identical with the first. This is accomplished by unwinding the parent helix; at the same time new nucleotides assemble and join, by means of hydrogen bonds, with the corresponding nucleotides of each of the old strands: thymine with adenine and cytosine with guanine. The DNA molecule of a bacterium such as *Escherichia coli* contains ap-

proximately 10,000,000 nucleotide pairs. It is an endless loop, about 1 mm. in length (see Figure 5–18). Replication takes place as indicated in Figure 9–3. The process starts at one point, called the *replicator,* where the helix opens, thus providing single strands that act as templates for the DNA replicating enzyme. The replicator also serves as a swivel, allowing the two old strands to unwind as new DNA strands form and join the old DNA strands.

The bacterial DNA molecule comprises a series of polynucleotide regions that make up the functional units commonly called genes. One of the principal activities of these units is to direct the synthesis of proteins,

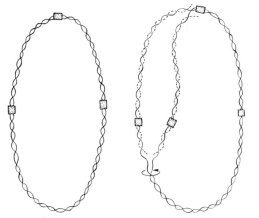

Figure 9–3. *Replication of the bacterial chromosome. Replicators are indicated by squares, and newly synthesized DNA is shown as broken lines. Replication starts at the top replicator, which serves as a swivel so that the old strands can unwind, and proceeds counterclockwise. (From Adelberg et al., Bact. Rev., 29:163, 1965.)*

both structural and enzymatic. The synthesis of a protein the size of hemoglobin (M.W. = 65,000) requires the information contained in approximately 2000 nucleotide pairs. The DNA in a bacterial cell therefore contains sufficient information for about 5000 different proteins.

The information in DNA is stored as a code in which the four nucleotide bases, adenine (A), guanine (G), thymine (T), and cytosine (C), represent "letters" of an alphabet, and each code word *(codon)* consists of three of these letters; e.g., −CAT−ATG−TTT−GTC−.

Each nucleotide triplet represents a specific amino acid used in the formation of a polypeptide chain. Twenty amino acids are found in proteins, and the four nucleotide bases provide 64 possible codons. As many as six codons correspond to a single amino acid. A sequence of nucleotide triplets that specifies the order of amino acids in a polypeptide is called a *cistron*. This is not necessarily equivalent to a gene, because some proteins are composed of more than one polypeptide chain; hemoglobin, for example, contains two chains, one of 141 amino acid residues and the other of 146 residues.

A group of genes and their regulatory agents that control a process or biosynthetic pathway and function together as a *unit of transcription* (see below) is called an *operon*. The *promoter* gene initiates transcription, after which the *operator* gene regulates the rate of transcription of succeeding structural genes that code for enzymes and other proteins.

To continue the analogy with communication of information, if a nucleotide is considered equivalent to a letter, a codon corresponds to a word, a gene to a sentence, and an operon to a paragraph.

The formation of a protein requires three principal events, which must be properly coordinated (Fig. 9–4): (1) *transcription* of the coded information of DNA into correspondingly coded RNA, (2) *translation* of the

Figure 9–4. *Structure of a tRNA molecule. Each solid circle represents a nucleotide. The single, small dots are hydrogen bonds (e.g., between adenine and uracil or guanine and cytosine, as in the Watson-Crick model of DNA) that maintain the structure of the molecule. X-X-X is a trinucleotide (anticodon) sequence that binds to a specific codon by hydrogen bonds in the same manner. The appropriate amino acid is attached to the C-C-A (cytosine-cytosine-adenine) end of the molecule.*

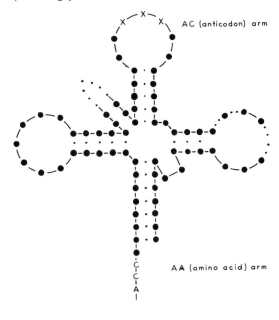

RNA instructions into the indicated amino acid sequence, yielding a polypeptide, and (3) *assembly* and folding of the latter into protein configuration.

Several forms of RNA participate in protein synthesis. *Messenger RNA* (mRNA) receives the code from DNA. This is a single-stranded molecule complementary to a limited region of one of the two strands of the nuclear DNA. RNA differs from DNA in one particular of base composition: it possesses uracil instead of thymine, which is always found in DNA. Uracil is paired with adenine, as guanine is paired with cytosine. The sugar component of RNA is ribose, while that of DNA is deoxyribose. As shown by electron microscopy, bacterial DNA is attached to the cell membrane or its mesosome, and mRNA is synthesized on DNA. One strand of DNA serves as template for the formation of mRNA. This is the process of *transcription.* It yields a molecule containing about 10,000 nucleotides, which represent sufficient information for several related proteins. Leaving the DNA, mRNA associates with the ribosomes, where the actual work of polypeptide formation *(translation)* takes place. This follows so quickly after transcription that the combined processes are termed "coupled transcription-translation."

Ribosomes are composed of two subunits, designated the 30S and 50S subunits. The letter S stands for the Svedberg unit, a measure of mass as indicated by the rate of sedimentation in a centrifugal field. The 50S subunit sediments more rapidly in the ul-tracentrifuge than the 30S subunit. The latter is depicted as a semilunar cap (see sketches in Figure 9–7), and the former as an ovoid, flattened on one end.

Whereas the ribosomes of animal cells are attached to the folded membranes of the endoplasmic reticulum, in bacteria they exist as unattached cytoplasmic granules composed of about 60 per cent RNA and 40 per cent protein. *Ribosomal RNA* (rRNA) is largely structural in function.

Transfer RNA (tRNA), or *soluble RNA*, serves as an amino acid carrier, bringing specific amino acids to the mRNA template associated with the ribosomes and holding them in the position specified by the coded instructions and properly oriented to be joined into a polypeptide structure. Each tRNA molecule contains about 80 nucleotides and its molecular weight is about 25,000. These molecules seem to be double stranded in part, presumably by means of the hairpin-like folding back of their linear structure. Where complementary bases (A–U, G–C) oppose one another, hydrogen bonds form and confer stability upon the molecule. At some region there is an exposed specific trinucleotide sequence corresponding to the amino acid with which the tRNA can combine. This provides the point of attachment with the complementary code triplet of mRNA, so that the amino acids are aligned in proper order for the specified polypeptide. It is not known what structures or forces permit the specific fit of amino acids to the various tRNA molecules.

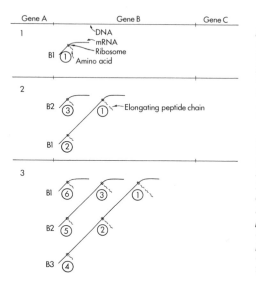

Figure 9–5. *Successive events [(1), (2), (3), . . .] in protein production. Transcription and translation of the information coded by DNA occur almost simultaneously, yielding mRNA and polypeptide chains. The process is initiated independently at the beginning of each gene (gene B in this figure), and many genes (e.g., A and C) are transcribed simultaneously. Transcription and translation of each codon occur repeatedly, so a great number of polypeptide chains form at the same time.*

As ribosome 1 moves along mRNA B1, the latter begins to peel off from the DNA of gene B and directs the selection of the first amino acid (i.e., tRNA-amino acyl complex). The next section of mRNA is then peeled off from the DNA and directs selection of the second amino acid, which is coupled to the first by the second ribosome. At the same time, transcription of the first codon of the gene occurs again, and another molecule of the first amino acid is selected to start a second peptide chain. These processes are repeated many times in succession. When the end of each gene is reached, the polypeptide is released and folds into an appropriate stable configuration or couples to other substances, according to the enzymatic and other conditions prevailing.

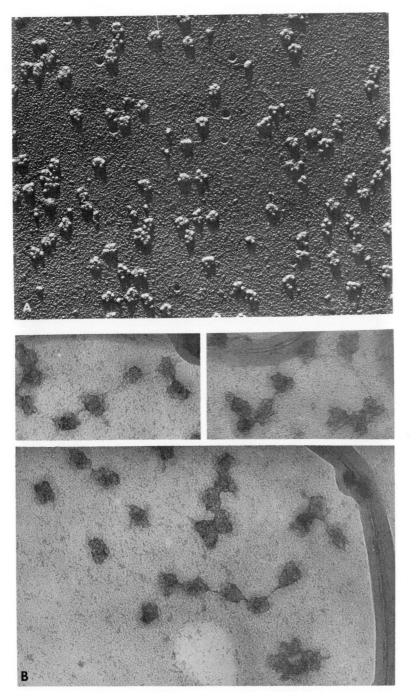

Figure 9-6. *Electron micrographs of polysomes. A, Platinum-shadowed preparation of human cell culture (HeLa) polysomes. B, Reticulocyte polysomes stained with uranyl acetate, showing thin strands (10 to 15 Å in diameter), believed to be mRNA, between ribosomes. (From Rich et al., Cold Spring Harbor Symposium on Quantitative Biology, 28:269, 1963.)*

The actual biosynthesis of a protein occurs in several steps, summarized diagrammatically in Figure 9–7. *Initiation* of the process involves attachment of mRNA to the 30S subunit of a ribosome, union with the formylated methionine-tRNA complex, and addition of the 50S subunit under the influence of initiation factors (proteins) and energy. *Recogni-* *tion* of subsequent codons in the mRNA chain leads to attachment of the corresponding amino acid–tRNA complexes, which is followed by *peptide bond formation* and *translocation* from the A (acceptor) site to the P (peptidyl) site. This positions the next codon in the A site, so the recognition step can be repeated, thus adding successive

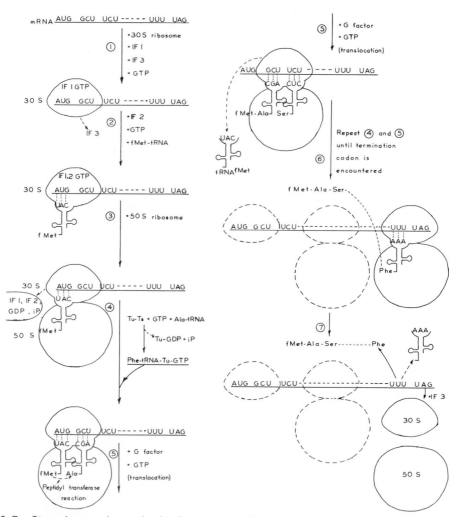

Figure 9–7. *Steps in protein synthesis directed by mRNA. (1–3). Formation of the initiation complex starts with binding of the 5'-OH end of the mRNA to a 30S ribosome subunit, aided by two small protein initiation factors, IF1 and IF3, and GTP (guanosine triphosphate). N-formylmethionine (fMet), coupled to tRNA and coded by AUG, is then added with the help of IF2 and GTP; IF3 is eliminated. This 30S-fMet-tRNA complex binds to a 50S ribosome subunit, and the initiation factors, GDP, and inorganic phosphate are released. (4–5). Formation of peptides and elongation of the chain occur by repeated addition of amino acyl-tRNA, directed by successive mRNA codons and removal of the tRNA molecules. Three protein elongation factors, Tu, Ts, and G, are essential, together with GTP and a peptidyl transferase enzyme, which is part of the 50S subunit. The incoming amino acyl-tRNA binds to a specific (acceptor) site on the ribosome. Peptidyl transferase then forms a peptide bond between the free -COOH of fMet or the growing peptide chain and the -NH2 of the new amino acid. This leaves a discharged tRNA at the second (donor) ribosome site. The new peptide translocates from the acceptor site to the donor site with the aid of GTP and G factor (a GTPase enzyme). (6–7). Peptide chain elongation continues until a termination codon (UAG, UAA, or UGA, for which there are normally no anticodon tRNA molecules) is encountered. Release factors then set free the mRNA, and IF3 dissociates the intact ribosome into its constituent 30S and 50S subunits, ready to be used again. (Modified from Joklik, W. K., amd Willett, H. P.: Zinsser Microbiology, 16th ed. New York, Appleton-Century-Crofts, 1976.)*

amino acids to the peptide chain. Finally, *termination* occurs when the termination codon, UAA, reaches the A site.

In an *inducible* enzyme system, the promoter gene binds RNA polymerase and initiates transcription, which is then controlled by an operator gene. The latter may itself be subject to control by a regulator gene, which codes for a protein that combines with the operator gene, turning off production of enzyme mRNA. When the substrate is present, it combines with the protein, releasing it from the operator gene, which in turn allows transcription of the structural gene to proceed.

Many sugars (e.g., lactose) are metabolized by inducible enzymes. It has been found that glucose can shut off synthesis of some inducible enzymes, even though the inducer (e.g., lactose) is present. This phenomenon is called *catabolic repression*. The repressor functions by reacting with the cyclic AMP binding protein and interfering with the activity of cyclic AMP (cAMP stands for cyclic adenosine-3′,5′-monophosphate). The latter is presumed to function in the lactose operon by acting at the promoter site (binding site of RNA polymerase) to facilitate initiation of transcription.

Sometimes the end product for a series of reactions will shut off production of the initial enzyme of a series. This is referred to as *end product repression*. In this case, the end product activates an inactive repressor.

BACTERIAL VARIATIONS

The variability of bacteria has troubled investigators since the early days of organized bacteriology. Morphologic and other changes in supposedly pure cultures led to the doctrine of *pleomorphism,* which maintained that any organism might appear in a variety of shapes. *Monomorphism,* on the other hand, was the belief that organisms of different shape were actually different species and not merely variant forms of the same species.

The conflict of ideas between pleomorphists and monomorphists produced two different approaches to the classification of bacteria. One school lumped together all reasonably similar forms into a few categories. The other established separate names for organisms differing from one another by only a single minor characteristic. Eventually it was found that bacterial properties vary with the age and growth conditions of the culture, and that mutations occur as in other living organisms. These normal variations must be taken into account in identifying and classifying bacteria.

Bacteria undergo both temporary and permanent variations. *Temporary variations* are morphologic or physiologic changes accompanying the normal development of a culture or induced by a specific environmental factor such as high osmotic pressure. Reversion to the normal form occurs quickly, either in the natural course of aging or upon return to usual conditions. *Permanent variations* consist of mutations—sudden, usually unpredictable, random genetic changes—and occur with relative infrequency. Reversion to the normal form occurs very rarely, and then only as a consequence of back-mutation.

TEMPORARY VARIATIONS

Morphologic Variations. The size and shape of bacteria vary continuously throughout the growth cycle (see page 113). This point is worth emphasizing because of the confusion students feel when attempting to identify their first "unknowns."

Old cultures of gram-negative rod bacteria such as *E. coli* consist almost entirely of cells so short that they appear coccoid. The same organism in a culture only a few hours old usually contains definite rods whose length is many times their diameter. In cases of doubt, therefore, it is often helpful to examine smears from young cultures; *true* cocci are rarely more than twice as long as they are broad at any stage of growth.

Another morphologic variation consists of irregular, bloated forms or bizarre shapes resembling the letters T, W, X, or Z. These odd shapes frequently occur in nature, whereas the same organisms appear as regular rods when growing upon suitable laboratory media. *Acetobacter* species in vinegar fermentation form long, bloated filaments. The root nodule bacteria, members of genus *Rhizobium* (Fig. 9–8), produce irregular shapes like the last letters of the alphabet when growing on the roots of legume plants. Both types of organism are ordinarily gram-negative rods in smears from agar cultures.

Certain bacterial structures are produced only under particular conditions of cultivation. *B. anthracis* forms *capsules* only within the animal body. *Streptococcus pneumoniae* is heavily encapsulated in the body but also produces capsules in culture media containing blood or serum and carbohydrate. Capsulation of saprophytic *Leuconostoc* species is favored by culture media containing sucrose.

Morphologic changes accompany *endospore* formation, and sporangia of *Clostridium* and other species are characteristically swollen. Spores usually form later in the growth cycle, but the time of sporulation varies from one species or strain of bacillus to another. Moreover, certain cultural conditions must usually be met; for example, aerobic bacteria sporulate only in the presence of oxygen.

The *staining* of bacteria is also subject to variation. This is fundamentally a physiologic rather than a morphologic characteristic. The gram reaction in particular varies according to the stage of development and other conditions within a culture. Old cells of gram-positive species frequently decolorize more readily than young cells; the same is true of cells that have produced an acid reaction in a medium containing fermentable sugar.

Physiologic Variations. Nongenetic enzymatic adaptation is an example of physiologic variation. Many bacteria produce a particular enzyme only when the specific substrate of the enzyme or some other inducer is present in the medium.

The formation of the lactose-splitting enzyme β-galactosidase by *E. coli* is an inducible process. Cells of this organism do not produce β-galactosidase in a glucose medium; when transferred to a medium containing lactose they are unable to attack this sugar for a short time, but within 30 minutes sufficient β-galactosidase is produced to

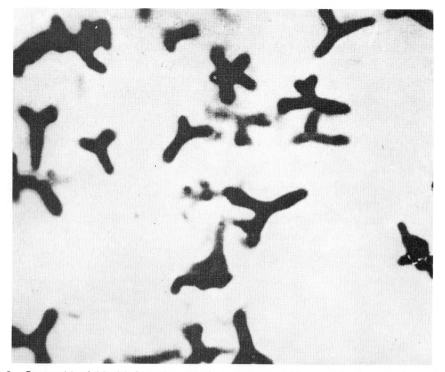

Figure 9–8. *Bacteroids of rhizobia from the root of a legume plant. These oddly shaped cells are found only in the nodule; they grow on agar in the laboratory as ordinary short rod bacteria. Magnification, about 2000X. (From Sarles et al.: Microbiology. New York, Harper & Brothers, 1956.)*

hydrolyze lactose vigorously. As soon as lactose disappears from the medium, the formation of the enzyme suddenly ceases. Lactose is the inducer and β-galactosidase is an *inducible enzyme*. Its formation presumably confers a competitive advantage on *E. coli,* because the monosaccharides of which lactose is composed can be utilized as sources of energy by the organism. It should be pointed out that there are variant strains of *E. coli* that do not produce this enzyme even in the presence of lactose and hence cannot ferment lactose. These are mutants, which appear following genetic variation.

Physiologic variation may occur in response to a change in the physical environment. *Lactobacillus plantarum* can grow in a medium lacking the amino acids tyrosine and phenylalanine at 26° C., but it cannot grow in the same medium at 37° C. unless tyrosine and phenylalanine are added. The protoplasm of this organism always contains the two amino acids; therefore, they are synthesized by cultures incubated at 26° C. from other substances, but they are not synthesized at 37° C., presumably because an essential enzyme is inactive at the higher temperature.

Inducible enzyme formation and the temperature dependence of enzyme activity are both under ultimate genetic control, but their demonstration in cultures depends upon particular environmental conditions, and reversion to "normal" occurs when the inducing condition is withheld.

MICROBIAL MUTATIONS

The DNA macromolecules composed of individual genetic units are amazingly stable. Nevertheless they change to other equally stable configurations once in 10,000 to 10,000,000,000 gene generations. These gene changes are mutations and are caused by alteration in the base pairs at specific loci; for example, an adenine-thymine base pair may be altered to give a guanine-cytosine base pair. This produces a change in the genetic code, which will be expressed as a change in the polypeptide whose structure is determined by this gene.

The rate at which mutations occur is enhanced by x-rays, ultraviolet light, nitrogen mustard, or other agents. Changes in the behavior of an organism are also caused by rearrangement of portions of a chromosome or translocation of portions between chromosomes. Individual genes are not affected by these processes, but their activities may be altered.

Most microbial mutations are *spontaneous;* those that take place at an increased rate in response to deliberately altered environmental conditions are called *induced mutations*. Both spontaneous and induced mutations are random; that is, it is impossible to predict which gene will be affected and hence what change in behavior will be observed.

In addition to mutations, there are types of genetic change over which it is possible to exert some control. Two of these are *transformation* and *transduction*.

SPONTANEOUS MUTATIONS

Colonial and Morphologic Mutations. A given bacterial species may produce more than one colonial form on agar media. The normal colony form (S) of many true bacteria is smooth, round, and glistening. Common mutant forms have rough (R) or mucoid (M) colonies. Rough colonies are dull, granular, or wrinkled, often with irregular margins; mucoid colonies are shiny, round, transparent, and slimy in consistency. In addition, various minute colonies have been described: G and L colonies contain granules or other elements that pass through filters used to retain normal bacteria, and dwarf (D) colonies often contain irregular club-shaped organisms (diphtheroid forms).

The normal smooth colony contains single cells in no particular arrangement. As multiplication occurs the individual cells separate and freely slide past each other upon the agar. This assumption of a random arrangement is doubtless facilitated by slippery cell constituents (e.g., slime). Rough colonies contain multicellular units consisting of chains of four or more cells. If they also lack the surface slime layer they cannot easily slip past one another to produce the random arrangement. Microscopic examination of the edge of such a colony often reveals parallel chains or threads of bacterial cells. Mucoid colonies produce excessive slime materials or even capsules.

Frequency of Colonial Mutations. The frequency of colonial mutations varies widely from one species to another and even within species. Mutation rates from one per 100 to one per 10 billion cell divisions have been reported; the average is about one per 10 million. This figure means that, on the average, 10 million cells must be produced in a culture before one mutant cell will arise.

Detection of Colonial Mutants. Detection of one rough mutant in a population of 10 million cells is a difficult problem unless some culture change suppresses the smooth form or enhances multiplication of the rough form. This is actually the case with *Brucella abortus,* which undergoes the S $\longrightarrow$ R mutation at the average one in 10 million rate. In an appropriate medium the amino acid d-alanine accumulates as a waste product and inhibits the smooth organisms but does not retard multiplication of the rough forms. The mutants therefore eventually predominate and are readily detected by plating.

Reverse Mutations. Reverse mutations from rough to smooth usually occur at a lower rate. In some cases an apparent reverse mutation yields a form having the same colony appearance as the parent but possessing some other genetic difference. Reversion of the R form of *B. abortus* produces a smooth colony that possesses much greater resistance to d-alanine than the original S form. The sequence of mutations is represented as S $\longrightarrow$ R $\longrightarrow$ S'. The S' form will eventually predominate over both S and R.

Associated Characteristics. Colony form is associated with other characterisitics of an organism. Smooth forms usually grow diffusely in liquid media, whereas rough forms tend to settle rapidly and produce a sediment. Smooth forms of pathogenic species are usually most virulent. A notable exception is *Bacillus anthracis,* but in this case the normal colony is rough, and avirulent mutant strains produce smooth colonies.

Resistance Mutations. Bacterial mutants can be obtained that are resistant to antibiotics or other chemicals, to bacteriophage, or to radiations. A normal population is exposed to the lethal agent for sufficient time to kill all sensitive cells. Subculture on appropriate medium then permits the growth of any surviving resistant cells. Mutations to the resistant forms occurs at a very low rate (e.g., 1/1 billion), and large populations must be exposed to the inhibitory agent for a few resistant cells to appear. Repeated trials are often necessary.

Resistance to Antibiotics. Resistance to high concentrations of streptomycin can be obtained in a single step. Mutants resistant to penicillin and most other antibiotics appear in a series of steps. A culture is exposed to a low concentration of the drug, and progeny of those organisms that survive are then exposed to a higher concentration; surviving mutants are then isolated. This process is repeated until organisms resistant to high concentrations are obtained.

The isolation of mutants resistant to antibiotics or other chemicals is facilitated by the *gradient plate technique.* Nutrient agar is poured into a tilted Petri dish and allowed to solidify. A second layer of nutrient agar containing the chemical is then added and allowed to solidify while the plate is level (Fig. 9–9). The antibiotic diffuses downward in proportion to the relative thicknesses of the two layers and establishes a uniform concentration gradient across the plate. The upper layer is inoculated heavily with the organism whose resistant mutants are desired. After incubation, dense growth is obtained at one side of the plate, in the region of tolerated low concentrations of the chemical. Higher concentrations toward the other side of the plate inhibit sensitive cells but permit colony formation by a few isolated resistant cells (Fig. 9–10).

Methods in which sensitive organisms are cultivated in contact with the inhibitory agent yield resistant mutants but do not demonstrate that the mutants arise spontaneously, even in the absence of the inhibitory agent. The Lederberg *replica plating technique* provides proof that such mutations do occur spontaneously. Plates of nutrient agar

(10 cm. diameter; 2 x 20 ml. agar)

Figure 9-9. *Preparation of a gradient plate. The lower layer consists of nutrient agar, which is poured into a slanted Petri dish. After the agar has hardened, the plate is placed in a horizontal position and nutrient agar containing antibiotic is added. Downward diffusion establishes a uniform concentration gradient of antibiotic across the plate, the highest concentration being at the left. (From W. Szybalski and V. Bryson, J. Bact., 64:489–499, 1952.)*

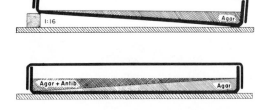

are inoculated on the surface with a normal or sensitive strain of the organism in question and incubated for a few hours or until colonies appear. Nutrient agar plates containing the inhibitory chemical are then inoculated from the first plate by means of sterilized velveteen wrapped tightly over the end of a wooden or metal cylinder slightly smaller than the Petri dish. The threads of cloth act as tiny inoculating needles, and when pressed against the agar upon which colonies are present they pick up 10 to 30 per cent of the cells in each colony. They are then pressed against the agar containing the inhibitory chemical and inoculate it. Several plates can be inoculated in succession. The colonies that appear are located exactly as on the original plate, but only resistant organisms can multiply (Fig. 9–11). The corresponding colonies on the original plate are readily identified, and isolation from these original colonies provides organisms that have never been in contact with the inhibitory

agents. They are found to be resistant to the chemical.

Mutants resistant to antibiotics appear *in vivo* with approximately the same frequency as they do *in vitro*. This fact has important consequences in the treatment of disease. The rate of mutation depends upon the antibiotic and the bacterial species, and in some cases is so high that therapy with a single antibiotic is virtually useless. However, simultaneous administration of two antibiotics is often successful, and resistant mutants do not arise. The explanation of this observation is that resistance mutations occur independently, and therefore organisms resistant to one antibiotic are killed or inhibited by the second, and vice versa. If bacteria resistant to each antibiotic appear at the rate of one per 10 million, for example, cells resistant to both antibiotics can be expected to appear at a rate of only one per 100 trillion (that is, $1/10,000,000 \times 1/10,000,000$).

Figure 9-10. *Growth of* Escherichia coli *on a gradient plate containing penicillin. A few resistant colonies are seen in the black area in the left half of the plate. Two such colonies were streaked out farther toward the left (toward the high concentration of penicillin); second step resistant colonies developed at the very left edge of the plate. (From W. Szybalski and V. Bryson, J. Bact., 64:489–499, 1952.)*

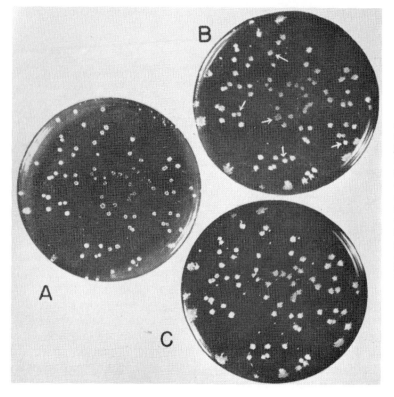

Figure 9–11. *Replica plating. Portions of colonies from the initial plate,* A, *were imprinted on plates* B *and* C *by means of sterile velveteen wrapped tightly over a circular support of wood or cork. Plate* B *contained the same medium as* A, *but* C *contained a medium on which certain mutants could not grow. The mutants are indicated by arrows on* B; *corresponding locations on* C *are vacant. (From J. Lederberg and E. M. Lederberg, J. Bact., 63:399–406, 1952.)*

Bacteriophage Resistance. Bacterial mutants resistant to bacteriophage can often be secured by continued incubation of a lysed culture. Resistant mutants can also be obtained by the replica plate technique. The success of this method demonstrates that phage resistance is acquired by mutation, because the bacteria are not exposed to bacterial virus during the isolation procedure. This mutation occurs at the very low frequency of about 1/1,000,000,000.

The *fluctuation test,* devised by Luria and Delbrück, can also be used to demonstrate that resistance to bacteriophage or other agents arises spontaneously by mutation at a certain determinable frequency, rather than by somatic adaptation of a constant proportion of the cells in a culture. If the latter situation were true, the percentage of resistant organisms in a population or in a series of parallel subcultures would always be the same. According to the mutation hypothesis, however, there is a small but constant probability *per generation* that each bacterium will become resistant, but when mutation has

occurred, *all* progeny of the mutant will be resistant. If mutation takes place early in the growth cycle of a culture, a large number of resistant cells will be present at a specified point in the history of the culture, whereas if mutation occurs late, few resistant cells will be present at the same point in the cycle of the culture. When several equal portions of a culture are used to seed subcultures, and the latter are examined after appropriate incubation, some subcultures are expected to contain many resistant organisms, others to contain fewer. The fact that a wide fluctuation in content of resistant bacteria actually occurs (Table 9–1) confirms the mutation hypothesis and supports the conclusion that the original resistant mutants arose prior to the exposure of the population to the phage or other inhibitory agent.

Radiation Resistance. Resistance to ultraviolet irradiation can be acquired by mutation. Sensitivity to ultraviolet light is quantitative rather than qualitative within the proper range of wavelengths; therefore intermittent or fractional irradiation is used to select pure

TABLE 9–1. Comparison of the Number of Phage-resistant Bacteria in Different Samples of the Same Culture with the Number in a Series of Parallel Cultures[*][†]

Number of Phage-resistant Bacteria	
10 Samples from the Same Culture	Samples from 10 Parallel Cultures
46	30
56	10
52	40
48	45
65	183
44	12
49	173
51	23
56	57
47	51

[*]Volume of samples = 0.05 ml.

[†]From data of S. E. Luria and M. S. Delbrück, Genetics, *28:*491, 1943.

cultures of resistant mutants. Sensitive cells are easily killed; increasingly resistant mutants survive and multiply.

E. coli mutants that are resistant to ultraviolet light are also resistant to hydrogen peroxide, potassium tellurite, crystal violet, and various other substances. Some of these chemicals are oxidizing agents, and it is now believed that resistance to radiation is associated with resistance to oxidation (see page 206).

Biochemical Mutations. The basis of all mutations is biochemical, but the term *biochemical mutation* is employed in a specific sense to designate mutations in which a nutritional requirement is altered or some easily recognized enzymatic process is affected.

Auxotrophic Mutants. The "normal" or usually encountered form of an organism is designated the "wild-type" or *prototrophic* form. Auxotrophic mutants are derived from a wild-type organism and require one or more growth factors. Prototrophs of many species can grow in simple, chemically defined media without addition of other growth factors. The simple medium in which the wild-type organism grows is called a *minimal medium,* whereas a medium containing all the likely nutritional requirements is known as a *complete medium* (e.g., nutrient agar and yeast glucose agar).

The isolation of auxotrophic mutants is facilitated by irradiation of the prototrophic population with ultraviolet light or x-rays (see page 206). The irradiated culture is plated on complete medium, and a large number of colonies is subcultured on complete medium and also on minimal medium. Growth on the complete medium but not on minimal medium indicates that the isolate may be an auxotrophic mutant. Several hundred colonies must often be tested before an auxotroph is obtained. Replica plating greatly facilitates the foregoing process and eliminates most of the labor of subculturing. Moreover, judicious selection of the culture medium permits one to pick any kind of auxotroph desired.

Auxotrophic mutants are useful tools in the hands of the biochemist interested in tracing natural pathways of synthesis of organic compounds. Each gene controls the activity of a single enzyme, and therefore any mutation interrupts a single biochemical process or step. Tatum and Bonner in 1944 found two mutants of the mold *Neurospora crassa* that could not synthesize tryptophan, although the parent strain could. Both mutants grew if supplied with indole, and one grew on anthranilic acid. In either case the amino acid serine was necessary. The steps in biosynthesis of tryptophan therefore seemed to be those outlined in Figure 9–12.

Both mutants possessed enzyme (*3*) and could join indole and serine to form tryptophan, but only one mutant possessed enzyme (*2*) and produced indole from anthranilic acid. Neither was able to make anthranilic acid from its unknown precursor *X*.

Auxotrophic mutants can be employed to identify and determine the concentrations of growth factors, such as amino acids, vitamins, and nucleic acid derivatives. The nutritional requirements of auxotrophic mutants are also being used as "markers" in the important field of microbial genetics, just as hair and eye color are used in human genetics.

A historically and biologically interesting

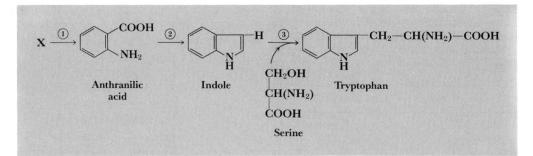

Figure 9–12. *Steps in the biosynthesis of tryptophan. An auxotrophic mutant that cannot carry out step 1 (conversion of an unknown precursor to anthranilic acid) must be supplied with anthranilic acid, indole, or tryptophan if it is to grow. A mutant that cannot perform step 2 must be supplied with indole or tryptophan, etc.*

mutation appears in S. typhi. This organism usually requires tryptophan when isolated from patients with typhoid fever, but some laboratory cultures will grow in media lacking this amino acid. It was originally believed that the organism gradually became "trained" by laboratory cultivation to synthesize its own tryptophan. Later, however, it was found that one cell in about 10 million mutates to a form able to grow without added tryptophan. This "tryptophan independent" mutant is not ordinarily detected in laboratory cultures, because the usual media contain sufficient tryptophan to permit luxuriant growth of the tryptophan-requiring parent. A medium containing enough tryptophan to support only 10 million to 100 million tryptophan-requiring cells eventually contains a considerable number of tryptophan independent mutants. If the medium contains even less tryptophan, visible growth appears only after prolonged lag, but most of this population consists of the mutant form.

The S. typhi mutation is of interest biologically because it may represent the reverse of what happens during acquisition of the parasitic mode of existence. A forebear of this species may have been a saprophyte that did not need an outside source of tryptophan. This organism was ingested by some animal or by man, established itself within the body, and mutated to a form that required tryptophan. The mutant was able to survive because an adequate supply of the amino acid was available in the host. Perhaps it multiplied a little more rapidly than the parent form, which was presumably completely eliminated and which disappeared except as the *back-mutation,* recognized in the laboratory, once more yields the form capable of synthesizing tryptophan.

Many species of parasitic bacteria can be cultivated only with great difficulty when first isolated, but ultimately they grow upon much simpler media. When these organisms are studied in sufficient detail to reveal their specific nutrient requirements situations similar to that in S. typhi will undoubtedly be found.

Fermentation Mutants. Fermentation mutants are forms that arise spontaneously in small numbers in populations consisting of cells with other fermentation properties, for example, the appearance of organisms that can ferment lactose or maltose in cultures that do not ordinarily possess these properties.

One of the best known fermentation mutants is derived from a variant of E. coli that fails to ferment lactose. This organism was at one time called E. coli mutabile. Young colonies of E. coli mutabile are white on E.M.B. agar (which contains lactose and other nutrients and an indicator system composed of eosin and methylene blue). Older colonies develop black secondary or daughter colonies known as papillae (Fig. 9–13). A black papilla may be subcultured and found to consist of stable, lactose-fermenting organisms. Replating the parent white colony continues to yield colonies that produce papillae. The lactose-fermenting mutants arise at the rate of approximately one per 100,000 cells.

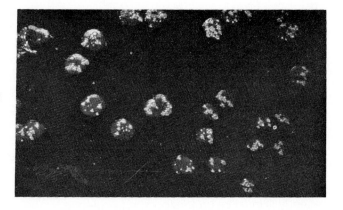

Figure 9–13. *Papillae of the lactose-fermenting mutant of* Escherichia coli mutabile *on colonies of the nonfermenting parent. (Parr.)*

The specific property affected in fermentation mutants is not known with certainty. It may be an enzyme necessary to attack the substrate in question, it may be permeability to the substrate, or some other factor.

Fermentation mutants arise at a higher rate than nutritional or resistance mutants: between one in 100,000 and one in 10 million cells. The prolonged incubation period before the appearance of papillae is accounted for by the time required for mutants to arise and the additional period during which they multiply sufficiently to produce papillae.

Pigmentation Mutants. *Mutations affecting pigment production* are easy to observe, but the actual biochemical changes are unknown in most cases. The mutants are almost completely stable, and are distinguished by complete loss of color or production of intermediate tints. *Serratia marcescens,* for example, normally produces a dark, blood-red pigment, but some strains produce stable, white mutants; rose red and pink variants are also fairly common. *S. aureus* often loses its golden pigment after extended cultivation on laboratory media. Many other staphylococci or micrococci are notorious for the frequency with which white colonies appear in otherwise yellow populations.

Nonmutational variations in pigmentation are frequently encountered, particularly among cocci and gram-negative rod bacteria. These temporary variations are caused by changes in nutrients, temperature of incubation, oxygen supply, and pH. Aerobic or facultative organisms are often most in-

tensely pigmented under aerobic conditions and at moderate or low temperatures.

Antigenic Mutations. The antigenic structure of bacteria is studied by techniques that will be described in Chapter 19. For the moment it will suffice to say that antigens are proteins or polysaccharides that stimulate the formation of modified blood proteins known as antibodies when injected into appropriate animals. Antibodies possess the ability to react detectably (e.g., by precipitation) with the corresponding antigenic substances. They are therefore *specific* reagents for the detection of antigens.

Bacterial cells contain many antigens, some associated with the cell bodies, some with the capsules, and others with the flagella. Their presence can be determined by proper tests with suitable antibody solutions (antisera).

Antigenic mutations occur with considerable frequency in bacteria and may involve any type of antigenic component. A motile organism that mutates to the nonmotile state and loses the power to produce flagella simultaneously loses its flagellar antigens. A mutation commonly encountered among *Salmonella* species results in the loss of some flagellar antigens and the acquisition of others. Mutations of this sort occur at a high rate: one mutant in about 10,000 cells is average, but rates as high as one in 250 cells have been reported.

Loss of cell body antigens by mutation is also a common occurrence. Freshly isolated *S. typhi* usually possesses a component known as the Vi antigen, but this disappears after only a few transfers on laboratory media. The transformation from smooth to rough colonial forms is accompanied by the loss of other cell body antigens. These are all mutational events. Reverse mutations occur with much lower frequency.

Mutations Affecting Virulence. Since the earliest days of medical bacteriology it has been noted that the virulence or disease-producing power of pathogenic bacteria may vary, but the mutational origin of this variation has been demonstrated only recently. It must be understood that virulence depends on numerous factors, some affecting the pathogen itself, others affecting the host. Virulence can be measured only by determining the effect of the pathogen on a host, and a resistant host can withstand large doses of the pathogen without adverse effect. Some pathogens cause disease by means of the toxins they secrete; the ability to produce toxin can be lost by mutation. Pathogenicity of certain species depends upon capsulation, and this property also is subject to mutational loss.

Studies with *Brucella abortus, Francisella tularensis, S. typhi,* and other bacteria have indicated a correlation between resistance to certain metabolic products and virulence, or between nutritional requirements and virulence. For example, mutants of *S. typhi* that required purines, *p*-amino-benzoic acid, or aspartic acid are less virulent for mice than the parent organism. The mutants multiply more slowly in the host, apparently because of limited availability of the purines, PABA, and aspartic acid in mice; the injection or in some cases even the feeding of these metabolites increases the pathogenic effect of the mutants.

Much remains to be done before factors affecting virulence are completely understood, and nutritional and resistance studies like the foregoing may be very fruitful. Mutations affecting virulence occur, but the ability of the mutant to establish itself depends upon chemical conditions within the host. These chemical conditions are influenced by the diet and general physiologic state of the host and by changes caused by the pathogen itself. The interaction between host and parasite is complex.

INDUCED MUTATIONS

Mutations resulting from exposure to *mutagenic agents* are known as induced mutations. Mutagenic agents increase mutation rates by a factor ranging from several hundred to as much as 100,000, but it is usually considered that they do not determine the type of mutation that occurs; they merely accelerate the rate of spontaneous mutation. The mutagenic agents that have been most widely used are x-rays, ultraviolet light, and nitrogen mustard. Other mutagens include peroxides, carcinogenic chemicals, and even some simple substances like manganese chloride.

Mutagenic agents are believed to damage or change a single genetic unit of DNA so that it possesses erroneous information. Spontaneous mutations are known to be the result of frameshifts; that is, deletion or addition of a single base at a given place in the DNA polynucleotide chain leads to a nonsensical situation because all subsequent bases are out of phase. Purines and pyrimidines occasionally undergo electron and proton rearrangements, which affect their ability to form hydrogen bonds, and hence they may attract different bases from the normal. The presence of base analogues may increase the frequency of such changes, as may a marked disturbance in the nature and concentration of nucleotide components. Excess replication of a given base may occur, as well as failure of replication, with the result that the information in the affected DNA region is altered or even "senseless."

There is some indication that oxidation plays a part; the presence of reducing substances in the culture medium decreases both the mutagenic and lethal effects of irradiation. Alkylating agents, nitrous acid, hydroxylamine, and other chemicals may add or change base radicals, thus producing different compounds that will react abnormally in the DNA strand.

All the progeny of a uninucleate cell that has undergone mutation display the mutated characteristic. Some bacterial cells are multinucleate, particularly during active growth phases, and only a portion of their progeny

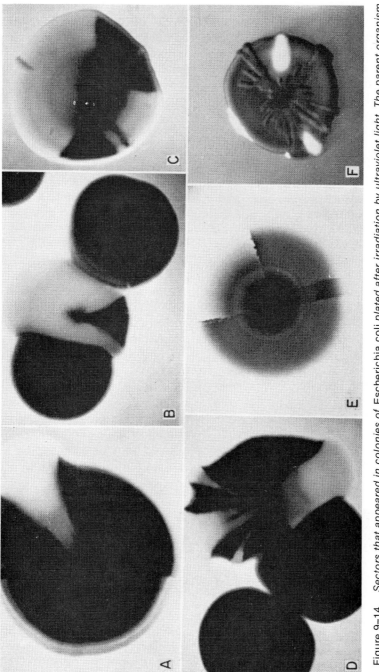

Figure 9–14. Sectors that appeared in colonies of Escherichia coli plated after irradiation by ultraviolet light. The parent organism ferments lactose and produces black colonies on the indicator-agar medium. Lactose negative mutant progeny produce white colonies or sectors. (Newcombe; from Braun: Bacterial Genetics, 2nd ed. Philadelphia, W. B. Saunders Co., 1965.)

consist of mutated forms, because the probability that more than one nucleus in the parent cell will be affected by the mutagen is extremely small. Mutation in a multinucleate cell can lead to the appearance of "sectored" colonies (Fig. 9–14), which show particularly well when the affected characteristic is pigmentation, colony smoothness or fermentation (of a sugar present in an indicator-agar medium). It should be pointed out that sectors of this sort may also arise when two or more individual cells are trapped upon the agar so closely together that their colonies merge; this is obviously not mutational sectoring.

Directed Genetic Changes

Genetic changes can sometimes be controlled specifically by the investigator. There are two principal methods of producing these changes: *transformation* and *transduction*. The techniques employed are dissimilar, but basically both consist of the transfer of DNA from one organism to another. They differ in the method by which transfer is accomplished.

Transformations. The first transformation discovered was the conversion of one antigenic type of pneumococcus into another. There are nearly 100 pneumococcus types, which are distinguished by possession of different capsular polysaccha-

rides. Antibodies against the polysaccharide of one type will not react with polysaccharides of other types.

Griffith in 1928 injected mice subcutaneously with noncapsulated, rough, type 2 pneumococci and capsulated type 3 pneumococci killed with heat. The animals that died were found to be infected with capsulated pneumococci of type 3. The type 3 organisms injected had been killed by heat; thus it was obvious that the rough, type 2 organisms had been transformed into type 3. The new type 3 culture maintained its type specificity through subsequent laboratory cultivation. Several other transformations of pneumococcal types were induced, and in each case the acquired property was that of the dead, capsulated organism injected.

Pneumococcus type transformations were later performed *in vitro* (Fig. 9–15). The substance in dead capsulated pneumococci that was responsible for transformation was eventually isolated and found to be deoxyribonucleic acid; its action was abolished by the depolymerizing enzyme, deoxyribonuclease.

Transformation is not limited to pneumococcal polysaccharides. Other transformations include transfer of penicillin resistance, streptomycin resistance, and resistance to other drugs from one strain of pneumococcus to another by means of DNA derived from resistant strains, and transfer of various hydrolytic, oxidative, and synthetic

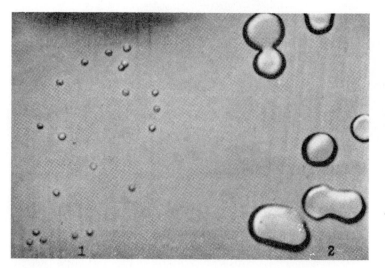

Figure 9–15. *Colonies of Streptococcus pneumoniae. Left, A rough, noncapsulated, type 2 strain. Right, The same strain after cultivation with active, cell-free transforming principle (DNA) from a smooth, capsulated, type 3 strain. These large, mucoid colonies are identical with those usually produced by capsulated type 3 pneumococci. (From O. T. Avery, C. M. MacLeod, and M. McCarty, J. Exper. Med., 79:137–158, 1944.)*

enzymes from one strain of organisms to another.

Transformation has been demonstrated in most instances between strains of the same bacterial species, but it has also been effected between certain different, closely related species. Some strains of a species are more readily transformed than others, and transformation can often be more readily accomplished in a particular medium. These facts indicate that both genetic and environmental factors control the phenomenon. Cells of a transformable strain acquire a physiologic state suitable for transformation when placed within a favorable medium for a relatively brief period; such cells are said to be *competent*. The nature of competence is poorly understood. Inasmuch as the first steps in transformation involve adsorption and penetration of exogenous DNA, it is assumed that competent cells either possess specific, receptive surface components or develop "holes" or "naked" areas in the cell wall, or both. There is evidence that competent cells possess as many as 75 adsorption sites, and to be taken up, the DNA must be of a certain minimum molecular size (M.W. = 100,000).

After penetration, the next steps are association, or "pairing," and integration of the transforming DNA with the corresponding segment of the recipient's DNA; replication of the chromosome with the integrated new information; and formation of a transformed cell population. The genetic change is permanent; transformed cells and their progeny continue to produce DNA having the same properties as that which caused the transformation.

Transduction. Transduction is genetic transfer mediated by bacteriophage. Zinder and Lederberg reported in 1952 the transfer of several characters between strains of *Salmonella.*

Certain strains of bacteria normally contain bacteriophage in a sort of commensal relationship. The bacteriophage lyses only a few cells of its normal host and perhaps a few of some other strains. The bacteriophage is a *temperate* strain, which ordinarily does not replicate unrestrictedly, as does a virulent phage. Instead, its DNA is inserted in that of

the bacterial host and replicates at the same rate as the host DNA. Temperate phage, associated with bacterial DNA and known as *prophage,* spontaneously acquires the ability to replicate vegetatively in about one bacterium in every million, whereupon the host cell lyses. The bacterial strain infected with prophage is described as *lysogenic.* Replication of prophage can be induced in a higher percentage of lysogenic cells by exposure to ultraviolet light, nitrogen mustard, or other agents. Phage particles released by lysis, either spontaneous or induced, can infect other bacteria of an appropriate strain, and they may attach to a suitable site on the chromosomal DNA. One view of the nature of attachment of prophage DNA to bacterial DNA and its subsequent separation is shown in Figure 9–16 (*A* and *B*).

Zinder and Lederberg found that when a lysogenic strain of *Salmonella* was cultivated with another *Salmonella* that possessed certain different properties but was susceptible to the same bacteriophage, some cells of the lysogenic organism acquired one of the characteristics of the second organism. For example, a lysogenic mutant of *S. typhimurium* that required tryptophan was cultivated with another strain that required histidine but not tryptophan. A few prototrophic cells were recovered that required neither tryptophan nor histidine. This is the phenomenon of *transduction.* It is characterized by the ability of a few infecting phage particles to incorporate a small amount of the host DNA and subsequently transfer it to a susceptible cell. The mechanism by which this may occur is illustrated in Figure 9–16 (*C* and *D*). Faulty attachment of phage DNA to bacterial DNA results in the incorporation of a small portion of the latter into the phage and vice versa, so that upon subsequent infection and lysis of another bacterium, DNA from the original host is transferred to the second and becomes incorporated into its genome (Fig. 9–17).

Zinder and Lederberg demonstrated that transduction does not require physical contact of the donor and recipient cells (and thereby differs from *conjugation,* to be described next); they accomplished this by

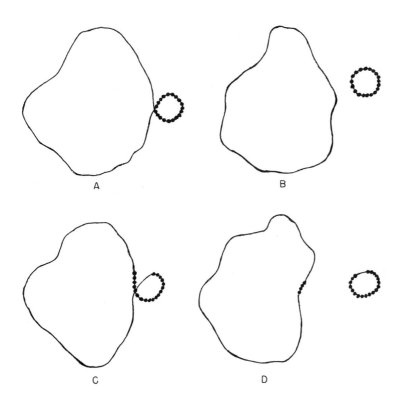

Figure 9–16. A, *Large, circular chromosome of an organism such as* E. coli *with a small loop of prophage DNA attached.* B, *Separation of the phage DNA.* C, *Faulty attachment of phage DNA to bacterial chromosome, so that* (D) *separation leaves a bit of chromosome DNA incorporated in the phage DNA and vice versa. Subsequent infection of another bacterium with this aberrant phage introduces the piece of foreign bacterial DNA into the recipient's chromosome, producing a genetic change. (From Braun: Bacterial Genetics, 2nd ed. Philadelphia, W. B. Saunders Co., 1965.)*

Figure 9–17. *Transduction. Bacteriophage normally present in cells of a lysogenic strain of bacteria (A) lyses a few cells (B). If this organism is cultivated in mixture with a related but not identical strain of bacteria, phage infection of a cell of the second strain may occur (C), and the phage may acquire a small bit of the bacterial DNA (D), which it then carries to another cell of the lysogenic strain (E and F). Reinfection is followed by introduction of the new piece of DNA into the bacterial chromosome and expression of a characteristic derived from the donor strain (G).*

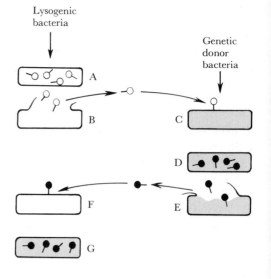

Figure 9–18. *U-tube with fritted glass filter separating the two arms used in a transduction experiment to demonstrate that genetic material is filterable. A lysogenic, tryptophan-requiring auxotrophic mutant of S. typhimurium (22A) was grown in the left arm, and a histidine-requiring auxotrophic mutant (2A) was grown in the right arm. Prototrophs that required neither amino acid were recovered from the left arm. (From Braun: Bacterial Genetics, 2nd ed. Philadelphia, W. B. Saunders Co., 1965.)*

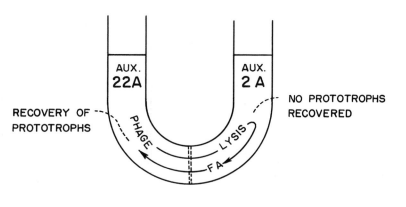

means of a U-tube with the two arms separated by a fritted glass disk through which bacteria cannot pass (Fig. 9–18). The lysogenic auxotrophic mutant of *S. typhimurium* (22A) requiring tryptophan was cultivated in one side of the U-tube, and the histidine-requiring auxotrophic mutant (2A) was cultivated in the other side. A few prototrophs that required neither tryptophan nor histidine were recovered from the first side. Phage from the lysogenic strain traversed the filter, infected strain 2A, picked up the genetic capability for producing tryptophan, lysed 2A, and then recrossed the filter, entering 22A and becoming incorporated into its genome.

Numerous traits have been transduced: ability to ferment various carbohydrates, resistance to antibiotics, and antigenic characteristics. Various well known types of *Salmonella* have been converted into other recognized types as well as into hitherto unknown types.

CONJUGATION AND GENETIC RECOMBINATION

The demonstration of bacterial transformation showed that genetic material could be transferred from one organism to another with subsequent alteration of the genetic characteristics of the recipient. This finding paved the way for an intensive search for indications of a natural sexual type of genetic exchange in bacteria such as exists in higher forms. There was little or no microscopic evidence that sexual fusion occurred in bacteria, so it was presumed that this event was extremely rare if it existed at all.

Lederberg and Tatum in 1944 demonstrated that a sexual type of genetic recombination can occur in *E. coli*. Normal strains of this organism require no added growth factors. Two mutant strains were secured, one of which required biotin (B) and methionine (M), and the other required threonine (T) and leucine (L). Partial genetic formulae for the two organisms may be written as follows: $B^-M^-T^+L^+$ and $B^+M^+T^-L^-$. The symbols B^- and M^- mean that the strain is unable to synthesize biotin and methionine; T^+ and L^+ signify that the organisms can synthesize threonine and leucine.

About 100 million cells of each type were mixed and incubated and then plated on a synthetic medium containing none of these four growth factors. Neither strain could grow alone on this medium, but several score colonies appeared when the mixed culture was plated. These were found to have the genetic composition $B^+M^+T^+L^+$; that is, they possessed the synthetic abilities of both parent strains.

The first demonstration of gene recombination in bacteria suggested that *E. coli* is capable of a sexual type of reproduction; it also showed that only about one cell in a million participates. The possibility of observing this phenomenon directly therefore appeared extremely remote.

Further experiments with other traits confirmed the probability of a sexual mode of reproduction. It was also demonstrated that bacterial genes are linked in a certain order and that recombination does not occur at random. This observaton implied that the genes are arranged upon a chromosome as in higher organisms. It was found further that some cells of *E. coli* behave as genetic donors, others as genetic recipients, and finally that conjugation involving a few donor strains produces an unusually high frequency of genetic recombination.

MECHANISM OF BACTERIAL CONJUGATION

Each genetic donor or male cell possesses a "sex factor" or *conjugon* that distinguishes it from a recipient or female cell. The sex factors are composed of DNA and are probably circular, self-replicating structures like bacterial chromosomes, but only about $\frac{1}{100}$ as large (Fig. 9–19). The sex factors have several functions, including that of providing the conditions necessary for establishing cell-to-cell contact, probably by directing the formation of a specific polysaccharide in the walls of male cells. During conjugation they are usually transferred to the female cells. Independent genetic structures of this description that are additional to the normal chromosome content of the cells they inhabit are also called *episomes*. If an episome is integrated into the chromosome, it replicates with the chromosome; if it is not integrated into the chromosome, it replicates autonomously.

Several sex factors are known. The first reported and most studied is the F (fertility) factor. Cells possessing this episome are called F+, and those lacking it F−. Others are the *resistance transfer factor* (RTF), which transfers resistance to certain antibacterial substances at the same time that it initiates conjugation, and the *colicine factor,* which is responsible for the production of certain polypeptide antibiotics.

In a mixture of F+ and F− cells, random collisions lead to the formation of specific mating pairs as F− cells encounter the polysaccharide attachment sites of F+ cells. Cellular connection, or conjugation, occurs and may (or may not) be followed by transfer of chromosomal material (Fig. 9–20).

Immediately before or simultaneously with this process, the chromosome of the F+ cell replicates, and a strand of DNA remains in the linear (noncircular) state, in which form it can traverse the conjugation bridge and enter the F− cell. DNA replication may provide the energy used in conjugation and transfer of genetic material. Usually only part of the male chromosome enters the female cell, because either the conjugation bridge or the polynucleotide chain breaks before transfer is complete. This can be detected by genetic analysis of the progeny. Under favorable conditions in certain strains of *E. coli,* transfer is completed in about 110 minutes.

The transferred genetic material integrates with that of the recipient, and replication and segregation take place. It should be noted that genetic transfer is not necessarily followed by recombination.

As stated earlier, certain bacterial strains participate in genetic recombination with unusual frequency. When normal F+ cells mate with F− cells, recombination occurs in about one cell in 100,000. Conjugation of strains designated *Hfr* (high frequency of recombination) with F− cells is followed by recombination in approximately one of every 100 cells. The conjugation-promoting powers of Hfr and F+ cells are about the same, but genetic recombination is 1000 times more frequent with the former.

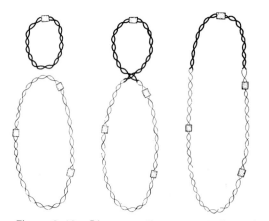

Figure 9–19. *Diagrammatic representations of a bacterial chromosome (light lines) and a sex factor (heavy lines). Integration of the sex factor into the chromosome by breakage and reunion is indicated in the center and right sketches. (From Adelberg et al., Bact. Rev., 29:164, 1965.)*

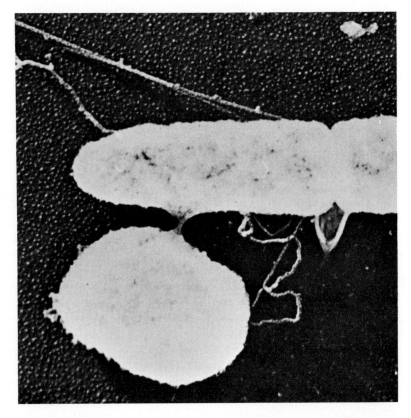

Figure 9–20. *Conjugation of E. coli. The elongated cell in the process of dividing (top) is an Hfr strain of K-12, and the F⁻ recipient (plump cell) is strain C. The long, thin structures attached to K-12 are flagella. The conjugation bridge is clearly shown. (From Anderson et al., Ann. Inst. Pasteur, 93:450, 1957.)*

The progeny obtained in F⁺ × F⁻ matings are always F⁺ and, in fact, when F⁺ cells are grown with F⁻ cells, most or all of the latter become F⁺ within an hour, owing to the rapid replication and spread of the F factor. The progeny of Hfr × F⁻ matings are nearly all F⁻, only a few Hfr cells being produced.

These facts are explained by differences in the location and behavior of F within F⁺ and Hfr cells (Fig. 9–21). The F factor is autonomous in F⁺ cells; that is, it is not normally associated with the chromosome and replicates independently of it. At conjugation it is often transferred alone, thus converting

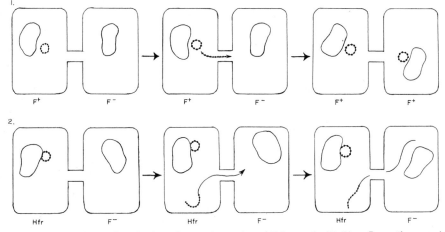

Figure 9–21. *Diagram illustrating the location and transfer of F factor in (1) F⁺ × F⁻ matings and (2) Hfr × F⁻ matings. Chromosomal DNA is indicated by plain lines, F by dotted lines, and replicated structures in the process of transfer by lines with arrowheads. Both chromosomes and F factors are circular except when being transferred. In F⁺ cells, F is ordinarily autonomous, whereas in Hfr cells it is integrated with the chromosome and is transferred last (if at all) in conjugation. (From Braun: Bacterial Genetics, 2nd ed., Philadelphia, W. B. Saunders Co., 1965.)*

an F⁻ cell to F⁺. F may, however, become associated with the donor chromosome, and in such a circumstance genetic material is also transferred to the F⁻ cell, and a low frequency (e.g., 1 in 100,000) of recombination occurs. The association between F and chromosome material involves no particular site, so that any trait (marker) can be transferred.

In Hfr cells, F is regularly associated with specific chromosome sites, which differ from one Hfr strain to another. At conjugation, the circular chromosome breaks at the site where F is located, and the linear DNA strand transfers to the F⁻ cell with the F factor at the tail end. Since complete transfer rarely occurs, F does not ordinarily enter the recipient, which therefore remains F⁻; only in the occasional instance of complete transfer does the F⁻ cell become Hfr.

The fact that F is associated with different chromosome sites in different strains of bacteria can be shown by interrupted mating experiments. Conjugation can be terminated and DNA transfer stopped at any time by violently agitating the culture in a Waring

Blendor to break the conjugation bridges. Genetic analysis of the recombinants then indicates what traits (markers) have transferred from the male cells to the female cells. The longer conjugation continues, the more traits are transferred. With a given Hfr strain, a certain marker always transfers first, followed by other markers in a particular order (e.g., A, B, C, D, E, . . .). When a different Hfr strain of the same species is used, another marker transfers first, and this is followed by markers in the same sequence (e.g., C, D, E, F, G, . . .). In some cases, markers transfer in the reverse sequence. These situations are illustrated in the circular genetic linkage map of *E. coli,* strain K-12 (Fig. 9–22). When Hfr substrain C is used in a conjugation experiment, marker *T6* is transferred first, followed by *ade, lac, pro,* etc., and finally *gal.* With substrain H, however, transfer begins with *fim, R_{try}, val-r,* etc., and ends with *pyr-1.*

Genetic recombination has been studied most intensively in *E. coli.* The phenomenon is not restricted to this organism, but the extent of its occurrence in other organisms is not yet known.

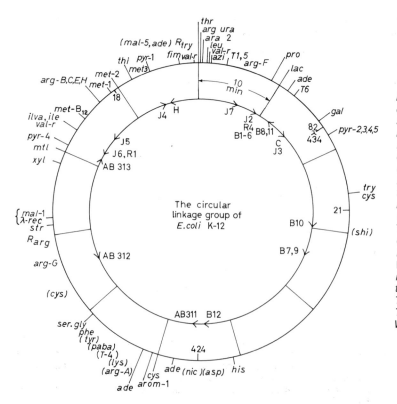

Figure 9–22. *Circular genetic linkage map of E. coli K-12. The double circle is divided into 11 sections, each representing a 10 minute transfer interval as determined from interrupted mating experiments. Around the outer circle are located various genetic markers: e.g., gal = galactose fermentation; str = streptomycin resistance. Prophage loci are indicated inside the outer circle. The various Hfr substrains of E. coli K-12 are shown inside the inner circle at the arrowhead (J4, H, J7, etc.), which mark the leading end and direction of the transfer of markers. (From Hayes: The Genetics of Bacteria and Their Viruses. New York, John Wiley & Sons, Inc., 1964.)*

THE RECOMBINANT DNA PROBLEM

Since 1973 techniques have been available for deliberately altering the genetic character of almost any form of life by transferring genes from cells of one organism to those of another. The basic facts that make this possible are the chemical and structural identity of DNA from all known forms and the universality of the genetic code.

The procedure is comparatively simple. DNA is isolated from cells of the donor species and is then cut into fragments by one of several possible enzymatic or physical methods. The fragments usually contain between 1 and 10 genes. The ends of the fragment chains are or can be made "sticky" by use of suitable enzymes so that they can recombine with other DNA chains or fragments, either of the same species or of any other species. The DNA to which they join need not be nuclear in origin—it may be derived from extranuclear bodies known as episomes or plasmids, or it may be viral DNA. Bacterial viruses were extensively used in early research because of the ease with which they are taken up by host bacterial cells. Plant or animal cells cultivated *in vitro* may also serve as hosts for plasmids or appropriate viruses and can be employed. In any case, the recombined DNA replicates within the host cell, either in phase with that of the host or independently. The new genetic material occasionally fuses with and permanently alters a host chromosome. The host cell becomes a chimera, since its genetic properties are those of two diverse species. The donor and host may be completely unrelated, even of different kingdoms, so it is possible to create an organism like no other in existence today.

Because of the ease with which bacteria can be cultivated and their speed of multiplication, they are the usual hosts into which recombinant DNA is introduced. A well-known strain of *E. coli,* K12, is most frequently employed. When the host cell multiplies, the recombinant DNA it contains also replicates, and it is therefore possible to prepare very large amounts in a short time. Moreover, since pure cultures of bacteria can readily be obtained by familiar cloning procedures, the recombinant DNA they contain can be purified more easily and completely than by any chemical or physical method.

The introduced genes may code for something beneficial to mankind, such as scarce hormones, enzymes that decompose environmental pollutants like petroleum, or enzymes that synthesize edible protein or some other nutrient substance. On the other hand, they may code for resistance to antibiotics, or for a carcinogen or a toxin or other agent that renders the host bacterium pathogenic. Accidental escape of such a bacterial host might cause a disaster if the organism finds a suitable niche (e.g., the human body) and multiplies and spreads unrestrainedly.

Realization of this possibility in 1974 caused concerned scientists to call a moratorium on certain types of recombinant DNA research until an international conference could be convened to discuss the problem. The Asilomar Conference, held in Pacific Grove, California, in February, 1975, agreed that some especially hazardous experiments should not be done at present, but that various others might be performed with suitable safeguards. In June, 1976, the director of the National Institutes of Health released guidelines for research on recombinant DNA molecules. These included reduction or elimination of hazards to laboratory workers, society, and the environment by both physical and biologic methods.

The minimal or P1 level of physical containment consists of standard microbiologic techniques and practices; no special equipment is necessary. P2 (low), P3 (moderate), and P4 (high) levels of physical containment are provided by increasingly rigid procedures and design features in the equipment and laboratory structures. A P4 facility is specially engineered to prevent escape of microorganisms to the environment.

Biologic containment is illustrated by the use of host organisms which, either naturally or through mutation, do not reproduce in the environment or under the conditions that prevail in the experiments.

SUPPLEMENTARY READING

Adelberg, E. A.: *Papers on Bacterial Genetics,* 2nd ed. Boston, Little, Brown and Company, 1965.
Braun, W.: *Bacterial Genetics,* 2nd ed. Philadelphia, W. B. Saunders Co., 1965.
Davis, B. D., et al.: *Principles of Bacteriology and Immunology.* New York, Harper and Row, Publishers, 1968.
Hayes, W.: *The Genetics of Bacteria and Their Viruses.* New York, John Wiley & Sons, Inc., 1964.
Stahl, F. W.: *The Mechanics of Inheritance.* Englewood Cliffs, N.J., Prentice-Hall, Inc., 1969.
Stanier, R. Y., Adelberg, E. A., and Ingraham, J.: *The Microbial World,* 4th ed. Englewood Cliffs, N. J., Prentice-Hall, Inc., 1976.
Stent, G. S.: *Molecular Genetics; An Introductory Narrative.* San Francisco, W. H. Freeman and Company, 1971.
Stent, G. S.: *Papers on Bacterial Viruses,* 2nd ed. Boston, Little, Brown and Company, 1965.
Strauss, B. S.: *An Outline of Chemical Genetics.* Philadelphia, W. B. Saunders Co., 1960.
Watson, J. D.: *Molecular Biology of the Gene,* 2nd ed. New York, W. A. Benjamin, Inc., 1970.

GROWTH AND DEATH 10
OF BACTERIA

METHODS OF STUDYING GROWTH AND DEATH

CELL GROWTH

Growth of individual cells can be observed by microscopic methods. As described in Chapter 5 (page 113), the chromatin body divides and the cell increases to several times its original size. Crosswalls then form, and the daughter cells become physiologically independent, often separating physically. The process of growth may be followed in unstained microcultures, but measurement of cell size is easier in stained preparations. However, drying and fixing shrink the cells in a smear, and the measurements obtained are only relative.

POPULATION CHANGES

Bacterial growth is more frequently studied by counting the organisms in a culture at intervals than by tediously determining the average cell size. Casual inspection of a broth culture every few hours after inoculation reveals gradually increasing turbidity, which roughly parallels the increase in population. Turbidity estimations are more precise when made with a photoelectric turbidimeter. The bacterial mass can be determined by removing the cells from the culture (by centrifugation) and weighing them; for some purposes quantitative analysis of a characteristic chemical constituent such as nitrogen suffices. These methods indicate the total amount of bacterial growth but do not necessarily reveal the number of cells because of the variation in average cell size during the growth cycle.

Direct (Microscopic) Count of Bacteria. Breed Method. Microscopic examination of stained smears can also be used to estimate bacterial populations. A known volume of culture or suspension is smeared uniformly over a definite area of a slide; this is stained and examined with a microscope that has been calibrated so that the diameter of the oil immersion field is known. The bacteria in several fields are counted, and the number per milliliter of the original suspension is calculated as follows:

$$\frac{\text{bact./ml. of}}{\text{suspension}} = \frac{\text{avg. no. of}}{\text{bact./field}}$$
$$\times \frac{1 \text{ ml.}}{\text{vol. of susp.}} \times \frac{\text{area of smear}}{\text{area of field}}$$

In the Breed method of counting bacteria in milk, 0.01 ml. of the sample is smeared over a 1 sq. cm. area of a slide; it is then stained and the individual cells or clumps are counted. If the organisms in an entire oil immersion field, whose diameter is 160 μm., are enumerated, the equation becomes:

$$\frac{\text{bact./ml. of}}{\text{suspension}} = \frac{\text{avg. no. of}}{\text{bact./field}}$$
$$\times \frac{1}{0.01} \times \frac{100,000,000 \text{ sq. } \mu\text{m.}}{20,100 \text{ sq. } \mu\text{m.}}$$
$$= \text{avg. no. of bact./field} \times 497,500$$

In other words, every cell seen in the microscope represents approximately 500,000 per milliliter in the suspension. This method is fairly accurate with large populations; with small populations, however, the experimental error is great unless many fields are examined.

Counting Chamber Method. Various microscope slides are available with chambers designed to contain a cell suspension above an accurately ruled area etched into the glass. The Petroff-Hausser chamber is so constructed that the depth of the suspension is 0.02 mm., each ruled square being 0.05 mm. on a side. After the chamber is filled with suspension and covered with a special, flat coverglass, the average number of bacteria in the ruled squares is determined. The calculation of the cells per milliliter is made by the following equation:

$$\frac{\text{bact./ml. of}}{\text{suspension}} = \frac{\text{avg. no. of}}{\text{bact./square}} \times \frac{1 \text{ mm.}}{0.02 \text{ mm.}}$$

$$\times \frac{1 \text{ mm.}}{0.05 \times 0.05 \text{ mm.}} \times 1000 \text{ cu. mm.}$$

$$= \text{avg. no. of bact./square} \times 20{,}000{,}000$$

This method is particularly useful with suspensions containing large numbers of cells. It is similar to the method used in counting blood cells.

Cultural Counts of Bacteria. All the methods so far mentioned indicate the *total* population, including dead as well as living cells. This information is sometimes desired, but usually the living population is of principal interest. Living cells can be counted or estimated only by a method that depends upon a detectable vital activity, such as cell multiplication, lactose fermentation, or cellulose digestion. Two culture procedures are in common use, the *dilution count* and the *plate count*. The basic assumption underlying both methods is that every viable (living) cell inoculated into fresh medium will multiply and produce easily detected evidence of growth: turbidity, acid or gas in broth, colonies on agar. This assumption is not always justified, particularly with very small numbers of bacteria, and constitutes a source of inaccuracy recognized by practicing bacteriologists. However, the results obtained by repeated examinations of a given specimen are reasonably reproducible, so that the methods have the virtue of consistency if not of strict accuracy.

Dilution Count. The dilution count provides an estimate of the number of bacteria in a population that are capable of multiplying in a liquid medium. Any medium may be used that will support the growth of the or-

ganisms of interest. To illustrate the method, assume a suspension containing 100 bacteria per milliliter. Serial dilutions (1:10, 1:100, 1:1000 . . .) are prepared by transferring 1 ml. into 9 ml. of sterile water and then mixing this dilution and transferring 1 ml. into another 9 ml. portion of sterile water, etc. (see Figure 10–1). The respective dilutions theoretically contain 10, 1, 0.1 . . . bacteria per milliliter. One milliliter portions of each dilution are then inoculated into the desired broth medium and these cultures are incubated (Fig. 10–2).

The anticipated results in the example just described are also indicated in Figure 10–1. Growth should occur in tubes inoculated with the undiluted specimen and with the 1:10 and 1:100 dilutions. Growth is not expected in broth inoculated with the 1:1000 dilution; that is, in nine trials out of 10 there would be no organism in a 1 ml. portion. The same results might be obtained if the specimen contained 500 or even 900 bacteria per milliliter, but there is obviously a greater probability that broth inoculated with the 1:1000 dilution will contain a viable organism and hence become turbid. It is usually *assumed,* however, that the highest dilution that yields growth in subculture contains one bacterium per milliliter; the minimum number of organisms in the specimen is therefore indicated by the reciprocal of this dilution.

The dilution count becomes more accurate if several tubes of broth are inoculated with 1 ml. aliquots of each dilution. Some aliquots of the critical dilution(s) then contain a viable organism and others do not. Tables have been calculated from which the most probable number of bacteria in the specimen can be read. One of these is Table 10–1, which shows the most probable number (MPN) of bacteria per milliliter of a specimen (culture, water or milk sample, etc.) as calculated from the observation of growth in duplicate broth cultures inoculated with 1 ml. aliquots of decimal dilutions of the sample. The figures representing positive results (i.e., growth) in the three successive dilutions may be termed the *significant number.* For example, if positive results are obtained from two tubes inoculated with an undiluted sample, one tube with a 1:10 dilution, and none with a

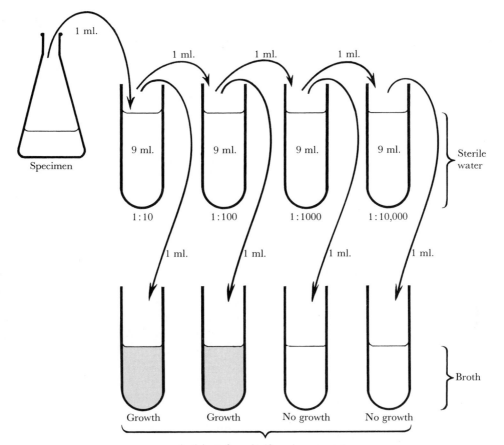

Figure 10–1. *Procedure for dilution count of a suspension containing 100 bacteria per milliliter with anticipated results. Arrows indicate transfer of 1 ml. of specimen or dilution.*

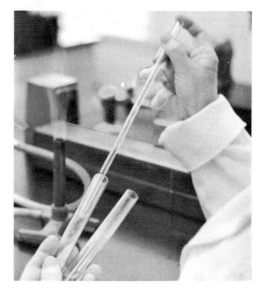

Figure 10–2. *Transferring bacterial specimen by pipette. The culture tubes are held as usual. Plugs are also held as usual by the little and ring fingers of the other hand. This hand also manipulates the pipette, which was taken sterile from the long can in the background. Note that the index finger is used to control the flow of liquid in the pipette.*

TABLE 10-1. Most Probable Number (MPN) of Bacteria per Milliliter of a Sample as Calculated from Observations of Broth Cultures Inoculated with 1 ml. of Three Decimal Dilutions*

No. of Positive Tubes from Each Dilution†			MPN of Bacteria per ml.
10^0	10^{-1}	10^{-2}	
2	2	2	100
2	2	1	70
2	2	0	24
2	1	2	21
2	1	1	13
2	1	0	6.2
2	0	2	9.5
2	0	1	5.0
2	0	0	2.3
1	2	2	3.7
1	2	1	2.9
1	2	0	2.1
1	1	2	2.8
1	1	1	2.0
1	1	0	1.3
1	0	2	1.9
1	0	1	1.2
1	0	0	0.6
0	2	2	1.9
0	2	1	1.4
0	2	0	0.94
0	1	2	1.4
0	1	1	0.92
0	1	0	0.46
0	0	2	0.90
0	0	1	0.45
0	0	0	0

*Two tubes were inoculated with each dilution.

†Exponential notation is commonly used to designate decimal dilutions: 10^0 = undiluted; 10^{-1} = 1:10; 10^{-2} = 1:100.

1:100 dilution, the significant number is 210. According to the table, this indicates that the sample most probably contains 6.2 bacteria per milliliter. The table can be used with any set of successive decimal dilutions by multiplying the indicated *most probable number* (in Table 10-1) by the reciprocal of the dilution corresponding to the first figure of the significant number. In the examples in Table 10-2 the significant number is indicated in boldface (heavy) type.

This method of counting viable populations is not very accurate, but it affords a means of estimating certain physiologic groups of organisms (e.g., indole producers and lactose fermenters) for which other procedures are not available or are cumbersome. Its precision is increased by inoculating more tubes with each successive dilution, five or 10, for instance.

Plate Count. The plate count is based upon the assumption that each bacterium trapped in or on a nutrient agar medium will multiply and produce a visible colony (Fig. 10-3). The number of colonies should therefore be the same as the number of viable bacteria inoculated into the agar. One milliliter of an undiluted specimen containing 100 bacteria per milliliter is expected to produce 100 colonies, a number that can be counted without difficulty.

A specimen containing 2,500,000 bacteria per milliliter would yield a plate far too crowded to count; moreover, metabolic products from the multiplying bacteria would inhibit the development of neighboring colonies. Such a specimen should be diluted, and 1 ml. of a 10^{-4} dilution (i.e., 1:10,000) should be plated. The number of colonies obtained (250 are expected in this case) has to be multiplied by 10^4, the reciprocal of the dilution, to find the number of bacteria per milliliter of the original specimen (Fig. 10-4).

Duplicate or triplicate plates are usually prepared from each of several dilutions in the anticipated critical range. Plates from the dilution that yields between 30 and 300 colonies can be counted easily and accurately. The average, multiplied by the reciprocal of the dilution, gives a result expressed as "bacteria per milliliter." Replicate plates show some discrepancies attributed to uneven distribution of bacteria, pipetting and counting errors, and other factors. The normal error in the plate count may be as great as 10 per cent.

The plate procedure provides a *minimum* count of viable bacteria. Some organisms normally occur in clusters that do not break up in the course of preparing dilutions, and hence each cluster produces a single colony. Moreover, individual bacteria may become trapped in the agar so close together

TABLE 10–2. Examples of MPN Determination by Use of Serial Decimal Dilutions

Sample	Positive Dilutions						MPN per ml.
	10^{-1}	10^{-2}	10^{-3}	10^{-4}	10^{-5}	10^{-6}	
A	2	1	0	0	0	0	62
B	2	2	2	0	0	0	2,400
C	2	2	2	1	1	0	13,000

that their colonies merge and are counted as one.

The dilution and plate count methods of estimating populations are widely used whenever the number of living bacteria is desired. It should be emphasized that no cultural procedure can be relied upon to provide an accurate estimate of the total number of organisms in a physiologically heterogeneous population, because no one culture medium or set of incubation conditions will permit the growth of all kinds of bacteria. Nutritionally fastidious organisms require special ingredients in the medium; certain bacteria are inhibited by substances that favor others; and strict aerobes will not grow under anaerobic conditions nor anaerobes under aerobic conditions. Each group must be counted separately.

Turbidimetry. For some purposes, the number of bacteria can be determined with sufficient accuracy, and certainly with greatest ease by estimating the degree of turbidity of the suspension. Optical nephelometers have been used for this purpose for many years; photoelectric turbidimeters provide greater accuracy, but any type of spectrophotometer can be used with equally great accuracy. There is an inverse relationship between the turbidity or optical density of a bacterial suspension and the amount of light that passes through it; the light transmitted can be measured accurately by a suitable photoelectric cell.

In practice, a spectrophotometer is usually employed and light transmission (T) or optical density (OD) is read on the galvanometer scale (OD = −log T). The value of T varies with the length of the light path through the suspension, with the solvent, the wavelength

Figure 10–3. *Pouring agar into a Petri dish. The mouth of the bottle or test tube of agar is flamed and inserted beneath the edge of the lid, which has been raised only enough to admit the agar container. The required medium is poured and the lid is carefully replaced.*

Figure 10–4. *Quebec colony counter. The magnifier and a darkfield type of illumination increase the accuracy of counting.*

of the light, and the nature and physical state of the organism, as well as with the number of bacteria, so it is necessary to eliminate or standardize all variables except the last. Matched test tubes or cuvettes are employed, and the instrument is adjusted to 100 per cent light transmittance through a blank consisting of the bacteria-free solvent, using a wavelength that will be maximally absorbed by the bacteria. This is determined by trial; it is usually between 450 and 650 nm. The bacterial suspension is then tested under identical conditions and its OD is determined (see Figure 10–5).

Turbidimetry is most effective with suspensions of moderate density. It indicates the total bacterial cells—dead as well as living. Results can be expressed in arbitrary units, or OD values can be standardized with a bacterial suspension of known cell content.

BACTERIAL POPULATION CURVES

Bacterial population curves are determined by inoculating a small number of organisms into a culture medium and counting the bacteria in aliquot samples at intervals thereafter (e.g., every hour). Counts made by microscopic examination represent the total number of cells, as shown in curve A of Figure 10–6. Dilution or plate counts indicate only the population of living cells, as shown in curve B. Logarithms are used to plot the numbers of organisms in order to present clearly the population changes when counts are low and yet show the maximum counts.

It will be noted that after a short latent period the population increases rapidly toward a maximum and then levels off.

Eventually the living population decreases

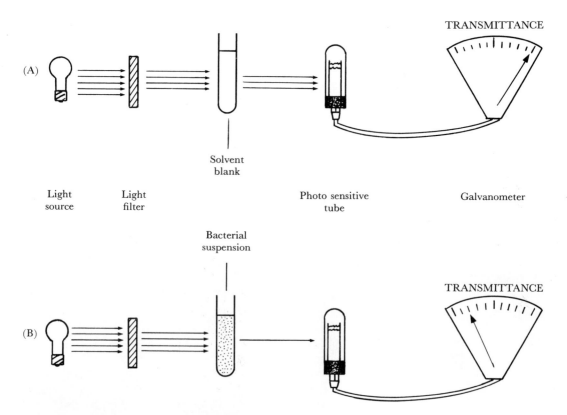

Figure 10–5. *Spectrophotometric determination of optical density* (OD). (A) *The instrument is adjusted so that the galvanometer indicates complete light transmittance (1.0 or 100 per cent) with a blank tube containing solvent alone.* (B) *A bacterial suspension is substituted for the blank and the galvanometer is read. As illustrated, a value of 0.2 is obtained, from which the* OD *is calculated as follows:*

$$T = 0.2$$
$$OD = -\log 0.2 = 0.7$$

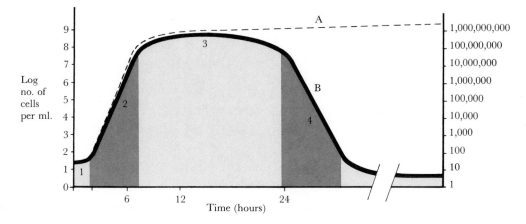

Figure 10–6. *Idealized population curves obtained by counting the bacteria in a culture at intervals after inoculation.* Curve A, *Total (living plus dead) cells;* curve B, *living cells. 1, Phase of cell enlargement; 2, phase of exponential multiplication; 3, maximum stationary level; 4, phase of exponential death.*

toward zero (curve B). The graph (A) of total population shows a slow but gradually diminishing rise after the period of most rapid cell division, which indicates that limited multiplication continues even though most cells are dying. Ultimately multiplication ceases and the total count becomes stationary.

These curves are not peculiar to bacteria. Curves of the same general form are characteristic of populations of any sort confined in a closed system and subject to starvation or to the harmful effect of toxic waste products, such as carbon dioxide or acids, as the population increases. A similar curve traces the rise and fall of ancient human populations in which geographic crowding, loss of economic productivity, disease, and war played their customary roles.

Curve B describes the normal viable microbial population changes in natural products, such as milk, cheese, sauerkraut, and silage, and in all kinds of decomposing animal and plant matter. Inasmuch as a "wild" or natural population usually consists of many species in intimate association with one another, the shape of the curve for each species is skewed one way or the other as the release of nutrients and other growth-promoting substances accelerates multiplication or as predation, antibiosis, and competition increase the death rate.

The term *population curve* better de-scribes the over-all changes depicted than the older expression "growth curve." The curve is commonly divided into four principal sections or phases for further discussion.

PHASE OF CELL ENLARGEMENT

The first portion of the population curve in a bacterial culture is the phase of cell enlargement, sometimes known as the "lag phase." The latter name is somewhat misleading because it implies inactivity or dormancy, which is contrary to the actual situation. It is true that the number of individual bacterial cells does not increase for a period of time that varies with the organism and the conditions, but protoplasm synthesis and cell growth begin almost immediately upon transfer to a fresh medium. As long ago as 1923, Henrici reported that the average size of bacterial cells increases markedly during the "lag phase," just before the period of active division (Fig. 10–7). Moreover, electron micrographs show that growth precedes cell division (page 113). The increase in size may amount to three or four times the initial cell volume, principally as a result of elongation.

Duration. The duration of the period of cell enlargement varies with conditions and with the species. A culture inoculated with cells that are already actively multiplying

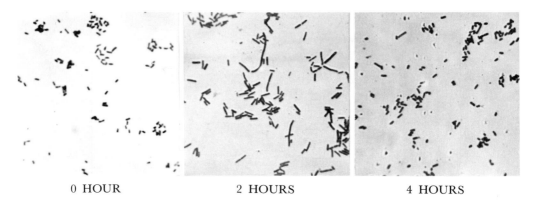

0 HOUR   2 HOURS   4 HOURS

Figure 10–7. *Cell enlargement during the period preceding the phase of logarithmic multiplication. Escherichia coli* *on nutrient agar at 37° C., observed at the time of inoculation (0 hour) and two and four hours later.*

displays little lag in population increase and hence little if any unusual cell enlargement. The phase of enlargement is prolonged, however, in a culture inoculated with "dormant" cells from an older culture. Bacteria that are capable of rapid multiplication under favorable conditions (e.g., *E. coli* or *P. vulgaris*) start to multiply within 30 minutes to two or three hours. Slowly growing organisms, such as *Mycobacterium tuberculosis,* do not start to divide until after a much longer interval.

Physiologic Activity. The "lag phase" was described as a period of intense physiologic activity before it was realized that protoplasmic growth begins almost at once. Measurements of oxygen utilization, carbon dioxide evolution, or deamination, when compared with the number of living cells, indicated great activity per cell. The same measurements compared with the total mass of protoplasm reveal that metabolic activity per unit of protoplasm is constant throughout this period.

The transition period between cell enlargement and active division is a critical interval. At this time the cells are unusually sensitive to unfavorable environmental conditions such as extremes of temperature (i.e., at either end of the growth range), high osmotic pressure, or disinfectant chemicals. Cells in the very early phase of enlargement or in the stationary periods of the population curve are more resistant.

The factors that prevent cell division during the phase of enlargement are unknown. It has been suggested that traces of toxic minerals inhibit division or that division requires certain proteins or other nitrogenous substances that are synthesized only slowly. Whatever the cause, cell division begins slowly and then rapidly accelerates during a period that varies from only a few minutes to an hour. The culture then enters the phase of exponential multiplication.

PHASE OF EXPONENTIAL MULTIPLICATION

Cell division occurs rapidly at a constant rate during this phase; that is, each successive division requires the same time as those preceding. A single cell of *E. coli,* for example, divides into two cells within 15 minutes. These cells divide and yield four cells in the next 15 minutes, and so forth. Each individual cell does not divide at exactly the same instant; some divide a little more rapidly, others a little more slowly, but the average division time is 15 minutes. Bacterial cell divisions in a culture may often be synchronized by alternately chilling and incubating the culture.

The population of a normal culture doubles in each consecutive time interval, so the number of bacteria increases in an exponential or logarithmic manner. This is illustrated in Table 10–3, which shows the number of

TABLE 10–3.   The Number of Cells in a Multiplying Bacterial Population as a Function of the Number Initially Present and the Generations Elapsed*

Generation	Number of Cells after Each Generation When Starting with:		
	1 Cell	5 Cells	a Cells
0	$1 = 2^0$	$5 = 5 \times 2^0$	$a = a \times 2^0$
1	$2 = 2^1$	$10 = 5 \times 2^1$	$2a = a \times 2^1$
2	$4 = 2^2$	$20 = 5 \times 2^2$	$4a = a \times 2^2$
3	$8 = 2^3$	$40 = 5 \times 2^3$	$8a = a \times 2^3$
4	$16 = 2^4$	$80 = 5 \times 2^4$	$16a = a \times 2^4$
5	$32 = 2^5$	$160 = 5 \times 2^5$	$32a = a \times 2^5$
n	2^n	5×2^n	$a \times 2^n$

* Numbers of cells expressed both arithmetically and exponentially.

cells after each generation when observations are made starting with one cell, five cells, or any given number of cells (*a*).

Bacterial Generations.   A *generation* is defined as a doubling of population. It is evident that the population after *n* generations is 2^n when the progeny of a single cell are enumerated. Five bacteria have 5×2^n progeny after *n* generations. The relationships are generalized in the expression $a \times 2^n$ for a case in which *a* bacteria undergo *n* cell divisions. It will be noted that the exponent is the same as the number of generations elapsed. The numeral 2 appears in the exponential expressions because each cell divides into two cells.

Generation Time.   The rate of cell division can be expressed in various ways. It is most commonly stated as the generation time, the interval required for one generation. The confusing feature of this method of expression is that small figures indicate rapid multiplication and large figures indicate slow multiplication. An equation for calculating the generation time can be developed from the general expression worked out in the preceding paragraph as follows:

Let *a* = number of bacteria at start of observation

b = number of bacteria after *n* generations

n = number of generations

g = time (minutes) for each generation

t = time (minutes) for *n* generations

Then
$$b = a \times 2^n$$
$$\log b = \log a + n \log 2$$
$$n = \frac{\log b - \log a}{\log 2}$$

But
$$n = \frac{t}{g}$$

So
$$\frac{t}{g} = \frac{\log b - \log a}{\log 2}$$

And
$$g = \frac{t \log 2}{\log b - \log a}$$

This relationship is used to calculate the average generation time from two bacterial counts during the phase of exponential multiplication. Generation time (*g*) can also be calculated from data obtained by spectrophotometric determination of *OD* values according to the equation:

$$g = \frac{t \log 2}{\log OD_b - \log OD_a}$$

in which OD_a and OD_b are optical densities at the beginning and end, respectively, of time interval *t*.

Factors Affecting the Rate of Cell Division.   Growth starts with *initiation* of DNA replication, and replication requires about 50 minutes. *Transcription* then yields mRNA, and *translation* of the mRNA results in protein synthesis; cell division may follow about 40 minutes after DNA replication. Meanwhile, DNA replication continues, often faster than the process of cell division, so the number of chromosomes may be greater than the number of cells for a time. At some point, energy becomes limiting and the organisms begin to use their endogenous reserves and even their rRNA as sources of energy. This destroys the ribosomes, whereupon protein synthesis ceases. The cells are then irreversibly committed to death. Other factors such as accumulation of toxic metabolic products also accelerate death of the culture.

The multiplication rate of bacteria varies from one species to another; i.e., it is genetically determined to some extent. Bacteria such as *E. coli* have a minimum generation time of 13 to 17 minutes in rich media. *M. tuberculosis,* on the other hand, divides no faster than once in 18 hours. It is not known why one species undergoes cell division at a different rate from another.

The rate of cell division during the exponential phase parallels the rate of synthesis of protoplasm; in other words, it is correlated with the chemical activity of the organisms. Any factor that affects the numerous chemical processes concerned in synthesis can therefore be expected to influence the rate of cell division.

Classical chemistry teaches that the rates of chemical reactions increase as the temperature is elevated. Bacterial multiplication also increases as the temperature is raised, but within the limits imposed by thermal inactivation or denaturation of the cell constituents (enzymes and other proteins) responsible for synthesis. The upper limit for many commonly encountered bacteria is 40° to 50° C.

The concentration of nutrients in the culture medium sometimes determines the rate of multiplication; it should be pointed out, however, that bacteria can grow in very dilute media. The nutrient broth used for routine cultivation of bacteria contains a great excess of nutrient materials. One milliliter of a 1 per cent peptone solution can support a population of 500,000,000 cells. These organisms weigh about 1 mg., of which 80 per cent is water. Only 2 per cent of the peptone therefore actually enters into the composition of the bacteria. The concentration of peptone can be varied within rather wide limits without affecting the rate of bacterial multiplication appreciably.

Growth and multiplication rates are increased by the addition of readily utilized sources of energy such as fermentable carbohydrates, growth factors normally present in minimal concentrations, and substances that an organism can synthesize slowly. The rate of multiplication is reduced by decreasing the concentration of nutrients, particularly an essential nutrient.

The duration of the exponential phase varies with conditions. It may be only 10 to 15 generations. The rate of cell division gradually diminishes, and increasing numbers of cells die.

Continuous Exponential Multiplication of Bacteria.   Bacteria transferred at intervals of a few hours to fresh medium remain in a state of logarithmic multiplication. A continuous flow apparatus called a *chemostat* has been devised to accomplish the same result, and by its use a culture can be maintained at a constant population and constant multiplication rate. The culture vessel is equipped with a nutrient inlet, air inlet, and overflow tube (Fig. 10–8). Each drop of fresh medium added displaces a drop of culture, so that the total population remains constant.

The medium used is adequate in all respects for the organism to be cultivated except that one essential nutrient (e.g., an amino acid) is provided in limited amounts. The rate of growth therefore depends on this ingredient and can be increased or decreased by regulating the flow of nutrient solution.

The chemostat is used in physiologic and genetic studies. The dependence of a mutant strain of *E. coli* on the amino acid tryptophan is shown in Figure 10–9; the generation time decreased sharply as the tryptophan level in the culture was increased to 1 μg. per liter and more gradually at greater concentrations.

MAXIMUM STATIONARY LEVEL

The maximum stationary level is a constant high population maintained by a balance between cell division and cell death; the net viable population does not change appreciably. Total (living plus dead) counts during

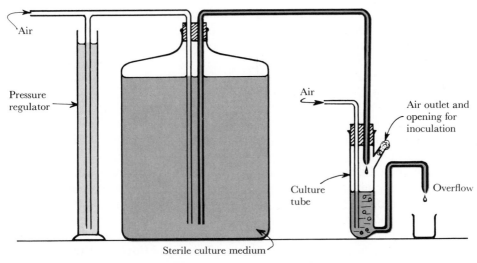

Figure 10–8. *Simplified form of a chemostat.*

this interval rise slowly and can be used to calculate the rate of death.

Two factors that limit bacterial growth have already been mentioned: nutrient supply and toxic wastes. At one time, in accordance with a hypothesis of Bail, it was believed that each individual organism required a certain amount of space. In dilute liquid media, the supply of nutrients or of any single essential nutrient may be limiting, whereas in more concentrated media this is not the case. For any species growing under constant conditions, a more or less uniform maximum population per unit volume is usually attained. If the medium is made free from bacteria by centrifugation and is then reinoculated, further growth occurs. Obviously, the medium

still contains nutrients despite the fact that it has already supported maximal growth; moreover, the amount of wastes present is below the lethal level. This observation led Bail to propose that for each organism there is a maximum attainable population or *M concentration.* It is implied that each cell requires a certain *biologic space,* and as the available space becomes occupied, growth decreases and eventually stops.

This hypothesis did not really explain anything, but ultimately led to a suggestion regarding the nature of biologic space. It has long been known that, although bacteria ordinarily grow only slowly if at all in very dilute media, considerable growth takes place if the same media are placed in small con-

Figure 10–9. *Effect of tryptophan concentration on the growth rate of a strain of* Escherichia coli *in a chemostat. (Replotted from data of A. Novick and L. Szilard: In E. J. Boell (Ed.): Dynamics of Growth Processes, Princeton, N. J., Princeton University Press, 1954.)*

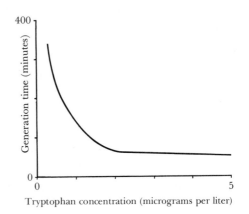

tainers or if glass beads, sand, or other solids are added. This phenomenon has been attributed to the adsorption of nutrients to inert surfaces and the development of local regions of increased concentration in which multiplication can occur. Evidently a certain minimal concentration of nutrient *per cell* must be available before sufficient nutrients can enter a cell to provide energy and building material for new cells. According to this concept, in a culture at the M concentration, the nutrient supply per cell may also be limiting. The removal of most of the bacteria increases the available nutrients per cell; consequently further multiplication can take place. It should be noted, however, that since the second crop is not as great as the first, other factors are also important. Among these is undoubtedly the nature and concentration of metabolic waste products.

It is readily demonstrated that the duration of the stationary level varies with the kind of waste products present, which, in turn, depends upon the composition of the culture medium. Fermentable carbohydrates, for example, are usually dissimilated to acids, which lower the pH and shorten the maximum stationary level; early death ensues. The incubation temperature also affects the height and duration of the maximum stationary level, particularly when very toxic wastes are formed. It will be shown later that chemical disinfection is promoted at elevated temperatures (page 253); the toxic substances that accumulate at high temperatures therefore quickly kill the bacteria that produced them, and the maximum population attained is lower.

The continued accumulation of wastes and the development of unfavorable conditions is followed by the complete cessation of multiplication. The culture then gradually enters the death phase.

DEATH PHASE

The death of cells in a culture often follows an exponential curve, half the surviving cells dying in each successive equal time interval. For example, a population decreases from 100,000,000 to 50,000,000 in the first hour, to 25,000,000 in the next hour, to 12,500,000 in the third hour, and so forth. Presumably the culture will be sterile when "less than one" survives.

Factors Affecting the Rate of Death. Some species are notorious for the rapidity with which they die in laboratory media; *Streptococcus pneumoniae* and *Neisseria gonorrhoeae,* the causes of lobar pneumonia and gonorrhea, respectively, die within a very few days. These organisms are particularly susceptible to autolysis, presumably the result of digestion by enzymes present in the bacterial cells.

The steepness and duration of the death curve depend in part upon the nature and concentration of toxic by-products elaborated during growth. A high concentration or unusually toxic substances produce a steep death curve.

The rate of death can be calculated by the same formula used to calculate the rate of exponential multiplication. The result in this case is a minus quantity and indicates the time required to diminish the population by 50 per cent. The same method can be used to compare the effectiveness of germicidal agents, a topic to be discussed in greater detail in the next two chapters.

SUPPLEMENTARY READING

Brock, T. D.: *Biology of Microorganisms.* Englewood Cliffs, N.J., Prentice-Hall, Inc., 1970.

Buchanan, R. E., and Fulmer, E. I.: *Physiology and Biochemistry of Bacteria*, Vol. I. Baltimore, The Williams & Wilkins Co., 1928.

Gunsalus, I. C.: Growth of Bacteria. *In* Werkman, C. H., and Wilson, P. W. (eds.): *Bacterial Physiology.* New York, Academic Press, Inc., 1951.

Lamanna, C., Mallette, M. F., and Zimmerman, L. N.: *Basic Bacteriology*, 4th ed. Baltimore, The Williams & Wilkins Co., 1973.

Maaløe, O.: Synchronous Growth. *In* Gunsalus, I. C., and Stanier, R. Y. (eds.): *The Bacteria: A Treatise on Structure and Function,* Vol. 4. New York, Academic Press, Inc., 1962.

Mandelstam, J., and McQuillen, K.: *Biochemistry of Bacterial Growth.* New York, John Wiley & Sons, Inc., 1968.

Porter, J. R.: *Bacterial Chemistry and Physiology.* New York, John Wiley & Sons, Inc., 1946.

Stanier, R. Y., Adelberg, E. A., and Ingraham, J.: *The Microbial World,* 4th ed. Englewood Cliffs, N.J., Prentice-Hall, Inc., 1976.

11 EFFECTS OF THE ENVIRONMENT ON BACTERIA

Bacteria are almost completely at the mercy of their environment. It is true, as previously noted (page 185), that some organisms possess limited means of regulating the pH of their surroundings and that motile organisms may be repelled by harmful substances or may possibly be attracted chemotactically to a source of food. Microorganisms in a compost heap, manure pile, or vinegar generator produce enough heat to raise the surrounding temperature several degrees. These, however, are unusual situations that depend upon the excellent insulating properties of the particular environments concerned.

Physical factors that affect microorganisms are discussed in this chapter. Sterilization and disinfection, particularly by chemicals, are discussed in Chapter 12.

TEMPERATURE

GROWTH TEMPERATURES

The Cardinal Points. The temperature of the culture medium determines the rates of growth, multiplication, and death of microorganisms. Each organism can grow only within a *growth temperature range* characteristic of the species and sometimes of the particular culture. The lowest temperature at which growth occurs is the *minimum growth temperature;* this point is difficult to determine exactly, because there is no sudden increase in generation time as the temperature gradually drops. The highest tem-

perature at which growth can take place is the *maximum growth temperature*. The *optimum growth temperature,* commonly defined as the temperature at which most rapid multiplication occurs, is often only a few degrees lower than the maximum. Optimum and maximum growth temperatures can usually be ascertained with considerable accuracy from multiplication curves determined at short temperature intervals. The maximum, optimum, and minimum growth temperatures are known as the *cardinal points*.

The effect of temperature on the rate of growth of two bacterial species is illustrated in Figure 11–1. It is coincidental that each organism attained approximately the same maximum multiplication rate; the temperature at which this occurred was about 40° C. in the case of *E. coli* and 50° C. in the case of *L. delbrueckii*. Maximum growth temperatures were only 5 to 10 degrees higher than the optimum growth temperatures, whereas minimum growth temperatures were approximately 30 degrees lower.

It is essential in determining the effects of temperature on microbial activity to specify exactly what is being studied, because the optimum temperature for growth is not always most favorable for other behavior. Certain by-products may be produced in greater yield or more rapidly at some other temperature. Moreover, the greatest total crop of cells is usually obtained at temperatures a few degrees less than the optimum growth temperature (Fig. 11–2), whereas the duration of the lag phase is greater. Tem-

230

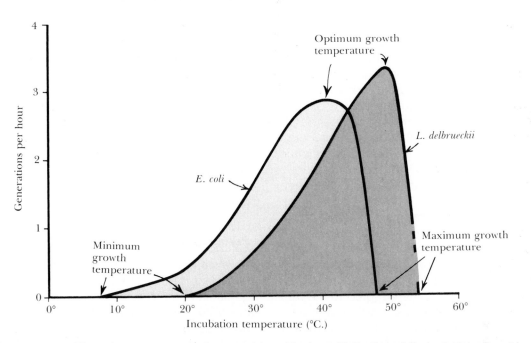

Figure 11–1. *Effect of temperature on the rate of logarithmic multiplication of* Escherichia coli *and* Lactobacillus delbrueckii. *Note in each case that the optimum growth temperature is much nearer to the maximum growth temperature than to the minimum growth temperature. (Plotted from data of J. L. Ingraham, J. Bact., 76:75–80, 1958; and A. Slator, J. Chem. Soc., 109:2T–10T, 1916.)*

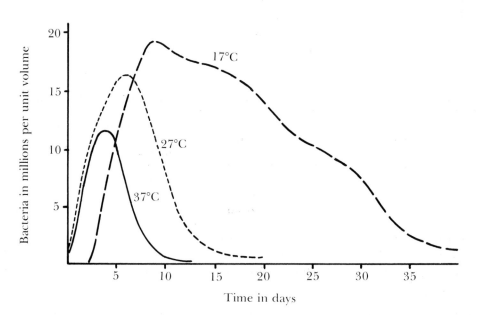

Figure 11–2. *The effect of incubation temperature on the "total crop" of* Staphylococcus aureus *in broth cultures. (After Graham-Smith.)*

perature also affects the rates of respiration and fermentation, spore formation, pigment production, and other processes. These effects can be attributed in part to the differing temperature optima of various enzymes, in part to the greater toxicity of acids and other wastes at higher temperatures. Obviously one must be careful to measure the growth *rate,* if this is the accepted function, rather than the *total yield* of cells or product after a single, arbitrarily selected interval of incubation.

Growth Temperatures and Habitats. The cardinal points of various bacterial species differ widely (Table 11–1). Such differences are a consequence of adaptation and natural selection. Highly parasitic organisms have become adapted to the temperatures of their natural hosts and through long association have lost the ability to grow at temperatures that depart widely from those of the hosts. These organisms therefore have a growth temperature range of only a few degrees; *Neisseria gonorrhoeae,* for example, grows only between 30° and 40° C. Many saprophytic bacteria, on the other hand, display a wide growth temperature range. In their natural habitats they are often exposed to extremes of temperature. *Bacilus subtilis* may grow within the range of 8° to 55° C.; it is found in soil and on plant materials, where it survives the freezing conditions of winter and the very warm conditions of summer. Other bacteria grow at subzero

temperatures in cooling brines; still others are found in hot springs.

Classification of Bacteria According to Growth Temperature. The total temperature range within which bacteria may grow extends from a few degrees below 0° C. to approximately 80° C. Bacteria are sometimes classified roughly into three groups according to their temperature preferences (Table 11–2). These groups are not sharply defined; the distinctions are arbitrary, but the classification has some practical utility.

Psychrophiles

Psychrophilic bacteria are ubiquitous and numerous. They are the predominant organisms in many uncultivated soils, and in many lake and stream water samples. They are also present in meats, in ice cream, and in various other materials. They can multiply at temperatures as low as −5° C. in brines or sugar solutions that do not freeze solid. Organisms that grow well at 0° to 5° C. are frequently encountered as causes of spoilage in refrigerated foods. Their rate of multiplication at such temperatures is often slow, and spoilage is not apparent unless storage is greatly prolonged.

Two groups of psychrophiles have been distinguished: (1) *obligate psychrophiles* cannot grow at temperatures above 19° to 22° C., whereas (2) *facultative psychrophiles* may grow at 30° to 35° C. The optimum tem-

TABLE 11–1.   Growth Temperatures and Habitats of Various Species of Bacteria

| Species | Habitat | Growth Temperature (°C.) | | |
		Minimum	Optimum	Maximum
Pseudomonas gelatica	Sea water	0	20–25	30–32
Pseudomonas syringae	Garden peas	7	28	37
Bacillus subtilis	Soil, decomposing vegetable matter	8	28–40	50–55
Escherichia coli	Intestines of vertebrates	8	37	47
Clostridium tetani	Soil, feces	14	37	43
Neisseria gonorrhoeae	Human infection	30	37	40
Lactobacillus delbrueckii	Fermenting vegetable and grain mash	20	50	55
Bacillus stearothermophilus	Soil, spoiled food products	33–37	50–65	70

TABLE 11–2.   Classification of Bacteria According to Their Growth Temperatures

A. Grow well at 0° C. (generation time less than 48 hours)	1. Psychrophiles
AA. Do not grow well at 0° C.	
B. Optimum growth temperature less than 45° C. (most species do not grow below 10° C. or above 52° C.)	2. Mesophiles
BB. Optimum growth temperature greater than 45° C.; in general, grow at 55° C. (*obligate* thermophiles do not grow at 37° C. or less, whereas *facultative* thermophiles can do so)	3. Thermophiles

perature of each group is usually within 5 degrees of the maximum. The chief feature that distinguishes a psychrophile is its ability to grow at 0° C.

Mesophiles

Most of the commonly studied bacteria are *mesophilic,* and these fall into two reasonably well defined subdivisions: (1) those whose optimum growth temperatures are from 20° to about 35° C., and (2) those whose optimum temperatures are between 35° and 45° C. The former group consists chiefly of saprophytes and plant parasites, whereas the latter are principally animal parasites or commensals (organisms that live in a host but do not harm it). Minimum and maximum growth temperatures vary correspondingly, but for the most part are within the range of 10° to 52° C.

Thermophiles

Thermophilic bacteria are interesting because they prefer temperatures intolerable to most forms of animal life. Water at 45° C. is hot to the touch—hotter than an ordinary bath; nevertheless, thermophilic bacteria grow best at temperatures above 45° C. Many will not grow below 40° C., and some will grow at 75° to 80° C. Bacterial multiplication has been demonstrated in natural hot springs at temperatures above 90° C.; generation times as low as 2 hours were recorded.* Vegetative cells of nonsporeforming mesophiles are usually killed within a

---

Bott and Brock, Science, 164:1411, 1969

few minutes at 60° C.; most other forms of life are also killed at this temperature.

In general, thermophilic bacteria are those that can grow at 55° C. *Obligate thermophiles* do not grow at 37° C., whereas *facultative thermophiles* can grow at both 37° C. and 55° C. It appears that facultative thermophiles carry on a mesophilic type of metabolism at temperatures below about 44° C. When they are warmed to 44° to 52° C. and allowed a few minutes to adapt, their metabolism shifts to the thermophilic type. Obligate thermophiles are incapable of adapting to the mesophilic type of metabolism.

Thermophiles obviously possess unusual proteins or an unusual physicochemical situation that permits, and even favors, physiologic activity at temperatures that denature the proteins of other organisms. The optimum temperature of various enzymes from thermophilic bacteria is known to be in the thermophilic range, and there is evidence that some enzymes are active at higher temperatures in the intact cell than after extraction from the cell.

FACTORS AFFECTING THE MAXIMUM GROWTH TEMPERATURE

The upper limits of growth temperature differ widely among the various forms of life (see Table 11–3). In general, procaryotic organisms can grow at higher temperatures than eucaryotic organisms, and among procaryotes, nonphotosynthetic bacteria can grow at higher temperatures than blue-green algae.

Metabolism and reproduction depend upon the integrity and proper functioning of

**TABLE 11–3.   Approximate Growth Temperature
Maxima of Various Groups of Organisms**

Organism	Maximum Growth Temperature (°C.)
Animals (including protozoa)	45–51
Eucaryotic microorganisms (certain fungi and the alga, *Cyanidium caldarium*)	56–60
Photosynthetic procaryotes (blue-green algae)	73–75
Nonphotosynthetic procaryotes (bacteria)	90

Adapted from Brock, T. D., Science, *158:*1012–1019, 1967

all essential cell components. The labile components of cells include enzymes and other proteins, and also the various membranes. The temperatures at which proteins are denatured vary, but the essential proteins in a healthy, active cell must be in their natural condition. The cell dies when the least stable of these proteins is denatured.

A relationship between maximum growth temperatures of bacteria and the heat resistance of their enzymes is suggested by the data in Table 11–4. The maximum growth temperature of four strains of *Bacillus cereus* var. *mycoides* is 40° C., and three respiratory enzymes of this organism are inactivated at 40° or 41° C. The same enzymes in species of higher maximum growth temperature withstand greater amounts of heat. Since the correlation between growth temperatures and enzyme inactivation temperatures is not perfect, it can be concluded that some of these enzymes are not vital to the growth of certain of the species. It has been shown that when an essential microbial enzyme is inactivated

by heat, the organism can be made to grow at an otherwise unfavorably high temperature by adding the normal product of that enzyme to the culture medium.

The importance of membranes is indicated by the observation that thermal death of psychrophiles seems to be associated with the loss of integrity of their cell membranes and subsequent cell lysis. Formation of holes due to rupture of relatively weak bonds at high temperature allows leakage of vital constituents. It may be noted that the types of organisms listed in Table 11–3 are arranged in order of decreasing number or extent of membrane structures. As discussed earlier, eucaryotes are distinguished from procaryotes by the membranous nature of their organelles. Among procaryotes, the blue-green algae possess membranous chloroplasts, where photosynthesis occurs, whereas the only significant membranes in nonphotosynthetic bacteria are the protoplasmic membranes and their associated structures.

**TABLE 11–4.   Relationships between Maximum Growth Temperatures of Bacteria and
Inactivation Temperatures of Their Enzymes**

Species	Maximum Growth Temperatures (Means)	Inactivation Temperatures (Means)		
		Indophenol Oxidase	Catalase	Succinic Dehydrogenase
Bacillus cereus var. *mycoides* (4 strains)	40° C.	41° C.	41° C.	40° C.
Bacillus cereus (21 strains)	45° C.	48° C.	46° C.	50° C.
Bacillus subtilis (10 strains)	54° C.	60° C.	56° C.	51° C.
Thermophiles (9 strains)	76° C.	65° C.	67° C.	59° C.

ALTERATION IN GROWTH TEMPERATURES

The growth temperature characteristics of bacteria may be changed by altering their genetic make-up. Two methods that have been successfully used* are (1) ultraviolet-induced mutations selected by cultivation at low temperature and (2) transduction with a phage grown on a psychrophile. The minimal growth temperatures of strains of the mesophilic species *Pseudomonas aeruginosa* were lowered from approximately 11° to 0° C. Concomitantly, the maximum growth temperatures decreased from 44° to 32° C., so the growth temperature ranges were not affected.

Thermoduric Bacteria

Thermoduric bacteria are unusually resistant to heat. The term can be applied to endospore-forming species, but is more commonly restricted to nonspore-forming types. Thermoduric organisms are often found in pasteurized milk or other products that have been heated but not sterilized. They include *Microbacterium* species, various micrococci and streptococci, certain lactobacilli, and some gram-negative rod bacteria.

MECHANISMS OF GROWTH CONTROL BY TEMPERATURE

The effect of temperature on microbial growth is complex; it can be expressed in its simplest terms as the resultant of two opposing activities. The rates of enzyme reactions, like those of all chemical reactions, vary directly with temperature. They are slow at low temperatures and accelerate as the temperature rises. Degradative processes, such as denaturation of proteins, including enzymes, are very slight at low temperatures and do not become marked until moderate temperatures are attained, but then they increase rapidly. The difference between beneficial enzyme activity and harmful denaturation is

*Olsen and Metcalf, Science, 162:1288, 1968

greatest at the optimum growth temperature, whereas at the maximum growth temperature destructive activity is so rapid that it just balances constructive processes and growth ceases.

The Killing of Microorganisms at High Temperature

TDP vs. TDT. The lethal effect of high temperatures upon microorganisms was formerly indicated by a number called the *thermal death point* (TDP): the temperature at which an organism is killed in 10 minutes. Numerous factors affect the apparent thermal death point, and it is difficult to obtain reproducible figures. These factors include the number and previous history of the organisms and the nature and pH of the suspending medium.

More recently, *thermal death times* (TDT) have been determined. The thermal death time is the time required to kill a given number of cells or spores under stated conditions of temperature, taking into account the nature of the suspending medium and all other pertinent information.

Survivor curves (which resemble the death phase of a population curve) may be determined by the plate method. The results can be expressed as negative generation times, or the time required to kill a stated percentage (e.g., 99.9 per cent) of the population, or a death rate constant that expresses the rate of death as a positive figure.

Factors Affecting the Death of Microorganisms by Heat. *Temperature.* The striking effect of temperature on the killing of bacteria is illustrated in Table 11–5. Six hours at the temperature of boiling water was required to kill spores of *Clostridium botulinum*. The sterilization time decreased markedly as the temperature was raised, so that only five minutes was required at 120° C. Not all strains of *C. botulinum* are as resistant as this. The thermophilic organism obviously possessed much higher heat resistance. It is presumably one of the most resistant encountered; yet even it was killed within 20 minutes at the highest temperature recorded. Hospital and laboratory sterilizers, commercial canning retorts, and home pressure

TABLE 11–5. Effect of Temperature upon the Thermal Death Times of Spores

Temperature	*Clostridium botulinum* (60,000,000,000 Spores Suspended in Buffer at pH 7)	A Thermophile (150,000 Spores per ml. of Corn Juice at pH 6.1)
	Minutes	
100° C.	360	1140
105° C.	120	
110° C.	36	180
115° C.	12	60
120° C.	5	17

cookers are often operated at 120° C., which is the temperature of steam at a pressure of approximately 15 pounds per square inch (Table 11–6).

Number of Organisms. Greater heat treatment is required to kill a large number of organisms than a small number (Table 11–7). Moreover, the cells or spores in a given suspension are not necessarily of equal resistance; a few highly resistant individuals may survive much longer than the majority. These few resistant cells are important in practical sterilization because they determine the total heat process required for preservation of canned goods, sterilization of surgical dressings, and preparation of culture media in the laboratory.

Species. The heat resistance of different bacterial species varies through a wide range. Any temperature above the maximum growth temperature is lethal to vegetative cells if applied for a sufficient time. *Treponema pallidum,* the cause of syphilis, is killed within one hour at 41.5° C. (Table 11–8). This species is unusually susceptible to heat. Most nonspore-forming bacteria are killed within 10 to 15 minutes at 60° to 65° C., but a few thermoduric organisms require a higher temperature or longer time. Vegetative cells of spore-forming bacteria are killed as readily as nonspore-forming bacteria. Spores are much more resistant but differ widely according to the species.

Medium: Composition and Viscosity. The nature of the suspending medium affects the results of thermal death time studies as well as practical sterilization procedures. Proteins, fats, and other substances offer some protection to bacteria. A highly viscous medium retards the distribution of heat by convection and limits heat transfer to the slower process of conduction. These factors are illustrated in Table 11–9, which shows thermal death point (at 10 minutes of heating) determinations of *E. coli* suspended in cream, whole milk, skim milk, whey, and broth. Cream is the most viscous and contains a high percentage of fat; whole milk is

TABLE 11–6. Relationship between Steam Pressure and Temperature in the Autoclave

Gauge Pressure (lbs./sq. in.)	Temperature (°C.)
0	100.0
5	109.0
10	115.5
15	121.5
20	126.5

TABLE 11–7. Effect of the Number of Spores of *Clostridium botulinum* on the Thermal Death Time at 100°C.

Number of Spores	Thermal Death Time (Minutes)
72,000,000,000	240
1,640,000,000	125
32,000,000	110
650,000	85
16,400	50
328	40

TABLE 11–8. Thermal Death Times of Vegetative Cells and Spores of Bacteria

Organism	Temperature (°C.)	Time (Min.)
Treponema pallidum	41.5	60
Most nonspore-forming bacteria	60–65	10–15
Staphylococcus aureus	65	30
Spore formers, vegetative cells	60–65	10–15
Spore formers, spores:		
Bacillus megaterium	100	16
Bacillus subtilis	100	180
Bacillus sp.	100	1200

next in viscosity and fat content. The remaining media contain little if any fat but decreasing amounts of protein.

pH. In general, the resistance of an organism to heat is greatest at a pH favorable for growth (often near neutrality), and death occurs more rapidly as the acidity or alkalinity of the medium is increased. The death time of *C. tetani* at 100° C. is 30 minutes at pH 7, but only 15 minutes at either pH 6 or 8. At 100° C. *C. botulinum* is killed in 50 minutes at pH 5.05 and in 15 minutes at pH 3.98. Most food spoilage bacteria grow best at approximately neutral reaction, and relatively little heating is necessary to sterilize acid foods (below pH 4.5). This explains why fruits are more easily preserved by canning than are meats, fish, and nonacid vegetables such as corn or peas.

TABLE 11–9. The Effect of the Medium upon the Thermal Death Point of *Escherichia coli**

Medium	Thermal Death Point (°C.)
Cream	73
Whole milk	69
Skim milk	65
Whey	63
Bouillon (broth)	61

* Heating time: 10 minutes.

Mechanisms of Killing by Heat. The lethal action of heat is doubtless exerted in several ways, including denaturation of proteins and inactivation of essential enzymes. There are also indications that the osmotic barrier may be affected. Increased permeability would permit toxic substances to enter, or vital components to be lost; decreased permeability would retard the entrance of nutrients or excretion of toxic wastes.

Effects of Low Temperatures

Low temperature does not necessarily kill bacteria. Multiplication ceases below the minimum growth temperature, but the organisms may remain viable in the dormant condition, often for long periods. Growth and multiplication resume when the temperature is again raised to the normal range.

The ability of bacteria to withstand freezing or subfreezing temperatures varies with the species. Endospores and certain cocci such as staphylococci are very hardy. Some gram-negative rod bacteria, e.g., *Pseudomonas aeruginosa,* are killed more readily at subfreezing temperatures. The mechanism of death by freezing is not known but has been attributed to denaturation of cellular proteins.

The rapidity of death caused by temperatures below the growth range depends upon the temperature; oddly enough, bacteria often resist very low temperatures better than temperatures only slightly below the minimum for growth. One experiment, for example, showed that only 4 per cent of the cells of *E. coli* remained alive after 11 days at −1° to −2° C., whereas 25 per cent survived storage at −20° C. for 163 days. Staphylococci can withstand the temperature of liquid air in practically unchanged numbers after the initial shock of freezing. The organisms in the experiment summarized in Table 11–10 were dried on garnets and then some were stored at room temperature, some in the refrigerator, and some at −190° C. Garnets were removed from the various storage conditions at intervals and the surviving bacteria were counted.

TABLE 11–10.   Effect of Low Temperature on Survival of Staphylococci Dried on Garnets

Time of Storage	22° C.	5° C.	−190° C.*
		Survivors	
1 day	90,800	88,800	65,900
8 days	11,400	37,700	60,700
32 days	300	550	67,900

* Liquid air

Preservation of Bacteria by Cold. The resistance of bacteria to low temperature provides a means of preserving many species for long periods without the necessity of frequent transfers. Hardy organisms frozen quickly on agar slants or, preferably, in dilute peptone or milk may be kept at −20° C. for a year or two without complete loss of viability. More delicate organisms like gonococci and meningococci, which ordinarily must be transplanted every two or three days, are preserved several days or weeks when properly frozen and stored. They survive even better if dried rapidly in a vacuum while frozen and kept at subfreezing temperatures.

Survival of Bacteria in Frozen Foods. The survival of bacteria at low temperatures is important in the food industry. Many years ago, several outbreaks of typhoid fever were reported that apparently were caused by bacteria in natural ice used several months after the ice was harvested. Laboratory tests later showed that typhoid bacteria can survive in ice as long as 22 weeks. Other intestinal pathogens possess similar survival power.

The tremendous expansion of the frozen food industry emphasizes the importance of bacterial resistance to low temperature. In one experiment, living typhoid bacteria were recovered from inoculated unsliced sweetened strawberries after storage at −18° C. for 14 months; the organisms died in berries stored for eight days at 5° C. or six hours at room temperature. Freezing is therefore not a means of destroying pathogens in food, and no food should be frozen that could not be eaten raw. Moreover, frozen food usually spoils more quickly after thawing than the original unfrozen food; freezing apparently damages the food tissues so that nutrient materials are released upon thawing and promote very active multiplication of surviving spoilage bacteria. Thawed frozen foods should therefore be used immediately or else discarded.

Toxin-producing bacteria, such as *C. botulinum* and *S. aureus,* may survive freezing, but if they did not multiply before freezing and if the food is used immediately it is safe. The harmful effects of these organisms are attributed to the toxins they produce while growing in food, not to the organisms themselves.

RADIATIONS

Investigators during the last quarter of the nineteenth century noted that sunlight is lethal to many bacteria, and the practice arose of cultivating microorganisms in the dark. Colored filters were used to test the effect of light of different colors; red, orange, and yellow did not harm bacteria, but blue-violet was markedly inhibitory. Later it was discovered that radiations outside the visible range were bactericidal; these were ultraviolet rays.

GERMICIDAL RADIATIONS

The electromagnetic spectrum is shown in Figure 11–3, which also indicates some of the radiations destructive to bacteria. Radiations of wavelengths greater than 300 nm possess little if any bactericidal power. These include visible light waves, infrared waves, and the much longer radio waves. Some of these radiations produce heat but are not germicidal per se.

Ultraviolet light at the wavelength 265 nm is especially effective against the majority of bacteria and some molds and viruses (Fig. 11–4). Most work has been done with mercury vapor lamps that emit radiations of 253.7 nm, well within the germicidal range. Many radiations shorter than ultraviolet are also germicidal, particularly the shorter

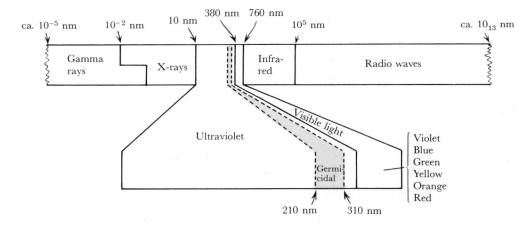

Figure 11-3. *The electromagnetic spectrum. The germicidal wavelengths of ultraviolet light are indicated.*

x-rays and gamma rays. Some of these are even more efficient than ultraviolet radiations but are more difficult or dangerous to use.

EFFECTS OF RADIATIONS

Radiations have two effects on microorganisms. The first is death, which appears to be exponential; that is, a constant percentage of the surviving cells is killed in each successive equal time interval. The other effect is the production of mutants in the surviving population. These genetically altered forms differ from the parent population, usually in one characteristic, such as resistance to bacteriophage, fermentative power, or the ability to synthesize a substance necessary for metabolic activity (i.e., an essential metabolite).

Mode of Action of Radiations. Lethal or mutagenic radiations are first absorbed by some portion of the irradiated cell. Ultraviolet light is known to be absorbed by nucleic acids, which are constituents of the DNA of nuclei. According to the "target" theory, the radiation makes a direct hit on a sensitive area like a nucleic acid molecule, and the absorbed energy alters it in some manner. It has been suggested that a lethal mutation occurs and the cell soon dies. X-rays seem to produce their effects by general bombardment, selective absorption by specific molecular structures playing little or no part. Instead, tracks or paths of ionization are produced, and OH^- or other ions cause lethal chemical reactions or genetic changes.

Killing by radiation depends upon the wavelength and the total radiation dosage, that is, the total amount of incident energy, but the dosage necessary to kill a given species varies with the wavelength. X-rays are more efficient than ultraviolet rays.

Photoreactivation. The question, When is a cell dead? is raised by the observation that bacteria exposed to ultraviolet light "revive" when exposed to visible light of 365 to 510 nm wavelength. This is known as *photoreactivation* (see Figure 11-5).

E. coli cells treated with ultraviolet light

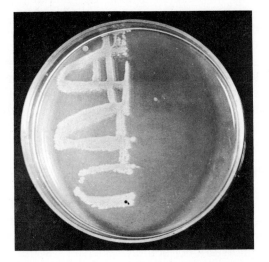

Figure 11-4. *Killing of* E. coli *by ultraviolet light. The bacteria were streaked uniformly on the agar in this dish. The left half was covered with black paper and the right half was exposed a few minutes to the light from a carbon arc. Only the protected bacteria on the left grew when the plate was then incubated.*

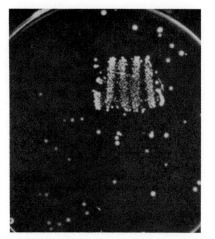

Figure 11–5. *Photoreactivation of* E. coli *irradiated with ultraviolet light. The bacteria were spread on the surface of a nutrient agar plate and irradiated. The plate was then covered with black paper except for a small square area, upon which the filament of a projection lamp was focused. After exposure to the visible light, the plate was incubated in the dark. (From Kelner, J. Bact., 58:512, 1949.)*

and then exposed to visible light survived at the rate of 120,000 cells per million irradiated. The survival rate of other suspensions, which were kept dark, was only 4.5 cells per million. Ultraviolet-treated cells survived only if exposed to visible light within three hours after irradiation.

The mechanism of photoreactivation appears to involve the use of visible light energy to activate an enzyme that breaks specific chemical bonds formed under the influence of ultraviolet light. During irradiation, DNA strands are distorted by the formation of dimers between two thymine residues on the same or complementary strands. Exposure to visible light at the blue-violet end of the spectrum activates an enzyme that hydrolyzes the thymine dimers.

The usual criterion of cell death is inability to multiply. According to this definition, nearly all the irradiated cells that were kept dark would have been considered dead. Most of the same cells, however, if exposed to visible light, would have proved to be capable of multiplication. This finding reemphasizes the fact that there is no sharp dividing line between life and death; in fact, the distinction is even more tenuous than previously believed.

APPLICATIONS OF IRRADIATION

Production of Mutants. Ultraviolet and x-irradiations cause mutations, and by the use of suitable techniques a desired mutant can be selected from the random population of induced mutants. This is a trial and error process; it is not yet possible to induce particular changes at will. Industrially important strains of microorganisms have been derived in this manner; for example, a culture of *Penicillium* that produces unusually large amounts of penicillin.

Sterilization. Ultraviolet irradiation is of some use for killing spoilage or disease bacteria. Practical sterilization is limited by the fact that glass, water, and organic matter absorb ultraviolet light and prevent its access to bacteria. It is therefore possible to sterilize only air and the surfaces of objects, such as surgical implements, glassware and chinaware, and the like. Ultraviolet lamps are used in hospital operating rooms to reduce surgical infection and in classrooms to reduce respiratory disease. The lamps must be arranged and shielded so that human skin is not exposed too long, lest serious damage be done. Ultraviolet irradiation has been used experimentally to inactivate viruses (e.g., influenza) for use in immunization.

The sterilization of food by gamma rays has been under investigation for a number of years. These rays penetrate deeply enough to be used with many products and sterilize very quickly, when used in sufficient dosage. Their advantage is that packaged as well as canned foods can be preserved and there is little rise in temperature to alter the nature of the product.

PRESSURE

Bacteria seem to possess considerable resistance to mechanical or hydrostatic pressure. The findings of different investigators are not entirely consistent; some of the discrepancies are undoubtedly the result of differences in procedure. The statement can be found that nonspore-forming bacteria, such as *Serratia marcescens* and *Streptococcus lactis,* are killed within five minutes at 85,000 to 100,000 pounds pressure per square inch. Another report indicates that 88,000 pounds per square inch applied for 14 hours is necessary to kill nonspore-forming bacteria, and that twice that pressure is required to kill spores.

Zobell and co-workers found that spore-forming and nonspore-forming bacteria failed to grow in broth when subjected continuously to hydrostatic pressure of 2940 to 8820 pounds per square inch. Temperature was a factor; several species grew at 40° C. under 8820 pounds pressure but did not grow at 20° C. Marine bacteria seemed to possess somewhat greater ability to grow at high pressures than terrestrial bacteria. Morphologic changes also occurred under pressure; some species lost motility and some cells grew but failed to divide.

SONIC VIBRATIONS

The human ear detects sound waves with frequencies between 32 and 32,000 vibrations per second. Supersonic waves range from the higher audible sound, 9000 vibrations per second, to inaudible sound of 200,000 vibrations per second. Vibrations over 200,000 per second are classified as ultrasonic waves. High frequency sound waves generated by nickel bars or quartz crystals stimulated electrically have been used extensively for experimentation in biology. Most work has been done at about 9000 vibrations per second.

EFFECTS OF SONIC VIBRATIONS

Supersonic vibrations drastically affect the cells of higher organisms, causing severe disturbance of the cellular contents and often eventual rupture of the cell walls and complete disintegration. Bacteria are much smaller and the effects of supersonic vibration cannot be observed as readily. Bacterial cultures, however, can be killed and lysed by high frequency sound. The sensitivity of different species to sonic energy varies greatly, as might be expected. *Neisseria gonorrhoeae* is easily sterilized, whereas most bacterial spores are unaffected.

Factors Affecting the Rate of Killing by Sonic Vibrations. The efficiency of sound waves depends more on their amplitude than on their frequency; that is, waves of relatively low frequency but of high intensity are more effective than high frequency waves of low intensity. The time required for sterilization is appreciable with intensities so far studied. For example, 99 per cent of a young (12 hour) culture of *E. coli* were killed in 20 minutes by high intensity waves at 8900 cycles per second. The cells of this culture died at a logarithmic rate. Older cultures contained a higher percentage of resistant cells. Logarithmic death of *Klebsiella pneumoniae* by ultrasonic waves is illustrated in Figure 11–6. Approximately 0.5 per cent of the initial population was still viable after 50 minutes.

According to the most favored hypothesis, death and lysis of bacteria by sonic vibrations are attributed to *cavitation*. It is presumed that the cell boundaries are bombarded by minute gas bubbles, which form in the suspending medium as a result of disturbance by the high frequency sound waves.

MICROBIOLOGIC APPLICATIONS OF SONIC VIBRATIONS

Sonic oscillators are used to study the composition of bacterial cells and to secure intracellular enzymes for biochemical inves-

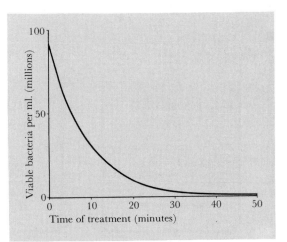

Figure 11–6. *The killing of* Klebsiella pneumoniae *by ultrasonic waves (700,000 cycles per second). (Plotted from data of D. Hamre, J. Bact., 57:279–295, 1949.)*

tigation. Solutions of microbial constituents can be obtained for use as immunizing agents and for serologic analysis.

MOISTURE

Water is the vehicle by which holophytic organisms like bacteria, yeasts, and molds secure food and eliminate waste products. Most bacteria and yeasts prefer media of very high water content. Ordinary nutrient broth (0.3 per cent beef extract and 1.0 per cent peptone) contains 98.7 per cent water; some bacteria grow better when 0.5 per cent sodium chloride is added. Molds require much less water; many grow on substrates containing 50 to 60 per cent sugar, or even upon leather goods and book bindings during humid weather.

OSMOTIC PRESSURE

Osmosis is the diffusion of solvent molecules through a semipermeable membrane. The *osmotic pressure* of a solution is directly dependent on the concentration of dissolved substances in the solution. Two solutions of the same osmotic pressure are said to be *isotonic*. If two solutions of unequal osmotic pressure are compared, the one of higher pressure is *hypertonic,* the other *hypotonic*.

The osmotic pressure of the protoplasm of normal bacterial cells is greater than that of the usual culture media; that is, it contains a higher concentration of salts, amino acids, and other organic and inorganic compounds. This creates a tendency for water to enter the protoplast by osmosis through the semipermeable cytoplasmic membrane (Fig. 11–7). Under customary conditions of cultivation the cells maintain a state of *turgidity*.

Cells of animals and higher plants undergo drastic changes when placed in media of very low osmotic pressure such as distilled water. There is a tendency for excessive amounts of water to pass into the protoplasm, thus creating very high pressure. Cells without extremely strong walls undergo *plasmoptysis:* The cell walls burst and protoplasmic materials are extruded. Most bacteria do not suffer serious harm when suspended for a short time in distilled water, because their walls are very strong and rigid. However, bacteria that have become acclimated to very hypertonic solutions such as pickling brines may undergo plasmoptysis when transferred to very dilute media. Many species of marine bacteria fail to grow in fresh-water media.

Plasmolysis occurs when normal cells are placed in solutions of very high osmotic pressure. Water passes from the protoplasm into the surrounding medium, and the protoplasm shrinks to a small mass within each cell. Metabolic activity ceases, and the cells become dormant and may even die. They may recover if they are returned soon enough

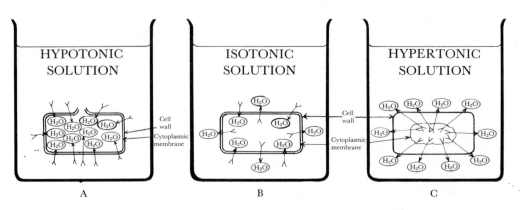

Figure 11–7. A, *Plasmoptysis (bursting) of a cell in a strongly hypotonic solution after water molecules have entered the cell;* B, *normal turgidity in an isotonic (or slightly hypotonic) solution;* C, *plasmolysis in a strongly hypertonic solution.*

to normal osmotic conditions. Gram-positive bacteria are more difficult to plasmolyze than gram-negative bacteria.

The morphology of cells grown in media of moderately high osmotic pressure may be atypical; long, filamentous, bloated, irregularly staining forms are frequently found.

OSMOPHILIC AND HALOPHILIC MICROORGANISMS

Microorganisms that have become adapted to high osmotic pressure are called *osmophiles*. Marine bacteria and other organisms that cannot grow in salt-free solutions are *halophiles*. Some halophilic bacteria can be readapted to normal media by successive transfers through media of lower and lower salt concentration. Less than 10 per cent of marine bacteria can grow in fresh water when first isolated, but about 75 per cent eventually can be adapted to solutions of low osmotic pressure.

Halophilic bacteria are interesting from the ecologic viewpoint, particularly those found in the sea and in other natural bodies of water of high salinity. The salt content of the major oceans is between 3.5 and 4.0 per cent; that of the Dead Sea is about 29 per cent. Some bacteria isolated from the Dead Sea have failed to grow in media containing less than 13 per cent salt. It is believed that marine bacteria are not separate species from terrestrial forms but actually are terrestrial organisms adapted to the marine environment. Organisms can be found in both environments with identical physiologic properties except for their osmotic requirements. They become indistinguishable after proper adaptation.

APPLICATIONS OF PLASMOLYSIS

Plasmolysis is an important method of controlling bacterial growth. It has been used throughout recorded history to preserve foods. Experience has shown that 10 to 15 per cent sodium chloride or 50 to 70 per cent sugar are satisfactory preservative concentrations.

Desiccation. Desiccation is a practical means of applying osmotic relations to the control of microbial activity. Drying any material increases its effective osmotic pressure so that microorganisms cease to grow and some eventually die. As long as the product is kept dry, spoilage will not occur.

Many bacteria survive for long periods in the desiccated condition; endospores and gram-positive cocci are particularly resistant. Foods preserved by this method should therefore not be considered sterile. They may contain a large, dormant microbial population, which can begin to grow and cause spoilage almost immediately after reconstitution with water.

Preservation of Bacteria by Desiccation. Desiccation is used to preserve bacterial cultures. The organisms are first frozen quickly and then dried under high vacuum while still frozen. *Lyophilization* is one of the best known and most widely used processes. Bacterial suspensions in small ampules are frozen in a bath of alcohol and dry ice ($-76°$ C.) and evacuated under very high vacuum. The water evaporates from the ice and the ampules are then sealed in a flame. If freezing and drying are performed rapidly, even sensitive species are preserved in a viable condition without change in their physiologic or pathogenic characteristics for months or years.

HYDROGEN ION CONCENTRATION

Microbial growth and activities are strongly affected by the pH of the medium, but there are wide differences between the pH requirements of the various species. These differences reflect the normal habits and habitats of the organisms. Each species can grow only within a certain pH range, and most rapid or luxuriant growth occurs in a narrow *optimum pH* zone (Fig. 11–8). Table 11–11 lists minimum, optimum, and maximum pH values for the growth of a few common microorganisms.

pH RANGES OF MICROORGANISMS

The intestinal bacteria tolerate greater acidity and alkalinity than most other animal

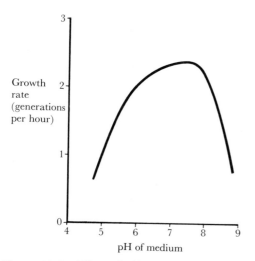

Figure 11–8. *Effect of pH on growth of* E. coli. *(Plotted from data of Gale and Epps, Biochem. J., 36:1942.)*

parasites; their optimum pH is near or slightly below neutrality. These organisms gain access to the intestine only after withstanding the acid of the stomach, and in the intestine they encounter bile, which is highly alkaline. Their adaptation to acid or alkaline conditions is not surprising. The pH of blood and tissues, on the contrary, is much more constant. Normal blood has a reaction of about pH 7.3. Most blood or tissue parasites are favored by neutral or slightly alkaline media.

Many plant and soil bacteria prefer relatively alkaline conditions. Yeasts and molds tolerate a wide range of pH, but usually grow best in acid media. Acid foods, such as fruits, pickles, and other fermented products, are more susceptible to yeast or mold spoilage than to bacterial spoilage.

TABLE 11–11.   Minimum, Optimum, and Maximum pH for the Growth of Various Bacteria and Fungi

Organism	Minimum pH	Optimum pH	Maximum pH
Animal parasites or commensals			
Intestinal			
Escherichia coli	4.3	6.0–8.0	9.5
Salmonella typhi	4.0	6.8–7.2	9.6
Shigella dysenteriae	4.5	ca. 7.0	9.6
Blood or tissue parasites			
Streptococcus pneumoniae	7.0	7.8	8.3
Streptococcus pyogenes	4.5	7.8	9.2
Neisseria meningitidis	6.1	7.4	7.8
Brucella melitensis	6.3	6.6–8.2	8.4
Vibrio cholerae	5.6	7.0–7.4	9.6
Mycobacterium tuberculosis	5.0	6.8–7.7	8.4
Corynebacterium diphtheriae	6.0	7.3–7.5	8.3
Lactobacillus	3.8–4.4	5.4–6.4	7.2
Plant pathogens or commensals			
Erwinia caratovora	4.6		9.3
Rhizobium leguminosarum	3.2–5.0		10.0–11.0
Agrobacterium radiobacter	4.5–5.0		11.5–12.0
Soil bacteria			
Bacillus subtilis	4.5	6.0–7.5	8.5
Nitrobacter	5.7	8.4–9.2	10.2
Nitrosomonas	7.6	8.5–8.8	9.4
Thiobacillus thiooxidans	1.0	2.0–5.0	9.8
Yeasts	2.5	4.0–5.8	8.0
Molds	1.5	3.8–6.0	7.0–11.0

pH CHANGES IN CULTURES

The nature of microbial metabolic activities is such that the pH of a culture medium does not ordinarily remain constant after growth begins. Degradation of proteins and other nitrogenous compounds frequently yields ammonia or other alkaline by-products; carbohydrate fermentations often produce organic acids. The nature of the organism and of the substrate therefore determines whether the pH of a medium rises or falls as growth proceeds. The reaction may continue to change until the maximum or minimum pH for the organism is reached, whereupon the culture dies.

Reversion of reaction occurs when some species grow in certain media. *Klebsiella pneumoniae* utilizes glucose vigorously and produces acid until the reaction falls to about pH 5.0. After the glucose is exhausted, the organism attacks its acid products and oxidizes them to carbon dioxide and water; the reaction therefore returns toward pH 7.0.

BUFFERS

Buffers are often added to culture media to retard pH change as acids or alkalies are formed. Most buffers used in media are either mixtures of weakly acidic and weakly alkaline compounds or they are amphoteric substances, that is, compounds that dissociate as either an acid or a base, depending on the pH of the solution.

Mixtures of potassium or sodium phosphate, or both, are most commonly employed because they are nontoxic in useful concentrations and they supply needed phosphorus. The pH of an equimolar solution of the acidic salt, KH_2PO_4, and the basic salt, K_2HPO_4, is 6.8. If an acid such as acetic acid is formed by bacteria in a medium buffered with phosphate, part of the basic salt is converted to the weakly acidic salt:

$$K_2HPO_4 + HC_2H_3O_2 \rightleftharpoons KH_2PO_4 + KC_2H_3O_2$$

and the pH of the medium falls only slightly. Conversely, a basic microbial product reacts with the acidic salt to form a dibasic compound that is only weakly alkaline.

Many culture media contain amphoteric substances such as proteins and peptides. These compounds possess both amino and carboxyl radicals, which can dissociate as basic and acidic groups, respectively:

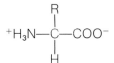

At proper pH, which differs from one substance to another, the positive charges balance the negative charges and the compound is electrically neutral. Addition of hydrogen ions depresses ionization of the carboxylate group, so the molecule acquires a net positive charge:

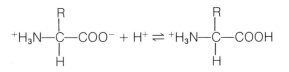

and, within limits, the pH of the medium does not change. Conversely, hydroxyl ions can be added without changing the pH of the solution.

Insoluble carbonates such as $CaCO_3$ and $MgCO_3$ are also added to media to prevent a drop in pH as acid is produced. Being insoluble, they have no direct effect on pH, but when acid is formed and the reaction falls below pH 7.0, the carbonate decomposes, CO_2 is evolved, and the acid is converted to its calcium or magnesium salt.

OXIDATION-REDUCTION POTENTIAL

The ability of an organism to grow when transferred to a fresh culture medium depends in part upon the oxidation-reduction (O-R) potential of the medium. Oxidation-reduction potentials are controlled by the oxidizing and reducing agents present and can be determined by measurement with appropriate electrometric apparatus. Strongly oxidizing substances produce positive potentials (e.g., +200 mv.), and strongly reducing substances produce negative potentials. The potential of hydrogen at 1 atmosphere is about −400 mv.

The various bacteria differ with respect to the O-R potentials at which they can begin to grow. Aerobic organisms tolerate higher potentials than anaerobes, which usually require media with negative potentials. The aeration of culture media produces positive potentials in the range of 200 to 300 mv., but the initiation of growth of many bacteria (not only anaerobes) is favored by a somewhat lower potential. Consequently, the boiling of media to expel dissolved oxygen, which lowers the O-R potential to −100 to +100 mv., or the incorporation of a reducing substance is often helpful in starting the growth of small inocula. As soon as the bacteria begin to metabolize, the O-R potential falls further, frequently to negative values. Incorporation of reducing agents such as sodium thioglycollate provides an O-R potential below −100 mv., in which most anaerobes will grow.

SUPPLEMENTARY READING

Buchanan, R. E., and Fulmer, E. I.: *Physiology and Biochemistry of Bacteria,* Vol. II. Baltimore, The Williams & Wilkins Co., 1930.

Ingraham, J. L.: *In* Gunsalus, I. C., and Stanier, R. Y., (eds.): *The Bacteria: A Treatise on Structure and Function,* Vol. IV. New York, Academic Press, Inc., 1962.

Lamanna, C., Mallette, M. F., and Zimmerman, L. N.: *Basic Bacteriology*, 4th ed. Baltimore, The Williams & Wilkins Co., 1973.

Mitchell, P.: Physical Factors Affecting Growth and Death. *In* Werkman, C. H., and Wilson, P. W. (eds.): *Bacterial Physiology.* New York, Academic Press, Inc., 1951.

Oginsky, E. L., and Umbreit, W. W.: *An Introduction to Bacterial Physiology.* 2nd ed. San Francisco, W. H. Freeman & Company, 1959.

Porter, J. R.: *Bacterial Chemistry and Physiology.* New York, John Wiley & Sons, Inc., 1946.

Salle, A. J.: *Fundamental Principles of Bacteriology,* 7th ed. New York, McGraw-Hill Book Co., Inc., 1973.

Thimann, K. V.: *The Life of Bacteria*, 2nd ed. New York, The Macmillan Company, 1963.

INHIBITION AND 12 KILLING OF MICROORGANISMS

NATURAL EQUILIBRIA

The vital activities of any organism or cell help to maintain a dynamic state or equilibrium condition. This is perhaps the most important feature differentiating a living organism from nonliving matter. The maintenance of intracellular equilibria is a function of the various enzymes. Each enzyme is under the control of a gene. Gene and enzyme activities are therefore affected by conditions within the cytoplasm: its composition, pH, oxidation-reduction potential, and other factors. Many essential enzymes are associated with the cytoplasmic membrane, which must be intact in order for the enzyme activities to contribute to the cell's maintenance. The disturbance of any cellular equilibrium by altering the factors controlling it may lead to death of the cell.

REVERSIBLE AND IRREVERSIBLE INJURY

A distinction is often made between agents whose effects on equilibria are reversible and those whose effects are irreversible. Reversibly injurious agents produce *stasis* or inhibition without immediately lethal action. *Bacteriostatic* agents inhibit bacteria, *fungistatic* agents inhibit fungi, and so forth. Irreversibly injurious agents cause fairly prompt death; *bactericidal, fungicidal,* and *virucidal* agents kill bacteria, fungi, and viruses, respectively.

It should be emphasized that there is no sharp distinction between bacteriostatic and bactericidal action: the difference is quantitative rather than qualitative. This can be demonstrated by a simple experiment with various concentrations of phenol in nutrient broth. Three-tenths per cent phenol prevents the growth of *E. coli* but does not kill all the cells within a test period of several days; 1 per cent phenol broth similarly tested contains no viable bacteria after one hour. The lower concentration of phenol is bacteriostatic for *E. coli,* whereas the higher concentration is bactericidal. Eventually the bacteria will die in 0.3 per cent phenol broth, but in the meantime the survival of some cells can be shown by transferring a small portion to a medium lacking phenol, whereupon growth resumes.

Organisms exposed to a bactericidal concentration of a chemical survive only a short time; an endpoint of one hour is often arbitrarily selected. The distinction between bacteriostatic and bactericidal action ultimately becomes a matter of definition. Nevertheless, the terms bacteriostatic and bactericidal are useful with reference to various agents in the concentrations or under the conditions *in which they are normally employed.*

TYPES OF CELLULAR INJURY

INJURY TO CELL MEMBRANES

Any agent that alters the permeability of the cell membranes either interferes with the intake of essential substances or the excretion of waste materials or else permits the entrance of toxic substances or the loss of essential cell components. Detergents, for example, break the osmotic barrier and allow leakage of metabolically active cellular components, such as nitrogen and phosphorus compounds. Some chemicals dissolve or remove the cell membranes and destroy the equilibria that maintain constant composition and osmotic pressure. Penicillin damages the amino acid transport mechanism of the cells of sensitive species and inhibits the terminal, cross-linking steps in synthesis of murein (see Figure 5–7, page 88).

INJURY TO THE NUCLEUS AND GENES

Certain agents have a particular affinity for nuclei or genes or they damage them specifically. Basic dyes such as crystal violet react strongly with the nucleic acids of nucleoproteins, presumably by salt formation. Dilute solutions are bacteriostatic; more concentrated solutions are bactericidal. Gram-positive bacteria are more sensitive than gram-negative bacteria, a difference probably associated with the more acid nature of the proteins of gram-positive cells. Heavy metals may react with the sulfhydryl (—SH) groups of nucleoproteins.

Any damage to genes is reflected by inhibition of the enzymes they control; if the enzyme is essential, growth will cease. Some enzymes are not essential if suitable alternate substrates are supplied or if another mechanism is available to form the required product.

INHIBITION OF ENZYMES

Enzymes are proteins and as such are denatured by alcohols, phenols, heavy metals in high concentration, surface active substances, and other active agents. Denaturation is more or less irreversible. Low concentrations of heavy metals and mild oxidizing agents form inactive compounds by reversible reactions with —SH, phenol, indole, and amino radicals of enzymes or coenzymes.

Competitive Inhibition. Enzyme inhibition by competition or antagonism between either the substrate or the enzyme and the inhibitor substance is shown diagrammatically in Figure 12–1. Competitive inhibition is illustrated by the behavior of sulfonamide drugs (Fig. 12–2). Para-aminobenzoic acid (PABA) is a constituent of folic acid. If the chemically related substance sulfanilamide is present, it may replace PABA and prevent the formation of folic acid. Folic acid is a coenzyme essential to the synthesis of amino acids, purines, and pyrimidines.

STERILIZATION AND DISINFECTION

DEFINITIONS

Sterilization is the destruction or removal of *all* forms of life, whether animal or plant; macroscopic, microscopic or submicroscopic; harmful or harmless. It can be accomplished by filtration, by fire, by heat, by radiations, and by chemicals.

The word *disinfection* was originally used, about 1600, to express the idea of removing an agent capable of causing infection or disease. It is impossible to limit the term *disinfectant* to agents that combat infections, because many apparently harmless organisms produce disease under appropriate conditions. Moreover, pathogenic organisms such as *B. anthracis* and *C. tetani* are among the most resistant microorganisms, and any

A. Enzyme + Inhibitor 1 ⟶ Enzyme-Inhibitor 1

B. Enzyme + Substrate ⇌ Enzyme-Substrate ⇌ Enzyme + Product(s)

C. Inhibitor 2 + Substrate ⟶ Inhibitor 2-Substrate

Figure 12–1. *Inhibition of enzyme action by chemical competition. The normal reaction is indicated at B; enzyme and substrate interact to form a product or products and the enzyme is released for further activity. At A, an inhibitor that is chemically related to the normal substrate reacts with the enzyme, thus blocking the enzyme from reacting with the usual substrate. At C, a different inhibitor reacts with the normal substrate and forms a compound that cannot react with the enzyme.*

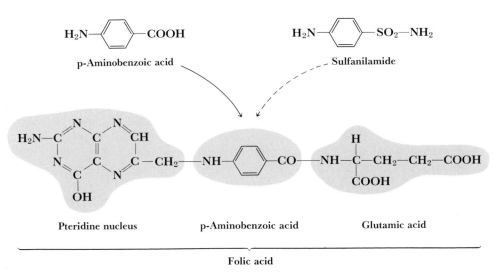

Figure 12–2. *Competitive inhibition. Para-aminobenzoic acid is an essential part of folic acid. If sufficient sulfanilamide is present, it replaces* p-*aminobenzoic acid, and folic acid is not formed.*

agent that kills them will also kill nearly all other species. Disinfectants are therefore essentially sterilizing agents.

Germicides are chemicals that kill germs. Most bacteriologists dislike the colloquial reference to harmful microorganisms as germs, but accept the term germicide as practically synonymous with disinfectant.

The word *antiseptic* originally designated an agent used to prevent *sepsis* or putrefaction. Sepsis is caused by growing microorganisms; hence, an antiseptic may inhibit multiplication without necessarily killing. Antiseptics are therefore bacteriostatic or fungistatic. In the United States Federal Food, Drug and Cosmetic Act, it was stated that an antiseptic is essentially the same as a germicide except in the case of an inhibitory drug to be used in prolonged contact with the body (e.g., a wet dressing or ointment).

Sanitization is the process of making an object sanitary or safe to use; the term implies freedom from esthetically objectionable material as well as harmful microorganisms. Proper dishwashing practices constitute a form of sanitization.

STERILIZATION

Removal of Microorganisms. *Filtration* is a practical method of removing microorganisms from liquids, both in the laboratory and in industry. It is used in the laboratory and in drug manufacturing plants to sterilize solutions that deteriorate when heated. Certain culture medium ingredients, for example, decompose at a high temperature; antisera and some other solutions for injection lose desired properties when sterilized by heat.

There are various types of filters (page 145). The Chamberland filter devised by a co-worker of Pasteur in 1884 is a hollow, unglazed porcelain candle. Berkefeld and Mandler filters are hollow cylinders of diatomaceous earth; Seitz filters are flat disks or pads of asbestos; sintered glass filters are made by heating finely powdered glass almost to the point of fusion; thin sheets or disks of cellulose-ester are used to remove bacteria from suspensions or to collect them for counting (in the latter application, the disk is moistened with a liquid medium and incubated a few hours, until colonies appear).

The effectiveness of a bacterial filter depends upon the size of the pores, the electric charge on the filter material, and other factors. Obviously an organism 1 μm in diameter cannot pass through a filter with pores only 0.5 μm in diameter, but it is not necessarily true that an organism 0.5 μm in diameter will pass through a filter whose pores are twice as large. The irregularities of the pores as they twist through a thick filter may retain bacteria of much smaller diameter than the average pore. More important, however, is the electric charge on the filter. Most bacteria are negatively charged in solutions of neutral reaction. They are therefore retained more effectively by a filter possessing a positive charge, even though the pores of the filter are large, than by a filter with a negative charge. Most of the filters just mentioned carry a negative charge.

Natural processes of filtration purify water as it percolates slowly through the soil. Microorganisms adsorb to soil particles, and subterranean water at considerable depths is sterile or nearly so. Water and sewage purification plants often use filtration through sand and gravel as a method of removing bacteria.

Sedimentation is another natural process by which microorganisms are removed from bodies of water. The specific gravity of most bacteria is slightly greater than 1.0; they therefore settle slowly in the fresh water of lakes, reservoirs, and slowly moving streams.

Sedimentation is hastened if the water contains soil or other particles of greater specific gravity to which the organisms adsorb. Natural sedimentation is a slow and uncertain method of purifying water, but it is reasonably effective over long periods.

Sedimentation is deliberately hastened in the laboratory by *centrifugation*. A centrifuge is a machine in which test tubes or bottles are whirled at high speed. The force on each particle within the container is several thousand times that of gravity, and bacteria can be thrown to the bottom of the container within a few minutes or hours. The supernatant liquid is rarely sterile, but its microbial population is tremendously reduced.

Sterilization by Heat.   *Incineration* is the most effective method of sterilization but is obviously limited in application. It is employed daily in the laboratory when needles and loops are flamed; the bacteriologist's first move upon entering his laboratory is to light his Bunsen burner. Contaminated swabs, paper materials, inexpensive clothing, and the bodies of discarded animals are often disposed of by incineration.

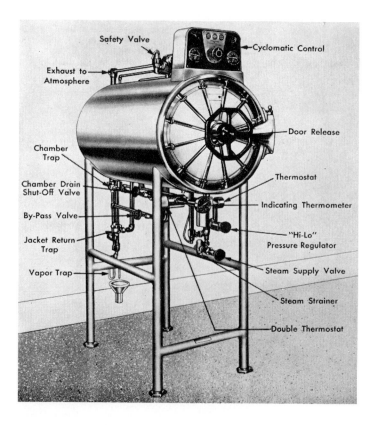

Figure 12–3.   *Autoclave with automatic controls used in bacteriologic laboratories. (Courtesy of American Sterilizer Company.)*

Boiling is a satisfactory method of disinfection when it is known that spore-forming bacteria are not a problem. The safety of water suspected of containing intestinal pathogens is ensured by boiling for 15 to 20 minutes. Boiling can never be relied on for sterilization unless adequate time is allowed. This may be several hours, as indicated in the preceding chapter.

Intermittent sterilization (also called tyndallization after Tyndall, who devised the method) is used when temperatures above 100° C. cannot be employed. The material to be sterilized is heated in flowing steam or boiling water for one-half hour on each of three successive days to kill vegetative cells. The germination of surviving spores is fostered by incubation at 30° to 37° C. between periods of heating.

Intermittent sterilization was formerly employed in the preparation of culture media that appeared to be adversely affected by temperatures higher than 100° C. Later, it was found that many of the same media can be sterilized satisfactorily and without serious damage at higher temperatures for shorter periods if the process is carefully controlled. Intermittent sterilization of home canned foods was practiced before pressure cookers became available, but it was never considered very safe. Pressure cooking is the only recommended procedure today.

Steam under pressure is the most widely used sterilizing agent in the hospital, laboratory, and food cannery. Laboratory media are usually sterilized in an autoclave at 15 pounds per square inch steam pressure (Fig. 12–3). The temperature is approximately 121° C. (Table 11–6), and sterilization is accomplished within 15 minutes if the materials are in small containers (test tubes) and properly distributed so that the entire contents of the autoclave can reach the sterilizing temperature. Large containers, such as flasks of culture medium, require longer sterilization. Hospital autoclaves loaded with surgical dressings also require longer treatment. The actual exposure time at 121° C. required to kill the most resistant bacterial spore is no longer than five or 10 minutes; the additional period of treatment permits steam to penetrate all parts of the autoclave chamber and

heat the contents of the containers to this temperature.

It is essential in operating an autoclave to be certain that the air has been completely exhausted. If steam at 15 pounds per square inch is introduced into an autoclave from which all air has been removed, a temperature of 250° F. (121° C.) can be attained (Fig. 12–4). When no air is removed, the final temperature reached is 212° F. (100° C.). Most autoclaves are so constructed that steam enters the chamber at the top and forces air out an exhaust port at the bottom. Only when pure steam issues uninterruptedly should the exhaust valve be closed to permit pressure to build up within the chamber. The 15 minute or longer period of sterilization is begun when a thermometer in the exhaust line indicates that the exhaust steam is at the sterilizing temperature (121° C.).

Laboratory glassware, such as pipettes

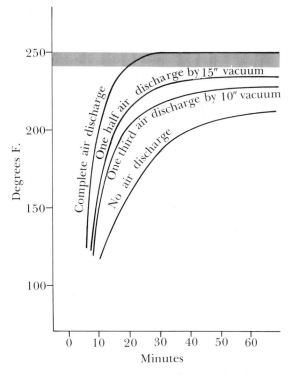

Figure 12–4. *Temperatures of steam-air mixtures in an autoclave. Steam at 15 lb. per square inch was introduced into a chamber completely or partially evacuated or unevacuated, and the temperatures within the chamber were determined at intervals. (After W. B. Underwood: A Textbook of Sterilization. Erie, Pa., American Sterilizer Co., 1941.)*

and Petri dishes, is usually sterilized dry in an oven. *Dry sterilization* requires a higher temperature for a longer time than moist sterilization: one to two hours at 160° to 180° C.

The presence or absence of moisture is important in sterilization by heat. Water presumably assists the denaturation of proteins. The effect of water on heat coagulation of egg albumin was shown many years ago (Table 12–1). It is probably not entirely a coincidence that dry egg albumin coagulated at the same temperature as that recommended for dry sterilization, whereas dilute solutions coagulated at a temperature that kills nonspore-forming bacteria within a few minutes.

FACTORS AFFECTING DISINFECTION

Killing microorganisms by any means, physical or chemical, is influenced by various factors. These should be known and understood, not only by those whose vocation is directly concerned with microorganisms, but also by every individual.

TIME

Time is one of the most important and most frequently overlooked factors in the control of microorganisms. Killing a microbial population is a gradual process, except in the case of incineration. The rate of death depends on

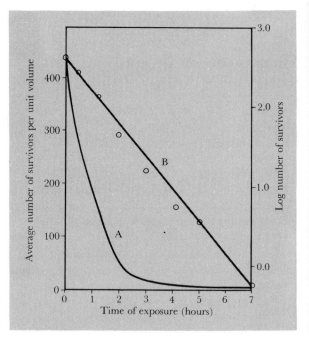

Figure 12–5. *Killing of spores of* Bacillus anthracis *by 5 per cent phenol at 33.3° C. A, Arithmetic plot;* B, *logarithmic plot of the same data. (Replotted from data of H. Chick, J. Hyg., 8:92–158, 1908.)*

the intensity or concentration of the killing agent as well as on various other factors which will be discussed.

The relationship between time and killing of spores of *B. anthracis* by phenol is shown in Figure 12–5. Curve *A* is an arithmetic plot of the surviving spores; the population decreased rapidly at first and more slowly thereafter. The same data plotted logarithmically lie close to the straight line *B*, which indicates that the rate of death was logarithmic; that is, half the surviving spores died in each successive equal time interval. The rate of death, expressed as −*g*, calculated according to the equation on page 225, is 43 minutes.

Any microbial population consists of cells of varying grades of resistance. Most cells are intermediate in resistance, a few are of low resistance, and a few of high resistance (Fig. 12–6). The most susceptible cells die first; those of greater resistance die in successive intervals until eventually the most resistant are killed. Young, actively growing cells are unusually susceptible to disinfectant agents, whereas mature or dormant cells are very resistant (Table 12–2).

TABLE 12–1. Effect of Moisture on the Coagulation of Egg Albumin by Heat*

Water (Per Cent)	Coagulation Temperature (°C.)
50	56
25	74–80
18	80–90
6	145
0	160–170

*Heating time, 30 minutes.

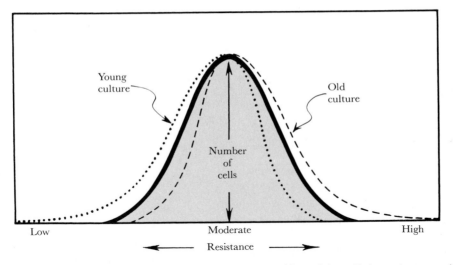

Figure 12–6. *The resistance to disinfection of cells in a culture. Most of the cells in an average culture* (solid line) *are moderately resistant. Young cultures* (dotted line) *contain a greater proportion of susceptible cells; old cultures* (dashed line) *contain more resistant cells.*

TEMPERATURE

Temperature markedly influences the efficiency of chemical disinfection. Killing is fundamentally a chemical process. The rates of chemical reactions increase with the temperature. It is to be expected, therefore, that disinfection will occur more quickly at high than at low temperatures. Table 12–3 shows that *S. aureus* is killed approximately five times as rapidly at 20° C. as at 10° C. by various concentrations of phenol. The factor 5, known as the *temperature coefficient* of disinfection (also called Q_{10}), indicates the effect of a 10 degree rise in temperature on the rate of killing. The Q_{10} of disinfection varies with the organism, the disinfectant, and other factors. It may be as low as 1.5 or as high as 50; average values are between 2 and 15.

If, as in the above instance, the temperature coefficient of disinfection is 5 between 20° and 30° C., the rate of destruction at 40° C. should be 25 times as great as at 20° C., and at 50° C. it should be 125 times as great. The practical importance of this fact is obvious; whenever it is possible to employ high temperatures it is highly advisable to do so. This explains why dishwashing machines, which can operate at 80° C., are recommended or

TABLE 12–2. Susceptibility of Young and Older Cells of *Escherichia coli* to Disinfection by 0.01 N NaOH at 30° C.*

Age of Culture (Hours)	Time to Kill 99.99% of the Cells (Minutes)
8	9.90
17	26.75

* From data of Watkins, J. H., and C.-E. A. Winslow, J. Bact., *24:* 243–265, 1932.

TABLE 12–3. Effect of Temperature on Killing of *Staphylococcus aureus* by Phenol*

Phenol (Per Cent)	Disinfection Time at	
	10° C.	20° C.
	Minutes	
1.82	17.5	5
1.66	40	7.5
1.54	70	12.5
1.43	100	20
1.33	150	30

* From data of Tilley, F. W., J. Bact., *43:* 521, 1942.

even required in eating establishments in preference to hand washing, which cannot be performed at temperatures much above 45° C.; the bactericidal efficiency of soap or detergents is enormously enhanced by the 35° temperature difference.

CONCENTRATION

The concentration of a disinfectant chemical profoundly influences the rate of killing of bacteria (Table 12–4). A moderate increase in concentration often multiplies the rate of killing by a large factor. Low concentrations exert no bactericidal action and still lower concentrations may actually stimulate microbial growth. This phenomenon is not peculiar to microbiology; small doses of poisonous substances are frequently used in medicine as stimulants (e.g., strychnine, epinephrine).

The concentration of a chemical to be employed for disinfection depends upon the chemical itself and upon the conditions in which it is to be used. The upper limit is usually dictated by the destructive or toxic action of the material and by economic considerations. Manufacturers' instructions are useful guides.

NATURE OF THE MEDIUM

The nature of the *medium* in or upon which the organisms are situated affects the efficiency of disinfection. Most disinfectants

TABLE 12–4. Effect of the Concentration of Phenol on the Killing Time of *Salmonella typhi**

Phenol (Per Cent)	Killing Time (Minutes)
1.11	∞
1.17	50
1.25	30
1.33	15
1.43	10
1.54	5

* From data of Tilley, F. W., J. Bact., 38: 499–510, 1939.

kill microorganisms by virtue of their ability to react with organic cell constituents. It is not surprising that extraneous organic matter also combines with the disinfectant and reduces its effective concentration.

pH

The pH influences the disinfection process by its effects on the organisms and on the disinfectant. Microorganisms are usually more resistant when suspended in media of a reaction satisfactory for growth, and killing by chemicals proceeds more rapidly as the pH departs from this value; pH also affects the degree of ionization of chemicals. Many disinfectants are more active in the undissociated state. Chlorine compounds decompose and liberate the effective constituent, chlorine, more rapidly in neutral or slight acid solutions than in alkaline solutions. It is apparent that the optimum conditions must be determined for each specific application.

NATURE OF THE ORGANISMS

The nature of the organisms to be killed, of course, cannot be controlled, but this factor must be taken into account. Spores of bacteria are more difficult to kill than vegetative cells; bacteria possessing capsules are usually more difficult to destroy than noncapsulated cells; acidfast bacteria, although not spore-forming, resist many disinfectant agents. The previous history, age, and cultural conditions of the test organisms are important factors in their susceptibility to killing agents.

CONTACT

A final factor affecting chemical disinfection is the opportunity for effective contact between the disinfectant and the microorganisms. A dry chemical placed in contact with bacteria will obviously be ineffective. Moisture is essential for the disinfection process. Surface tension depressants such as soaps and synthetic detergents improve the contact between chemicals and microbial cells by concentrating the chemicals at the cell surfaces. An alcoholic solution of iodine (i.e., tincture of iodine) owes much of its usefulness to the alcohol that brings the chemical into effective contact with skin bacteria. Alcohol is an excellent wetting agent.

CHEMICAL DISINFECTION

EVALUATION OF DISINFECTANT CHEMICALS

There are many methods of evaluating the disinfectant power of chemicals. No one method is satisfactory for all substances or for all conditions under which they may be used. Several methods will be described briefly.

Phenol Coefficient. One of the oldest procedures is the phenol coefficient method. This was originally intended to be used only for comparing compounds that act on bacteria in the same manner as phenol, that is, cresols and other derivatives or higher homologues of phenol. The procedure was later applied to chemicals that act in other ways, with the result that much useless or misleading information was secured and publicized. More recent awareness of this fact has led to the introduction of other methods of testing.

The phenol coefficient is a figure comparing the dilutions of phenol and of another chemical that possess equivalent killing power for a specified test organism, either *S. aureus* or *Salmonella typhi*. Test tubes containing various dilutions of phenol and of the chemical of unknown potency, X, are inoculated with constant amounts of the test organism (Table 12–5). Five, 10, and 15 minutes later, subcultures are made from each solution in tubes of special nutrient broth, which are then incubated. Results like those in the table indicate that a dilution of 1:80 phenol is strong enough to kill *S. aureus* within five minutes; a dilution of 1:90 phenol does not kill in five minutes but does kill within 10 minutes; a dilution of 1:100 phenol kills the test organism within 15 minutes but not in 10. Similar results are obtained with chemical X in different dilutions. Those dilutions that kill the test organism within 10 minutes but not in five minutes are compared, and the phenol coefficient is the number obtained by division, as indicated.

The phenol coefficient indicates that chemical X can be diluted five times as much as phenol and still possess equivalent killing power for *S. aureus* under the conditions of the laboratory experiment. The results cannot necessarily be applied to any other organism; in fact, phenol coefficients with *S. typhi* frequently differ from those with *S. aureus*. If chemical X has the same general chemical or biologic properties as phenol, the results are a little more significant than if the two chemicals are entirely different. A

TABLE 12–5. Phenol Coefficient Determination

Chemical	Dilution	Growth of *S. aureus* in Subculture after Exposure for		
		5 Min.	10 Min.	15 Min.
Phenol	1:80	−	−	−
	1:90	+	−	−
	1:100	+	+	−
X	1:400	−	−	−
	1:425	−	−	−
	1:450	+	−	−
	1:475	+	+	−
	1:500	+	+	+

Phenol coefficient of X = $\frac{1}{90} \div \frac{1}{450} = 5$

substance of different properties may be affected quite differently by extraneous organic matter, or by variations in pH, temperature, etc.

MIC Test. A similar measure of the antimicrobial activity of a chemical is the MIC (minimum inhibitory concentration) test. The MIC is determined by preparing a series of dilutions of the agent in a culture medium, inoculating each with an equal number of a suitable test organism, and observing for growth as shown by turbidity after appropriate incubation. The lowest concentration that prevents appearance of turbidity is the MIC. This value is affected by the nature and number of bacteria used in the test, the composition and pH of the medium, and the temperature and time of incubation. It is useful for comparing the effectiveness of different chemicals against a given organism or for comparing the sensitivity of different organisms to the same chemical. This test is broader in its applicability than the phenol coefficient test.

Agar Plate Tests. Agar plate methods are widely used to test the inhibitory power of chemicals, ointments, antibiotics, and other substances against various organisms. Appropriate nutrient agar inoculated heavily with the test organism is poured into Petri dishes and allowed to harden. Open-ended cylinders pressed into the agar are filled with the test chemical; filter paper disks saturated with the solution, or drops of ointment, are placed upon the agar. After the dishes are incubated, zones of inhibition of bacterial growth surround those substances that possess bactericidal or bacteriostatic action (Fig. 12–7). The diameter of the zones of inhibition depends upon the diffusibility of the chemicals and their antibacterial potency. This method does not necessarily measure killing effectiveness, because the chemical remains in constant contact with the test organism. However, it is useful for comparing the inhibitory powers of a variety of chemicals against any single organism or group of organisms; it is also convenient for testing the sensitivity of various bacterial species to any given inhibitory substance. The test is easily performed, and results are often available within a few hours.

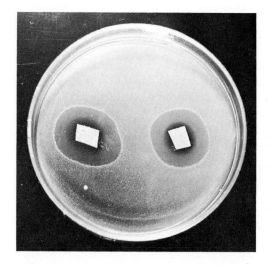

Figure 12–7. *Inhibition of bacterial growth by a common household antiseptic. The agar in the dish was inoculated with bacteria. Pieces of paper were soaked in the antiseptic and laid on the agar. The chemical, diffusing out of the paper, prevented growth of bacteria in the immediate vicinity.*

Death Curves. The most laborious but for some purposes the best method of testing disinfectant chemicals is the death curve determination. Mixtures of the chemical solution and the test organism are prepared, and counts of the surviving viable bacteria are made at suitable intervals. The temperature, composition, and pH of the test solution, the number and previous history of the organism, and other factors can and must be controlled in order to secure consistent and significant results. The information obtained constitutes a complete study of the dynamics of disinfection by the chemical in question.

No laboratory test is as satisfactory as actual experience. The laboratory can, however, reduce the number of trials and errors that will be necessary in practice. The ultimate test of a surgical disinfectant is the occurrence or nonoccurrence of surgical infection; the test of a dairy disinfectant is the microbial quality of the milk.

Toxicity Tests. One other type of test applied to antiseptics or disinfectants used upon the human body should be mentioned. Any chemical used for personal hygiene must be nontoxic. Many antibacterial chemicals kill or damage human cells. The dividing line between toxicity for microbe and for man is often narrow.

Several methods are available to assay the toxicity of chemicals for human or other animal cells. Test cells or tissues include guinea pig leukocytes, the chorioallantoic membrane of the developing chick embryo, and tissue cultures of chick heart fragments. The toxicity of a germicide for animal cells is compared with its bactericidal power, and a figure known as the *toxicity index* is obtained. A chemical with a low toxicity index is less toxic to tissue than to bacteria, and is preferred for personal use if its germicidal activity is satisfactory.

Chemicals that Inhibit or Kill Bacteria. A few chemicals (Table 12–6) that are used to inhibit or kill bacteria will be discussed according to their principal mode of action (see page 247). Some will be mentioned more than once because their behavior is complex.

Chemicals that Injure Cell Membranes. *Soaps* are markedly surface active; that is, they tend to concentrate upon surfaces. These molecules orient themselves upon the surfaces of bacterial cells so that their hydrophobic fatty acid constituents adsorb to the relatively nonaqueous surface, and the cation is attracted to the surrounding aqueous medium. The mechanical strain thus established disrupts the cell membranes, and part of the protoplasm leaves the cells, or toxic materials enter. The extent of damage determines whether the effect is bacteriostatic or bactericidal.

The cleansing action of soaps is important because it reduces the number of organisms to be killed or removed by other means. The surgical scrub is an effective method of reducing "adventitious" skin bacteria: those that gain access from the environment.

Synthetic detergents, like soaps, contain both hydrophobic and hydrophilic radicals and are usually subdivided according to the nature of the hydrophilic group: anionic, cationic, or nonionic. Anionic and cationic detergents disrupt cell membranes and permit nitrogen and phosphorus compounds to leak out of the cells. Probably they combine with membrane lipids and proteins, and also denature proteins.

Strong solutions of *acids* and *alkalies* quickly digest any form of organic matter and destroy bacterial cell walls and membranes. The bactericidal action of mineral acids appears to depend upon the pH of their solutions rather than their normality. Solutions at pH 2 or less are bactericidal against nonspore-forming bacteria. Poorly ionized organic acids, such as acetic and benzoic, are bactericidal at higher pH values; their effectiveness is attributed either to specific action of the anion or to the undissociated molecule.

Alkalies owe their bactericidal properties to the hydroxyl ion concentration of their solutions. Sodium and potassium hydroxides are germicidal at pH 12 or higher. Ordinary household lye (NaOH) is the most practical disinfectant for barns, stables, and chicken houses. Quicklime in the form of *fresh* whitewash is also used to disinfect chicken houses. Trisodium phosphate combines cleansing action with alkalinity and is useful in cleaning dairy equipment and in dishwashing machines. Its effect is greatly enhanced at high temperatures.

Chemicals that Inhibit Enzyme Activity and/or Denature Proteins. Most bacteriostatic or bactericidal chemicals inhibit or kill microorganisms by reacting with proteins, which are distributed generally throughout the cells as structural and enzymic components. Some chemicals denature proteins, others oxidize radicals such as the sulfhydryl group (—SH), and still others form substitution or addition products.

Chemicals that *denature* proteins include alcohols, phenols, and heavy metals and their salts. Early reports of the disinfectant action of *ethyl alcohol* unduly emphasized a peculiar effect of concentration that has been handed down in textbooks and medical practice for many years without effective challenge. Some experiments, particularly with dried bacteria, indicated that the most effective disinfection occurred with solutions containing only 50 to 70 per cent alcohol in water; absolute or even 95 per cent alcohol had practically no killing effect. The belief therefore arose that reliance could not be placed upon the usual 95 per cent solution for practical disinfection. Later observations indi-

Text continued on page 261

TABLE 12–6. Properties and Uses of Representative Bacteriostatic or Bactericidal Chemicals

Group	Examples	Significant Properties	Mode of Action	Uses
Chemicals that injure cell membranes				
Soaps		Na or K salts of long chain fatty acids; surface tension depressants; bactericidal action improved by high temperature; effective against pneumococci	Disrupt cell membranes and increase permeability	Cleansing, mechanical removal of microorganisms
Detergents		Hydrophobic hydrocarbon chain, sterol, etc., and a hydrophilic anion or cation; activity better at high temperature; inhibited by soaps	Disrupt cell membranes, probably by combining with lipids and proteins; N and P compounds leak out of cells	Cleansing and bactericidal action
	Sodium lauryl sulfate	Anion: carboxyl, sulfate, sulfonate, etc.; more active in acid solution		Selective culture media; kill gram-positive bacteria
	Quaternary ammonium halides	Cation: substituted ammonium; more active in alkaline solution		Skin disinfection, dairy sanitation; kill gram-positive and gram-negative bacteria
Acids	Mineral acids: H_2SO_4	High concentration of H^+ ions	Destroy cell walls and membranes	Limited by corrosiveness mainly to laboratory use
	Organic acids: acetic, benzoic, boric, etc.	Poorly ionized	Anions or undissociated molecules combine with protoplasmic constituents	Preservation, mild antisepsis
Alkalies	Lye (NaOH), quicklime [$Ca(OH)_2$], Na_3PO_4	OH^- ions	Destroy cell walls and membranes	Disinfection of barns, chicken houses; dishwashing and dairy sanitation (Na_3PO_4)

Chemicals that inhibit enzyme activity and/or denature proteins

Alcohols	In order of increasing activity: methyl, ethyl, propyl, butyl, amyl		Bacteriostatic and bactericidal; denature and coagulate proteins	Skin disinfection (70-95% C_2H_5OH), used in tinctures to increase "wetting" power of other chemicals; kill vegetative cells, little effect on spores
Phenols	Phenol, cresols, lysol, hexylresorcinol, etc.	Surface tension depressants; activity improved by high temperature, acid	Bactericidal; denaturation and precipitation of proteins	Disinfection of laboratory equipment, instruments, bench tops, garbage pails, toilets
	Bis-phenols (hexachlorophene)		Bacteriostatic	Deodorants in soaps; inhibit gram-positive bacteria
Oxidizing agents	Halogens: Cl, Na or Ca hypochlorite	Inhibited by extraneous organic matter	Bactericidal; oxidize $-SH$, $-NH_2$, or indole nucleus of enzymes or coenzymes	Purification of water, dairy disinfection, restaurant sanitation
	Iodine			Skin disinfection, especially as tincture
	H_2O_2	Unstable; decomposes to H_2O and O_2	Bacteriostatic, mildly bactericidal	Antisepsis of cuts, minor wounds
Heavy metals	$HgCl_2$	Very toxic; combines with organic matter	Highly bacteriostatic; react with $-SH$ groups of enzymes or coenzymes; precipitate proteins	Laboratory disinfectant
	"Merthiolate," "Metaphen," phenyl-mercuric nitrate	Not greatly affected by organic matter; relatively nontoxic		Skin antisepsis, preservation of biologicals (sera, etc.)
	$AgNO_3$, silver proteinate (Argyrol)			Antisepsis of mucous membranes of throat and eyes (to prevent ophthalmia neonatorum)

TABLE 12-6. Properties and Uses of Representative Bacteriostatic or Bactericidal Chemicals
(Continued)

Group	Examples	Significant Properties	Mode of Action	Uses
Formalde-hyde	HCHO	Reacts with —NH$_2$ and —OH groups; relatively unaffected by extraneous organic matter; effective against spores	Bactericidal in high concentration; reacts with enzymes, nucleic acids; coagulates proteins	Disinfection of contaminated laboratory and surgical instruments and equipment
Sulfon-amides	Sulfanilamide, sulfadiazine, etc.	Chemically related to p-aminobenzoic acid	Bacteriostatic; compete with PABA in synthesis of folic acid	Chemotherapy of human and animal infections; inhibit pneumococci, meningococci, streptococci, *Haemophilus influenzae*
Antibiotics	Penicillin, streptomycin, chloramphenicol, tetracyclines, etc.		Bactericidal or bacteriostatic; interfere with one or more essential enzyme reactions	Chemotherapy of human and animal infections
Chemicals that injure nuclei and genes				
Basic dyes	Crystal violet, brilliant green	React with acid radicals	Bacteriostatic; probably form salts with nucleic acids	Selective culture media, skin and oral antisepsis; inhibit gram-positive bacteria

cated that *moist* nonspore-forming bacteria of many species are killed within five minutes by ethyl alcohol in concentrations between 40 and 99 per cent.

Alcohol is one of the most widely used chemicals for the destruction of bacteria on the skin, for example, in surgical practice. Any concentration between 70 and 95 per cent destroys or removes approximately 90 per cent of the normal resident skin flora within two minutes.

Viruses are apparently not as readily inactivated or killed by alcohols as are bacteria. Outbreaks of serum hepatitis have been attributed to virus on hypodermic needles or lancets that had been cleansed or stored in alcohol between use on successive patients. Resterilized or disposable sterile needles and lancets are now recommended.

Phenols and phenolic compounds are highly bactericidal in proper concentration. Some phenols are poorly soluble in water but can be emulsified in dilute soap solutions to provide effective disinfectants. Lysol is such a preparation containing ortho-, meta-, and para-cresols. Phenol and Lysol are too irritating or caustic in germicidal concentrations to be used for personal disinfection, but they are often used as laboratory and household disinfectants.

Oxidizing agents, such as halogens, hydrogen peroxide, and ozone, are bacteriostatic or bactericidal and exert their inhibitory or lethal action by oxidizing chemical groups such as —SH, —NH$_2$, or the indole nucleus, which are essential for the activity of some enzymes or coenzymes. Halogens probably also damage cell membranes and form substitution or addition products with proteins.

Halogens and their derivatives were among the earliest chemicals used to control microbial activities. In 1854, before the role of microorganisms in putrefaction was known, chloride of lime was added to the sewage of London as a deodorant. The first large scale use of *chlorine* for water purification in the United States was in 1908 in Chicago. Chlorine is ordinarily used in the form of gas or as one of the hypochlorites, but in either case the reaction with water yields hypochlorous acid, HCIO, some of which

decomposes and liberates nascent oxygen. The disinfectant action is believed to be caused by the ClO$^-$ ion and by nascent oxygen.

Care must be taken in using chlorine compounds to ensure an adequate concentration for the desired purpose, because of the lability of chlorine and its great capacity for combining with extraneous material. The concentration of chlorine necessary to kill nonspore-forming bacteria under favorable conditions is extremely small. Water that is free from organic matter is rendered safe for drinking within a few minutes by only 0.1 to 0.2 parts of available chlorine per million parts of water (0.1–0.2 μg./ml.).

Iodine is probably the most widely used skin disinfectant. It is strongly bactericidal against many kinds of microorganisms, including *M. tuberculosis* and bacterial spores, and the concentration required to kill does not vary greatly among different species. Moreover, iodine solutions are rapidly germicidal and relatively nontoxic. It should be cautioned that rapid germicidal action does not necessarily mean immediate germicidal action. Rubbing or swabbing an area of skin for two minutes or applying a wet solution for five minutes is usually recommended.

Chemicals that form *substitution* or *addition products* with proteins may interfere with cellular metabolism by reacting with enzymes. Heavy metals, such as mercury and silver, for example, combine with the essential —SH radicals of certain enzymes and coenzymes and produce compounds lacking the normal activity. Formaldehyde and the halogens may also react with —NH$_2$, —OH, and other radicals of proteins and nucleic acids.

Heavy metals and their salts are bactericidal only in relatively strong solutions, but they are almost unbelievably bacteriostatic in dilute solution. The mercuric ion reacts particularly with —SH radicals, and —S—Hg—S— linkages form. When low concentrations of mercury salts are used, the foregoing reaction is readily reversed by addition of H$_2$S or other compounds containing the —SH radical (e.g., thioglycollic acid,

$CH_2(SH) \cdot COOH$). Organisms exposed to low concentrations of mercury salts appear to be dead when subcultured in a medium such as nutrient broth, which contains few sulfhydryl compounds; nevertheless they multiply readily when tested in thioglycollate broth. These solutions are bacteriostatic rather than bactericidal. Higher concentrations of mercury produce irreversible damage, perhaps by protein precipitation, and hence are bactericidal.

Silver and copper are bacteriostatic in extremely small concentration. A coin pressed lightly into the surface of nutrient agar inoculated heavily with *S. aureus* dissolves sufficiently to inhibit growth of the bacteria within a zone of several millimeters. The growth of *E. coli* is inhibited by one part of silver in 5 billion parts of broth, and the same species is apparently killed within a few hours by 10 to 100 parts of silver per billion parts of distilled water. This phenomenon is known as the *oligodynamic action of metals.* Bacteria, yeasts, and trypanosomes are killed by solutions calculated to contain only 100,000 to 10,000,000 silver ions per cell, a figure of the same order of magnitude as the estimated number of protein molecules in each cell. It can therefore be assumed that only one metallic ion per protein molecule suffices to kill the cell.

Formaldehyde inactivates most but not all enzymes. It reacts with —NH_2 and —OH radicals of enzymes, proteins, and nucleic acids, and is also a protein coagulant. It is bacteriostatic in low concentration and bactericidal in high concentration and is capable of killing spores of both aerobic and anaerobic bacteria. Formaldehyde disinfects contaminated surgical and laboratory implements after an exposure of several hours, despite the presence of considerable organic matter.

Chemicals that Injure Nuclei and Genes. A few chemicals with particular affinity for nucleic acids are bacteriostatic or bactericidal. Formaldehyde and other chemicals react with amino and hydroxyl radicals of nucleoproteins as well as other proteins. Basic dyes are known to stain nuclear material of higher forms intensely; they also stain bacteria deeply, presumably by combining with the nucleic acids. Crystal violet, brilliant green, and malachite green are powerfully bacteriostatic basic dyes, especially against gram-positive organisms; gram-negative bacteria are affected relatively little.

The hypothesis that basic dyes react with nucleic acids is supported by the observation that dye inhibition of bacteria is counteracted by adding nucleic acid or nucleotides. Inhibition can also be reversed by transferring bacteria from a medium containing dye to one lacking it, whereupon the organisms resume growth. The reversibility of inhibition by either method is evidence that the action of basic dyes is bacteriostatic rather than bactericidal.

Basic dyes are used in culture media to promote the selective growth of gram-negative bacteria. Drinking water is considered unsafe if it contains gram-negative rod bacteria of intestinal origin; these organisms are frequently mixed with gram-positive cocci and spore-formers, which confuse and prolong the analytical procedure. Brilliant green or crystal violet is added to culture media to inhibit the latter, while permitting growth of the gram-negative bacteria.

CHEMOTHERAPY

Chemotherapy is the treatment of disease by chemicals that inhibit or kill the infectious agent but do not harm the host in the concentrations employed. Since the composition and physiologic activities of most parasites are fundamentally like those of their hosts, relatively few chemicals are effective without being harmful. This is one aspect of the specificity of chemotherapeutic action.

Specificity is further demonstrated by the selective antimicrobial behavior of chemotherapeutic substances. Chemicals effective against some kinds of organisms are useless against others. This characteristic is attributed to the fact that different chemicals interfere with different physiologic processes and, although all organisms perform much the same basic activities (respiration, synthesis, etc.), they may differ in the individual steps or pathways utilized, as was pointed out in Chapter 8.

SYNTHETIC DRUGS

Following Ehrlich's long successful search for a "magic bullet" that would kill or inhibit a disease-producing organism without harming man, a quarter century elapsed before the next great advance in chemotherapy. Domagk reported in 1935 that Prontosil, a red dye synthesized and patented three years previously, was highly effective against hemolytic streptococci in animal experiments. Finally, administered to humans, it miraculously cured the usually fatal streptococcal septicemia known colloquially as "blood poisoning."

Prontosil was soon found to break down within the tissues to sulfanilamide. Eventually, as already pointed out (see page 248), it was shown that sulfanilamide competes with p-aminobenzoic acid in the synthesis of folic acid, which is essential as a coenzyme in catalyzing the formation of amino acids. Mammalian tissues are not poisoned by the drug because animals secure preformed folic acid in their diets.

Sulfanilamide was not a cure-all. It was found to be bacteriostatic, inhibiting some but not all bacteria and permitting normal body defenses to combat infection. Moreover, the drug might be excreted rapidly, the speed of elimination depending partly upon the route of administration. In order to combat a wider variety of infections, caused by agents localized in different parts of the body, a search was made for derivatives of sulfanilamide that would be more effective or more useful by virtue of different solubilities, excretion rates, and other characteristics. Several thousand compounds were synthesized and tested, and a few turned out to be useful: sulfapyridine, sulfathiazole, sulfadiazine, sulfaguanidine, etc. Structural formulas for these and the parent compounds are shown in Figure 12–8.

The active portion of each compound is the aminobenzene ring; the sidechains affect solubility and other accessory properties. Sulfaguanidine and sulfamethoxypyridazine (Kynex), for example, are absorbed more slowly from the intestine than the other compounds and hence are more effective in treating bacillary dysentery. Sulfisoxazole (Gantrisin) is highly soluble and therefore does not form crystals in the urine, as do some other sulfonamides.

There is little qualitative difference in antibacterial properties among these compounds. In general they are active against streptococci, staphylococci, pneumococci, gonococci, meningococci, plague, and dysentery bacteria, but are relatively ineffective against the typhoid organism and other salmonellae and rickettsiae. They are also useful in urinary infections caused by gram-negative rods and in the prevention of rheumatic fever, bacterial endocarditis, and wound infections.

Other products of the chemistry laboratory with *in vivo* antibacterial activity include p-aminosalicylic acid (PAS), which is effective in treating tuberculosis. This was synthesized as an analogue of salicylate, because the latter can be utilized by *M. tuberculosis*. It was later shown to behave as an analogue of PABA in the metabolism of the organism. PAS is bacteriostatic, and its effectiveness is greatly enhanced by use in conjunction with isoniazid. Isoniazid resembles nicotinamide and pyridoxamine, and inhibits various enzymes for which pyridoxamine phosphate is a cofactor. The effectiveness of isoniazid is limited to *M. tuberculosis*, against which it is bactericidal even when the bacteria are situated intracellularly.

Nitrofurans are bactericidal against various gram-negative and gram-positive bacteria. Furadantin is used in chronic infections of the urinary tract.

Nalidixic acid is useful in urinary tract infections, particularly those caused by gram-negative bacteria.

ANTIBIOTICS

The term *antibiosis* was apparently first used by Vuillemin in 1889 to describe the competitive nature of biologic societies in which only the strongest or most fit survive. The same word was applied to microbial antagonism a few years later.

Antagonistic activity among microor-

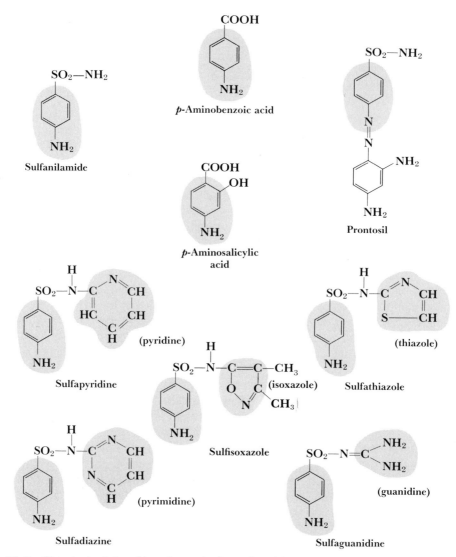

Figure 12–8. *Chemical relationships of* p-*aminobenzoic acid,* p-*aminosalicylic acid, sulfanilamide, Prontosil, and other common sulfonamides.*

ganisms had been known for several years. It has long been observed that plate cultures from the soil or air frequently contain "spreaders," whose growth is interrupted or "punctuated" by colonies of other bacteria surrounded by zones in which the spreader is inhibited. The suggestion had been made that products of these inhibitory organisms might be useful for treating disease. Emmerich and Löw in 1899 found that pyocyanase from cultures of *Pseudomonas aeruginosa* was highly bactericidal to many organisms, both gram-positive and gram-negative; unfortunately it is also toxic to animals.

Antibiotics are products of microbial activity that specifically inhibit or kill other microorganisms. Following Fleming's discovery of penicillin in 1929 (see page 37), Dubos (1939) found an antibiotic principle, tyrothricin, in cell-free filtrates of cultures of *Bacillus brevis*. This was later shown to contain two active crystallizable polypeptides, gramicidin and tyrocidine. Gramicidin is effective against hemolytic streptococci, pneumococci, and tetanus and gas gangrene bacteria, whereas tyrocidine is active against various gram-positive bacteria *in vitro,* but is ineffective *in vivo* because it is inactivated by body fluids such as blood and

TABLE 12-7. Sources and Properties of Some Common Antibiotics

Antibiotic	Date	Microbial Source	Chemical Nature	Specific Uses
Penicillin	1929	*Penicillium notatum*	Dipeptide (see Fig. 12-9)	Active against gram-positive bacteria, gonococci, meningococci, syphilis spirochetes.
Tyrothricin (gramicidin and tyrocidine)	1939	*Bacillus brevis*	Polypeptides	Active against gram-positive bacteria; local treatment of wounds and upper respiratory infections.
Streptomycin	1944	*Streptomyces griseus*	Basic glucoside (see Fig. 12-10)	Active against tularemia, some *Salmonella* and other gram-negative bacteria; treatment of tuberculosis.
Bacitracin	1945	*B. licheniformis*	Polypeptide	Active against gram-positive bacteria; mildly toxic.
Chloramphenicol (Chloromycetin)	1947	*Str. venezuelae*	Nitrobenzene derivative (see Fig. 12-11)	Active against gram-positive and gram-negative bacteria, rickettsiae, large viruses, spirochetes; specific for typhoid fever.
Polymyxin (Aerosporin)	1947	*B. polymyxa*	Polypeptide and fatty acid	Local treatment of mouth, throat, and wound infections; treatment of systemic *Pseudomonas* infection.
Chlortetracycline (Aureomycin)	1948	*Str. aureofaciens*	Tetracycline (see Fig. 12-12)	Active against gram-positive and gram-negative bacteria, rickettsiae, large viruses.
Neomycin	1949	*Str. fradiae*	Aminoglycoside	Active against gram-positive and gram-negative bacteria; for topical application and GI antisepsis.
Oxytetracycline (Terramycin)	1950	*Str. rimosus*	Tetracycline (see Fig. 12-12)	Active against gram-positive and gram-negative bacteria, rickettsiae.
Nystatin (Mycostatin)	1951	*Str. noursei*		Active against *Candida albicans* and other fungi.
Erythromycin	1952	*Str. erythraeus*	Macrolide (large lactone ring with attached sugars)	Activity like penicillin but including penicillinase-producing bacteria; more toxic than penicillin.
Tetracycline (Achromycin, etc.)	1953	Hydrogenation of chlortetracycline	Tetracycline (see Fig. 12-12)	Active against gram-positive and gram-negative bacteria, rickettsiae.
Novobiocin	1955	*Str. niveus*	Aromatic lactone, sugar, substituted phenol	Activity like penicillin but effective against penicillin-resistant staphylococci and some gram-negative bacteria.
Kanamycin	1957	*Str. kanamyceticus*	Aminoglycoside	Active against *S. aureus* and most gram-negative bacteria (except *Pseudomonas*); short-term treatment of urinary infections.

serum. Both drugs are toxic when injected into experimental animals and can be used in man only by topical application or instillation into certain body cavities.

The specificity and limited usefulness of the early antibiotics led to immediate search for other substances, particularly compounds effective against gram-negative bacteria and *M. tuberculosis*. Waksman and his associates isolated streptomycin from a soil actinomycete, *Streptomyces griseus,* in 1944, and many other antibiotics were discovered in the following years. Some of them are listed in Table 12–7.

An antibiotic must meet certain requirements in order to be suitable for use in treat-

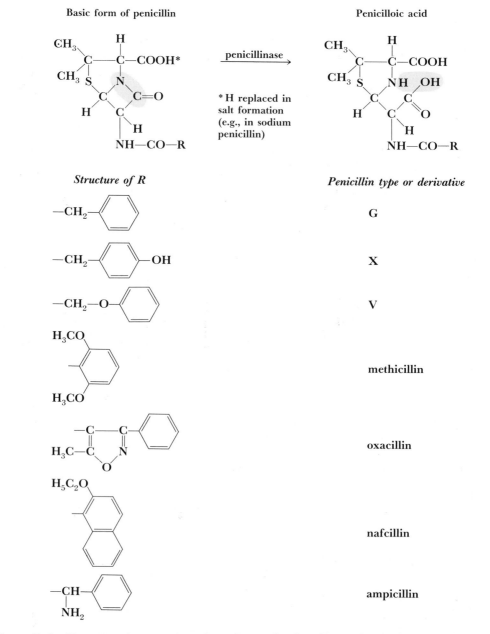

Figure 12–9. *Structures of some commonly used natural and semisynthetic penicillins and of penicilloic acid, which is formed upon hydrolysis of a C—N bond of penicillin (shaded area) by penicillinases from various bacteria. Methicillin, oxacillin, and nafcillin are resistant to penicillinase. Ampicillin has a broader antibiotic spectrum than penicillin G, but it is sensitive to penicillinase.*

ing human disease: (1) it must possess high antimicrobial activity *in vivo;* (2) it must not produce harmful or undesirable reactions in the patient (that is, it must not be toxic or hemolytic or produce histamine-like effects or allergic reactions [see page 384], or precipitate body proteins); (3) it should be soluble in water, saline, and body fluids; (4) it should be stable. Many substances meet some of these requirements, but only a few meet them all. Natural, modified, and semisynthetic penicillins and streptomycin are most widely employed; the tetracyclines (Aureomycin, Terramycin, tetracycline) and chloramphenicol are of next importance, followed by tyrothricin, bacitracin, and polymyxin.

The chemical structures of penicillin, streptomycin, chloramphenicol, and the tetracyclines are shown in Figures 12–9 through 12–12. Many of the other antibiotics are polypeptides. It is obvious that no one type of chemical structure is responsible for antibiosis, and therefore it is not surprising that inhibition or killing is attributed to a variety of mechanisms.

Nature and Uses of Common Antibiotics. *Penicillin.* There are actually many penicillins (Fig. 12–9), which share a common structure and possess different "R" radicals that comprise the second portion of the dipeptide. Penicillins G, X, and V are most active *in vivo.* Penicillin G is considered the

most practical and is used more often than the other forms. Penicillin V is resistant to the acids of the stomach, and when given orally it is absorbed from the intestine and yields high blood concentrations. Sodium, calcium, potassium, and procaine salts are prepared; the potassium and procaine compounds are most common clinically.

Crystalline salts of penicillin are stable for several months, particularly when kept cold; aqueous solutions are unstable and must be refrigerated. Penicillin is inactivated by heat, alkali, acid (except penicillin V), and penicillinase, which is produced by many species of bacteria. The hydrolytic action of penicillinase is indicated in Figure 12–9.

Some modified penicillins are manufactured biosynthetically by adding an excess of a compound containing the desired R group to the culture medium. Others are made semisynthetically: when the *Penicillium* is deprived of an R group donor substance, the mold produces a penicillin precursor (6-aminopenicillanic acid), to which a variety of R groups can easily be attached by simple chemical reactions. Semisynthetic penicillins, such as methicillin, oxacillin, and nafcillin, are particularly interesting because the structural alteration caused by introducing large inert ring compounds near the relatively weak C—N bond makes them resistant to penicillinase.

The amino group added to the —CH_2—

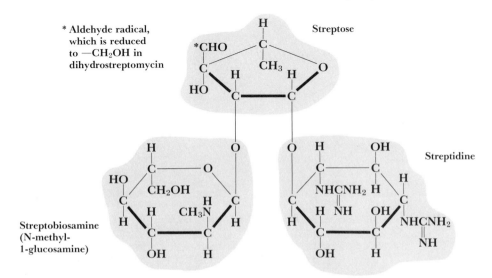

Figure 12–10. *Structure of streptomycin.*

portions of the R sidechain of penicillin G in manufacturing ampicillin markedly extends the antibiotic spectrum of ampicillin; although its activity against gram-positive bacteria is only about half that of penicillin G, it is effective against many gram-negative bacteria. As might be expected from the fact that the R group is small, ampicillin is sensitive to penicillinase.

Penicillin is soluble in water, ethyl alcohol, ether, and acetone. It is effective in the presence of blood, pus, and body fluids, and is remarkably nontoxic. However, an increasing number of individuals develop allergic hypersensitivity on continued use of the drug; reactions range from urticaria with itching hives and fever to acute, fatal anaphylactic shock.

Penicillin interferes with one of the final steps in the synthesis of cell wall material in sensitive bacteria, as pointed out earlier (see page 248). This occurs only under conditions that would normally permit growth of the organisms. The cells elongate and may swell

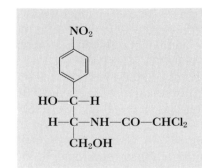

Figure 12–11. *Structure of chloramphenicol. This antibiotic is now made synthetically.*

but do not divide. Susceptible cells lose their walls and lyse, but if they are protected by 0.2 M sucrose, they convert to protoplasts (naked cells, bounded only by a plasma membrane). Animal cells are not affected by penicillin because they do not have cell walls containing cross-linked glycosaminopeptide (Fig. 5–7), and hence do not normally perform the critical reaction that would be inhibited.

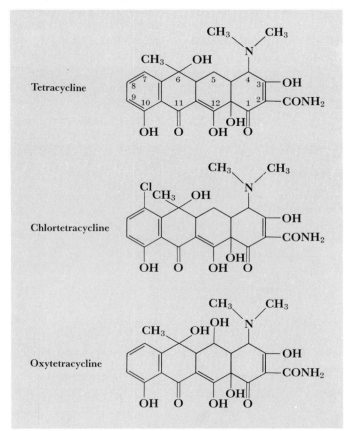

Figure 12–12. *Structures of the tetracyclines.*

Streptomycin. Streptomycin is a complex three-part basic glucoside (Fig. 12–10). Dihydrostreptomycin is produced from streptomycin by the catalytic hydrogenation of an aldehyde radical. It is not as toxic as streptomycin but has essentially the same antibiotic activity. The toxicity of these compounds results in damage to the eighth cranial nerve and vestibular apparatus; prolonged therapy may cause vertigo, tinnitus, and deafness.

Streptomycin is effective against numerous gram-negative bacteria, some gram-positive bacteria, and the tuberculosis organism. The usefulness of this drug is limited by the readiness with which many bacteria become resistant to it. Susceptibility to streptomycin varies from strain to strain and must be tested in individual cases. Development of resistance is particularly significant in the case of *M. tuberculosis,* because treatment of tuberculosis is prolonged. Combination therapy with isoniazid (isonicotinic acid hydrazide, INH) or *p*-aminosalicylic acid (PAS) greatly reduces the opportunity for the appearance of resistant mutants.

Streptomycin kills susceptible bacteria a few minutes after contact. Protein synthesis in a culture is inhibited at about the time that the viable count begins to decrease. Later, DNA and RNA synthesis and respiration (the oxalacetate-pyruvate condensation step in the Krebs cycle) are inhibited, and the cell membrane is damaged to such an extent that it becomes increasingly permeable to both intracelluar and extracellular substances. The site of action of streptomycin appears to be the ribosomes; streptomycin causes misreading of the genetic code, so that an mRNA codon on the ribosome binds an aminoacyl tRNA that would be expected to bind to a different codon. It is not yet clear how this explains the observed effects on the bacterial cell.

Chloramphenicol. Chloramphenicol, first isolated from cultures of *Streptomyces venezuelae* and later from cultures of other *Streptomyces* species, is now prepared by synthesis. It is the only antibiotic synthesized on a large scale. This compound is a nitrobenzene derivative.

Chloramphenicol is a broad spectrum antibiotic. It is effective against a greater variety of microorganisms than streptomycin or penicillin, including gram-positive and gram-negative bacteria, rickettsiae, and some of the larger viruses. It is particularly effective against the typhoid fever organism and is recommended for treatment of this disease. It is relatively ineffective against bacillary dysentery.

Chloramphenicol blocks the formation of peptide bonds on bacterial ribosomes and therefore inhibits protein synthesis. It is relatively nontoxic and is readily absorbed from the gastrointestinal tract; it can therefore be taken orally. There have, however, been a number of cases of anemia following its use.

Tetracyclines. The tetracyclines include the important broad spectrum antibiotics, chlortetracycline (Aureomycin), oxytetracycline (Terramycin), and Tetracycline. These compounds are very closely related chemically and possess similar antimicrobial properties. They inhibit many gram-positive and gram-negative bacteria and rickettsiae and are relatively nontoxic.

Chlortetracycline presumably inhibits protein synthesis and phosphorylation of mitochondria and adenine nucleotides, a reaction which suggests interference with the Krebs cycle. Oxytetracycline blocks the utilization of RNA reserves that are lost by depolymerization to mononucleotides. It also seems to interfere with the transfer of amino acids from mRNA to polypeptides and hence inhibits protein formation.

Antimicrobial Spectra. Antibiotics are characterized and identified by their antimicrobial spectra. The antimicrobial spectrum is determined by testing the ability of the chemical to inhibit a variety of microorganisms (Table 12–8). Penicillin, for example, inhibits many gram-positive bacteria and very few gram-negative organisms; polymyxin, on the contrary, inhibits a variety of gram-negative bacteria and is practically inactive against gram-positive organisms. Streptomycin, chloramphenicol, and the tetracyclines inhibit most representatives of both groups.

When a new antibiotic is obtained, it is easier to determine its probable identity by testing its ability to inhibit 15 or 20 species of bacteria than by chemical analyses. A sub-

TABLE 12–8. Antimicrobial Spectra of Some Commercially Available Antibiotics*†

Organism	Penicillin	Streptomycin	Chloramphenicol	Tetracyclines	Gramicidin	Bacitracin	Polymyxin
Gram-positive							
Streptococcus pyogenes A	+	+	+	+	+	+	−
Staphylococcus aureus	+	+	+	+	−	±	−
Bacillus subtilis	+	+	+	+	+	−	+
Clostridium perfringens	+	−	±	+	±	±	−
Corynebacterium diphtheriae	+	+	+	+	+	+	−
Mycobacterium tuberculosis	−	+	−		+		−
Gram-negative							
Neisseria meningitidis	+	+	+	+	−	+	+
Escherichia coli	−	+	+	+	−	−	+
Salmonella species	−	+	+	+	−	−	+
Pseudomonas species	−	+	−	±	−	−	+
Haemophilus influenzae	+	+	+				+
Brucella species	+	+	+				+

* + = inhibition.

† Adapted from Welch: *Principles and Practice of Antibiotic Therapy.* New York, Medical Encyclopedia, Inc., 1954.

stance that inhibits the same organisms as bacitracin, for example, is considered to be identical with bacitracin, or at least to possess no advantage over it as far as inhibitory properties are concerned.

Resistance to Antibiotics. The development of resistance to antibiotics by previously sensitive strains of bacteria was discussed in Chapter 9. It was pointed out that perhaps once in every million or 10 million cell divisions a mutant might appear that is resistant to a given antibiotic. If this mutation occurs in a patient under treatment with the antibiotic, the mutant will possess greater survival value than the patient's normal bacteria and will shortly outnumber them. Further treatment with the same antibiotic will be of no advantage. Moreover, any subsequent cases of the disease arising from exposure to the first patient will be caused by bacteria resistant to the antibiotic.

Since 1945, penicillin-resistant strains of *Staphylococcus aureus* have become widespread. This organism is present on the skin and mucous membranes of many normal individuals and can be transferred readily from one person to another. It is capable of infecting wounds, sinuses, lungs, bones, kidneys—any part of the body. Most *S. au-*

reus infections were successfully treated by penicillin when the drug was first available. By 1955, however, there had been a tremendous increase in penicillin-resistant staphylococci, particularly in hospitals. Physicians, nurses, and other attendants apparently developed a staphylococcal flora resistant to penicillin, partly by constant association with and use of penicillin, and partly by contamination from patients under treatment with penicillin. The result is that patients entering a hospital for any cause are likely to become infected with penicillin-resistant staphylococci in the hospital. Physicians and hospital administrators recognize that the solution to the problem is no different now than it was a century ago: strict aseptic technique.

Control of Resistance to Antibiotics. The development of antibiotic resistance by pathogens within the body of a patient can be decreased by administering doses of antibiotic large enough to eliminate the infectious agent at once. This is possible only in the case of drugs of little or no toxicity used in patients not allergic to the drug and infected by organisms that do not develop resistant mutants rapidly.

Combined drug therapy also decreases

the appearance of resistant organisms. Resistance to different, unrelated antibiotics develops independently. Cells resistant to one antibiotic may arise once in a million cell divisions, and cells resistant to a second antibiotic may also arise once in every million divisions. The probability that a cell resistant to both antibiotics will occur is therefore one in one million million, an almost infinitesimal chance.

The use of antibiotics is obviously not an unmixed blessing. Reliance on antibiotics to control bacterial infection doubtless led some physicians to relax aseptic techniques and to treat various infectious diseases without adequate diagnosis. This practice was often successful before antibiotic-resistant bacteria became common. Moreover, some antibiotics were made available without prescription, and their overuse in treating trivial ailments incited allergic hypersensitivity in many persons. These individuals could not then be given the same agent to treat serious illness without risk of a fatal anaphylactic reaction.

Sensitivity Tests. It is advantageous to test the sensitivity to antibiotics of bacteria isolated from a patient before instituting treatment. The physician can then select the drug that is most likely to be effective in each specific case. There are several types of sensitivity tests. Pure cultures of the isolated organism from the patient are grown in broth containing dilutions of the available antibiotics. Choice of an antibiotic to use in treating the infection is limited to those that inhibit the organism in low concentration (i.e., when highly diluted). This method requires considerable labor and expense, but the results are likely to be significant.

The agar cup plate technique is easier and quicker. Open cylinders of glass, porcelain, or stainless steel are placed on Petri dishes containing an agar medium inoculated with the organism responsible for the patient's infection. The cylinders are filled with dilutions of the antibiotics and incubated for a few hours. The organism is usually most susceptible to the antibiotic that produces the largest zone of inhibition. A simpler modification of the same technique makes use of filter paper disks, either singly or in-

Figure 12–13. *Test of the sensitivity of an organism isolated from a patient to various antibiotics and synthetic drugs by means of a filter paper "multidisk." The tip of each arm is impregnated with a chemical, which diffuses into the culture medium. Antibiotics F, B, and CA markedly inhibit the test organism illustrated. (From Microbial Testing for Susceptibility to Chemotherapeutic Agents. Chicago Heights, Ill., Consolidated Laboratories, Inc., 1958.)*

corporated in a "multidisk" (Fig. 12–13), instead of the cylinders; each disk, impregnated with a different antibiotic or concentration of antibiotic, is placed upon the surface of the inoculated agar and incubated for a few hours. Crude, unpurified clinical specimens (e.g., blood, throat swab) can often be used, and this will save one or two days.

It should be emphasized that the results of sensitivity tests are not always confirmed by clinical experience, and the physician should utilize the laboratory findings only as a guide. Studies of resistance are especially useful in patients undergoing long-term treatment with antibiotics, because the laboratory observations indicate when it is time to shift from one antibiotic to another.

Antiviral Chemotherapy. At first thought it would appear hopeless to attempt chemotherapeutic control of virus diseases, because viruses replicate only within and through the (enforced) cooperation of host cells, and any agent that interferes with this process might be expected also to damage the host cell. However, at four points in the multiplication cycle the virus is critically dependent upon biochemical processes that are not essential for the cell:

1. The initial attachment of virus to cell, which occurs at specific sites on each.

2. Transcription of mRNA from the viral nucleic acid, which requires a virus-specific polymerase.

3. Translation of viral mRNA into viral protein, which is inhibited by translation-inhibitory protein (TIP). TIP can recognize the difference between viral and mammalian mRNA. Its production is induced by interferon, another cell-coded protein, whose formation is triggered by viral or other foreign nucleic acid. Interferon inhibits the synthesis of a wide variety of viruses.

4. Replication of viral nucleic acid, which requires virus-coded enzymes that are of no use to the cell. RNA viruses may be particularly susceptible to interference with synthesis of these enzymes.

Drugs acting at each of these points have been found effective in laboratory experiments, but only two, isatin-β-thiosemicarbazone and iododeoxyuridine, are useful clinically. One problem is that in most viral

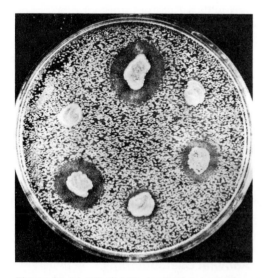

Figure 12–14. *Antibiotic-producing bacteria may be found in the human intestine. This plate was inoculated over its surface with one of the dysentery bacteria, and then spot inoculations were made with coliform bacteria picked directly from a stool culture. Four of the six organisms inhibited the dysentery bacteria. The antibiotics produced by coliform organisms are called colicines. (Photograph by S. P. Halbert; A.S.M. LS-215.)*

diseases virus multiplication is nearly over before symptoms appear. Moreover, laboratory diagnosis often requires several days or longer. Another problem is the emergence of mutant viruses resistant to the drugs, which seems to occur as frequently as it does with bacteria.

Attachment of virus to host cell is inhibited in cultured cells and experimental animals by a bacterial enzyme, neuraminidase, which destroys cell receptors for myxoviruses and some paramyxoviruses, and by a three-ringed symmetrical amine, adamantanamine, which prevents engulfment of influenza A_2 virus. Neither agent appears useful at present in treating human disease.

Rifampicin, an antibacterial drug that *inhibits transcription* by binding to bacterial DNA-dependent RNA polymerase, blocks the multiplication of poxviruses in cultured cells. Its mode of action in the virus is not clear, and so far it has not been used in human smallpox.

Translation is inhibited by TIP in response to the presence of interferon. Therapeutic administration of interferon is not practical because of the tremendous amounts required. The alternative is to stimulate the host to produce it. This can be done by administration of a synthetic double-stranded polynucleotide such as polyinosinic:cytidylic acid (poly I:C). This procedure is effective against a variety of viruses in cultured cells and in animals, but it has had only limited application in man (herpes simplex infections of the lip and eye), because poly I:C is toxic when given systemically.

Methylisatin-β-thiosemicarbazone (Marboran) has been used successfully in treating smallpox. It reduced the attack rate in several Indian epidemics by 75 to 95 per cent. A disadvantage was the occurrence of severe vomiting.

Replication of nucleic acid is prevented in cultured cells infected with picornaviruses by guanidine and by benzimidazoles. The drugs apparently inhibit synthesis of the viral RNA polymerase. They are not effective in animals, probably because drug-resistant mutants develop rapidly.

Iododeoxyuridine (IUdR) is useful when applied topically in early herpes simplex infections of the eye. The drug is incorporated into newly synthesized viral DNA in place of thymidine and produces a nonfunctional molecule.

Antiviral chemotherapy is obviously in its infancy. The foregoing points of vulnerability in the virus cycle suggest possible approaches. Methods of treatment are urgently needed for respiratory virus diseases and latent, slow, and oncogenic viral infections.

SUPPLEMENTARY READING

American Society for Microbiology: *Anti-Microbial Agents and Chemotherapy–1961.* Bethesda, Maryland.

Barber, M., and Garrod, L. P.: *Antibiotic and Chemotherapy.* Edinburgh and London, E. & S. Livingstone Ltd., 1963.

Buchanan, R. E., and Fulmer, E. I.: *Physiology and Biochemistry of Bacteria,* Vol. II. Baltimore, The Williams & Wilkins Co., 1930.

Davis, B. D., and Feingold, D. S.: Antimicrobial Agents: Mechanism of Action and Use in Metabolic Studies. *In* Gunsalus, I. C., and Stanier, R. Y. (eds.): *The Bacteria: A Treatise on Structure and Function,* Vol. IV. New York, Academic Press, Inc., 1962, pp. 343-397.

Lamanna, C., Mallette, M. F., and Zimmermann, L. N.: *Basic Bacteriology,* 4th ed. Baltimore, The Williams & Wilkins Co., 1973.

Lennette, E. H., Spaulding, E. H., and Truant, J. P.: *Manual of Clinical Microbiology,* 2nd ed. Washington, American Society for Microbiology, 1974.

Oginsky, E. L., and Umbreit, W. W.: *An Introduction to Bacterial Physiology,* 2nd ed. San Francisco, W. H. Freeman & Company, 1959.

Porter, J. R.: *Bacterial Chemistry and Physiology.* New York, John Wiley & Sons, Inc., 1946.

Salle, A. J.: *Fundamental Principles of Bacteriology,* 7th ed. New York, McGraw-Hill Book Co., Inc., 1973.

Schnitzer, R. J., and Hawking, F. (eds.): *Experimental Chemotherapy,* Vols. I–V. New York, Academic Press, Inc., 1963.

Thimann, K. V.: *The Life of Bacteria,* 2nd ed. New York, The Macmillan Company, 1963.

Wyss, O.: Chemical Factors Affecting Growth and Death. *In* Werkman, C. H., and Wilson, P. W. (eds.): *Bacterial Physiology.* New York, Academic Press, Inc., 1951.

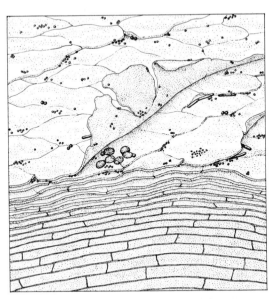

SECTION
THREE

HIGHER
PROTISTS

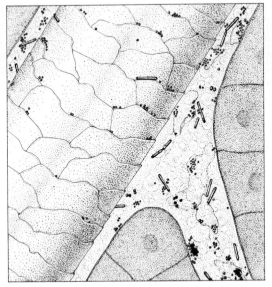

From *Life on the Human Skin* by Mary J. Marples. Copyright © 1969
by Scientific American, Inc. All rights reserved.

PROTOZOA   13

Protozoa are the smallest of the animals. They range in size upward from something less than the diameter of a red blood corpuscle (about 7.5 μm). Most of them are visible to the naked eye.

Apart from their role as parasites and causes of disease in animals of interest to man, protozoa are of concern as nuisances and as helpful organisms of decomposition. They are widespread in water and may affect its color, taste, or odor. Some species produce green or red colors; others liberate aromatic oils that cause fishy or cucumber odors, bitter or spicy tastes.

Protozoa are abundant in the soil and contribute to its fertility. One gram of garden soil may contain 10,000 to 100,000 protozoa. They digest particles of insoluble organic matter and liberate soluble waste materials that can be utilized readily by plants or other microorganisms. Many protozoa feed on bacteria and other microorganisms. This is one means of preserving the balance of Nature.

Protozoa display several signs of irritability; that is, responsiveness to external stimuli. Certain protozoa are attracted by weak light and repelled by intense light. Temperature affects the rate of motility and of various physiologic processes, but optimum temperatures vary from one species to another. Protozoa react violently to mechanical stimuli such as contact with hard objects. An ameba, for example, withdraws when irritated by a sharp needle. Chemicals attract or repel protozoa, depending upon their potential usefulness as food. Most free-living protozoa are aerobic and swim toward a supply of gaseous oxygen.

Many protozoa "withdraw from the world" when conditions become unfavorable and enter into a dormant state. The process, known as encystment, occurs when food or water supplies become inadequate, too little oxygen is available, or when the environment becomes otherwise unsuitable. Cysts are more resistant than vegetative cells (trophozoites) to drought, extremes of temperature, harmful chemicals, and starvation. Their function is survival rather than reproduction, but not all cysts return to active life when conditions become normal. Apart from normal mortality, certain nutrient and other conditions particularly favor excystment.

SUBPHYLA OF PROTOZOA

Approximately 25,000 species are recognized within the phylum Protozoa. They are grouped into four subphyla, principally according to their methods of locomotion (see Figures 13–1 and 13–2). The *Sarcodina* move about by flowing into protoplasmic extrusions called *pseudopodia*. The *Mastigophora* are propelled by one to four (occasionally more) long, whiplike flagella. *Ciliata* are usually covered with thousands of very short, hairlike cilia, whose rhythmic, coordinated, back-and-forth movement provides locomotion. Adult *Sporozoa* are usually nonmotile; they also differ from other protozoa by undergoing sporulation as a stage in their reproductive cycle.

SARCODINA

Morphology.   Members of the Sarcodina lack a cell wall and hence do not have a definite shape [Fig. 13–3(2)]. Pseudopodial extrusions of protoplasm serve the function of locomotion and they also, by flowing around particles of food, provide the means

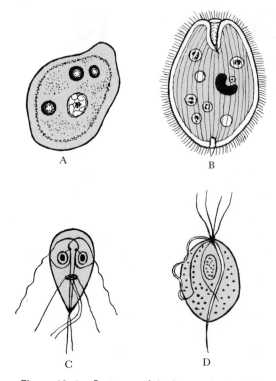

Figure 13–1. *Protozoa of the human body:* A, En-tamoeba histolytica, *cause of amebic dysentery;* B, Balantidium coli, *an intestinal ciliate that sometimes produces a dysentery-like inflammation;* C, Giardia intestinalis, *a flagellate cause of tropical diarrhea;* D, Trichomonas vaginalis, *a flagellate of the vagina that may produce vaginitis. (Magnifications: A = 1200X, B = 450X, C = 900X, D = 1300X.)*

water intracellularly. This collects in small vacuoles, which coalesce until they reach a critical size, whereupon they discharge to the surface.

Reproduction. Members of Sarcodina, such as *Amoeba,* reproduce sexually by binary fission. The nucleus divides mitotically, the cell elongates and constricts in the middle, and the two daughter cells separate (Fig. 13–6).

Physiology. Sarcodina respond chemotactically to the proximity of a food particle by extruding pseudopodia toward it and engulfing it. Typical foods include other protozoa, small algae, and bacteria. The particle is enclosed within a vacuole, into which digestive enzymes are secreted. Hydrolysis of the particle yields building materials for the synthesis of new protoplasmic ingredients and structures, the energy required

of ingestion (Fig. 13–4). In some forms, such as *Amoeba,* there is no structural differentiation at the outer edge of the cytoplasm; other forms, such as the Foraminifera, have a well-developed shell with openings through which the pseudopodia are extruded, and the Heliozoa and Radiolaria have skeletons of various materials (Fig. 13–5).

The cytoplasm is usually differentiated into an inner portion and an outer portion—the *endoplasm* and *ectoplasm,* respectively. The endoplasm of most species contains a single nucleus, food vacuoles, and various granules. Contractile vacuoles are always present in freshwater forms but absent from marine and parasitic forms. They help maintain a proper osmotic balance within the cell. Freshwater protozoa live in a medium having a much lower osmotic pressure than their protoplasm and therefore tend to accumulate

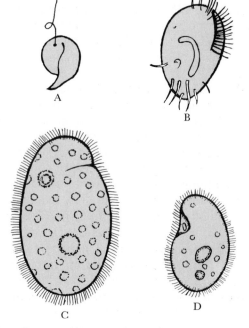

Figure 13–2. *Fresh water protozoa:* A, Phacus, *a small flagellate, and three ciliates,* B, Euplotes; C, Nassula; D, Colpidium. *(Magnification: 250X.)*

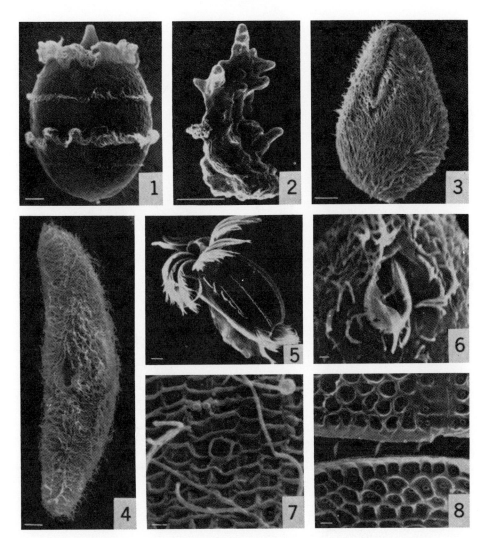

Figure 13–3. *Scanning electron micrographs of protozoa.* (1) Didinium nasutum, *showing bands of cilia (scale, 10 μm).* (2) Amoeba proteus *(scale, 100 μm).* (3) Nyctotherus ovalis *covered with cilia (scale, 10 μm).* (4) Paramecium multimicronucleatum *covered with cilia (scale, 10 μm).* (5) Uronychia *sp. with cilia fused into cirri and membranelles (scale, 10 μm).* (6) Tetrahymena pyriformis, *showing oral membranellar cilia (scale, 10 μm).* (7) P. multimicronucleatum *pellicle denuded of cilia, showing contractile vacuole pore (scale, 1 μm).* (8) Ceratium hirudinella, *exoskeleton and transverse girdle (scale, 1 μm). (From Small and Marszalek, Science, 163:1064, 1969.)*

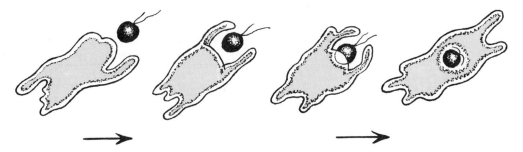

Figure 13–4. *An ameba feeding. It encounters a small flagellate and puts forth pseudopodia, which enclose the smaller organism. A vacuole forms around the food particle, and enzymes digest it.*

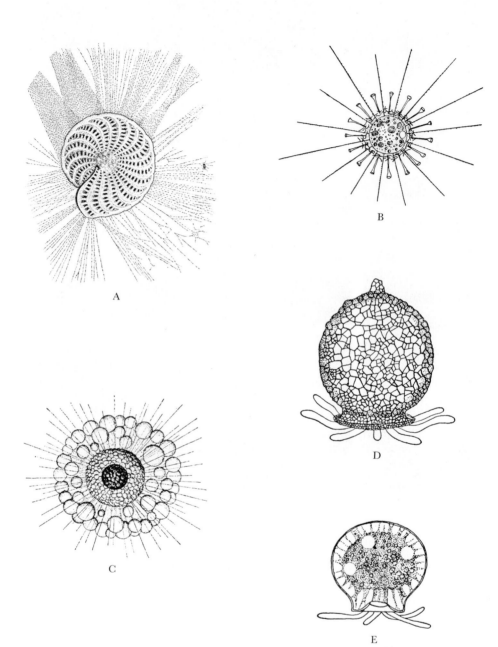

Figure 13–5. *Ameboid organisms possessing a shell or skeletal structure.* A, Elphidium, *one of the Foraminif-era (25X).* B, Raphidocystis, *a member of the Heliozoa (500X).* C, Thalassicolla, *a radiolarian (15X).* D, Difflugia *(130X).* E, Arcella *(140X). (From Kudo: Protozoology, 4th ed. Springfield, Ill., Charles C Thomas, 1966.)*

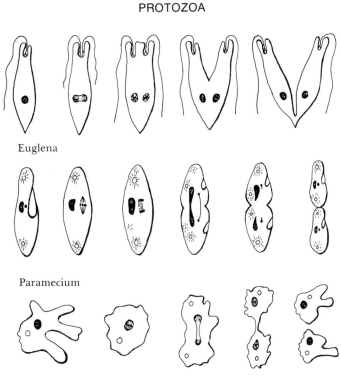

Euglena

Paramecium

Ameba

Figure 13–6. *Asexual reproduction in three common protozoa. The nucleus divides mitotically; all other cell parts are then replicated, and finally cell division occurs. (From Villee, Walker, and Smith: General Zoology, 3rd ed. Philadelphia, W. B. Saunders Co., 1968.)*

being derived from oxidation-reduction and other chemical reactions. Undigested materials and waste products such as acids and CO_2 are excreted from the cell surface.

Significant Groups. The most familiar group of Sarcodina are the free-living forms such as *Amoeba*. These may be as large as 600 to 800 μm, and are found in fresh or salt water and in damp soil.

A second group includes the Foraminifera (Fig. 13–5). Many of these free-living forms possess a multi-chambered calcareous shell, somewhat similar to a snail's shell. They are comparatively large organisms (as great as 1 mm. in diameter) and drift about in fresh or salt water, settling to the bottom, where they move sluggishly. Pseudopods extend through pores in the shell for locomotion and to secure food. The white cliffs of Dover are composed of these shells.

Radiolaria are also marine organisms. Most of them possess shells of silica. Some species are as large as 1 or 2 mm. in diameter.

Among freshwater ameboid organisms resembling the foraminiferans are the Heliozoa, which are covered by a gummy substance or siliceous needles. Most species are somewhat less than 50 μm in diameter. *Difflugia* and *Arcella* construct protective shells, the former of cemented sand grains and the latter of a chitinous material (Fig. 13–5). They vary in size from 30 to about 150 μm.

The third group of ameboid protozoa of particular interest includes the human intestinal parasites, *Entamoeba, Endolimax,* and *Iodamoeba* (Fig. 13–7). *Entamoeba* species vary from 10 to 35 μm in diameter, and are irregularly round or ovoid. During encystment, the cells become nearly round, and nuclear division occurs, sometimes yielding 4 or 8 or more nuclei.

Entamoeba histolytica is the cause of amebic dysentery, a very serious disease in many tropical and subtropical areas. The problem in laboratory examination of fecal specimens is to distinguish the cysts from

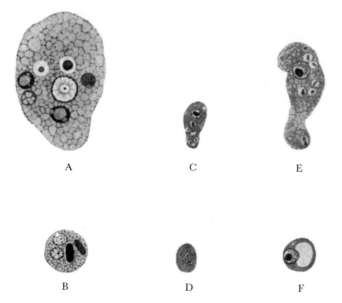

A   C   E

Figure 13–7. *Trophozoite (A, C, E) and cyst (B, D, F) stages of human intestinal amebae (1150X)*. (A, B) Entamoeba histolytica, (C, D) Endolimax nana, (E, F) Iodamoeba butschlii. *(From Kudo: Protozoology, 4th ed. Springfield, Ill., Charles C Thomas, 1966.)*

B   D   F

those of normal ameboid intestinal inhabitants. *E. histolytica* cysts are 5 to 20 μm in diameter and contain four nuclei; the cytoplasm is greenish yellow. *Entamoeba coli* cysts are somewhat larger (10 to 30 μm), yellowish brown, and contain one to eight nuclei. *Endolimax nana* is a small organism; its cysts are 5 to 12 μm in diameter, greenish yellow, and contain one to four nuclei. *Iodamoeba butschlii* cysts are 6 to 15 μm in diameter; they contain a single nucleus, yellowish cytoplasm, and possess a large glycogen body, which stains characteristically with iodine.

MASTIGOPHORA

Morphology. The majority of flagellates are elongate organisms with a moderately strong outer layer of cytoplasm, so the shape of the cells is practically constant. Species that can creep as well as swim change shape as they move over a surface, and others, whose movement is associated with rotation about the longitudinal axis, are spiral in shape. Swimming forms tend to be pointed at one end (see Figure 1–14). Movement occurs in the direction of the flagellate end, and the oral opening, or cytostome, if present, is usually near the anterior end. Most forms are uninucleate. The cytoplasm contains vacuoles and granules of various kinds, and some groups contain chloroplasts.

Reproduction. Asexual reproduction occurs by longitudinal binary fission. In organisms with a single anterior flagellum, the flagellar apparatus divides first, followed by the remainder of the organelles and the cell body (Fig. 13–6).

Colonial Protozoa. Many flagellates form colonies, owing to incomplete separation of daughter individuals. *Volvox* is an example of a colonial form consisting of several thousand individual cells more or less intimately joined (see Figure 1–17, page 17). It is a flagellate, closely related to *Euglena*, and illustrates an early stage in specialization or the acquisition of specific functions by certain cells. A colony is composed of a single layer of cells surrounding a cavity filled with fluid. Each cell possesses two flagella and a nucleus and is connected by bits of protoplasm with the six surrounding individuals. Most of the cells grow and multiply and assist in propelling the colony through the water. A special function has been acquired by a few individual cells, however, which develop into germ cells or sexual cells and bear the burden of reproducing new colonies.

Physiology. The flagellates are sometimes divided into two groups, based on presence or absence of a photosynthetic pigment. The "phytoflagellates" are motile, single-celled, algalike, eucaryotic organisms such as *Euglena;* they are on the borderline between the plant and animal kingdoms. They are photosynthetic, but they can also live saprophytically during periods of poor illumination if they have access to sufficient organic matter in solution. Most "zooflagellates" also live saprophytically, securing inorganic and organic nutrients from the medium and obtaining energy by oxidation of organic matter. Some are equipped with an oral opening or cytostome and can ingest solid food particles such as bacteria, algae, and smaller protozoa.

Significant Groups. The *dinoflagellates* comprise a heterogeneous group of organisms; they possess two distinguishing features that are recognizable even in forms that have evolved far from the original unicel-

lular, alga-like progenitors. Dinoflagellates possess two flagella, one lying in a groove or girdle that encircles the cell, the other extending posteriorly away from the cell (see Figure 13–8). They also have a single, massive nucleus with an unusual manner of division in which the chromosomes are often visible during interphase.

Many dinoflagellates are photosynthetic and possess dark yellow, brown, or greenish chloroplasts. Some of these are partially covered by a series of plates, between which pseudopodia can protrude to ingest food particles. There are also forms that have lost the photosynthetic pigments and rely solely on ameboid ingestion, and still others secure nutrients by simple diffusion of soluble organic and inorganic substances. The interesting interrelationships of these forms are illustrated in Figure 13–8.

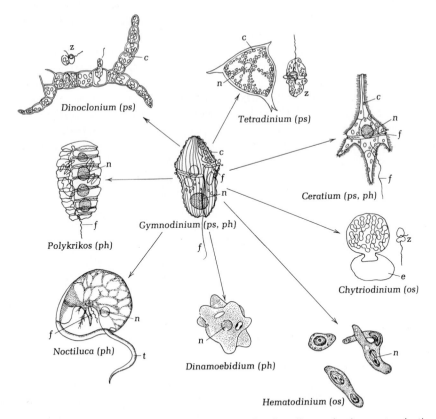

Figure 13–8. *Evolutionary trends shown among the dinoflagellates. At the center is the comparatively unspecialized photosynthetic* (ps) *Gymnodinium, which can also feed phagotrophically* (ph). *Evidence of dinoflagellate ancestry is shown in the more highly specialized genera by the formation of zoospores* (z) *with typical morphology or characteristic dinoflagellate nuclei* (n). os, *osmotrophic nutrition;* f, *flagellum;* c, *chloroplast;* t, *tentacle;* e, *parasitized invertebrate (host) egg. (From Roger Y. Stanier, Michael Doudoroff, and Edward A. Adelberg: The Microbial World, 3rd ed.,* © *1970, p. 108. By permission of Prentice-Hall, Inc., Englewood Cliffs, N.J.)*

Dinoflagellates occasionally arouse popular interest when they "bloom" in coastal waters and produce a so-called *red tide*. They wash ashore following prolific multiplication in tropical or subtropical areas or in temperate areas late in the summer, and they may also kill fish and crustacea. The lethal effect has been attributed to depletion of dissolved oxygen by the decomposing dinoflagellates, but more important are the highly toxic substances produced by a number of species.

Several genera of flagellates are common in man. Most are part of the normal fauna, but some produce disease. *Trichomonas* species (see Figure 13–1D) are ovoid or pear-shaped with four flagella attached to blepharoplasts at the anterior end, the flagella being about as long as the cell itself. A fifth flagellum is attached to an undulating membrane, which also has a thick, backbone-like *axostyle* extending from the anterior end, through the cell, and projecting from the posterior, tapering to a point. *Trichomonas* species vary in length from 6 to 23 μm, exclusive of flagella and axostyle. *T. tenax, T. hominis,* and *T. vaginalis* are frequently encountered in the human mouth, intestine, and genitourinary tract, respectively. Under some conditions the latter causes inflammatory disease in both males and females.

The *trophozoite* (i.e., vegetative) stage of *Chilomastix* is asymmetric and pear-shaped, 10 to 15 μm long (Fig. 13–9). It has a spiral groove around the cell toward the tapered posterior end. Two short and three long flagella project from the anterior pole, and another lies along the large cytostome. The cyst is about 8 μm long, and is easily identified by its lemon shape, no other species having a cyst of this type. *Chilomastix* is an enteric protozoon.

Giardia lamblia, another intestinal inhabitant, is bilaterally symmetrical, and its paired nuclei, axostyles, and flagella give the trophozoite a characteristic facelike appearance (see Figure 13–1C). The cells are 12 to 20 μm long, but only 3 to 4 μm thick, flattened dorsally and convex ventrally. The ovoid cysts are slightly smaller than the trophozoite; they contain paired structures, which correspond to the axostyles and flagella, and two or four nuclei.

Flagellate protozoa found free in the human bloodstream or intracellularly are called *hemoflagellates* and are members of the family Trypanosomidae. They are presumed to have evolved from intestinal parasites of insects, and those that infect man are distributed by blood-sucking insects.

Several morphologic stages occur in the life cycles of these protozoa, as illustrated in Figure 13–10. In the trypanosomal stage a flagellum arises from the posterior end of the cell and passes forward, attached to an undulating membrane, and continues as a free flagellum at the anterior end of the cell. The crithidial stage is similar, except that the flagellum and undulating membrane start about the middle of the cell, in front of the nucleus. In the leptomonad stage there is no undulating membrane, and the flagellum starts near the forward end of the cell. The leishmanial stage is ovoid and has no flagellum. The flagella are attached to blepharoplasts, and these are found even in the leishmanial form.

Seven species of the Trypanosomidae occur in man. *Trypanosoma gambiense*, the etiologic agent of African sleeping sickness, can be seen swimming amongst the erythrocytes in the blood and lymph of early cases. The cells are 15 to 40 μm long—two to five times the diameter of red blood cells. Infection takes place via the bite of a tsetse fly at least 20 days after it has fed on a previously infected host. The organisms multiply in the

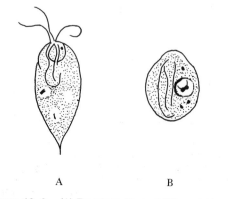

A   B

Figure 13–9.   (A) *Trophozoite* and (B) *cyst stages of* Chilomastix, *an intestinal flagellate.*

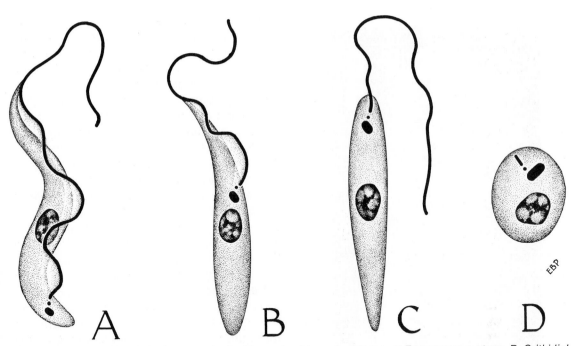

Figure 13–10. *Morphologic stages in the life cycle of hemoflagellates. A, Trypanosome stage. B, Crithidial stage. C, Leptomonas stage. D, Leishmania stage. (From Burrows, Textbook of Microbiology, 20th ed. Philadelphia, W. B. Saunders Co., 1973.)*

crop, stomach, and intestines of the fly and finally reach the salivary glands. The incubation period varies from two weeks to several months. The initial systemic infection produces irregular fever; anemia, wasting, cardiac injury, and edema may follow. The parasites then become abundant in the cerebrospinal fluid, and the sleeping sickness state ensues, with somnolence, apathy, and weakness; death usually occurs in several months. African sleeping sickness is limited to areas where the tsetse fly is found: tropical Africa and parts of South Arabia. Man is the usual source of infection.

The association between termites and their intestinal flagellates provides an interesting example of another type of interrelationship between organisms. This illustrates a form of symbiosis (Greek: *syn*, with + *bios*, life) known as *mutualism*, a relationship in which each partner in the association receives some benefit from the other.

Termopsis angusticollis is a large species of termite that feeds on wood. Its intestine contains four different flagellate protozoa. The termite can be kept in an atmosphere of pure oxygen for 72 hours without harm, but its intestinal protozoa are destroyed by this treatment. The termite continues to eat wood after its return to a normal atmosphere but cannot digest the wood and dies of starvation within three to four weeks. Reinfection of such a termite from another termite restores its digestive powers. One of the four intestinal protozoa, *Leidyopsis,* is essential for digestion of wood within the termite; the other three protozoa contribute nothing vital. *Leidyopsis* is completely dependent on the termite and can live nowhere else.

CILIATA

Morphology. The free-swimming members of the Ciliata are usually spherical or ovoid; creeping forms are flattened. There is great variation in size, from less than 10 μm to about 2 mm. in length.

Each cilium arises from a basal structure or kinetosome. The kinetosomes are interconnected in longitudinal, oblique, or spiral rows by fibrils, which coordinate the activity

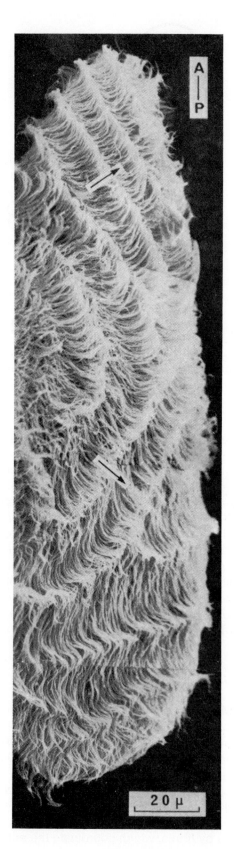

of the cilia so that they beat rhythmically (see Figure 13–11). Cilia in the cytostomal region propel food particles toward and into the oral opening. In some organisms, cilia fuse to form cirri, or undulating membranes.

The cytoplasm is differentiated into ecto-plasm and endoplasm. The ectoplasm contains the ciliary basal granules and is covered by a pellicle. The endoplasm contains nuclei, food and contractile vacuoles, crystals, and other materials.

The cytostome is usually situated toward the anterior end of the cell, and the anal opening, or cytopyge, is at the posterior. In *Paramecium*, the cytostome is not at the anterior region, but a ciliated oral groove leads into it from the anterior end.

Reproduction. Asexual reproduction is usually by transverse binary fission (see Figure 13–6). In *Paramecium,* for example, there are two nuclei, a *macronucleus* and a *micronucleus*. The former, which is necessary for binary fission, is sometimes called the "vegetative" nucleus. The process begins with elongation of the macronucleus in the direction of the long axis of the cell. The cytoplasm undergoes structural reorganization, a second cytostome and associated structures develop, the cell invaginates near the center and eventually separates into two daughter cells.

The micronucleus is required for sexual reproduction or conjugation. Some strains can multiply indefinitely in the absence of a micronucleus, but in its presence the sexual cycle occurs at intervals; after conjugation, a new macronucleus is derived from a micronucleus. The process of conjugation and genetic exchange is complex. As shown diagrammatically in Figure 13–12, the macronucleus disintegrates at the beginning and plays no part in succeeding events.

Some ciliates also reproduce asexually by *budding*, a process in which a small bit of cytoplasm containing nuclear material

Figure 13–11. *Scanning electron micrograph of* Opalina, *showing waves of ciliary action. Arrows indicate directions of transmission of waves of activity and of the effective stroke. A—P, anterior-posterior axis. (From Horridge and Tamm, Science, 163:817, 1969.)*

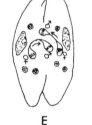
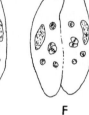

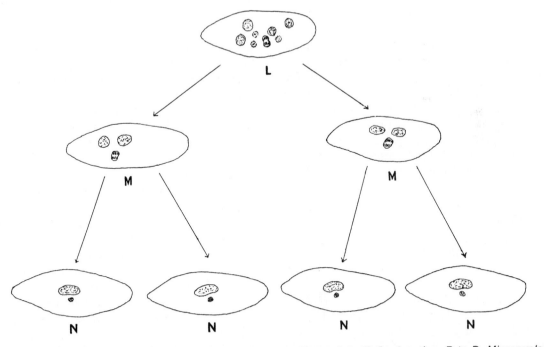

Figure 13–12. *Sexual reproduction in* Paramecium caudatum. *A to F, Conjugation. B to D, Micronuclei undergo three divisions, the first two of which are meiotic. E, "Male" micronuclei are exchanged. F, They fuse with the stationary micronucleus of the opposite conjugant. G, Exconjugant with macronucleus and zygote micronucleus; other micronuclei have been resorbed. H to K, Three divisions of zygote, forming eight micronuclei; old macronucleus is resorbed. L, Four micronuclei form macronuclei; three are resorbed. L to M, Remaining micronucleus divides twice in course of two cytoplasmic divisions. Resulting daughter cells each receive one of the four macronuclei in L. N, Normal nuclear condition restored. (From Villee, Walker, and Barnes: General Zoology, 4th ed. Philadelphia, W. B. Saunders Co., 1973.)*

pinches off from a cell and enlarges until it can separate and maintain independent existence.

Physiology. Most ciliates ingest food in the form of particles—bacteria, algae, protozoa, and other small animals. These are enclosed within food vacuoles, which circulate around in the semifluid cytoplasm while digestion occurs. Undigested residues and wastes are excreted through the anal pore. Contractile vacuoles, which collect fluid, also discharge at intervals through the cell surface.

In some cases algae become established as symbionts within the cytoplasm of protozoa in a mutualistic relationship. This is illustrated by the association between *Paramecium bursaria* and the green alga *Zoochlorella*. The algae are ingested by the paramecium like any other particle of food but for some reason are not digested. They grow and multiply within the paramecium and are then passed on to the next generation when the protozoon divides, so the association continues indefinitely. The alga contributes to the partnership by taking carbon dioxide and other protozoan waste materials and by supplying oxygen and carbohydrates, both of which are products of photosynthesis carried out within the alga. The paramecium provides a protected environment for the plant, a constant supply of carbon dioxide, and transportation to areas of light where photosynthesis is possible.

SPOROZOA

Morphology. All members of the Sporozoa are parasitic and produce spores at some stage in their life cycle. There is great variety in size and shape, both among genera and species and among stages in the life cycle of a given species. In general, they are nonmotile, but some move by pseudopodia when immature, and gametes are often motile.

Reproduction. Many species have both asexual and sexual reproductive cycles. The asexual cycle consists of repeated fission or budding of intracellular trophozoites or *schizonts;* this process, called *schizogony,*

produces great numbers of individuals. Sexual reproduction by fusion of male and female gametes often initiates *sporogony* (spore formation). These steps will be illustrated later in a description of the life cycle of *Plasmodium vivax.*

Physiology. As parasites, the Sporozoa are largely saprozoic; that is, they secure soluble nutrients from the body fluids, tissue fluids, or cell fluids of their host. Moreover, they are protected from unfavorable environmental agents as long as they do not destroy their host or are able to find a means of transfer from one host to another. This is the *sine qua non* of successful parasitism.

Plasmodium. The malaria parasite, *Plasmodium vivax,* is a notorious protozoon whose reproductive cycle can be completed only by invasion of two hosts (Fig. 13–13). The infective stage, the *sporozoite,* is stored in the salivary glands of certain *Anopheles* mosquitoes. These spindle-shaped cells are about 0.01 mm. in length and are deposited in the blood stream of man when the mosquito feeds. The sporozoites develop in tissue cells for about nine days and then enter the red blood corpuscles, where they appear first as *ring forms;* later, as the parasite grows at the expense of the erythrocyte contents, they transform into *ameboid trophozoites.* Nuclear divisions occur until 12 to 24 nuclei are present; this is the *schizont* stage. Cytoplasm and wall material surround each nucleus in the *segmenter* stage. Disintegration of the erythrocyte releases the spores or *merozoites.* These may enter fresh red cells and repeat the asexual cycle.

After several asexual cycles, some trophozoites develop into large, uninucleate, sexual stages known as *macrogametocytes* (female) and *microgametocytes* (male). These are terminal forms and degenerate unless they are ingested by a mosquito. In the stomach of the mosquito the macrogametocyte leaves the blood cell and becomes a macrogamete. The microgametocyte produces four to eight long whiplike microgametes at its surface. One of these spermatozoonlike structures fertilizes a macrogamete, and the resulting *zygote* finds its way to the outer surface of the stomach. Nuclear division occurs and a large *oocyst*

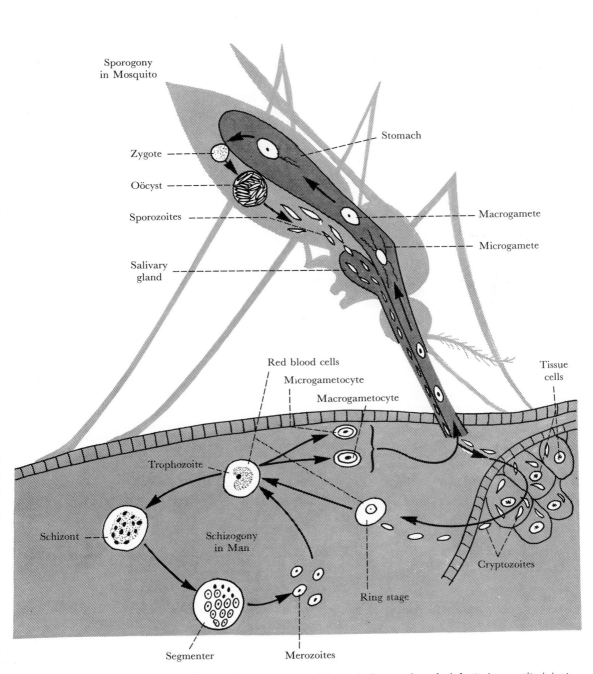

Figure 13–13. *The life cycle of* Plasmodium vivax, *one of the malaria parasites. An infected mosquito injects sporozoites into the bloodstream of a human, from which they enter tissue cells and undergo segmentation. Some of the resulting cryptozoites enter red blood cells, where asexual cycles of schizogony take place. Eventually sexual stages, the macro- and microgametocytes, form. If they are ingested by an appropriate mosquito, a sexual cycle ensues, ultimately yielding sporozoites. (Modified from Villee, Walker, and Barnes: General Zoology, 4th ed. Philadelphia, W. B. Saunders Co., 1973.)*

forms. It contains as many as 10,000 nuclei, which become surrounded by cytoplasm and form the spindle-shaped sporozoites. Rupture of the oocyst permits the sporozoites to spread all over the body of the mosquito, and those that enter the salivary gland may be injected into man, thus completing the life cycle of the parasite. The events in the body of the mosquito require one to two weeks.

Chills and fever occur in man when a sufficient number of parasitized corpuscles break down simultaneously. The usual period of incubation before this takes place is about two weeks. In malaria caused by *P.*

vivax the cycle of schizogony requires 48 hours, so chills and fever recur regularly every other day. This form of the disease is known as *tertian* malaria.

Malaria is also caused by other species of *Plasmodium*. In general, the series of events in man and mosquito are alike, differing only in details. The asexual cycle of *P. malariae* in man requires 72 hours, and the fever, therefore, recurs every third day. This is *quartan* malaria. *P. falciparum* behaves somewhat differently, paroxysms occurring daily *(quotidian)* or every other day *(tertian)*.

SUPPLEMENTARY READING

Buchsbaum, R.: *Animals Without Backbones,* rev. ed. Chicago, University of Chicago Press, 1948.
Cheng, T. C.: *The Biology of Animal Parasites.* Philadelphia, W. B. Saunders Company, 1964.
Hegner, R.: *Big Fleas Have Little Fleas, or Who's Who Among the Protozoa.* Baltimore, The Williams & Wilkins Co., 1938.
Jahn, T. L.: *How to Know the Protozoa.* Dubuque, Iowa, W. C. Brown Co., 1949.
Kudo, R. R.: *Protozoology,* 4th ed. Springfield, Ill., Charles C Thomas, 1954.

ALGAE 14

As discussed in Chapter 5, the bacteria and blue-green algae are now classified in a separate kingdom, Procaryotae. The plant kingdom includes three principal major divisions, one of which, Thallophyta, includes organisms described in this chapter and Chapters 15 and 16. The remaining divisions of the plant kingdom are Bryophyta (liverworts and mosses) and Tracheophyta (vascular plants).

The Thallophyta are either simple, unicellular forms or more complex organisms that are multicellular but not differentiated into roots, stems, or leaves; their plant body is called a *thallus*. The first subdivision includes organisms containing chlorophyll: diatoms and the higher algae (green, red, brown, etc.). The second subdivision consists of nonphotosynthetic organisms: yeasts, molds, mushrooms, etc. Molds and yeasts will be discussed in Chapters 15 and 16.

The blue-green algae are procaryotes, whereas all other algae are eucaryotes. The former presumably evolved from ancestral photosynthetic bacteria. It has been postulated that other algae and higher plants developed from an amebo-flagellate eucaryote by symbiotic acquisition of photosynthetic plastids from early photosynthetic procaryotes.

OCCURRENCE

Algae can be found almost anywhere on Earth from the tropics to arctic regions. They are principally aquatic and hence occur in lakes and streams, seas, and oceans. Some fresh water forms adapt to salt water and are found even in the salt lakes of the southwestern United States. Many species live in damp soil; some grow on rocks and aid in the slow decomposition by which rocks are eventually converted into soil; still others can be found on the north side of trees where they are protected from the midday sun. Some species are adapted to very cold climates and grow on snow and ice in polar and mountain regions. Others have become adapted to high temperature and flourish in hot springs at 85° to 90° C. (for example, in Yellowstone National Park).

MORPHOLOGY

Many species of algae are microscopic single cells. Their shapes vary from spheres to rods, clubs, spirals, and irregular forms. Multicellular species exhibit great variations in form and complexity. Some species are not truly multicellular but consist of simple aggregations of single, apparently identical cells held together by a slimy, gelatinous outer coat. Colonial forms, on the contrary, contain cells with special functions, such as reproduction, or rootlike holdfasts that anchor the plant to rocks or other solid surfaces. Some algae are motile by means of flagella, others (diatoms) by an unknown mechanism that may involve cytoplasmic streaming. Many algae are not independently motile but drift in the water. Some normally nonmotile species possess motile reproductive cells.

REPRODUCTION

Algae reproduce by a variety of methods. Sexual and asexual processes occur, and many species have complicated life cycles with both sexual and asexual stages.

Asexual reproduction includes simple cell

division or fission in which a single cell enlarges and divides. Fragmentation of the filaments of some algae serves as a method of propagation. Asexual spore formation also occurs; nonmotile spores or motile *zoospores* are produced by specialized cells called *sporangia*.

There are three types of sexual reproduction. (1) One of the simplest is *isogamy,* the fusion or conjugation of two sex cells that are indistinguishable from each other. Gametes from two haploid plants (+ and − strains) fuse and produce a zygote, which germinates to form a diploid plant. This yields zygospores (i.e., flagellated spores), which undergo reduction division, after which haploid plants of the two strains are produced. (2) A more advanced form is *anisogamy,* the fusion of smaller and/or more active (i.e., flagellate) haploid male gametes with larger and/or less active haploid female gametes. The resulting zygote produces a diploid sporophyte, on which tetraspores form and undergo reduction division, yielding spores from which haploid plants grow and develop male and female gametes. (3) Finally, some species possess large, nonmotile ova and small, motile sperm cells, the union of which is known as *oogamy*. The haploid gametes, produced on minute filaments, fuse and produce a zygote while the ovum is still attached to the mouth of the oogonium. A small embryo develops before connection with the parent plant is lost. A large, diploid sporophyte grows and the zoospores it produces undergo reduction division and become haploid male or female gametophytes.

CLASSIFICATION OF ALGAE

Algae are classified according to the nature of the pigments they produce, the kinds of products they synthesize and store, and their methods of reproduction. Six major groups are described in Table 14–1. Some authorities include the dinoflagellates among the algae. They have already been included with the protozoa, but they might equally well be discussed as algae because, like the Euglenophyta, they are borderline organisms with properties of both plant and animal kingdoms.

The Cyanobacteria or blue-green algae are the most primitive algae in existence; as was pointed out in Chapter 1, they bear many resemblances to bacteria (Fig. 14–1) and are classified with them as procaryotes. The nuclear body of the blue-green algae, like that of bacteria, is not bounded by a definite membrane (Fig. 14–2). Similarly, photosynthetic activity is not confined to membrane-enclosed structures or chloroplasts, as it is in eucaryotic algae and in higher plants. Instead, it takes place in a continuous leaflike lamellar system containing chlorophyll and carotenoid pigments. It may be noted that photosynthetic bacteria differ from the blue-green algae in their possession of small, vesicular structures, the chromatophores, within which some photochemical activity occurs, but no fixation or reduction of carbon dioxide. Blue-green algae owe their color to the pigment phycocyanin, which is often sufficiently intense to mask the chlorophyll. Some species possess a red pigment, phycoerythrin, in addition, and these forms appear reddish or purple. The product of photosynthesis that is stored is the animal-like polysaccharide, glycogen. Reproduction is by asexual fission, as in bacteria.

The Euglenophyta lack cell walls, are flexible like protozoa, and possess eyespots and contractile vacuoles. They are unicellular flagellates and reproduce by longitudinal fission. In addition to the pigmented photosynthetic genera, there are unpigmented nonphotosynthetic genera that feed saprophytically, that is, by absorption of soluble organic and inorganic nutrients. These are sometimes known as *leucophytes* (Greek: *leukos,* white + *phyton,* plant). It has been noted that *Euglena* loses the ability to produce chlorophyll when cultivated in a medium containing penicillin, and thus becomes indistinguishable from a protozoan form.

The Chlorophyta, or green algae, comprise a large and heterogeneous group with many characteristics similar to those of higher plants (Fig. 14–3). Colonial types are

TABLE 14–1. Major Groups of Algae

Group	Size, Structure, etc.	Reproduction	Habitat
Procaryotes			
Cyanobacteria (blue-green algae)	Usually microscopic; multicellular or unicellular; contain carotene, carotenoid, phycocyanin and/or phycoerythrin pigments; store glycogen	Asexual fission	Fresh water and soil
Eucaryotes			
Euglenophyta (*Euglena*)	Microscopic, unicellular; store fat and paramylum	Longitudinal fission; simple sex cells	Fresh water
Chlorophyta (green algae)	Microscopic, a few macroscopic (e.g., a few inches); unicellular or multicellular; cellulosic walls; store starch	Asexual fission and zoospores; primitive sexual fusion	Fresh water, soil, tree bark
Chrysophyta (diatoms, etc.)	Microscopic, mostly unicellular; walls of 2 overlapping halves, often siliceous; contain carotenoid pigments; store oils	Usually asexual	Fresh and salt water (some in arctic), soil, higher plants
Phaeophyta (brown algae; seaweeds, kelp, etc.)	Multicellular, large (up to several hundred feet); walls of cellulose and algin; contain carotenoid pigments; store mannitol, laminarin (a polysaccharide), fat	Sexual; some asexual zoospores	Salt water (cool)
Rhodophyta (red algae; seaweeds)	Unicellular or multicellular and macroscopic (up to 4 feet); cellulosic walls; contain phycobilin pigment; store starch	Sexual by well-differentiated male and female germ cells; asexual by spores	Salt water (warm)

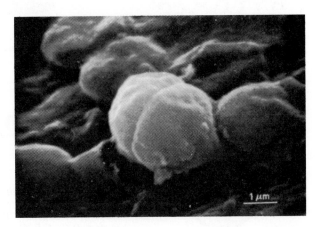

Figure 14–1. *Scanning reflection electron micrograph of the blue-green alga,* Anacystis montana. *(From Echlin, P., in Gibbs and Shapton (eds.): Identification Methods for Microbiologists, Part B. New York, Academic Press Inc., 1968.)*

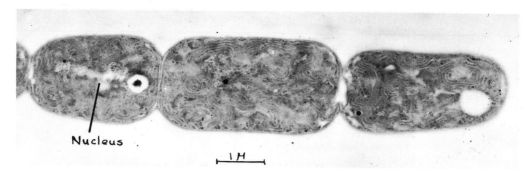

Figure 14–2. *Anabaena, a blue-green alga. The low density material is nuclear substance, but note the absence of a nuclear membrane. The membranous structures in the cytoplasm are the equivalents of chloroplasts. (Electron micrograph by G. B. Chapman.)*

common, and the product of photosynthesis is starch. Reproduction is by asexual fission, formation of asexual zoospores, and also by primitive sexual fusion.

The Chrysophyta include diatoms, microscopic and mostly unicellular organisms with shells composed of silica (Fig. 14–4). These organisms are widely distributed in both fresh and salt water, even in arctic regions, and their shells have accumulated in certain places to depths as great as 3000 feet. This material is known as diatomaceous earth; it is a filtering agent and also a very fine abrasive, and as such it is used in toothpaste. The chrysophytes produce and store

oils and are believed to have been important sources of petroleum deposits.

The remaining two groups of algae, Phaeophyta (brown algae) (Fig. 14–5) and Rhodophyta (red algae) (Fig. 14–6), are large, multicellular organisms familiar as seaweeds, kelp, and the like. Their colors are attributed to brown, red, and other pigments, which mask their chlorophyll. They are marine forms, the brown algae preferring cool waters of higher latitudes, the red algae being favored by warm, subtropical waters. Sexual reproduction is well developed among them, but asexual reproduction by means of spores also occurs.

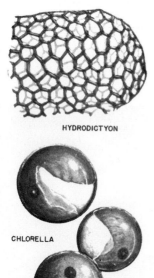

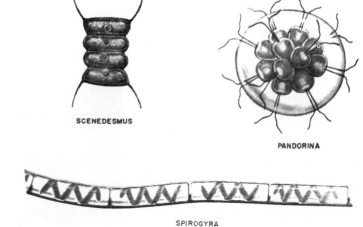

HYDRODICTYON

SCENEDESMUS

PANDORINA

CHLORELLA

SPIROGYRA

Figure 14–3. *Green algae, representatives of the Chlorophyta: Hydrodictyon (9X), Scenedesmus (900X), Pandorina (445X), Chlorella (4500X), Spirogyra (110X). (From Palmer: Algae in Water Supplies. Washington, D.C., U.S. Dept. of Health, Education and Welfare, 1962.)*

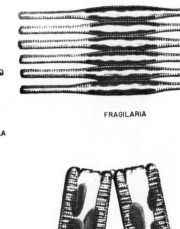

FRAGILARIA

ASTERIONELLA

NAVICULA

DIATOMA

Figure 14–4. *Diatoms, members of the Chrysophyta:* Asterionella *(225X),* Fragilaria *(885X),* Tabellaria *(1350X),* Navicula *(1350X),* Diatoma *(1350X). (From Palmer: Algae in Water Supplies. Washington, D.C., U.S. Dept. of Health, Education and Welfare, 1962.)*

TABELLARIA

Figure 14–5. Fucus, *a brown alga, commonly found attached to rocks along the seashore. The bulbous structures along the leaflike body are air bladders. (Courtesy General Biological Supply House, Inc.)*

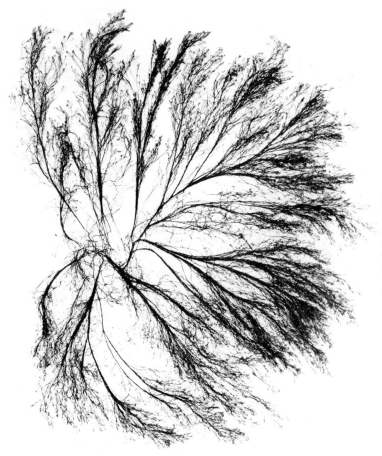

Figure 14–6. *A red alga, one of the Rhodophyta. These delicately branched, lacy forms occur in quiet, deep waters. The blue and violet rays of the sunlight can penetrate to the depths in which the Rhodophyta live and are utilized by the red pigment, phycoerythrin, characteristic of these algae. (From Villee: Biology, 5th ed. Philadelphia, W. B. Saunders Co., 1967.)*

PHOTOSYNTHESIS BY ALGAE

Photosynthesis by algae is essentially the same as in any other plant. Light is absorbed by chlorophyll or by carotenoid or other pigments. When it is absorbed by pigments other than chlorophyll, the energy obtained from the light is transferred to chlorophyll within the plant. The chlorophyll is converted to an excited state, and its extra energy then oxidizes water with the reduction of NADP:

$$12 \ H_2O + 12 \ NADP + 12 \ ADP + 12 \ Pi$$

$$\xrightarrow[\text{(chlorophyll)}]{\text{(light)}} 6 \ O_2 + 12 \ NADPH_2 + 12 \ ATP$$

The oxygen is released, electrons pass along the photosynthetic electron transport chain and reduce the excited chlorophyll,

and the hydrogen is transferred to NADP, yielding $NADPH_2$. At the same time, ADP is phosphorylated to ATP.

Meanwhile, a phosphate-containing five-carbon carbohydrate, ribulose diphosphate, has reacted with the CO_2 to produce a three-carbon compound, phosphoglyceric acid (Fig. 14–7). This is reduced to triose phosphate at the expense of energy from ATP and the hydrogen of $NADPH_2$. Ribulose monophosphate forms from triose phosphate, together with a mole of hexose phosphate, as the product of photosynthetic CO_2 fixation. The ribulose monophosphate is phosphorylated by ATP to the diphosphate, thus completing the cycle. This ATP may be derived from cyclic photophosphorylation:

$$6 \ ADP + 6 \ Pi \xrightarrow[\text{(chlorophyll)}]{\text{(light)}} 6 \ ATP$$

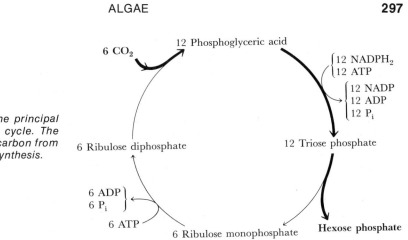

Figure 14-7. *Outline of the principal steps in the photosynthetic cycle. The heavy line traces the path of carbon from CO_2 to the product of photosynthesis.*

SIGNIFICANCE OF ALGAE

Algae are present in great numbers in water, both fresh and salt. The brown and red seaweeds are familiar to anyone who has walked along the ocean shore. They contain useful chemicals, such as iodine, bromine, and potassium, and also considerable protein. In some countries millions of tons are used each year for fertilizer.

The microscopic, single-celled algae are very abundant in natural waters. The ocean may contain several hundred thousand individuals per cubic foot in the surface layers where sunlight can penetrate. Algae are important primary photosynthesizing organisms in this location, and without them all oceanic animal life would soon cease. Microscopic algae constitute the basic food material for aquatic animals. Even the mammoth blue whale, which may reach 100 feet in length, feeds upon algae indirectly; one of its principal sources of food is a crustacean, which in turn feeds upon algae, particularly diatoms.

The distribution of algae in water is limited by the penetration of light. In northern oceans algae are not usually found much below 180 feet, whereas in the clear water of southern latitudes they may be found as deep as 600 feet. The pigments of algae play an important role in their absorption of light. The red phycoerythrin permits red algae to live in deeper water than most other algae because it absorbs blue light particularly well, and the blue or short wavelengths of sunlight penetrate more deeply into water than the longer red wavelengths.

Algae are abundant in soil, particularly moist soil. Their numbers in the upper few inches reach several tens of thousands per gram. The total volume of algae may be three times that of the soil bacteria. Organisms as abundant as these necessarily exert great influence upon the soil chemistry. As photosynthetic organisms they increase the organic matter of the soil. Moreover, at least 40 species in 14 genera of blue-green algae can combine or "fix" atmospheric nitrogen. When this occurs in soil, it may markedly increase fertility. Since a large proportion of the Earth's surface is covered by water, algal fixation of nitrogen in the oceans also contributes a large but at present unknown amount of combined nitrogen to the world economy.

Algae are not generally pathogenic, although a few species produce toxic substances that reach man by way of shellfish, particularly during the warm, summer months (those whose names lack the letter R.)

The oil droplets stored by diatoms are rich in vitamins A and D. These vitamins are transferred through crustaceans to fish and are extracted by man from the livers of the fish.

Carrageenin is a cell wall polysaccharide complex derived from species of *Chondrus*, one of the Rhodophyta. It gels in the presence of potassium and is used to stabilize emulsions and suspend solids in the pharmaceutical, food, brewing, textile, and leather industries. *Chondrus* is abundant in the Canadian Maritime Provinces. The Nova Scotia harvest increased from about 25 tons in 1941 to over 8500 tons in 1958.

Another group of thickeners and emulsifying agents derived from algae are alginic acid and its derivatives. Alginic acid is similar in structure to cellulose (Fig. 14–8). It occurs in the middle lamella and primary walls of numerous genera of the Phaeophyta—*Laminaria, Macrocytis, Lessonia, Eklonia*—from Japan, Australasia, South Africa, and California. The harvested algal fronds are washed, shredded, digested with sodium hydroxide, and purified as calcium alginate or alginic acid. Soluble alginates are hydrophilic colloids and find many uses in the food, textile, cosmetic, rubber, paper, and ceramic industries.

Agar, or "agar-agar," was probably originally derived from an East Indian member of the Rhodophyta, *Eucheuma,* but most agar is now secured from *Gelidium,* although many other genera can serve as sources of this polysaccharide complex. The washed and dried seaweed is extracted, clarified, filtered, purified, and dried in flakes or a powder. Not only is it used as a solidifying agent in microbial culture media, but it is employed in industry in the same way as carrageenin and alginates.

Kieselguhr, or diatomaceous earth, consists of the siliceous cell walls of diatoms sedimented during Tertiary and Quaternary times. Deposits in California, England, and elsewhere are scooped up with huge earth-moving equipment and purified for use as a filter aid in many industries, and as a filler in paints, paper products, and insulation. Thousands of tons of diatomaceous earth are used per year.

Algae are important indicators of pollution in streams and bodies of water. The number of algae and the particular species present are determined by the nature of the water: its temperature, oxygen content, pH, and amount of dissolved organic and inorganic matter (Table 14–2). A highly polluted water containing a large concentration of decomposing organic matter and no dissolved oxygen will support only an anaerobic bacterial flora, often with sulfur bacteria predominating. As purification proceeds, *Euglena* and certain blue-green algae *(Oscillatoria)* grow along with the sulfur bacteria. After further purification, little organic matter remains because most of it has been oxidized to CO_2, nitrates, and sulfates, which stimulate certain algae to such an extent that they are a nuisance because of the thick "blooms" of growth that develop, especially during warm seasons. This is a common occurrence

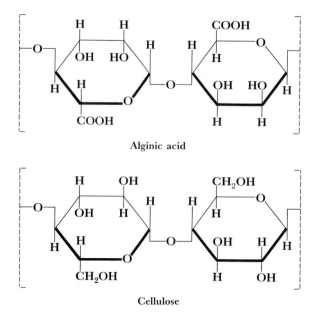

Alginic acid

Cellulose

Figure 14–8. *Structures of the disaccharide units of which alginic acid and cellulose are composed.*

TABLE 14–2. Effect of Degree of Pollution on the Flora of Water

Pollution Zone	Characteristics	Flora
1	Extreme pollution by unpurified sewage; high concentration of organic matter; reducing conditions; no dissolved O_2	Sulfur bacteria (no algae)
2	Partial purification; moderate concentration of organic matter; some dissolved O_2	Sulfur bacteria; *Euglena;* Cyanophyta (*Oscillatoria*)
3	Nearly complete purification; low concentration of organic matter; moderate concentration of inorganic matter (nitrates, sulfates, etc.); dissolved O_2	Cyanophyta (*Oscillatoria, Phormidium*); Chlorophyta (*Ulothrix, Cladophora, Stigeocloneum*); Rhodophyta (*Batrachospermum, Lemanea*)
4	Complete purification; very little organic matter	Cyanophyta (*Phormidium*); Chlorophyta (*Vaucheria, Draparnaldia*); Chrysophyta (*Meridon*); Rhodophyta (*Batrachospermum, Lemanea, Hildenbrandia*)
5	Minimal organic and inorganic matter	Cyanophyta (*Calothrix, Chamaesiphon*); Chlorophyta (*Draparnaldia, Chlorotylium*); Rhodophyta (*Hildenbrandia, Chantransia*)

where treated sewage is discharged into a stream that empties into a small lake. The algal growth ruins the lake for fishing, swimming, and all other recreational uses.

Certain algae are used for food. The green alga, *Chorella,* has been cultivated experimentally for this purpose for several years. It has not found wide acceptance as human food but has been used for animals. An interesting recent application of its biologic activities consists of experiments designed to provide oxygen to support human life in closed vehicles such as space ships. Carbon dioxide from human respiration is bubbled vigorously through illuminated tanks of *Chlorella* in a mineral salt solution. Photosynthesis by the algae produces the familiar reaction in which carbon dioxide is used and oxygen is set free.

In parts of Scotland and Ireland seaweeds such as *Laminaria* and *Fucus* are used throughout most of the year as feed for sheep. *Sargassum* species are employed as fodder in China.

SUPPLEMENTARY READING

Palmer, C. M.: *Algae in Water Supplies*. Washington, D.C., U.S. Department of Health, Education, and Welfare, 1962.

Round, F. E.: *The Biology of the Algae*. London, Edward Arnold Ltd., 1965.

Smith, G. M.: *The Fresh-water Algae of the United States,* 2nd ed. New York, McGraw-Hill Book Co., Inc., 1950.

Taylor, W. R.: *Marine Algae of the Northeastern Coast of North America*, rev. ed. Ann Arbor, Mich., The University of Michigan Press, 1957.

Villee, C. A.: *Biology,* 7th ed. Philadelphia, W. B. Saunders Co., 1977.

Villee, C. A., and Dethier, V. G.: *Biological Principles and Processes*, 2nd ed. Philadelphia, W. B. Saunders Co., 1976.

Weatherwax, P.: *Botany,* 3rd ed. Philadelphia, W. B. Saunders Co., 1956.

15 MOLDS

The Thallophytes are commonly subdivided into three groups: (1) algae, which have just been discussed, (2) fungi, and (3) lichens. Lichens are composite organisms consisting of symbiotic associations of algae and fungi; they will not be discussed further.

Fungi are frequently filamentous, at least in microscopic structure, but are sometimes single-celled (e.g., many of the yeasts). There are five major classes: Myxomycetes, Phycomycetes, Ascomycetes, Basidiomycetes, and Fungi Imperfecti, or Deuteromycetes. The Myxomycetes, or slime molds, resemble protozoa as much as fungi. The remaining forms, sometimes called "true fungi," number over 80,000 species. They include a wide variety of organisms called by common names including molds, yeasts, mushrooms, puffballs, rusts, and smuts. Molds and yeasts are found among the Phycomycetes, Ascomycetes, and Fungi Imperfecti. The terms *mold* and *yeast* have no taxonomic significance; they are colloquial designations for forms that cannot be accurately defined. This chapter is concerned with molds; yeasts will be discussed in Chapter 16.

GROSS STRUCTURE OF MOLDS

Molds are usually described as filamentous, multicellular fungi in which the filaments, known as *hyphae,* branch and sometimes rejoin to form a tangled mass, any large portion of which may be referred to as *mycelium*. Two types of hypha can often be distinguished. The entire *vegetative* portion of the mold (as contrasted with the reproductive structures) constitutes the mold *thallus*.

Vegetative hyphae penetrate the substrate or lie along its surface and secure water and nutrients for the plant; branching, rootlike *rhizoids* perform this function for certain groups of molds and also anchor the plant in place. *Fertile* hyphae usually extend into the air and bear the reproductive bodies or spores.

Many mold hyphae are divided by crosswalls or *septa* into definite cells, each containing one or more *nuclei*. Hyphae of other molds possess few septa and are essentially long tubes containing protoplasm with numerous nuclei scattered throughout. Active protoplasmic streaming can often be observed. These are *nonseptate* or *coenocytic* hyphae.

GROWTH AND REPRODUCTION OF MOLDS

CELL STRUCTURE

Individual mold cells possess rigid walls surrounding the protoplasm. The cell walls are composed of chitin, the substance that makes up the shells of crabs and lobsters, or of a chitin-cellulose complex. The protoplasm is held within a semipermeable cytoplasmic membrane and contains one or more small nuclei together with various vacuoles, granules, and droplets.

Mold cells are often cylindrical and vary greatly in size. The diameters of cells of some common molds range from 2 to 5 μm, and they may be two to four times as long as their diameter. Mold nuclei are not easily demonstrated; special staining methods must be employed.

VEGETATIVE GROWTH OF MOLDS

Mold growth is initiated by germination of a spore or enlargement of any hyphal fragment that falls upon a suitable substrate under proper conditions of temperature, moisture, and aeration. Apical growth and extension occur, and in a septate mold the terminal cells increase in length and crosswalls form to establish new cells. Developing hyphae branch repeatedly and produce a tangled mycelial mass.

Growth is rapid under favorable conditions. An orange or lemon may be completely covered by the bluish green mycelium of *Penicillium* in a day or two, and in the laboratory the same mold will produce colonies an inch or more in diameter on a suitable medium within the same period. Some species grow more slowly and require one or two weeks to produce colonies one-half inch in diameter. Certain molds grow very luxuriantly, producing a cottony mycelium that can fill any given container, such as a bread box, within a few days.

Growth of mold mycelium can continue indefinitely *under favorable conditions*. Older hyphae die and disintegrate, and new hyphae develop. Alexopoulos stated that colonies of fungi have been known to continue growing for over 400 years, and it is likely that some mycelia (but not the individual cells of which they are composed) are thousands of years old.

REPRODUCTION; SPORE FORMATION

Molds reproduce by spore formation. Sexual and asexual spores are produced. Both types are formed by members of the Phycomycetes and Ascomycetes; only asexual spores are found in the Fungi Imperfecti.

Asexual Spores. The spores most frequently observed are asexual. They are produced by the mycelium without nuclear fusion. There are several kinds of asexual spores, and the type of spore is more or less characteristic of the species of mold.

Sporangiospores. These are contained within a swollen structure, the *sporangium,* on the end of the fertile hypha of a nonseptate mold or Phycomycete (Fig. 15–1). The tip of the hypha enlarges, protoplasm and nuclei flow into the enlarged end, and walls form around each nucleus with its surrounding protoplasm. A fertile hypha bearing a sporangium is called a *sporangiophore;* its rounded or club-shaped end, which supports and is partially surrounded by the sporangium, is the *columella.*

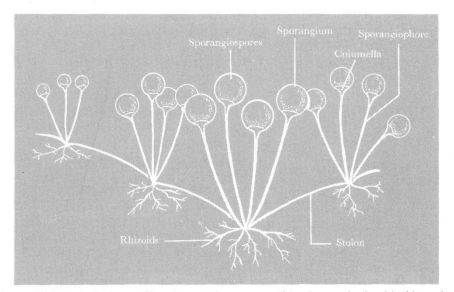

Figure 15–1.   Rhizopus nigricans. *The plant anchors to a solid substrate by its rhizoids and spreads by stolons.*

A fully developed sporangium may burst or be broken by mechanical contact and will then broadcast the thousands of spores it contains. These spores are somewhat more resistant to drying and other unfavorable conditions than are vegetative cells and can survive for some time while being blown about in the air. They germinate when they encounter a suitable moist medium, and a whole new plant may start from each spore.

Conidia. The Ascomycetes and many Fungi Imperfecti bear exposed or unprotected spores, *conidia,* upon fertile hyphae called conidiophores. The tips of the conidiophores differentiate into special cells or swellings from which the conidia form. The cell that immediately supports the conidia is a *sterigma*. A sterigma pinches off a conidium and then repeats the process several times, pushing the preceding conidia ahead to form a chain with the oldest conidium at the end. In some molds a swelling or bud pinches off from the terminal conidium; the bud then becomes the terminal conidium and later it too produces a bud. In this case the conidium at the end of the chain is the youngest.

Chlamydospores. Two or three additional types of asexual spores are found (Fig. 15–2). *Chlamydospores* may be produced by all molds. A thick wall develops around any cell in the mycelium. This cell, containing reserve food, is fairly resistant to drying, can remain dormant for considerable periods, and will germinate to produce new growth when it encounters a favorable environment.

Blastospores. These are asexual spores produced by budding.

Oidia. Hyphae of certain septate molds break up or fragment under proper conditions into their component cells, which are then known as *oidia* or *arthrospores*. These cells do not seem to be reproductive bodies in the same sense as sporangiospores or conidia and possess little if any greater resistance to drying than any other mold fragment. However, they are capable of initiating new growth under favorable conditions and are considered to be a growth form of the organisms.

Asexual spores, particularly sporangiospores and conidia, are produced in thousands by each mold plant. Common household molds, such as *Aspergillus* and *Penicillium,* ordinarily begin to produce spores within a day or so after inoculation onto fresh culture medium. The spores are usually colored, often quite brilliantly. The striking colors of mold colonies are associated almost solely with the spores on the older mycelium. The edge of an actively growing colony is colorless or grayish white where vegetative growth is taking place.

Sexual Spores. These spores are produced following nuclear fusion. Their manner of formation is used as the principal basis for classification. Sexual spores are less frequently observed than asexual spores, and some particular habitat or environmental condition is often necessary to induce sporulation.

Ascospores. Sexual spores of Ascomycetes are known as *ascospores* (Fig. 15–3). Two neighboring cells, either from the same mycelium or from two separate mycelia, send out tubelike processes, which meet and fuse. The two nuclei unite, and the single nucleus that forms then divides to produce daughter nuclei. Two more divisions may occur, yielding eight nuclei. Each nucleus is surrounded by a layer of dense protoplasm and covered by a spore wall. The spores are retained within the original wall that resulted from the union of the two cells, which is now known as an *ascus* or sac.

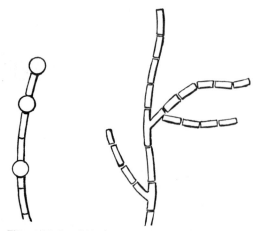

Figure 15–2. *Chlamydospores and oidia of molds.*

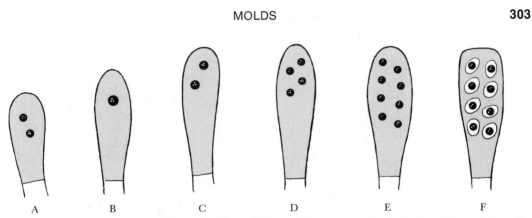

Figure 15-3. *Ascospore formation in a mold. The cell shown in A was produced by fusion of two cells from the same or different mycelia. Its two nuclei unite and then divide repeatedly until eight nuclei have formed (E). In F, the spore walls have been produced. The cell containing the spores is called an ascus.*

Oospores. Some Phycomycetes produce oospores by the union of small male cells with large female cells formed on neighboring hyphae. Oospores possess thick walls and are very resistant to drying; they can remain dormant for considerable periods.

Zygospores. On other Phycomycetes, zygospores are formed by the union of two apparently identical cells from the same or different plants (Figs. 15-4 and 15-8E-H).

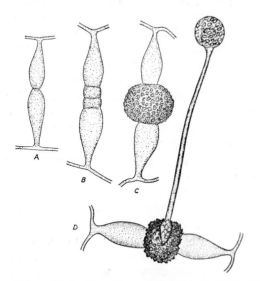

Figure 15-4. *Zygospore formation in* Rhizopus. *Cells from the same or different mycelia meet (A) and produce special sex cells (B), which fuse and form a zygospore covered with a thick, black, protective wall (C). The zygospore remains dormant until proper conditions arise; it then germinates and produces a typical asexual fruiting body. (From Frobisher: Fundamentals of Microbiology, 7th ed. Philadelphia, W. B. Saunders Co., 1962.)*

The two hyphae join, nuclear fusion occurs, and a thick wall surrounds the resulting zygospore. Zygospores, like oospores, can remain dormant for a long time.

Basidiospores. The reproductive cells of Basidiomycetes are formed from and are situated upon special cells called *basidia*. Basidia are binucleate, the two nuclei having been derived from the union of two uninucleate vegetative cells but without fusion of their nuclei. Production of basidiospores begins with enlargement of the basidium and fusion of its two nuclei, followed by meiosis of the zygote and formation of four haploid nuclei (see Figure 15-5). Sterigmata form at the end of the basidium, their tips enlarge, and the four nuclei enter them. They then complete development into basidiospores.

Significance of Sexual and Asexual Sporulation. Both sexual and asexual spores are important to the survival of a species, but in different ways. Sexual reproduction is of great importance in the process of evolution because it permits the recombination of genes and the appearance of progeny with combinations of characteristics different from those of either parent. Selective environmental factors eliminate the least well adapted forms, and only those with great survival powers persist. Asexual spores, particularly conidia and sporangiospores, are especially significant in multiplication and distribution. Their somewhat increased resistance, together with their very great number, favor survival of the species during unfavorable conditions.

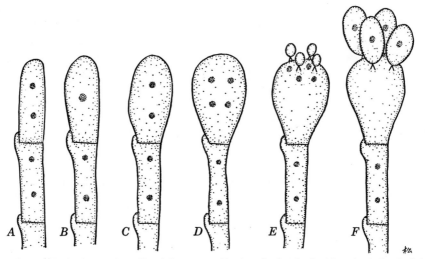

Figure 15–5.   *Stages in the formation of basidiospores. Nuclear fusion in the binucleate hyphal tip cell* (A,B) *is followed by meiosis* (C,D), *development of sterigmata with young basidiospores* (E), *migration of the four nuclei, and maturation of the basidiospores* (F). *(From Alexopoulos: Introductory Mycology, 2nd ed. New York, John Wiley & Sons, 1962.)*

PHYSIOLOGY OF MOLDS

The multiplication of any organism depends upon the availability of proper nutrients and suitable environmental conditions. Promotion or prevention of the growth of molds requires knowledge and control of these factors.

NUTRIENTS

Because of their rigid chitinous cell walls, molds secure nutrients only by diffusion or transport of soluble matter and hence are limited to relatively simple foods. Most molds secure carbon and energy from carbohydrates, especially glucose, and some can utilize alcohols or organic acids. Carbon can also be secured from proteins or, in the absence of a more readily available supply, from products of protein digestion. Very few species utilize fats extensively.

Sources of nitrogen include organic compounds, such as peptones, peptides, and amino acids. Some species can utilize ammonia or nitrates.

Many molds can synthesize all the vitamins and other growth factors they require, but some must be supplied with preformed thiamin or biotin or their precursors.

Molds vary greatly in their ability to utilize complex materials. Some, such as *Penicillium* and *Aspergillus*, excrete digestive enzymes which enable them to grow on nearly any substrate that contains organic matter; they are found on jams and jellies, dates, tobacco, fabrics, and bicycle saddles, to mention only a few places. Others are parasitic and in nature may be restricted to a few host species or even a single species or variety. Some molds parasitize man and other animals. Many are parasites of plants, among which they cause serious epidemics, such as wheat rust.

The metabolic activity of molds yields energy, which is used in the synthesis of the proteins, polysaccharides, and lipids that comprise the various cell structures and protoplasmic constituents. Excess carbohydrate and lipid, stored in the form of glycogen and oil, are readily available whenever needed as sources of energy. By-products of mold metabolism include citric, oxalic, gluconic, glycolic, and other organic acids; various aldehydes and esters; cholines; alkaloids, hydroxylamine, and other nitrogenous compounds; pigments; and antibiotics.

MOISTURE AND OSMOTIC PRESSURE

Mold growth is favored by a moist environment, but these organisms do not need as much water as do yeasts and bacteria, which require an almost completely aqueous medium. Growth upon such dry materials as dried fruits, grains, cloth, tanned leather, and furniture requires only a humid atmosphere. Mildew appears on books and shoes, for example, during prolonged periods of "muggy" weather or in damp climates.

The osmotic pressure of a solution varies with the molecular concentrations of dissolved substances and their degree of ionization. Distilled water has very low osmotic pressure, whereas a concentrated sugar or salt solution possesses high osmotic pressure. The osmotic requirements of different mold species vary widely. Some species are inhibited by 10 per cent salt or 15 to 20 per cent sugar; others grow luxuriantly in 50 or even 75 per cent sugar. Jams and jellies are readily spoiled by molds even though they are usually 50 to 60 per cent sugar.

pH

Molds tolerate changes in pH better than most other forms of life. Common species that grow on bread, citrus fruits, or milk can multiply between pH 2.2 and 9.6, although pH 5 to 6 is most favorable for many of these organisms. Certain unusual forms have been found growing in strong organic or inorganic acids, such as 1N acetic acid or even 2N sulfuric acid. The nutrients that support the growth of these organisms are impurities present in trace amounts in the acid solutions. The predilection of molds for acids explains why mold spoilage is so common in acid fruits and preserves, but their wide range of pH tolerance accounts in part for the universal observation that nearly any organic household object may develop mold growth at some time.

TEMPERATURE

Few mold species will grow at temperatures below 0° C. or above 42° C. Lowest,

TABLE 15–1. Temperature Relations in Common Mold Species

Temperature to kill spores within 30 min.	60–63° C.
Maximum growth temperature	30–40° C.
Optimum growth temperature	22–32° C.
Minimum growth temperature	5–10° C.

highest, and most favorable temperatures for growth of many common species are listed in Table 15–1.

Death occurs at any temperature above the maximum for growth, and the rate of killing accelerates as the temperature rises. Vegetative mycelium is, in general, easily killed; spores are more resistant, but there is wide variation between spores of different species. Almost all, however, are killed within 5 to 30 minutes at 60° to 63° C. For comparison, it may be noted that milk is commonly pasteurized by heating for 30 minutes at about 62° C.

OXYGEN

Practically all molds are highly aerobic and require abundant oxygen. Only a few species, such as that used in Roquefort cheese, will grow satisfactorily under conditions of reduced oxygen tension. Even in this case, however, growth is promoted by stabbing the cheese with wires to provide air holes through the ripening curd.

The necessity for oxygen limits mold spoilage to the surfaces of materials. Molds are found only on the top of the preserves in an imperfectly sealed jar or the jelly in a glass whose paraffin covering has been disturbed so that air can enter.

Despite the requirement for free oxygen, mold metabolism does not yield solely oxidized products. Complete oxidation of glucose and other carbohydrates produces only carbon dioxide and water; the acids and alcohols formed by molds result from incomplete oxidation.

Knowledge of the types of nutrients preferred by molds, together with their pH, temperature, and oxygen requirements, explains most common instances of mold spoilage. The same information, intelligently applied, indicates how to prevent undesired mold growth.

CLASSES OF FUNGI

Most fungi of domestic, industrial, or medical significance are members of the classes Phycomycetes, Ascomycetes, and Fungi Imperfecti (Deuteromycetes). The Basidiomycetes are important principally in agriculture. Some characteristics and common or interesting representatives of the four classes will be described.

PHYCOMYCETES

It will be recalled that fungi are classified primarily according to their methods of sexual reproduction. The Phycomycetes are heterogeneous in this regard, but spores are usually produced after sexual fusion or union, and these structures function as resting or resistant bodies. Most members of the class are also coenocytic, and the more primitive forms produce flagellated cells (zoospores or gametes) during one or more stages in their life cycles.

Chytrids. Chytrids are considered to be the simplest and least highly developed fungi. They are distinguished from all other fungi by production of motile cells (zoospores or planogametes), each with a single, posterior, whiplash flagellum. (A whiplash flagellum is composed of two principal portions: a long rigid basal segment, and a shorter flexible terminal portion.) The thallus of a chytrid is coenocytic and varies from a rudimentary structure to a rhizoid holdfast.

Chytrids are typically microscopic, aquatic forms, but some species grow in soil or on dead leaves and other organic matter, and some parasitize algae, other fungi, mosses, pollen grains, and flowering plants.

Rhizophidium couchii is a moderately developed chytrid that is parasitic upon the green alga, *Spirogyra*. A zoospore that settles on a filament of the alga produces a rhizoidal process that penetrates the wall and enters the host cell protoplast (Fig. 15–6A). The zoospore enlarges and forms a cyst, which continues to grow and develops into a sporangium containing a number of clearly outlined zoospores. When mature, the sporangium wall bursts, permitting the spores to emerge and swim away (Fig. 15–6B, C).

The sexual cycle also starts with zoospores, one of which becomes a female gametangium, and another zoospore attaches to it (Fig. 15–6D). The second zoospore remains small and empties its contents into the female gametangium through a short fertilization tube (Fig. 15–6E, F.) The fertilized gametangium, or oogonium, develops a thickened wall and becomes a resting sporangium (Fig. 15–6G). It is presumed that karyogamy (nuclear fusion) occurs at some point in this process and yields diploid nuclei. The steps in germination of the resting spore probably include several nuclear divisions and formation of a multinucleate structure, from which uninucleate zoospores emerge and start a new vegetative cycle.

Saprolegnia. *Saprolegnia* is one of the group that is sometimes called *water molds*. Most species are found in clear, fresh water, but many are also inhabitants of the soil. The great majority are of little economic importance, since they live on dead organic materials, but a few are parasitic on fish and fish eggs and cause considerable loss in fish hatcheries.

Isolation of *Saprolegnia* is simple. To a pint of pond water in a quart jar are added a few dead flies and split boiled hemp, wheat seeds, or boiled corn grains; after a few days colonies of profusely branched coenocytic mycelium should cover the bits of decaying plant or animal tissue.

The life cycle of a *Saprolegnia* species is diagrammed in Figure 15–7. The thallus of the plant consists of two kinds of hypha: rhizoids, which enter the substrate (the dead fly or plant seed) and attach to and absorb substrate materials, and branched hyphae on the outside of the substrate (Fig. 15–7A), which support the reproductive structures. In the asexual, vegetative cycle, long tapering

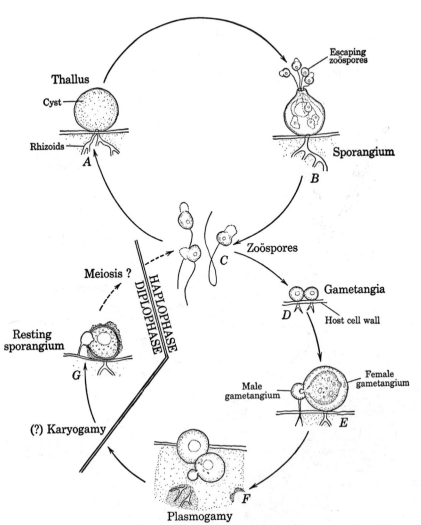

Figure 15–6. *Life cycle of a chytrid,* Rhizophidium couchii. *(See text for explanation.) (From Alexopoulos: Introductory Mycology, 2nd ed. New York, John Wiley & Sons, 1962.)*

sporangia grow at the tips of somatic hyphae and a large number of nuclei flow into them. A septum forms at the base of each sporangium, separating it from the rest of the hypha. Each nucleus serves as a center for a spore (Fig. 15–7B). An opening develops at the end of the sporangium, and the primary zoospores escape into the surrounding water, where they swim and finally come to rest as cysts. A short time later, each cyst opens and a zoospore with two lateral flagella emerges. After a brief swarming period this also encysts. It later puts forth a germ tube, which develops into a hypha and starts a new colony (Fig. 15–7G,H). The thallus continues to grow by proliferation and branching, new

sporangia form, and several asexual generations ensue.

Under appropriate conditions, a sexual reproductive cycle occurs. Oogonia and antheridia (Fig. 15–7I) develop, either on the same hypha or on different hyphae. Several uninucleate oospheres form in each oogonium. The antheridia are smaller than oogonia and are multinucleate. When fully developed, they attach to the oogonia, and fertilization tubes penetrate the oogonial wall and reach the oospheres (Fig. 15–7J). Nuclei now migrate from the antheridium to the oospheres; one nucleus fuses with each oosphere nucleus and forms a diploid zygote nucleus (Fig. 15–7K,L). A thick wall that

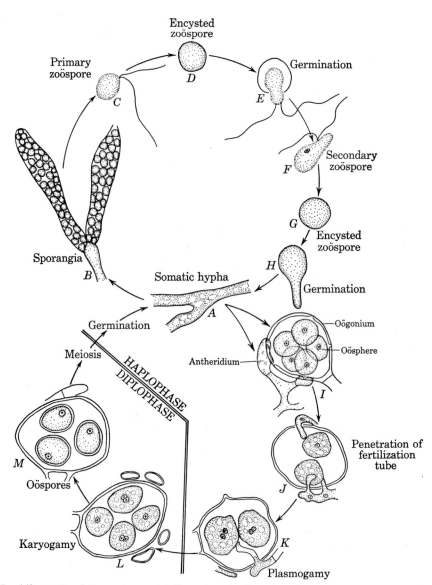

Figure 15–7. *Life cycle of the water mold,* Saprolegnia. *(See text for explanation.) (From Alexopoulos: Introductory Mycology, 2nd ed. New York, John Wiley & Sons, 1962.)*

grows around each oosphere after fertilization converts it into an oospore (Fig. 15–7M). After a long resting period, oospores are liberated from the oogonial wall. They germinate and produce hyphae on which sporangia form and complete the sexual cycle.

Rhizopus. *Rhizopus nigricans* is commonly known as the bread mold because it so frequently occurs on this food. It produces a cottony mycelium and grows rapidly and luxuriantly. Its structure, illustrated in Figure 15–1, resembles that of a strawberry plant.

Vegetative reproduction begins when a bit of mycelium makes contact with a suitable substrate. A rhizoid forms and serves as point of attachment and means of securing water and nutrients, and sporangiophores grow upward. The sporangiophore tips enlarge, and cytoplasm from the sporangiophore flows into it and brings nuclei, which are eventually surrounded by protoplasm and walls and mature into sporangiospores (Fig. 15–8B). The spores within the sporangium are gray or black and make the

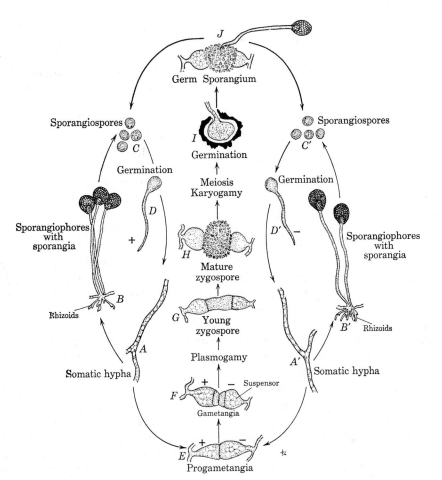

Figure 15–8.   *Life cycle of* Rhizopus nigricans. *(See text for explanation.) (From Alexopoulos: Introductory Mycology, 2nd ed. New York, John Wiley & Sons, 1962.)*

mature mycelium appear to the naked eye to be dotted with black specks.

A hypha similar to the runner or stolon of the strawberry plant burrows through or extends over the substrate and produces another rhizoid, from which more fertile hyphae arise. Growth through and over the substrate can continue almost indefinitely.

When the sporangia rupture, the sporangiospores are expelled; if they are carried to a suitable substrate, they germinate, producing somatic hyphae, and an asexual cycle is completed (Fig. 15–8C,D,A).

The sexual cycle begins with fusion of hyphal branches called progametangia from two physiologically distinct but compatible mycelia, designated + and − (Fig. 15–8E). These make contact with one another, and their ends become walled off from the adja-

cent suspensor cells. The gametangia fuse and form a zygospore, which matures with the formation of a thick, black, and warty wall (Fig. 15–8F–H). The zygospore remains dormant for one to three months, after which it cracks open and a sporangiophore emerges, develops a sporangium containing sporangiospores, and the vegetative cycle can be initiated once more.

ASCOMYCETES

The distinguishing feature of Ascomycetes is the formation of ascospores contained in a saclike structure, the *ascus,* as a result of sexual reproduction. Ascomycetes also possess septate mycelium, most species produce a fruiting body enclosing the asci, and they do not possess any type of flagellated cell. In addition to vegetative multiplication by hyphal growth, Ascomycetes have two reproductive phases, the sexual or ascus stage and the asexual or conidial stage. Ascomycetes include yeasts, some common green and black molds, powdery mildews, cup fungi, morels, and truffles.

Aspergillus and *Penicillium* will be discussed in this chapter. The classification of organisms resembling *Aspergillus* or *Penicillium* poses problems to the systematic mycologist, because sexual cycles have not been observed for all species that possess asexual and vegetative growth patterns like those of ascus-producing representatives. There seems to be a tendency, however, to include the forms that lack asci in the respective genera. From the practical, day-to-day laboratory viewpoint, this is the most satisfactory solution to the problem, since ascospore formation by these organisms sometimes requires several weeks, and most species can be identified from their cultural characteristics and asexual reproductive habits. It should be remembered, however, that many members of these genera are actually Fungi Imperfecti as at present defined.

Aspergillus. *Aspergillus* species are widely distributed and cause numerous kinds of spoilage. They rot figs and dates, decay tobacco and cigars, spoil nuts and bread, and grow on leather and clothing in humid weather. *Aspergillus* colonies are more limited in size than those of *Rhizopus:* colonies on malt extract or yeast glucose agar may be 40 to 50 mm. in diameter. Mature colonies are velvety in texture and dark brown or black, the color being attributed to the conidia.

The microscopic structure of an *Aspergillus* is illustrated in Figure 15–9. Septate vegetative hyphae grow in or over the substrate. At intervals a cell branches and sends a fertile hypha or conidiophore into the air; the top of the conidiophore enlarges to form a *vesicle,* upon which numerous sterigmata are attached radially. Each sterigma supports a chain of conidia. Microscopic study of an undisturbed conidiophore shows only the hypha surmounted by an indistinct black body, because the conidia are so tightly packed and opaque en masse that the internal structure of the "fruiting body" cannot be discerned. The sterigmata and vesicle are seen in preparations that have been disturbed mechanically to remove the conidia.

Formation of the conidia takes place within the tips of the sterigmata, which are actually tubes. A bit of protoplasm containing nu-

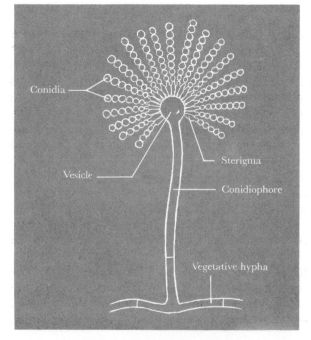

Figure 15–9. Aspergillus. *At intervals in the vegetative hypha a cell puts forth a fertile hypha, upon which conidia form.*

cleus is walled off, becomes rounded, and is pushed outward by formation of another conidium beneath it. The chain of spores lengthens as the sterigma protoplasm grows behind it. Individual spores are colored, and since they are produced in tremendous abundance, the colony takes their color (black, brown, yellow, green, etc.).

It is considered likely that many species of *Aspergillus* have actually lost their ability to reproduce sexually. Evidence of sexual degeneration has been seen in species that do produce asci, and different species vary greatly in their sexual patterns. In one group, antheridia (♂) and ascogonia (♀) are formed close together on vegetative hyphae. They are long, helical, multinucleate, and wind about each other. Nuclei from the antheridium may enter the ascogonium and pair with the ascogonial nuclei. Special hyphae are produced by the ascogonia, and they branch and bear asci at their tips within the developing ascocarp (fruiting body). Each ascus usually contains eight ascospores. The ascocarp is enclosed in one or more layers of cells, and the entire structure is called a *cleistothecium*. When ascospores germinate, they produce germ tubes that develop into mycelia.

Penicillium. Species of *Penicillium* are also widely distributed and occur on decaying organic matter. They are found on apples, pears, grapes, and citrus fruits in storage, on paper, and on textile fibers. They are used in the manufacture of Camembert and Roquefort cheese. The colonies are velvety and greenish or blue-green and may be 30 to 40 mm. in diameter on laboratory media. (See Plate IIIA).

The microscopic structure of *Penicillium* is somewhat similar to that of *Aspergillus*. Septate vegetative hyphae penetrate or grow over the substrate, and septate fertile aerial hyphae arise at intervals. These are sometimes branched and support the conidia, which pinch off from sterigmata. The sterigmata in turn arise from metulae. Several parallel chains of conidia are formed and superficially resemble the bristles of a brush (Figures 15–10 and 15–11). (The name *Penicillium* is derived from a Latin word meaning *little brush*.)

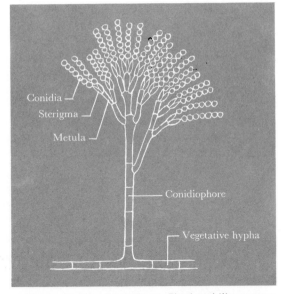

Figure 15–10. Penicillium. *The brushlike appearance of the parallel chains of conidia is characteristic of this genus.*

Sexual reproduction in *Penicillium* is less well known than in *Aspergillus*. In some species, antheridia and ascogonia are still functional, but in others the antheridia no longer play a direct role; in these, the ascogonial nuclei pair and the ascospores apparently develop by meiosis. The ascocarp and cleistothecium develop in much the same manner as in *Aspergillus*.

BASIDIOMYCETES

The Basidiomycetes include mushrooms, toadstools, bracket fungi, rusts, and other fungi. Their spores, designated *basidiospores,* are borne on the outside of a special structure, the *basidium,* as described previously (page 303). Basidiospores have a single, haploid nucleus, and usually four are produced on each basidium.

The vegetative mycelium of Basidiomycetes consists of septate hyphae that penetrate into and through the substrate. Although each hyphal filament is microscopic, the entire mass is easily seen by the naked eye. The mycelium is common in damp locations such as rotten logs and wet, dead leaves, where a white, orange, or bright yellow fan-shaped growth can be seen.

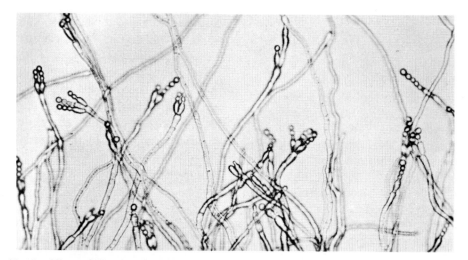

Figure 15–11. *Micrograph of a* Penicillium *showing twisted, branching hyphae and a few chains of conidia. (From Frazier: Food Microbiology. New York, McGraw-Hill Book Co., Inc., 1958.)*

Basidiomycete mycelium develops through three stages. (1) A basidiospore falls upon a suitable, damp substrate, the germ emerges and grows, and *primary mycelium* forms. This may be multinucleate at first as the spore nucleus undergoes rapid division, but septa soon divide the protoplasm into uninucleate cells. (2) The *secondary mycelium* develops next; it is characteristically binucleate as a result of fusion of uninucleate cells, which then branch, after which the two nuclei divide and the four sister nuclei separate into the two daughter cells. (3) Continuation of the process of cell fusion and nuclear and cellular division gives rise to the *tertiary mycelium,* the binucleate cells of which comprise the sporophores or basidia.

The basidia of higher Basidiomycetes are produced in complex fruiting bodies such as mushrooms, puffballs, and bracket fungi. It should be noted that the main mass of the fungus is not the fruiting body but is actually the extensive mycelium, which usually escapes notice.

Basidiospores are of various shapes, and may be colorless or green, orange, yellow, brown, black, or other colors. In mushrooms and toadstools they are often borne in pores or on the surface of gills or plates on the bottom of the fleshy, umbrellalike sporophores. In some species, the spore is forcibly discharged from its location at the tip

of the sterigma after the proper stage of maturity has been reached. When a spore comes to rest on a suitable, moist substrate, it germinates by producing germ tubes, which develop into uninucleate mycelia.

Asexual reproduction occurs in various ways: (1) by budding, (2) by fragmentation of the mycelium, (3) by formation of conidia (e.g., by pinching off from basidiospores or mycelia), or (4) by formation of arthrospores or oidia. Oidia can germinate and produce primary mycelium directly or they can function as *spermatia*.

Most groups of Basidiomycetes, with the exception of the rusts, lack sex organs *per se;* vegetative hyphae and oidia perform the sexual function. The fusion of two protoplasts is a method of securing the binucleate state from uninucleate cells, and this can be accomplished by union of two vegetative cells or by transfer of the contents of a spermatium into a receptive structure (e.g., a vegetative hyphal cell).

Wheat Rust.   Black stem rust of wheat caused by the fungus *Puccinia graminis* is found in nearly all parts of the world and affects barley, oats, rye, and certain grasses, as well as wheat. Damage to the wheat crop in the central United States and Canada is extensive despite an active program to combat the infection.

Two hosts are required for the complete

life cycle of the fungus (Fig. 15–12). The asexual phase occurs on wheat and other grains and grasses. Reddish blisters appear on the stems in late spring, shortly before maturity (Fig. 15–13). These blisters contain fungus mycelium and red spores (uredospores), which are blown to other wheat plants and reproduce the infection. An interval of 10 days is required for the production of a new crop of uredospores. These spores resist drying and may be scattered hundreds of miles by the wind. They are not particularly resistant to cold weather, but under suitable conditions the same mycelium produces a second type of spore (teliospore) in similar blisters; this spore is black or dark brown and the blisters appear black. Teliospores remain dormant over the winter in the straw or on the ground. They germinate in the spring and produce a third type of spore (basidiospore).

Basidiospores are blown about by the wind but do no damage unless they fall upon barberry leaves (Fig. 15–14), where mycelial growth of two different "mating types" is formed. Insects transfer material from one type of mycelium to the opposite type, whereupon fertilization occurs, followed by nuclear division and formation of aeciospores. Aeciospores, when carried by the wind to wheat or other grains or grasses,

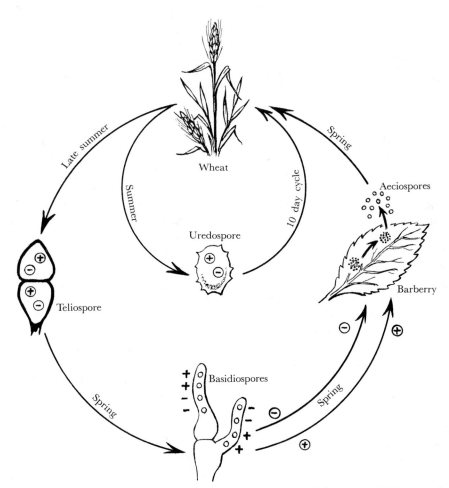

Figure 15–12.   *Life cycle of wheat stem rust. The fungus* Puccinia graminis *grows on the stems of wheat plants during the late spring and summer, producing uredospores, which are distributed to other plants in a 10 day cycle. Under favorable conditions teliospores are produced. These resist cold weather and survive until the next spring, when they germinate and form sexual basidiospores, which are then carried to the common barberry. Insects effect fertilization of cells of the (+) strain by nuclei from cells of the (−) strain, and aeciospores are liberated. When they are blown to wheat plants they germinate and initiate the summer cycle again.*

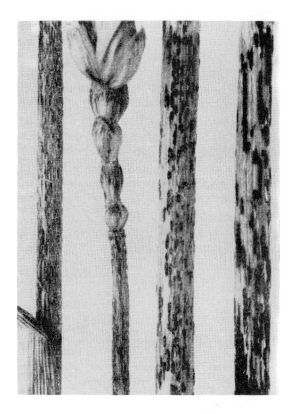

Figure 15–13. *Wheat stem rust. The dark, elongated patches on the stems are the blisters, which contain fungus mycelium and spores (3X). (From Weatherwax: Botany, 3rd ed. Philadelphia, W. B. Saunders Co., 1956.)*

germinate and produce the mycelium upon which uredospores are once more formed.

Epidemic spread of wheat stem rust is promoted by several factors: (1) large fields of wheat, such as those found in the central plains of North America from Texas to Canada; (2) mild winters, which permit uredospores to survive; (3) barberry plants on which the sexual cycle can take place; (4) humid or rainy weather in the spring, which favors uredospore germination; (5) vigorous succulent growth of the crop, in which the fungus is most active; and (6) continuous winds to spread the spores.

Wheat rust has been combatted by elimination of barberry plants. The first known attempt to control a plant disease by legislation occurred about three centuries ago when officials in Rouen, France, ordered the destruction of barberry bushes. Barberry eradication laws were also enacted in Connecticut, Massachusetts, and Rhode Island between 1726 and 1766. It is of interest in this connection that the life cycle of *P. graminis* was not known until about 100 years ago (1865).

Another approach to the problem of eliminating wheat rust is that of breeding resistant varieties of wheat. This has been partially successful, but complete success is hampered by the readiness with which new varieties of the fungus appear. Sexual recombination during the stages in the barberry permits the development of new varieties, some of which possess enhanced survival power even in resistant wheat.

FUNGI IMPERFECTI

Many fungi with septate mycelium apparently reproduce only by conidia, that is, they lack a sexual (perfect) phase. These are the Fungi Imperfecti, or Deuteromycetes. The conidial stages of many species are similar to those of the Ascomycetes, and it is presumed that they are actually Ascomycetes whose sexual (ascospore) stage has never been found or has been lost during evolution. In some cases sexual stages were discovered years after the organisms were originally classified in the Fungi Imperfecti. Most

Figure 15-14. *Wheat stem rust; an infected barberry leaf. Aeciospores produced in these enlarged areas on the lower surface of the leaf are capable of infecting the wheat plant. (From Weatherwax.)*

of these were Ascomycetes, but a few were Basidiomycetes. The Fungi Imperfecti may therefore be defined as the conidial forms of Ascomycetes, or occasionally of Basidiomycetes, whose sexual stages have not been observed. They reproduce by conidia, by oidia, by budding, or by purely vegetative processes. It will be recalled that conidia are asexual spores, not contained in a sac, and generally borne at the tips or along the sides of hyphae called conidiophores. The actual fruiting structure may take various forms, and this is used in part as a basis for classification. Shape, color, septation, and size of conidia are also considered, together with the use of other methods of reproduction, such as budding.

Some Fungi Imperfecti are normal residents of the human body, and a number of species are notorious pathogens. The dermatophytes, which cause superficial mycoses, are parasitic. The group that produces deep-seated mycoses are primarily soil organisms.

Candida. Species of *Candida* are normal inhabitants of the mouth and throat, the digestive tract, the vagina, and the skin. Members of this genus are described as *dimorphic* fungi because they produce yeastlike cells under certain conditions and form branching filaments with a mycelial habit of growth under other conditions. *Candida* species also appear in an intermediate, pseudomycelial phase in which elongated, budding cells cling together in chains. In addition, some species produce heavy-walled, resistant, asexual chlamydospores.

Dermatophytes. Three principal genera of dermatophytes are found on the skin with considerable frequency: *Microsporum, Trichophyton,* and *Epidermophyton.* The dermatophytes invade only the superficial, keratinized epithelium—skin, hair, and nails; they are parasitic and are rarely found elsewhere. In the skin and nails only mycelial fragments are observed. Filamentous colonies are produced on Sabouraud's glucose agar at room temperature. (See Plate III).

Differentiation of dermatophytes is based primarily upon cultural and morphologic characteristics on laboratory media: (1) gross appearance and pigmentation of mycelium in giant colonies; (2) form and arrangement of the macroconidia (as large as 160 μm in some species); (3) number and arrangement of microconidia (circular or club-shaped, 2 to 3 μm in diameter); (4) special structures such as helical coils or racket-shaped hyphal ends; and (5) chlamydospores and arthrospores.

Colonies of *Trichophyton* are granular to powdery, cottony to velvety, heaped, wrinkled, and folded, with a velvety to smooth or waxy surface; they are pink, red, purple, violet, brown, yellow, or light buff. A variety of macro- or microconidia are formed. *Epidermophyton* colonies are velvety to powdery and greenish-yellow. Only macroconidia are produced. *Microsporon* produces microconidia on its hyphae, and also large, spindle-shaped macroconidia, similar to but larger than those of *Epidermophyton*.

Deep-seated Mycoses. The fungi that cause deep-seated mycoses are usually soil saprophytes. When these are inhaled, they produce chronic infections that progress slowly at first but accelerate and finally lead to death.

Blastomyces dermatitidis. *B. dermatitidis* is dimorphic; growth is mycelial on laboratory media at room temperature but yeastlike at body temperature and in infected tissue. It is found in soil, but less commonly than other pathogenic fungi. It causes North American blastomycosis.

Coccidioides immitis. *C. immitis* is also dimorphic, but it differs from yeastlike fungi in that it does not reproduce by budding. In tissues it reproduces by a process of spore formation. Within a multinucleate, spherical body, 50 to 60 μm in diameter, many uninucleate spherical spores form. These are released when the spherical body ruptures. Each small spore then grows and develops once more into a large, multinucleate, spherical body. On artificial media at room temperature, or 37° C., under aerobic conditions, *C. immitis* has a mycelial type of growth.

Histoplasma capsulatum. This dimorphic fungus appears as yeastlike cells when observed in the mononuclear cells of the peripheral blood and in macrophages in bone marrow and spleen, but grows in a mycelial form on laboratory media at room temperature. The yeast form is found also on anaerobically incubated blood agar slants at 37° C. Histoplasmosis is a widespread and very serious disease.

Cryptococcus neoformans. *C. neoformans* is monomorphic and yeastlike, that is, it does not produce mycelial growth under any conditions. It is also distinctive in being heavily encapsulated, both in tissue or spinal fluid and in water suspensions from laboratory media. The capsules are strikingly demonstrated in India ink wet mounts. *C. neoformans* causes a form of meningitis.

SIGNIFICANCE OF MOLDS

The widespread distribution and nutritional versatility of molds fit them well for a role as spoilage agents, and it is this activity that is probably of most popular concern. A comparatively restricted number of species cause human disease. Aside from these unpleasant or harmful activities, molds are becoming increasingly valuable for the commercial production of chemicals. Citric acid is produced by strains of *Aspergillus niger;* this same organism can be used to produce oxalic acid. Gluconic, lactic, and other acids are also produced by certain molds. Industrially valuable enzymes (e.g., amylase, which hydrolyzes starch) are secured from various molds, and certain species are used in the manufacture of cheeses and other foods.

Penicillin is probably the most widely known commercially manufactured mold product. Originally produced by *Penicillium notatum,* it is now manufactured by use of *P. chrysogenum,* and annual production is about 1000 tons in the United States, with a market value of approximately $100,000,000. Growth of the industry over a period of only 18 years is indicated in Table 15–2. The 180-fold increase in production from 1945 to 1963 was accompanied by a drop in the cost per pound to less than $\frac{1}{70}$ of its earlier figure.

Another, quite different but equally significant use of molds is illustrated by the use of various species of *Neurospora* in the study of biochemical genetics. *Neurospora,* com-

TABLE 15–2. Production and Value of Penicillin in the United States Since 1945

Year	Production (Tons)	Market Value per Pound
1945	6	$3870
1950	215	266
1955	284	117
1960	430	84
1963	831	56

monly known as red bread mold, is an ascomycete. Its vegetative growth resembles that of most filamentous fungi; the mycelium consists of branching, multinucleate hyphae, and asexual conidia form at the ends of aerial conidiophores. Sexual spores are produced in asci after interaction of the two mating types, A and a. The fruiting body, or perithecium, contains numerous asci, which are elongated sacs, each containing a linear array of eight ascospores derived equally from the parental strains. Fertilization of the female element can be accomplished by almost any hyphal element of the male strain, including a conidium.

The usefulness of Neurospora in genetic research derives partly from the fact that mutations can be induced by irradiating conidia with x-rays and then using the treated spores to fertilize normal material of the opposite mating type. The sexual spores thus produced are separated by micromanipulation, germinated, and tested for mutation-induced characteristics.

Normal, "wild-type" Neurospora will grow on a simple "minimal" medium containing mineral salts, sugar, and biotin. Various mutants produced as just described cannot grow on the minimal medium until it has been fortified by other vitamins, amino acids, purines, or pyrimidines. These are biochemical mutants, and by appropriate crossing with the normal or wild-type strain they can be used to study the genetic basis of the mutations.

Relatively few of the thousands of species of molds are pathogenic, particularly in temperate climates. Almost all of the common spoilage molds are harmless and could be eaten without danger. It is a reflection of human inconsistency that Penicillium roqueforti and P. camemberti are eaten with relish when they occur upon Roquefort and Camembert cheeses, but most people consider the same molds undesirable on other foods and discard the moldy portions. The chief danger from moldy food is that unsatisfactory methods of preservation or storage may have permitted harmful microorganisms as well as molds to enter or survive.

Under natural conditions, molds are kept in check to some extent by the bacteria with which they are associated. Many kinds of bacteria grow more rapidly than molds and produce waste materials that inhibit the multiplication of molds; suppression or destruction of the bacteria therefore permits the molds to grow. This phenomenon is illustrated by the incidental results of treating humans with antibiotics. The bacterial flora of the intestine can be almost completely eliminated by suitable antibiotics, but when this occurs various fungi (e.g., Monilia) replace the bacteria, and the physical characteristics of the intestinal contents change markedly.

SUPPLEMENTARY READING

Ainsworth, G. C., and Sussman, A. S. (eds.): The Fungi; an Advanced Treatise. Vol. 1. The Fungal Cell. New York, Academic Press, Inc., 1965.

Alexopoulos, C. J.: Introductory Mycology, 2nd ed. New York, John Wiley & Sons, Inc., 1962.

Funder, S.: Practical Mycology. Manual for Identification of Fungi, 2nd ed. Oslo, A. W. Brøggers Boktrykkeri A/S, 1961.

Raper, K. B., and Fennell, D. I.: The Genus Aspergillus. Baltimore, The Williams & Wilkins Co., 1965.

Raper, K. B., and Thom, C.: A Manual of the Penicillia. Baltimore, The Williams & Wilkins Co., 1949.

Skinner, C. E., Emmons, C. W., and Tsuchiya, H. M.: Henrici's Molds, Yeasts, and Actinomycetes, 2nd ed. New York, John Wiley & Sons, Inc., 1947.

16   YEASTS

There is no clear-cut distinction between yeasts and molds. Yeasts have been defined as fungi whose usual and dominant growth form is unicellular. This is perhaps as good a definition as can be stated. The close relationship between yeasts and molds is indicated by the observation that large colonies of many yeast species develop hyphal filaments upon a suitable medium and under certain cultural conditions. Moreover, many yeasts produce sexual spores (ascospores) by a process similar to that of various molds.

MORPHOLOGY OF YEASTS

SHAPE AND SIZE

Yeast cells are spherical, elliptical, or cylindrical. Their sizes are highly variable. Cells of *Saccharomyces cerevisiae* range from 2 to 8 μm in diameter by 3 to 15 μm in length. Cells of some species attain a length of 100 μm.

Cell Structures

The protoplasm of a yeast cell is enclosed by a cell wall and cytoplasmic membrane, and contains a nucleus, a large vacuole, and numerous granules and fat globules (Fig. 16–1). No flagella or other organs of locomotion are present. Electron micrographs indicate that the cell wall is composed of an outer dense layer about 0.05 μm thick and a less dense layer of about 0.2 μm. The inner part of the wall may in turn be subdivided into about three layers. The cell wall is composed of polymers of glucose and mannose with smaller amounts of protein, lipid, and chitin.

The nucleus is less than 1 μm in diameter; practically nothing is known of its internal structure. The protoplasm of young, actively growing cells is fairly free from granules and globules of reserve substances, but large numbers accumulate later as growth ceases. Fat globules gradually coalesce and may form one large globule. Carbohydrate granules, principally glycogen, appear in yeasts as in more complex fungi. They stain deep reddish brown with iodine. Fine granules of protein have also been demonstrated within yeast protoplasm.

The large vacuole near the nucleus contains a solution or suspension of *volutin,* also found in higher fungi and in bacteria. Volutin is a complex material consisting of RNA, polyphosphates (polymers of phosphate), and lipoprotein. It may be absent from very young cultures, is abundant in old cultures, and disappears during spore formation.

Figure 16–1. *Sketch showing structures of a "typical" yeast, such as* Saccharomyces cerevisiae. *The vacuole is prominent in unstained yeast preparations; the nucleus and lipid globules are demonstrated by special staining or by electron micrography of ultrathin sections.*

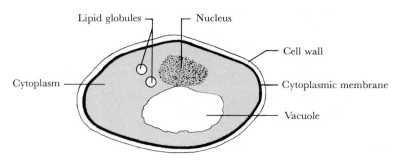

REPRODUCTION OF YEASTS

Yeasts are divided naturally into two groups on the basis of their methods of reproduction. One group reproduces by both budding and spore formation; the other reproduces only by budding. The spore-forming group is classified with the Ascomycetes and may represent a primitive form of this class. Yeasts that reproduce by budding only are included in the Fungi Imperfecti, because they lack a sexual phase. A third small group of yeasts is related to the Basidiomycetes.

BUDDING

Budding is an asexual process and is well illustrated in the series of electron micrographs in Figure 16–2. A small bulge develops in the cell wall and gradually enlarges. The cytoplasm of the mother and daughter cells remains continuous for some time, but eventually the opening between the two cells closes. A double crosswall forms,

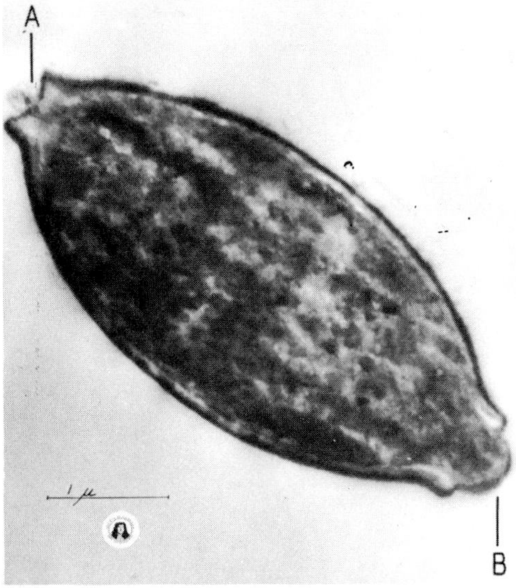

Figure 16–3. *Birth scar* (A) *and bud scar* (B) *on a cell of* Saccharomyces cerevisiae. *(Photograph by H. D. Agar and H. C. Douglas; A.S.M. LS-336.)*

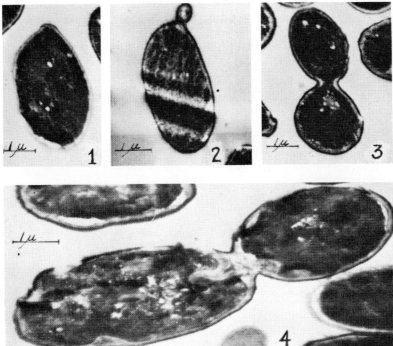

Figure 16–2. *Successive stages in budding of the yeast,* Saccharomyces cerevisiae. *A small bulge* (1) *marks the site of future bud formation. The bulge enlarges* (2) *and eventually reaches the size of the mother cell* (3); *the cytoplasm of the two cells is still continuous. Formation of cell wall material* (4) *makes possible independent existence of the two cells. (From H. P. Agar and H. C. Douglas, J. Bact., 70:427–434, 1955.)*

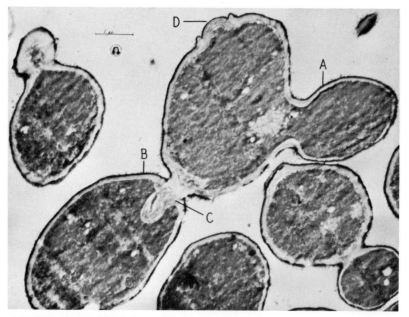

Figure 16–4. *A budding yeast cell. A young bud is shown at A; its cytoplasm is still connected to that of the mother cell. B is an older bud, now physiologically separated from the mother cell by cell wall substance (C). The mother cell has budded previously, as indicated by the bud scar (D). (Photograph by H. D. Agar and H. C. Douglas; A.S.M. LS-335.)*

whereupon the two cells are physiologically distinct and may separate. A convex bud scar remains on the mother cell after the daughter has separated, and the daughter cell retains a corresponding concave birth scar (Figs. 16–3 and 16–4). Nuclear division presumably preceded or accompanied development of the bud (Fig. 16–5).

Daughter cells do not always separate immediately but may remain attached while

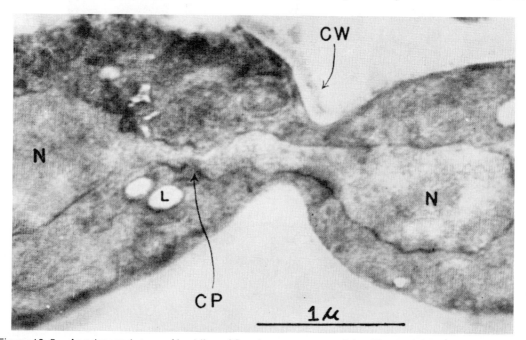

Figure 16–5. *An advanced stage of budding of* Saccharomyces cerevisiae. *The nuclei are still not separated. N, nucleus; CW, cell wall; L, lipid granule; CP, constriction point, where the two nuclei will eventually separate. (From Hashimoto, Conti, and Naylor: J. Bact., 77:344–354, 1959.)*

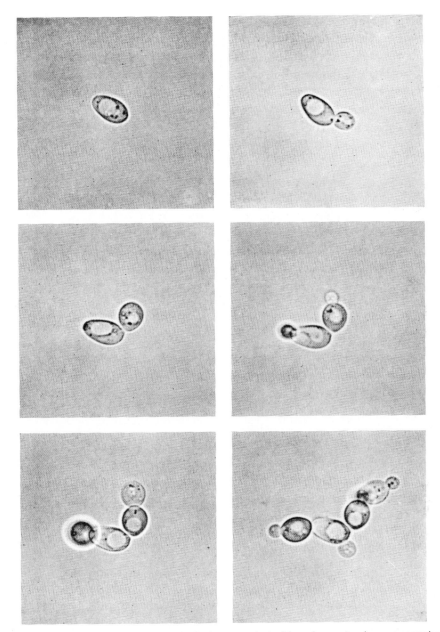

Figure 16–6. *Micrographs taken at intervals during a period of four hours to show successive budding of rapidly growing* Saccharomyces cerevisiae. *The original cell increased to eight cells. (From Sarles et al.: Microbiology. New York, Harper & Brothers, 1956.)*

one or more additional buds form on the mother cell (Fig. 16–6). Moreover, daughter cells themselves may undergo budding while still attached. The result is a pseudo-mycelial mass composed of as many as 64 connected cells.

BINARY FISSION

Binary fission is another form of asexual reproduction that occurs in a few yeast gen-era. This process is similar to that in bacteria and consists of elongation of the cell followed by formation of a crosswall and possible separation of the two cells.

SEXUAL REPRODUCTION OF YEASTS

There are three major life cycle patterns in yeasts, and they differ with regard to the stage in which union of haploid nuclei oc-

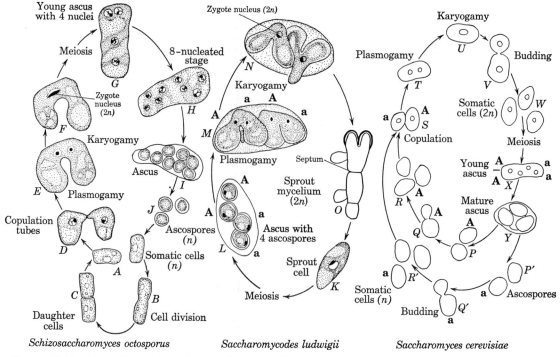

Figure 16–7. *Major life cycle patterns of yeasts. (From Alexopoulos: Introductory Mycology, 2nd ed. New York, John Wiley & Sons, 1962.)*

curs; these are illustrated in Figure 16–7. *Schizosaccharomyces octosporus* reproduces vegetatively by fission (A–C), and sexually by fusion of two cells and their nuclei (D–F). This is followed by three nuclear divisions, yielding eight ascospores (G–J), which can germinate and divide by fission. *Saccharomycodes ludwigii* cells undergo meiosis and produce four ascospores, two of each mating type (K–L); these fuse by pairs and form zygotes (M–N). Germ tubes grow through the ascus wall and produce sprout mycelium, which buds and yields separate yeast cells (O–K). *Saccharomyces cerevisiae* reproduces vegetatively by budding (P–R and U–W). Haploid cells of different mating types (R, R′) copulate and nuclear fusion produces large, diploid cells (S–U), which bud repeatedly until meiosis occurs, and two pairs of haploid ascospores of opposite mating types form (X–Y). Liberation of these paves the way for repetition of the sexual portion of the cycle.

HYBRIDIZATION OF YEASTS

An interesting outgrowth of theoretical studies of sexual reproduction by yeasts is the development of hybrids. Single spores are dissected from an ascus using microtools and a micromanipulator. Individual spores are allowed to germinate, and the characteristics of their progeny are determined. Two spore cultures, each possessing one or more characteristics desired for a certain process, such as brewing, are mixed and allowed to grow together. A form of sexual copulation occurs and a zygote forms. The progeny may possess the desired qualities from both parent spore cultures. This method might be used, for instance, to secure strains that produce a particular flavor and aroma together with increased yield of alcohol.

PHYSIOLOGY OF YEASTS

Nutrient Requirements. Yeasts require the same chemical elements as other forms of life: carbon, hydrogen, oxygen, nitrogen, phosphorus, potassium, sulfur, magnesium, iron, zinc, manganese, copper, and molybdenum. The last five metallic elements listed are required in minute quantities as components or activators of enzymes. They are often present in sufficient amounts as impurities in the water or other ingredients of culture media.

Carbon is ordinarily secured from sugars, organic acids, aldehydes, or glycerin. Part of the carbon is utilized in the synthesis of protoplasmic constituents, but the greater portion is oxidized with the release of energy for synthetic and other vital processes. Nitrogen is secured from products of protein hydrolysis, such as proteoses, peptones, amino acids, and ammonia, or from urea or amides. Ammonium sulfate, phosphate or chloride is often used as a source of nitrogen in culture media and in industrial processes. Phosphorus is essential for growth and plays an important part in carbohydrate metabolism; it is usually supplied as a phosphate salt.

Growth Factors. Yeasts require certain vitamin-like *growth factors*. As early as 1901, Wildiers noted that small inocula did not grow in simple, chemically defined media in which large inocula multiplied rapidly and luxuriantly. When large amounts of culture were transferred, sufficient growth-promoting substance was carried over from the original culture to initiate growth, but this did not occur when small amounts were used. The addition of a little yeast extract permitted multiplication of the organisms in small inocula; after growth was under way, the yeast manufactured its own growth-promoting substance. The unknown factor was called "bios" and was later found to consist of at least six substances: thiamine, biotin, pyridoxine, inositol, pantothenic acid, and niacin.

These growth factors are active in extremely low concentrations. For example, a yeast that fails to grow in the absence of biotin may grow maximally when biotin in a concentration of only 1:1,000,000,000 to 1:1,000,000 is provided. Inositol is much less active than biotin, but it stimulates the growth of yeasts when it is present in concentrations of 1:100,000 to 1:20,000. Biotin is significant in nitrogen metabolism; pyridoxine, thiamin, and niacin are precursors of certain coenzymes; and inositol is apparently built into the cell structure.

Water. In general yeasts require somewhat more water than molds but less than most bacteria. It should be emphasized, however, that great variation exists among yeasts; some species grow in media containing as little as 40 per cent water, for example, honey and jellies or jams. Organisms that grow in solutions of such high osmotic pressure are called *osmophilic*.

pH. Yeasts grow over a wide range of pH, although their requirements are more restricted than those of molds. Many species can multiply in solutions as acid as pH 3 and as alkaline as pH 7.5. The optimum reaction is usually between pH 4.5 and 5.0.

Temperature. Growth cannot be expected at temperatures much below freezing, nor does it occur above 47° C.; maximum temperatures for some species are lower. The most favorable temperature is usually between 20° and 30° C. Incubation at 30° C. is generally satisfactory. Highest yields of cells or of ethyl alcohol are obtained at somewhat lower temperatures, because the inhibitory effect of the toxic waste products that accumulate in the medium increases with temperature. This factor is of concern in the commercial manufacture of yeast.

Yeast spores are resistant to low temperatures; they can survive through the winter, frozen in the soil. Laboratory tests indicate that there is rapid death early during freezing, but the number of viable spores soon levels off and remains relatively constant. Yeasts have been found alive after storage for 160 weeks at −13° to −15° C.

Vegetative cells of most species are killed within five to ten minutes at 52° to 58° C. Spores are more resistant but are killed within a few minutes at 60° to 62° C. The medium in which the organisms are suspended affects the sterilization time and temperature; survival is often better in solutions containing high concentrations of sugar or salt.

Oxygen. Yeasts were the first organisms found to grow in the absence of atmospheric oxygen. Pasteur was greatly impressed by this fact, and he observed that anaerobic utilization of sugar yielded principally alcohol and carbon dioxide, whereas aerobic products were carbon dioxide and water. Yeast multiplication is more rapid and the yield of cells is greater under aerobic than under anaerobic conditions; an abundance of oxygen is therefore provided in the manufacture of commercial yeast, but oxygen is excluded when the desired product is alcohol (e.g., in brewing or wine making).

These observations are most readily understood by reference to the equations that express empirically the overall reactions of complete oxidation and of alcoholic fermentation of a simple sugar:

$$(1)\ C_6H_{12}O_6 + 6\ O_2 \longrightarrow$$
$$6\ CO_2 + 6\ H_2O + 688{,}000\ cal.$$
$$(2)\ C_6H_{12}O_6 \longrightarrow$$
$$2\ C_2H_5OH + 2\ CO_2 + 54{,}000\ cal.$$

These show that complete, aerobic oxidation of a sugar such as glucose can be expected to yield a theoretical maximum of 688,000 calories of energy, whereas anaerobic fermentation of the same sugar to alcohol and CO_2 makes available no more than 54,000 calories of energy. Since the manufacture of protoplasmic constituents and structures, budding, and all other vital activities require energy, it is obvious that aerobic conditions favor more rapid and extensive cell growth.

CLASSIFICATION OF YEASTS

Yeasts are divided into two groups on the basis of their ability to produce ascospores. Those that form ascospores are assigned to the class Ascomycetes and are sometimes colloquially designated "true yeasts." Yeasts that do not produce ascospores but reproduce chiefly by budding are classified among the Fungi Imperfecti and are sometimes called "false yeasts" or *torulae*. The name torula in this connection does not have taxonomic significance.

There are about 30 genera of yeasts. The best known and the most important industrially is *Saccharomyces*. *S. cerevisiae* is the species used in brewing and baking. The *ellipsoideus* variety of *S. cerevisiae* is commonly used in the manufacture of wine.

False yeasts appear in the white or grayish scum that develops on acid foods like pickles left in an opened container. Some species utilize the organic acid, which nor-

mally acts as a preservative, oxidizing it to carbon dioxide and water, thus paving the way for further spoilage.

CLASSIFICATION BY PHYSIOLOGIC CHARACTERISTICS

Protozoa, algae, and molds are classified principally on the basis of their morphology. Organisms as small as yeasts display so few distinctive morphologic features that additional characteristics must be found if fine subdivision is considered desirable. Properties useful in subdividing morphologically similar groups are based upon physiologic behavior, that is, what the organisms do rather than how they appear.

Physiologic classification is illustrated by three species of *Saccharomyces*: *S. cerevisiae*, *S. carlsbergensis*, another beer yeast, and *S. fragilis*, found in various fermented milks. These organisms produce oval cells and resemble one another morphologically; moreover, all ferment the simple sugar glucose ($C_6H_{12}O_6$) and produce large quantities of gas (CO_2). They can be distinguished, however, by their behavior in culture media containing two other carbohydrates, the disaccharides lactose and melibiose (Table 16–1). These compounds have the empirical formula, $C_{12}H_{22}O_{11}$, but their monosaccharide components are joined differently. *S. cerevisiae* ferments neither lactose nor melibiose. *S. carlsbergensis* ferments melibiose, and *S. fragilis* ferments lactose.

IMPORTANCE OF YEASTS

Spoilage. Yeasts are almost universally present in soil and from this source are

TABLE 16–1. Fermentation of Sugars by Three Common Species of Yeast

Species	Glucose	Lactose	Melibiose
Saccharomyces cerevisiae	+	–	–
Saccharomyces carlsbergensis	+	–	+
Saccharomyces fragilis	+	+	–

widely disseminated by insects; they also travel on dust particles in the air. Yeast spores are not as resistant as spores of bacteria and molds, but in the dried state they are known to survive at least four years. Hence, it is obvious that they may easily find access to substances capable of supporting their growth.

Yeasts have a particular predilection for acid foods that contain sugar, from which they produce ethyl alcohol and a large quantity of gas. Fruits are especially subject to this type of spoilage; untreated fruit juices are almost certain to undergo alcoholic fermentation. It is believed that insects inoculate fruits with yeasts that are commonly present in or upon their bodies. Since yeasts, unlike molds, can grow in the absence of oxygen, sealing a food container does not prevent spoilage. However, even moderate heat (e.g., 60° C. for a few minutes) destroys yeasts and provides practical preservation.

Fermentation. Yeasts are very important in industry. Alcoholic fermentation has been practiced on a trial and error basis for thousands of years; scientific principles have been applied only within the past century. Strains of *S. cerevisiae* are used to make beer and ale; this organism produces 4 to 6 per cent alcohol. "Bottom yeasts" are usually employed in making beer; they get their name because they gradually settle to the bottom of the fermenting solution. "Top yeasts" are used in making ale; they rise to the top of a fermentation tank, swept upward by the rapid evolution of gas. Strains of *S. cerevisiae,* var. *ellipsoideus,* are commonly present on grapes and other fruits and produce as much as 16 per cent alcohol in wine fermentation. Yeasts are found in fermented milks such as kefir, koumiss, and matzoon, beverages that are popular in eastern European countries. The alcoholic fermentation is a necessary preliminary to the manufacture of vinegar, in which bacteria of the genus *Acetobacter* oxidize alcohol to acetic acid. Yeasts also participate in the production of certain cheeses.

Commercial manufacture of ethyl alcohol is a large industry. Numerous by-products are formed, including carbon dioxide, which can be compressed into solid form as dry ice. Before World War I, it was discovered that sodium bisulfite modified the yeast fermentation so that glycerin was formed; the Germans used this procedure during the war to produce glycerin for explosives. An American process employs alkali for a similar purpose. Compressed yeast is used not only in baking but also as a source of vitamins and of enzymes useful in the manufacture of syrups and confectionery products.

SUPPLEMENTARY READING

Alexopoulos, C. J.: *Introductory Mycology,* 2nd ed. New York, John Wiley & Sons, Inc., 1962.

Cook, A. H. (ed.): *The Chemistry and Biology of Yeasts.* New York, Academic Press, Inc., 1958.

Ingram, M.: *An Introduction to the Biology of Yeasts.* London, Sir Isaac Pitman and Sons, Ltd., 1955.

Lodder, J., and Kreger-van Rij, N. J. W.: *The Yeasts, A Taxonomic Study.* New York, Interscience Publishers, Inc., 1952.

Skinner, C. E., Emmons, C. W., and Tsuchiya, H. M.: *Henrici's Molds, Yeasts, and Actinomycetes,* 2nd ed. New York, John Wiley & Sons, Inc., 1947.

SECTION
FOUR

ECOLOGY OF INFECTIOUS DISEASE

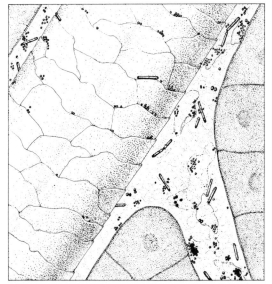

NORMAL AND 17
PATHOGENIC
BODY FLORA

Ecology is the study of the relationships of living organisms with their environment. Complete understanding of any given ecologic situation requires thorough knowledge of the populations concerned, the interrelationships among these organisms and between them and their environment, and the biochemical bases for these relationships.

The sum total of the interacting elements, both living and environmental, in a limited universe defines an ecosystem. In Nature, an ecosystem is usually an open system in a steady state, that is, production of each element of the system is balanced by its consumption. The adult human body comprises an ecosystem; although cells and tissues die, they are replaced continuously; food enters, wastes are excreted, and the system remains essentially unchanged.

The ecology of infectious disease is especially interesting because one element of the system, the host, which serves as the environment for interrelated nonpathogenic and pathogenic microorganisms, is itself affected by the external environment, e.g., climate, radiations, and nutrients. Familiarity with the normal body flora is an obvious prerequisite to an understanding of the ecology of infectious disease and to recognition of the abnormal or pathogenic flora. However, it should be emphasized that many, perhaps all, of the microorganisms in or on the normal individual may cause disease under appropriate circumstances.

GROUPS OF MICROORGANISMS

The microbial flora and fauna of the human or animal body include many species of bacteria, viruses, fungi, and protozoa.

Bacteria

Cocci. The predominant gram-positive cocci on the body are staphylococci and streptococci. Species of the gram-negative *Neisseria* and *Veillonella* are also common, as are members of the gram-positive genus *Micrococcus;* the latter are usually adventitious, being derived from soil and air.

Staphylococci. *S. epidermidis (albus)* is probably the most abundant of these organisms in the human body. This and *S. aureus* are aerobic or facultatively anaerobic, and were originally differentiated in terms of their pigmentation (see Table 17–1); at present, however, the distinction is based primarily on production of coagulase, an enzyme that clots blood plasma. Anaerobic staphylococci, including the genus *Peptococcus,* which are also present, have not been studied and characterized adequately.

Most staphylococcal infections are caused by *S. aureus.*

Streptococci and Gram-positive Diplococci. Streptococci and related bacteria abound in the respiratory and digestive tracts. Classification is based primarily on their behavior on blood agar. The majority of normal throat streptococci produce a zone of greenish (occasionally brownish) discoloration around their colonies. This is called the alpha (α) reaction, sometimes designated

TABLE 17–1. Differential Characteristics of *Staphylococcus* Species

Species	Pigment	Coagulase	Hemolysis
S. aureus	Golden	+	+
S. epidermidis (albus)	White	−	−

TABLE 17–2. Differential Characteristics of Streptococci and Gram-positive Diplococci of the Human Body

Species	Hemolysis	Bile solubility	Growth in 6.5% NaCl broth
S. mitis	α	–	–
Enterococci	α, γ	–	+
S. pneumoniae	α	+	–
S. pyogenes	β	–	–

TABLE 17–3. Differential Characteristics of *Neisseria* and *Branhamella*

Species	Pigment	Glucose	Maltose	Sucrose
N. gonorrhoeae	–	A	–	–
N. meningitidis	–	A	A	–
N. sicca	–	A	A	A
N. flava	+	A	A	*V
B. catarrhalis	–	–	–	–

*V = varies from strain to strain

alpha-hemolysis, but this term is not strictly correct since the erythrocytes are usually not dissolved. The most highly pathogenic streptococci produce a zone of clear hemolysis, the beta (β) reaction, in which the blood cells are completely lysed. A third group of organisms causes no change in the erythrocytes; this is called the gamma (γ) reaction. (See Plate IV.)

Four major normal and pathogenic groups can be differentiated according to properties listed in Table 17–2. Alpha-type cocci found most commonly in the throat are strains of *S. mitis*, sometimes also called *S. viridans*. Species from the intestine, the enterococci, produce either an alpha or gamma reaction, but can be distinguished readily by their ability to grow in broth containing 6.5 per cent NaCl. Hemolytic streptococci, often called *S. pyogenes*, and pneumococci are recovered from healthy individuals with sufficient frequency to be considered part of the indigenous flora, particularly during the winter months, when they may be present in as many as 60 per cent of apparently normal persons. Both are considered pathogenic and cause disease when certain predisposing conditions obtain or when they gain access to susceptible tissue. Pneumococci are easily distinguished from other alpha streptococci by their solubility in bile.

Anaerobic streptococci, like anaerobic staphylococci, have not been adequately studied. They comprise the genus *Peptostreptococcus*.

Neisseria and Branhamella. Gram-negative diplococci are common in the normal upper respiratory tract. *Neisseria* and *Branhamella* species are aerobic and produce an oxidase detectable by treating colonies with dimethyl- or tetramethyl-*p*-phenylene diamine; a blue-purple color is produced. They are typically coffee bean-shaped cells arranged in pairs with the adjacent sides flattened. *N. sicca*, *N. flava*, and *Branhamella catarrhalis* are representative of the normal throat population (see Table 17–3); they can be cultivated on ordinary nutrient agar and will grow on a variety of culture media at temperatures from 22° to 37° C. The pathogenic species, *N. gonorrhoeae* and *N. meningitidis*, are more fastidious, requiring blood or other complex organic nutrients and a temperature closely controlled at 37° C. They cause gonorrhea and cerebrospinal meningitis, respectively. During meningitis epidemics, a high percentage of the normal population may be carriers of *N. meningitidis*.

Veillonella. *Veillonella* consists of small anaerobic gram-negative cocci that occur in pairs or clusters. They do not ferment sugars vigorously, but they attack peptides, releasing CO_2 and H_2 and producing a rise in pH.

Gram-positive Rod Bacteria. Gram-positive rods indigenous to the human body include the genera *Lactobacillus* and *Corynebacterium*. Anaerobic spore formers, i.e., *Clostridium* species, are frequently present in the intestinal tract. Members of the genus *Bacillus* are not uncommon but are usually adventitious; they are derived from the external environment and are easily removed by normal cleansing operations.

Lactobacilli. Lactobacilli are catalase-negative, nonmotile, nonspore-forming, and are usually described as microaerophilic. They grow slowly under aerobic conditions and produce minute surface colonies. Some species grow best between 37° and 45° C., others are thermophilic, and still others grow best at about 30° C. Lactobacilli are actively saccharolytic. Two main subdivisions are distinguished; *homofermentative* lactobacilli produce lactic acid from glucose with only traces of other compounds; *heterofermentative* species produce ethyl alcohol, CO_2, acetic acid, and often other acids, in addition to lactic acid. Some species produce sufficient acid to lower the pH of the culture to 2.

Corynebacteria. *Corynebacterium* species are catalase-positive, nonmotile, non-spore-forming, and often pleomorphic. Club-shaped, irregularly staining, or granular cells are common, characteristically arranged in V or W patterns or in a parallel array called a "palisade." *C. pseudodiphtheriticum* and *C. xerosis,* the aerobic diphtheroids of mucous membranes, are classified according to their ability to ferment glucose and sucrose.

C. diphtheriae, the principal pathogen in the genus, causes disease through the activity of a toxin produced by strains that harbor a specific temperate bacteriophage.

Propionibacteria. Anaerobic or microaerophilic diphtheroid bacteria are now assigned to the genus Propionibacterium. *P. acnes* is abundant on the skin and is also found in the intestine and in wounds, blood, pus, and soft tissue abscesses. It is one of the most common contaminants in anaerobic cultures.

Clostridia. Anaerobic, spore-forming bacilli frequently present in the human intestine include *C. perfringens,* one of the gas gangrene group. *C. tetani* is much less common in man, although it is often present in other animals. Either of these organisms, introduced deeply into tissues sufficiently damaged by trauma to provide nutrients and a favorably low O–R potential, causes highly fatal disease.

Gram-negative, Nonspore-forming Rod Bacteria. These bacteria include the numerous members of the Enterobacteriaceae, the most thoroughly studied of the indigenous flora of man, and species of *Pseudomonas, Alcaligenes, Haemophilus,* the poorly defined and comparatively unrecognized *Acinetobacter-Moraxella* group,

TABLE 17–4. Differential Characteristics of Selected Enterobacteriaceae

Genus	Glucose	Lactose	V.P.	Urea	Motility
Escherichia	AG	AG	–	–	+
Enterobacter	AG	AG	+	–	+
Klebsiella	AG	AG	+	–	–
Proteus	AG	–	–	+	+
Salmonella	AG*	–	–	–	+
Shigella	A	–	–	–	–

*S. typhi: **Glucose A**

the abundant but often undetected *Bacteroides,* and *Fusobacterium.*

Enterobacteriaceae. Motile Enterobacteriaceae possess peritrichous flagella. All species reduce nitrate to nitrite and all produce acid from glucose; further classification is based upon fermentation and other biochemical reactions (see Table 17–4). The normal flora of man includes *Escherichia* and *Enterobacter (Aerobacter)* and species of *Klebsiella* and *Proteus*. Pathogenic members are principally *Salmonella* and *Shigella,* the causes of typhoid and paratyphoid fevers, gastroenteritis, and bacillary dysentery.

Pseudomonas. This genus includes aerobic, oxidase-positive, nonspore-forming straight rods, usually motile by means of polar flagella. Many strains produce water-soluble (i.e., diffusible) green, blue, or other color pigments (see Table 17–5). Acid production from glucose is an oxidative process in which gluconic acid or a similar product is formed. Nitrate is frequently reduced to nitrite, ammonia, or free nitrogen. *Pseudomonas* species are principally soil and water forms or plant pathogens, but some species (e.g., *P. aeruginosa*) are often

TABLE 17–5. Differential Characteristics of Selected Gram-negative Rod Bacteria of the Human Body

Genus	Motility	Pigment	Oxidase	Glucose
Pseudomonas	+	+	+	A
Alcaligenes	+	–	±*	–
Acinetobacter	–	–	–	A
Moraxella	–	–	+	–

* Weak + or –

associated with the intestinal tract and epidermis of man, and produce wound or burn infections.

Alcaligenes. *Alcaligenes* species, like the Enterobacteriaceae, are motile by peritrichous flagella. However, they do not utilize carbohydrates, and typically produce an alkaline reaction in litmus milk. They are common in the intestinal tract of vertebrates.

Acinetobacter-Moraxella. This group of bacteria consists of strictly aerobic, nonmotile, nonpigmented organisms that appear as diplococci or diplobacilli. They are easily confused with members of the genus *Neisseria* and with *Alcaligenes* and nonpigmented *Pseudomonas* strains. Classification of the group presents a problem to systematic and diagnostic bacteriologists (see Table 17–5).

Haemophilus. The genus *Haemophilus,* originally named because its members seemed to require blood media, is now defined to include aerobic, gram-negative, minute, coccobacillary to rod-shaped bacteria that require either hemin (X factor) or phosphopyridine nucleotide (V factor) or both. Either NAD or NADP can serve as V factor. *H. influenzae* and a hemolytic species, *H. haemolyticus,* which require both X and V factors, are often present in the healthy mouth and upper respiratory tract, as are corresponding forms that require only the V factor, *H. parainfluenzae* and *H. parahaemolyticus.*

If *H. influenzae* finds its way to the central nervous system in young children, it causes a highly fatal form of meningitis.

Bacteroides. Bacteroides are anaerobic, gram-negative, nonspore-forming, pleomorphic rods, occurring singly or in pairs or short chains. Some require enriched media. They usually ferment glucose but do not reduce nitrate to nitrite. They are the most abundant bacteria in feces.

Fusobacterium. *Fusobacterium* species are aerobic or microaerophilic, straight or curved, gram-negative rods with tapering ends. They are nutritionally fastidious and require special enrichment. Most species ferment glucose. They are common in the mouth, especially around the teeth. Tremendous numbers (i.e., 10^{10} per gram) of fusiform bacteria have been found in the intestinal contents of mice.

Spirochaetes. Two genera of anaerobic spirochaetes are indigenous to the mucous membranes of the mouth, intestine, and genitalia. They are flexible, highly motile, but nonflagellated.

Treponema. *Treponema* cells are 3 to 18 μm in length. They stain with difficulty and are best observed by darkfield microscopy. They consist of a filament about 0.25 μm thick wound in a fine coil about 1 μm in diameter. *In vitro* cultivation and even maintenance of viability are difficult. The most notorious pathogen is *T. pallidum,* the cause of syphilis. *T. vincentii* is a synergistic partner in the etiology of trenchmouth.

Borrelia. Cells of this genus consist of a slender filament, about 0.3 μm in diameter, loosely wound in a few irregular coils. They stain easily with ordinary aniline dyes and thus differ from many other spirochaetes. Some species have been cultivated *in vitro.* Pathogens include *B. recurrentis,* the cause of relapsing fever.

Mycoplasmatales. The order Mycoplasmatales, which consists of pleomorphic organisms that lack a rigid cell wall, was described in Chapter 6 (page 139). Most species are found in lower animals, but a few comprise part of the normal flora of the human genitourinary tract, the mouth, and the lower end of the digestive tract. *Mycoplasma pneumoniae* is a cause of primary atypical pneumonia.

Viruses

Viruses, as obligate parasites, are usually found in pathologic situations rather than as part of the normal flora. However, some viruses have been recovered from apparently healthy individuals under circumstances that indicate ability to survive and to replicate without causing observable harm to the host tissues; therefore they can be presumed to be part of the normal flora. One of the most outstanding examples is the "orphan virus" group. The ECHO (enteric cytopathogenic human orphan) viruses were first isolated in tissue cultures from the feces of presumably normal individuals. Specimens were inoculated onto cultured mammalian cells growing in a single layer (a monolayer) on agar, and virus was detected by production of plaques or areas of cytopathogenicity, from which isolations were subsequently made. A considerable group of viruses was thus secured, and many of them have never been shown to produce disease.

Viruses have also been isolated from other undiseased sources. For example, CELO (chicken embryo lethal orphan) virus is present and apparently harmless in the eggs from many flocks of hens. When it is inoculated into embryonated eggs from virus-free birds, it kills the embryos. The Coxsackie viruses were first isolated from the feces of individuals suspected of having poliomyelitis, and other identical or similar viruses were isolated from persons without illness. Some, if not all, of this group therefore appear to be indigenous to man. Various adenoviruses from the tonsils and adenoids of normal individuals have not been shown to produce disease.

Herpes simplex virus is present in cells about the mouth and nose of many normal persons. Primary infection during the first few months of life causes formation of vesicles inside the mouth and on the gums, and sometimes a severe, generalized, febrile illness occurs. Even though antibodies develop, the virus then persists in a latent condition throughout the remainder of life. It can be activated by stress (e.g., fever, sunburn, menstruation, allergic reactions) whereupon coldsores appear at mucocutaneous junctions such as the lips.

The SV40 (simian virus 40) virus seems to be harmless in its natural hosts, rhesus and cynomolgus monkeys, but when transferred to cell cultures from green monkey kidneys, it produces cytopathogenic effects. This oncogenic virus is of potential significance in man because poliomyelitis vaccines are prepared in tissue cultures of monkey kidneys. Since recognition of the problem, suitable precautions have been taken to avoid the presence of this virus in vaccines.

Fungi

Most fungi associated with man are naturally distributed throughout the environment, so it is difficult to decide whether a given organism is indigenous to man or is merely transient. A few are so common on the human body that they appear to be part of the normal flora, and under special circumstances they produce disease. Among these are *Candida* species, *Pityrosporum ovale,* and the dermatophytes.

Candida. Species of *Candida* are found in the mouth and throat, digestive tract, vagina, and on the skin. Under certain conditions they produce disease of any of these areas and can cause systemic infection with brain, meningeal, or heart involvement.

Pityrosporum. *P. ovale* is a yeast present on the skin, cultivated particularly from dandruff. It requires fat, fatty acids, or glycerol for continued growth, and the presence of these substances in the skin suggests that this organism is indigenous to the human body.

Dermatophytes. *Trichophyton* and *Epidermophyton* are found on the skin of the feet with sufficient frequency to indicate that they are part of the indigenous flora. They are associated with tinea pedis, or athlete's foot, but the distribution of the various species during outbreaks points to the possibility of endogenous rather than exogenous infection.

Protozoa

Protozoa found normally on man are members of the Mastigophora and the Sarcodina. Several genera occur with considerable frequency: the flagellates *Trichomonas, Chilomastix, Giardia, Trypanosoma,* and *Leishmania,* and the amebae *Entamoeba, Endolimax,* and *Iodamoeba.* Several of these were discussed in Chapter 13.

Leishmania. *L. donovani,* the cause of kala-azar in man, occurs as 3 to 5 μm, nonflagellated, leishmanial cells, situated intracellularly in macrophages. They multiply by fission until the host cell is crowded and then escape and infect new cells. Although they predominate in internal organs, some infect skin macrophages, from which they are ingested by the intermediate hosts, sandflies of the genus *Phlebotomus.* They then transform into the leptomonad stage, growing to a length of 14 to 20 μm, and multiply in the gut. The insect becomes infectious in a week or more.

Kala-azar is a chronic visceral disease; symptoms begin one to four months after infection, with high temperature and irregular fever. The spleen and liver enlarge greatly, and wasting, emaciation, edema, and often dysentery occur. Death is the usual outcome in untreated cases. The disease is found in China, India, the Mediterranean area, Africa, and parts of South and Central America.

Cultivation of Protozoa. Protozoa are usually identified by the morphologic and staining characteristics of their vegetative forms and cysts, if any, rather than by cultural methods. Cultivation in pure culture is limited by the difficulty of separating them from the bacteria, yeasts, and other microorganisms that constitute their usual food. However, some protozoa have been purified and maintained in. media containing complex animal and plant products and growth factors, together with penicillin, streptomycin, or a combination of these. Protozoa from the human body are almost entirely anaerobic, and cultivation requires substances such as ascorbic acid and cysteine that lower the O–R potential. Glucose also helps to maintain reducing conditions and supplies energy. Whole blood, serum albumin, and egg proteins are used in media, and purines, pyrimidines, ribose, and B vitamins satisfy growth factor requirements.

DISTRIBUTION OF MICROORGANISMS

The predominant normal flora of the human body is summarized in Table 17–6. Obviously any area such as the skin, upper respiratory tract, or mouth that has constant contact with the external environment contains microorganisms derived from the environment. Most of these are transients, and are easily destroyed or removed. The indigenous "resident" population consists of organisms that secure needed nutrients, growth factors, protection, and other conditions in a particular location.

TABLE 17–6.   Predominant Normal Flora of the Human Body

	Skin	Upper respiratory tract	Mouth	Lower intestine	Urogenital tract
Cocci					
Staphylococcus	+	+	+	–	+
Strep. mitis	–	+	+	–	–
Enterococci	–	–	–	+	+
Neisseria	–	+	+	–	–
Gram-positive rods					
Lactobacillus	–	–	+	+	+
Corynebacterium	+	+	+	–	+
Clostridium	–	–	–	+	–
Gram-negative rods					
Coliforms	–	+	+	+	–
H. influenzae	–	+	+	–	–
Bacteroides	–	+	+	+	+
Fusobacterium	–	–	+	–	–
Spirochaetales	–	–	+	–	–
Mycoplasmatales	–	–	+	+	+
Viruses	–	+	–	+	–
Fungi					
Yeastlike	+	–	+	+	+
Dermatophytes	+	–	–	–	–
Protozoa					
Flagellates	–	–	+	+	+
Amebae	–	–	+	+	–

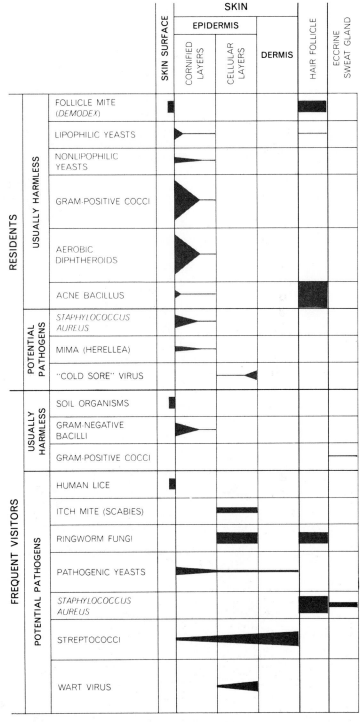

Figure 17–1. *Distribution of microorganisms in the various layers and structures of the skin. Some organisms are abundant in the outer cornified (horny) layers but are absent from deeper layers, and others apparently penetrate via hair follicles and sweat glands (see Figure 17–2). (From Life on the Human Skin by Mary J. Marples. Copyright © 1969 by Scientific American, Inc. All rights reserved.)*

The flora of the skin, for example, includes chiefly coagulase negative staphylococci, corynebacteria, and fungi (Fig. 17–1). All of these organisms are favored by fatty acids and other lipoidal substances that are available in the sebaceous glands (Fig. 17–2).

The upper respiratory tract contains a somewhat more diverse flora, supported by the mucous secretions of the nasopharyngeal, oropharyngeal, and bronchial membranes. The proteins and carbohydrates of these secretions help to satisfy the nutri-

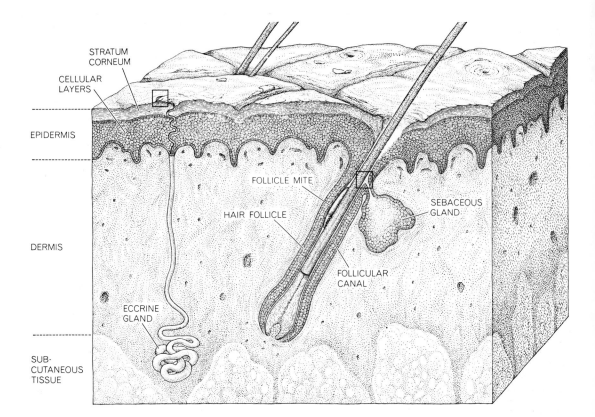

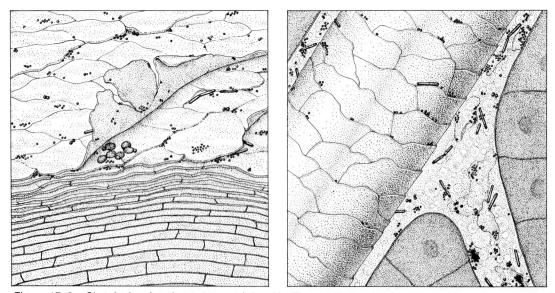

Figure 17–2. *Sketch showing the structure of the skin. The top illustration is a section of skin enlarged 25X. The two areas marked by small squares are shown below enlarged 500X. At the left is represented the upper layer of skin, the stratum corneum, consisting of flat, scaly "squames" that curl and flake off. Bacteria and yeastlike fungi reside around and under the scales. The hair follicle at the right houses many bacteria. (From Life on the Human Skin by Mary J. Marples. Copyright © 1969 by Scientific American, Inc. All rights reserved.)*

tional demands of *Streptococcus, Neisseria, Haemophilus,* and *Bacteroides* species, and cells of the adenoids and tonsils of some individuals are hosts to adenoviruses.

The flora of the mouth is large and varied. With the possible exception of the hands, it is the area most susceptible to contamination throughout life, and in addition the salivary and mucous secretions are enriched by ingested food. Viridans streptococci and *Neisseria* are universally present. Colonization begins during delivery of the baby, and bacteria are regularly demonstrable by the sixth to the tenth hour. Streptococci are among the first bacteria to appear, soon followed by micrococci, lactobacilli, enterococci, and coliform bacteria. The total population in the adult mouth is in the order of 150,000,000 per ml. of saliva, or 50,000,000 per mg. of material from gingival crevices. Of these, about one-third are aerobic and two-thirds are anaerobic. The anaerobic forms presumably find favorable conditions in tonsillar crypts, in crevices between teeth, and in association with aerobes.

The acidity of the stomach and the bile of the small intestine suppress the microbial flora of these parts of the digestive tract, but the lower intestine provides an environment suitable for a flora almost as varied as that of the mouth; the population attained is estimated to be approximately 10^{12} per gram of wet feces. By far the most numerous are anaerobes, especially *Bacteroides;* lactobacilli are also abundant. Coliforms number between 10^6 and 10^9 per gram, and enterococci average somewhat less.

The flora of the anterior urethra is essentially the same in men and women, and consists of coagulase negative staphylococci, alone or with enterococci. Diphtheroids may also be present.

The vaginal flora varies with the stage of sexual development, the presence of glycogen in the vaginal epithelium, and the presence of acidogenic microorganisms such as aerobic lactobacilli and corynebacteria. Before puberty, glycogen is absent and the vaginal flora is varied. During the reproductive period glycogen is present, and the flora consists largely of acid-tolerant *Lactobacil-lus, Candida,* and *Corynebacterium* species. Some of these produce acid from glycogen and other carbohydrates, and the acid presumably inhibits other microorganisms. At menopause, reversion to an alkaline secretion and varied flora occurs.

The blood and tissues (e.g., muscles, internal organs) of healthy individuals are normally sterile.

INTERRELATIONSHIPS BETWEEN THE HOST AND ITS FLORA

A host and its indigenous flora are normally in a state of dynamic balance. The host provides favorable living conditions for its microbial populations and does not react too violently to them; the microorganisms likewise do not harm the host and may, in fact, benefit it. At the same time the various groups of organisms interact, either stimulating or inhibiting each other.

Valuable information about the relationships between an animal and its microbiota has been secured from study of germ-free or gnotobiotic animals (Greek: *gnotos,* known + *bios,* life). Germ-free mammals, such as mice, rats, and guinea pigs, are delivered by caesarian section in sterile chambers and reared under aseptic conditions. Gnotobiotic animals are germ-free animals deliberately inoculated with one or more known organisms.

One of the most obvious ways in which a microorganism can benefit its host is by synthesis of vitamins and other growth factors and nutrients, which become available to the host. There is evidence that the intestinal flora of various animals contributes significant amounts of biotin, pyridoxin, pantothenic acid, and vitamins B_{12} and K. Germ-free rats must be supplied dietary vitamin K, whereas conventional rats derive it from their intestinal organisms.

Microorganisms greatly change the physical and chemical nature of the contents of the intestine and assist in digestive processes. The cecums of germ-free rats and mice are two to three times as large as those of conventional animals, and their content is four to six times as great. Upon introduction of normal bacteria into the digestive tracts of such animals, the cecal content begins to decrease within a few hours and returns to normal in less than a week; the cecal sac also decreases in size.

One of the best studied examples of beneficial microbiologic processes is the activity of microorganisms in the rumen of cattle, sheep, goats, and other animals that have a special stomach in which cellulose and other polysaccharides are digested (see Figure 17–3). The rumen is a large organ (100 to 200 liters in cattle) that receives ingested food. It provides a remarkably constant microbial environment. Its temperature is 39° C., with only minor fluctuation. The reaction of its contents is pH 5.8 to 6.8, the O–R potential is −0.35 volt, and the atmosphere is completely anaerobic. Intermittent but continuous ingestion maintains the concentration of organic matter at 10 to 18 per cent, which includes a population of at least 10^9 to 10^{10} bacteria per ml., 10 to 20 per cent of which are viable, and about 10^6 protozoa per ml. The bacteria include species of *Bacteroides, Ruminococcus, Butyrivibrio, Clostridium, Methanobacterium, Veillonella, Streptococcus, Peptostreptococcus,* and other genera. More than 100 species of protozoa have been found, chiefly ciliates. Although their total number is less than that of the bacteria, their cell mass is about the same.

The principal products of microbial decomposition of cellulose and other polysaccharides are CO_2 and CH_4 and acetic, propionic, and butyric acids. These acids are absorbed through the rumen wall and enter the bloodstream; they are oxidized and provide energy for the animal. The microbial cells, which comprise about 10 per cent of the rumen contents, are digested and furnish a considerable portion of the animal's amino acid and vitamin requirements.

The normal flora is also significant in the development of mechanisms of immunity. The amount of lymphatic tissue in germ-free animals is markedly depressed, particularly in regions, such as the intestinal tract, that are normally in contact with microorganisms. Lymphoid cells are the principal sites of manufacture of antibodies, proteins of globulin nature whose role in immunity will be discussed in Chapter 19. It has also been shown that the total globulin in the serum of germ-free animals is less than half that in conventional animals. Antibodies against microorganisms of the normal flora are absent, or their appearance is greatly delayed. It is likely that in conventional animals these antibodies protect against the normal flora. Germ-free animals exposed to living microorganisms of the indigenous biota may succumb rapidly to infection, while similar animals that are injected with killed organisms or fed a diet that includes killed organisms produce antibodies, and at the same time their lymphatic tissues undergo marked development. These animals can then resist exposure to the living organisms.

INTERRELATIONSHIPS BETWEEN MICROBIAL POPULATIONS

Two populations of microorganisms residing in the same locality may have no effect on each other (*neutralism*), but usually they interact in one way or another. When they are

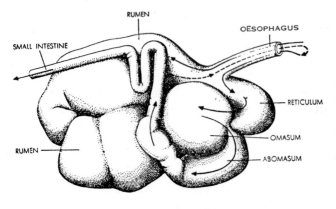

Figure 17–3. *Schematic diagram of the relationships between the four parts of the ruminant stomach. Ingested food is regurgitated, remasticated, and swallowed again and held in the rumen and reticulum until it attains a fine consistency, when it passes to the omasum, abomasum, and the small intestine. (From Annison and Lewis: Metabolism in the Rumen. New York, John Wiley & Sons, Inc., 1959.)*

PLATE III

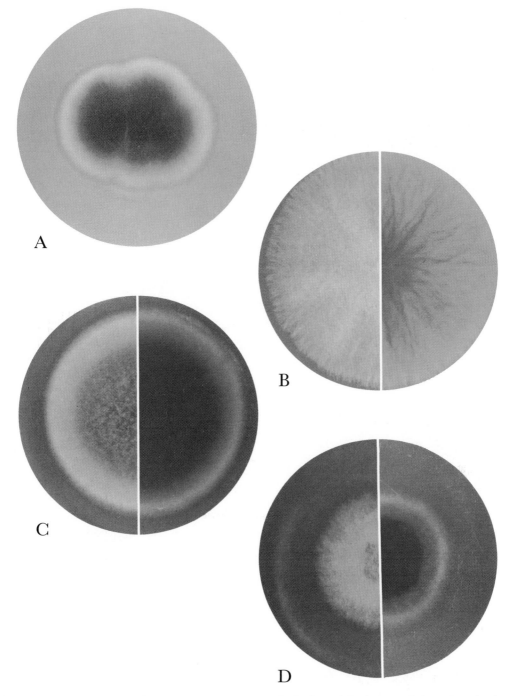

Giant colonies of molds on Sabouraud dextrose agar. In B, C, and D, the top of the colony is shown at the left, and the bottom of the colony, viewed through the agar, is shown at the right.

A, Penicillium notatum, *one of the sources of penicillin.*

B, Fusarium *species. This genus includes some plant pathogens, and some species produce toxins that cause illness in humans who eat moldy grain.*

C, Cladosporium pedrosoi, *a cause of chromoblastomycosis, a chronic skin or subcutaneous infection with warty or tumor-like lesions, sometimes ulcerative, usually confined to the extremities and the head.*

D, Microsporum audouinii, *a frequent cause of tinea capitis, a fungus infection of the scalp and hair, in children; it may also cause tinea caporis, or ringworm of the smooth skin.*

PLATE IV

A

D

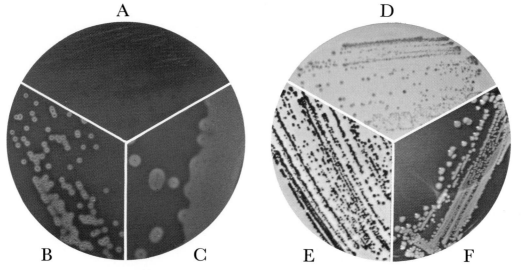

B

C

E

F

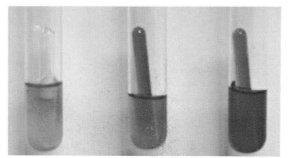

G

H

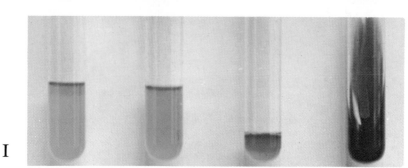

I

A–C, *Colonies on sheep blood agar.* A, *minute colonies of* Streptococcus viridans *surrounded by zones of greenish discoloration.* B, *S. pyogenes, a variety that produces a small zone of complete hemolysis.* C, *S. pyogenes, a strain that produces a large hemolytic zone.*

D–F, *Colonies on eosin methylene blue (E.M.B.) agar.* D, Salmonella typhi. E, Escherichia coli. F, Klebsiella pneumoniae.

G, *Cultures in Durham fermentation tubes containing a sugar broth with bromcresol purple indicator. Left, acid (yellow) and gas;* middle, *acid without gas;* right, *neither acid nor gas.*

H–I, *IMViC (indole-methyl red-Voges-Proskauer-citrate) tests with coliform bacteria.* H, E. Coli *(IMViC ++−−);* I, K. pneumoniae *(IMViC −−++).*

situated in or on the body, the result of the interaction may directly or indirectly affect the host. One or both populations may benefit from interaction, or one or both may suffer. *Mutualism* is an association in which the growth of both partners is enhanced, and, in fact, survival under natural conditions depends upon the relationship. One example—the association between termites and their intestinal flagellates—has already been cited (page 285). There are probably many mutualistic pairs of microorganisms within the human or animal body. These include, for example, organisms with different growth factor requirements that are not satisfied in a given site on the animal body. When two organisms grow together, however, each supplies missing factors to the other.

A second type of beneficial association is *cooperation* (some authorities prefer the term *proto-cooperation,* since cooperation implies an act of intelligence). Both organisms may benefit from the association, but the relationship is not obligatory. Cooperation is illustrated by "enzyme sharing," which is a feature of some examples of *synergism.* When various bacteria, such as *Streptococcus faecalis* or *S. aureus,* are inoculated simultaneously with *Proteus* species into a broth medium containing the disaccharide lactose, the sugar is utilized and gas is produced, whereas neither organism alone is able to produce gas from lactose. In these cases, the streptococcus or staphylococcus produces an extracellular lactose-hydrolyzing enzyme, and the *Proteus* ferments one or both products of hydrolysis of lactose with the formation of gas. A still unexplained synergistic association exists between *F. fusiforme* and *T. vincentii,* mentioned earlier (page 332), and results in the disease trenchmouth.

Commensalism is a relationship in which one organism benefits and the other remains unaffected. It is illustrated by the associated growth of aerobic and anaerobic microorganisms, the aerobic microorganism utilizing atmospheric oxygen and creating a low enough O–R potential to permit growth of the anaerobic microorganisms. It is assumed that this must be the mechanism that permits attainment of the high populations of *Bacteroides* and *Propionibacterium* species that are found in such aerobic locations as the skin and the mouth.

Satellitism is a form of commensalism in which one partner produces a substance that is utilized or even required by the other and therefore enhances or permits growth of the second organism. It will be recalled that *Haemophilus influenzae* and *H. parainfluenzae* require phosphopyridine nucleotide (V factor). This growth factor is produced by many organisms, such as *S. aureus* and yeast. Since it is soluble and diffusible, its action can be demonstrated by superimposing a single spot inoculation or a cross-streak of one of these organisms on a blood agar plate previously seeded with *Haemophilus.* Growth of *Haemophilus* is greatly enhanced within a few millimeters of the *S. aureus* or yeast.

One form of antagonism is *competition* of two species for essential growth factors, nutrients, or living space. Both organisms are adversely affected.

Amensalism is a relationship between two organisms in which one is inhibited but the other is not affected. This very common type of association is relatively easy to study. Usually the inhibitor or effector organism produces a substance that inhibits or kills one or more organisms. If only one or a very few species are affected, the substance is said to be highly specific, whereas if many species are inhibited or killed, it is nonspecific.

Acidity and hydrogen peroxide are nonspecific inhibitors. The low pH produced by lactobacilli and corynebacteria on mucous membranes limits the flora of the membranes to acid-tolerant microorganisms. Hydrogen peroxide liberated by pneumococci and streptococci in the mouth and upper respiratory tract kills *Neisseria* and *Corynebacterium* and other organisms that lack a peroxide-decomposing enzyme such as catalase.

Many organisms produce antibiotics, some of which are relatively specific, whereas broad-spectrum antibiotics affect a wide range of organisms, as discussed in Chapter 12. Species of the Actinomycetales

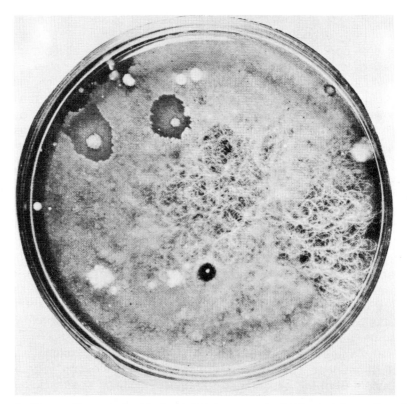

Figure 17-4. *Antibiosis. The agar in this Petri dish is covered with bacterial growth, principally of the* mycoides *variety of* Bacillus cereus. *Several colonies of other kinds of bacteria have inhibited the spreading organism by producing antagonistic substances, which diffuse a considerable distance through the agar. (From Grant: Microbiology and Human Progress. New York, Rinehart & Co., Inc., 1953.)*

and numerous fungi are noteworthy producers of antibiotics (see Figure 17-4).

Bacteriocins are bactericidal proteins produced by strains of a number of species of bacteria, particularly Enterobacteriaceae and other gram-negative rods. Each is active against only a few organisms closely related to the one that produced it. Bacteriocins from *E. coli,* for example, which are known as colicins, kill certain other strains of *E. coli* or of other species of Enterobacteriaceae.

E. coli, K. pneumoniae, and *Proteus* produce substances antagonistic to the yeast *Candida* and help suppress it in the intestine. When drastic chemotherapy used to combat intestinal infection reduces the normal bacterial flora sufficiently, *Candida* species multiply and dominate the flora and sometimes cause different disease symptoms.

The other mechanisms of antagonistic action are *parasitism* and *predation;* in both of these one organism directly attacks and is dependent upon the other. A parasite is smaller than the host, whereas a predator is larger. Most antagonistic microorganisms are, of course, parasites, when considered in relation to higher animals, but in their relationships with each other some microorganisms are parasites and some are predators. Many protozoa, for example, are predators of bacteria, and some bacteria are parasites of protozoa. As already pointed out, even bacteria are subject to parasitic attack by bacteriophages. Comparatively few of these parasite-predator interrelationships among microorganisms have been worked out, but it is obvious that they directly or indirectly affect the ecologic picture within the body of a larger host.

FACTORS AFFECTING THE BODY FLORA

Some of the many factors that affect the body flora have already been mentioned.

Others include the maturation and aging process during which chemical changes in mucus and other secretions influence the microbial populations of the various membranes.

There are also associated changes in hormone production, which indirectly affect the microbial flora of organs such as those of the female reproductive tract.

Since adrenal cortical hormones have been used therapeutically in large amounts, it has been found that a side effect of this treatment is activation of latent or avirulent microbial species and production of infectious disease. Cortisone also suppresses the inflammatory response, and this, as will be shown, is an important means of defense.

Infectious or noninfectious disease may lead to change in microbial flora. Viral influenza is frequently followed by pneumonia due to *S. pneumoniae,* which was present as part of the normal flora. Similarly, typhoid fever and other bacterial diseases may be followed by fatal pneumococcal pneumonia. In these instances, the actual mechanism may be suppression of host resistance, which permits proliferation of the pneumococci.

Organic disease such as diabetes markedly increases susceptibility to microbial infections. *Candida* infections of the mucous membranes of the mouth and gums, and also staphylococcal infections such as boils and cellulitis, are common in diabetes.

Massive x-irradiation of experimental animals induces marked change in the intestinal flora and decreases resistance to bacterial invasion. Such animals frequently succumb to *Proteus* infection. X-rays destroy neutrophiles, the principal phagocytic cells of the blood, and also decrease antibody production.

Use of antilymphocyte serum and other immunosuppressant drugs in an attempt to counteract graft rejection has led to a new problem: loss of protection against infectious agents. Many patients in whom a successful organ transplant has been performed die of pneumonia or other infection because the treatment necessary to permit retention of the graft damaged or destroyed their immunologic mechanism.

OPPORTUNISM

Opportunism is the phenomenon of disease production by normally harmless organisms of the indigenous flora as a result of fortuitous events affecting the host rather than the microorganism. Two examples will be described.

Subacute bacterial endocarditis is often caused by nonhemolytic streptococci or other members of the normal oral flora, which find their way to a previously damaged heart valve. A common clinical sequence begins with rheumatic fever, which leaves an endocardial scar, often undetected. At some subsequent date surgical or other manipulation, such as tonsillectomy or tooth extraction, provides access of normal mouth streptococci to the bloodstream and in the course of the transient bacteremia they become established in the endocardial scar, often protected by thrombi from phagocytosis and other blood-clearing mechanisms. The vegetative growth of the streptococci may lead to death by occlusion of the affected valve, or through other vascular accidents or embolism. Persistent bacteremia may also occur.

Candida albicans becomes invasive under special circumstances and produces disease. Situations in which this occurs include unrelated disorders such as severe diabetes and malignant lymphomas, intensive treatment with broad-spectrum antibacterial drugs, and immunosuppressive therapy. These conditions or agents, which interfere with normal inhibitory mechanisms, permit a variety of diseases to occur. (1) Thrush or oral candidiasis consists of white patches of yeast cells and hyphae on the mucous membranes of the mouth and pharynx. (2) Vulvovaginal candidiasis—invasion of the vaginal mucous membranes—sometimes occurs during pregnancy and in diabetes. (3) Skin infections may occur in those whose occupation requires long immersion in water, or in areas of the body that are continuously moist, such as the perineum and inframammary folds. (4) Intestinal infection with *Candida* often follows therapy with antibiotics sufficient to eliminate the antagonistic normal flora. (5)

Bronchial and pulmonary tissues are occasionally invaded when a chronic bronchial obstruction (e.g., cancer) has impaired drainage of secretions.

SUPPLEMENTARY READING

Brock, T. D.: *Principles of Microbial Ecology.* Englewood Cliffs, New Jersey, Prentice-Hall, Inc., 1966.

Marples, M. J.: Life on the human skin. *Scientific American, 220*(1):108–115, January, 1969.

Odum, E. P.: *Fundamentals of Ecology,* 3rd ed. Philadelphia, W. B. Saunders Co., 1971.

Rosebury, T.: *Microorganisms Indigenous to Man.* New York, McGraw-Hill Book Co., Inc., 1962.

Rosebury, T.: Microorganisms Indigenous to Man. *In:* Dubos, R. J., and Hirsch, J. G. (eds.): *Bacterial and Mycotic Infections of Man,* 4th ed. Philadelphia, J. B. Lippincott Co., 1965.

INFECTION AND 18
DISEASE

THE NATURE OF INFECTIOUS DISEASE

Disease is any departure from a state of health, that is, from the typical condition of an individual. *Infectious* diseases are those caused by microorganisms: protozoa, fungi, bacteria, viruses. All other diseases are *noninfectious*.

Infection is the process whereby a microorganism becomes established upon or within an individual and produces injury. Damage to cells or tissues upsets their physicochemical behavior; cellular metabolism and respiration are disturbed, and intermediate metabolic products may accumulate instead of being completely oxidized. The usual signs of disease—malaise, inflammation, fever, pain—call attention to the fact that an infection is in progress. The specific response in a given instance depends on the nature of the infectious agent and on the tissues or physiologic processes affected.

PARASITISM AND DISEASE PRODUCTION

Disease as an Accident. Disease is a biologic accident. The cells of a healthy individual are in a state of continuous chemical activity, transforming energy and synthesizing cellular constitutents.

Normal cellular chemical activity is disturbed by various types of accident. The ability of DNA to direct the formation of normal enzymes is altered by accidental variation in its replication (e.g., frameshift) or variation induced by chance exposure to ionizing radiation or a mutagenic chemical. Failure of the supply of essential nutrients, minerals, or vitamins seriously impairs the chemical behavior of body cells, producing deficiency or metabolic disease.

Infection is another kind of accident that upsets the chemistry of a living organism, because most infectious agents possess some degree of parasitic tendency. It has been pointed out previously that parasitism is an association in which one organism lives upon or within another (larger) living organism at the expense of the latter but without compensation for the advantages received. Parasitism is therefore a method of securing food or shelter, or both. The specific nutrients required vary from one parasite to another and include individual amino acids, growth factors, vitamins, and nucleotides.

The expression "without compensation for the advantages received" implies possible damage to the host. Damage, if it occurs, is an incidental consequence of the parasitic mode of existence. It is ultimately as harmful to the parasite as to the host, because the most successful parasites are those that cause the least disturbance. Violent host response may result in extermination of the parasite, whereas mild response leads eventually to prolonged association and perhaps commensalism (see page 339).

Treponema pallidum, the cause of syphilis, and its human host have apparently developed a high degree of compatibility by adaptive modifications during the past several centuries. Early reports described syphilis as an acute disease, but constant association has so modified the host or parasite or both that the disease is usually benign and chronic; a fatal outcome is delayed for many years, and patients frequently die of other, unrelated ailments.

Gradations in Parasitism. Parasitism is quantitative rather than qualitative, but there is great variation among parasitic microorganisms. Moulder divided parasites into four groups according to their behavior *in the living host* (Table 18–1).

Facultative parasites are organisms that live either as saprophytes or as parasites. They can multiply indefinitely outside a living host, but on occasion they enjoy a parasitic existence. *Pseudomonas, Escherichia, Proteus,* and *Clostridium* are often cited as examples. *E. coli* grows luxuriantly in the laboratory, even on synthetic media, although its normal habitat is the human or animal body. In its natural environment, it does not produce disease and cannot even be considered parasitic, because its source of food is only the body wastes of the intestinal contents. The same organism, however, occasionally gains access to the kidneys, the bladder, or to subcutaneous tissues and produces infections that are distressing even if not necessarily fatal. In this circumstance it leads a temporarily parasitic existence.

Obligate parasites are organisms that multiply *under natural conditions* only within a living host. Although they may persist for a time in nonliving material in a natural environment, and many can be cultivated indefinitely in laboratory media, they tend to die out in nature when not passed from one host to another. Most unicellular parasites fall into this category, and they are further distinguished according to the character of their relationships with their hosts.

Obligately extracellular parasites multiply only outside body cells, within tissue spaces or body cavities. They cannot invade living cells, and they cease to multiply or are destroyed when ingested by phagocytes.

Facultatively intracellular parasites can multiply either extracellularly or intracellularly within the cytoplasm or nucleus of host cells. Organisms of this kind generally produce severe and prolonged disease, such as gonorrhea, brucellosis, and tuberculosis, and are usually found within phagocytic cells rather than cells of other types. Apparently they are not able to invade host cells actively, but when ingested they find the cellular environment suitable for multiplication.

Obligately intracellular parasites multiply

TABLE 18–1. Relationship Between Parasitic Habit and Nutritional Requirements of Representative Microorganisms*

Probable Nutritional Requirements During Parasitic Life	Parasitic Habit			
	Facultative Parasites	Obligate Parasites		
		Obligately Extracellular	Facultatively Intracellular	Obligately Intracellular
Simple carbon and energy sources, inorganic nitrogen	*Pseudomonas* *Escherichia*	S. typhi V. cholerae		
B vitamins, amino acids, nitrogen bases, etc.	*Clostridium* *Proteus*	Staphylococcus S. dysenteriae B. anthracis		
Complex natural materials: blood, serum, etc.		Trypanosomes Streptococcus S. pneumoniae C. diphtheriae Leptospira	N. gonorrhoeae N. meningitidis Brucella M. tuberculosis	PPLO Bartonella
Unknown; satisfied only within living cells		T. pallidum		Viruses Psittacosis group Rickettsiae Malaria parasites

*Modified from Moulder, J. W.: *The Biochemistry of Intracellular Parasitism.* Chicago, The University of Chicago Press, 1962.

in nature only within living host cells. Many kinds of body cells may be invaded and parasitized, even those without phagocytic properties. Most of these organisms do not grow in nonliving media, for example, viruses, rickettsiae, and malaria parasites.

Table 18–1 indicates also that the nutritional complexity of parasites living *in vivo* correlates well with their parasitic tendency. That is, the facultative parasites require only simple carbon and energy sources and inorganic nitrogen *in vitro* and presumably also in the animal body. Obligate parasites in general require more complex nutrients, together with vitamins, nitrogenous bases, blood or serum; the obligately intracellular parasites can be cultivated only in living host cells, which provide ingredients or conditions not encountered elsewhere (e.g., enzymes, high energy compounds). Noting that in general parasites and host cells are closely similar in chemical and enzymic makeup and in biochemical activity, Moulder concluded that "the unique problem in the multiplication of obligate intracellular parasites . . . is why the intracellular parasites grow so well within suitable host cells and so poorly outside them."

COMMUNICABILITY AND THE ESTABLISHMENT OF INFECTION

Infectious diseases are subdivided according to the ease with which they are transmitted from one individual to another. *Contagious* or *communicable* diseases are readily transmitted by direct or indirect contact or through the air. Venereal disease, typhoid fever, and the common cold are highly contagious. *Noncontagious* diseases are not readily transmitted from one individual to another of the same species. They are often acquired only by direct inoculation; the living organisms of tetanus must be introduced into the body tissues via a wound. It should be understood that the terms contagious and noncontagious do not describe sharply distinct categories of infectious disease; the communicability of infectious diseases varies continuously from those that are highly contagious to those that rarely, if ever, are transmitted from one individual to another. Many diseases are only moderately or mildly contagious; repeated or constant exposure appears to be necessary for transmission of human tuberculosis.

Factors Determining Infection. Several factors determine the readiness with which infectious disease is established in an individual. Some of these were indicated by Theobald Smith in the following relationship:

$$P = NV/R$$

in which *P* is the probability that disease will result from a given exposure to a pathogen, *N* is the number of microorganisms in the infecting dose, *V* represents their virulence, and *R* designates the resistance of the host. This expression states that the probability that disease will result from a given exposure to a pathogenic agent depends on the number of organisms and their virulence but varies inversely with the resistance of the host.

Virulence and Pathogenicity. *Virulence* is the capacity of a *given strain* or pure culture of a microbial species to produce disease. The terms virulence and pathogenicity are sometimes loosely used in the same sense. *Pathogenicity* should properly be employed only with reference to the ability of a *group* of organisms (species, genus, etc.) to produce disease. The distinction implies that avirulent strains of pathogenic species may exist; this is, in fact, the case. Avirulent strains of *Mycobacterium tuberculosis, Corynebacterium diphtheriae, Brucella abortus,* and many other species are known.

Virulence and some factors in the transmission of microorganisms and the pathologic response of the host will be discussed in this chapter. The resistance the host offers will be discussed in the next chapter.

VIRULENCE

Virulence, the disease-producing capacity of a microorganism, is attributed to two factors: toxigenicity and invasiveness. *Toxigenicity* is the ability to produce toxic or poisonous substances, that is, substances that can directly damage host tissues. *Invasiveness* is the capacity of the organism to establish itself within a host. Most nonanimal parasites such as bacteria, fungi, and viruses possess no means of active penetration; entrance into the host body is therefore largely a matter of chance. The invasiveness of these organisms depends on their ability to withstand the shock of initial contact with the host and then to multiply within the host.

INVASIVENESS

The adaptability of a microorganism to a parasitic existence depends on the production of chemical components, metabolic products, and enzymes that can counteract normal body defenses or assist dissemination from the original site of infection.

Inhibition of Phagocytosis.   One of the most important active defenses against infection is *phagocytosis*. Phagocytic cells are present throughout the body. They include certain white blood cells or leukocytes, various wandering small cells within the tissues, and many fixed cells that line the capillaries and are particularly abundant in the liver. The process of phagocytosis is similar to ameboid ingestion; foreign particles are engulfed by the phagocytic cells and are usually digested by intracellular enzymes. Efficient operation of the phagocytic system stops many an infection before it can get started.

Establishment of the parasite is promoted by substances such as *leukocidins,* which kill white blood cells by some mechanism not yet understood; they are produced by some staphylococci, streptococci, and pneumococci.

The presence of *capsules* is associated with virulence of certain bacterial strains. Virulent forms of *Streptococcus pneumoniae, Klebsiella pneumoniae, Bacillus anthracis,* and *Haemophilus influenzae* possess capsules; noncapsulated variants are avirulent. Capsules possess some chemical or physical property that makes them resist phagocytosis. Enzymatic removal of the capsule from a virulent strain of *S.*

pneumoniae destroys the virulence of the organism for mice.

Enzymes.   Proteolytic and other *enzymes* contribute to the invasiveness of pathogenic bacteria. *Clostridium histolyticum* is actively proteolytic in gangrenous tissue; it digests muscle vigorously and spreads rapidly from the initial site of injury. A similar action has been attributed to hyaluronidase, which hydrolyzes hyaluronic acid, an important constituent of the intercellular cement that binds tissues into their normal structure. This enzyme is produced by some hemolytic streptococci, pneumococci, gas gangrene and other bacteria. Collagenase destroys collagen, an important structural supporting substance in muscle, bone, and cartilage; it is produced particularly by *C. perfringens,* one of the gas gangrene bacteria.

Hemolysins are enzymes or toxins that dissolve red blood cells. There are many hemolysins of different properties, and they are produced by nonpathogenic as well as pathogenic bacteria. The hemolysin of *C. perfringens* is known to be a lecithinase; it hydrolyzes lecithin, a constituent of erythrocytes and of many other body cells. Some hemolysins also possess leukocidin activity. Staphylococci and hemolytic streptococci are among the common pathogens that produce hemolysins.

Streptokinase, formerly called fibrinolysin, is produced by virulent hemolytic streptococci and activates a normal plasma protease that dissolves fibrin clots. Similar enzymes are produced by other species. Hemolytic streptococci also produce streptodornase or deoxyribonuclease, which liquefies purulent exudates and viscous material containing DNA. The detection of streptokinase and streptodornase is of some diagnostic significance, but the importance of these enzymes in virulence is debated. Streptococci are among the most highly invasive bacteria and spread rapidly all over the body from a local infection. One of the first responses of the body to local infection is the formation of a fibrin clot around the area. It may be more than coincidence that pathogenic streptococci are strongly fibrinolytic.

TOXIGENICITY

Toxins are normal cellular components or metabolic products that damage or interfere with the activity of tissue cells of other animals or plants. They are produced by higher plants (e.g., mushrooms) and animals (e.g., snakes) as well as by microorganisms.

Endotoxins. Bacterial endotoxins are complexes of polysaccharide, protein, and lipid; they comprise part of the cell wall structure and are released by autolysis of the dead cells. Their potency is low: 1 mg. usually contains not more than 10 lethal doses for the mouse. Endotoxins are found in gram-negative bacteria and are present in nonpathogens as well as in pathogens. The endotoxins of *E. coli* and *Serratia marcescens,* for example, are as potent as those of the typhoid and dysentery bacteria. Symptoms observed when these substances are injected into laboratory animals include fever, diarrhea, paralysis of the limbs, disturbed carbohydrate metabolism, and degeneration of blood vessels. All endotoxins produce the same general symptoms, no matter from what species of microorganism they are obtained.

The protein portion of endotoxin is not essential for toxicity. Most of the lipid can be removed from the remaining lipopolysaccharide without great loss of toxic activity, but preparations completely free from lipid are inactive. There are indications from experiments with germ-free animals that toxicity is a function of repeated interaction of the body with endotoxic material. Animals reared without the usual gram-negative intestinal flora are not normally susceptible to endotoxin, whereas those with typical flora are susceptible. Inasmuch as the intestinal population consists largely of bacteria with endotoxin, it appears likely that "natural antibodies" (see page 371) to them may develop slowly, and when the animal is infected or injected with endotoxic bacteria or material, some sort of generalized shock reaction ensues.

Very small doses of endotoxin actually enhance nonspecifically some of the defensive processes of the body, especially phagocytosis and antibody production.

Larger doses initially depress these processes. Enhancement of the antibody response is known as adjuvant action. Lipopolysaccharide from *Bordetella pertussis,* the cause of whooping cough, has been used in experimental studies of this phenomenon.

Exotoxins. Exotoxins are soluble protein poisons secreted by the living cells of a few bacteria, plants, and animals. Some of the bacteria are semiparasitic, but the most notorious exotoxin-producing species are saprophytic.

Toxin production by *C. botulinum* in foods will be discussed in Chapter 21. The toxin in affected food resists gastric acids and passes quickly through the stomach wall to the bloodstream.

C. tetani multiplies in deep wounds from which oxygen is excluded, utilizing devitalized tissue as source of nutrients. The potent toxin it produces is transported to the central nervous system and induces typical paralysis. This organism is considered a saprophyte because it does not grow in normal living tissue but prefers devitalized tissue, as found in the area of a wound.

Corynebacterium diphtheriae establishes itself within the human upper respiratory tract—throat, nostrils, larynx—but cannot invade the body more deeply. Most of the disease symptoms are attributed to the toxin, which is absorbed through the mucous membranes and distributed to the body by the bloodstream.

Properties of Exotoxins. Exotoxins are extremely powerful. The toxin of *C. botulinum* type A is one of the most potent poisons known; 1 mg. of the purified material is sufficient to kill 31,000,000 mice. Most exotoxins are converted to the nontoxic form, toxoid, by aging, heating at 60° C., or treatment with formaldehyde or various other chemicals. Toxoid retains nearly all the immunizing power of exotoxin but lacks poisonous properties; therefore, it can be used with relative impunity for artificial immunization.

Specificity of Exotoxins. Most exotoxins are highly specific with respect to their site and mode of action. *C. botulinum* toxin causes paralysis especially of respiratory muscles, apparently by interfering with the release of acetylcholine at the endings of peripheral motor nerves. Tetanus toxin causes a similar paralysis and, in addition, produces spasm of voluntary muscles by stimulating neuromuscular junctions. This toxin has great affinity for brain and spinal cord tissue. Diphtheria exotoxin produces local necrosis, adrenal hemorrhage, paralysis, and degeneration of heart muscles, kidneys, and liver. It seems to interfere with tissue synthesis of cytochrome *b,* which is important in biologic oxidations.

VARIATIONS IN INVASIVENESS AND TOXIGENICITY

Invasiveness and toxigenicity are separate properties of a microorganism, and each varies independently of the other. A species which is both highly toxigenic and highly invasive will be likely to produce serious disease. Hemolytic streptococci are of this nature and were greatly feared until the advent of sulfonamide drugs and antibiotics. *T. pallidum* is highly invasive but not highly toxigenic; syphilis is not rapidly fatal, but symptoms may appear in any organ. *C. botulinum* has no invasiveness whatsoever and yet is highly toxigenic.

TRANSMISSION OF INFECTIOUS DISEASE

Knowledge of the modes of transmission of infectious diseases is important in their control. Most infectious diseases would be eradicated if transmission of their causative agents were completely interrupted.

There are four principal methods by which infectious agents are transmitted: (1) direct contact, (2) in water and food, (3) by direct inoculation (bites of arthropods and lower animals, wounds), and (4) via air. Diseases transmitted by these various routes will be discussed in some detail in Chapters 20 to 23.

PORTALS OF ENTRY

Pathogenic microorganisms gain access to the human or animal body through the respiratory tract, the digestive tract, broken skin, and possibly through unbroken skin. One portal of entry is usually more effective than any other for a given microorganism, because infection is more easily established in specific tissues. The pneumococcus, for example, multiplies luxuriantly in the bronchi and alveoli of the lungs and is usually acquired by the respiratory route. Dysentery bacteria establish themselves only in the intestinal tract and are therefore effectively acquired by the oral route. Some bacteria are versatile and invade the body by several routes, producing a variety of diseases. Hemolytic streptococci, for example, enter by the respiratory route and produce pharyngitis, scarlet fever, or respiratory infections, or they traverse broken skin and produce wound infections, septicemia, erysipelas, etc. The possibility that bacteria may pass through unbroken skin is debated, but the high incidence of laboratory infections by the organisms of brucellosis and tularemia, even among experienced workers, suggests that they may be able to do so.

ROUTES OF EXIT

Respiratory pathogens leave the infected patient by the oral or nasal route in saliva and respiratory exudates. Intestinal pathogens are excreted in the feces and possibly in the urine. Microorganisms that cause wound infections are not ordinarily highly contagious, although some can be transmitted by pus or other secretions. It was not uncommon in the days before aseptic surgery for hemolytic streptococci transmitted from a wound by the hands or instruments of the physician to infect a woman in labor and produce fatal puerperal sepsis (childbed fever). The pus from an open venereal sore also transmits infectious agents. Certain pathogens leave a host via insect bites. Mosquitoes, body lice, ticks, and other insects acquire the pathogenic agent in the blood or other ingested material and pass it along to another host, with or without further development in the body of the insect.

SURVIVAL OUTSIDE THE HOST

Successful transmission of a microbial cause of disease is accomplished only if the microorganism survives outside the donor long enough to reach another host. This is not much of a problem if the pathogen is not highly parasitic. The typhoid fever organism withstands widely varying conditions of temperature, oxygen, and nutrient supply, and can even survive in water or ice for several weeks. It is therefore capable of producing disease many miles away from the individual who excreted it and many weeks later. The bacteria that produce venereal diseases, on the contrary, are highly parasitic; they die quickly if cooled more than a few degrees and fail to multiply unless supplied with complicated foods found in body fluids. These organisms are rarely transmitted by any route other than direct contact.

TYPES OF INFECTION

LOCAL INFECTION

The course of events following a minor injury, such as introduction of a contaminated splinter into the finger, will be described to illustrate various types of infection. The physical act of jabbing a splinter into the finger damages or kills some of the tissue cells. The materials of these damaged cells provide an excellent culture medium for the bacteria that were on the splinter or were carried along from the surface of the skin. These organisms multiply and produce further irritation.

Irritated tissues liberate polypeptides and globulin proteins, which cause a series of reactions. Increased permeability of capillaries in the immediate vicinity permits blood plasma to pass into the surrounding tissues and to form a fibrinous clot that inhibits spread of the infectious agent. The number of circulating white blood cells gradually increases, and actively phagocytic polymorphonuclear leukocytes are attracted to the site of infection. The pus that forms is composed largely of these cells, together with plasma and bacteria. Cellular metabolism in the infected area is disturbed as a result of interference with normal circulation, and acids accumulate; temperature control is

also disturbed, and there is local fever. Inflamed areas are often warm to the touch.

Inflammation is the universal response to irritation of any sort—physical, chemical, microbiologic. The five attributes of inflammation—swelling, pain, redness, heat, and loss of function—call attention to the source of irritation, often in time for proper treatment to be instituted.

Most local infections never progress beyond this stage. The infection remains confined, and eventually phagocytes ingest and destroy the infecting microorganisms. The condition of local acidity attracts large leukocytes or macrophages, which replace the polymorphonuclear leukocytes; they dispose of the debris resulting from infection and help repair damaged tissue. A few days later nothing remains but a scar.

GENERALIZED INFECTIONS

A local infection sometimes generalizes and spreads to other parts of the body. This can happen if the organisms are unusually virulent, the local defenses unusually weak, or if amateur surgery is attempted before phagocytosis has eliminated all living microorganisms. Some of the bacteria escape from the local, walled-off site in cellular or inflammatory lymph. Lymph from the finger drains into the cubital lymph nodes at the elbow, thence to the axillary nodes in the armpits, and finally into the bloodstream by way of the thoracic duct.

Lymph nodes are composed of single layers of cells separating lymph and blood and function as filters to prevent passage of particulate matter into the blood. Many infections that spread from a local site are stopped at the lymph nodes, because these organs contain phagocytic cells, which can ingest bacteria. Microorganisms may produce secondary inflammation of a lymph node; a well known example is the swollen parotid gland of mumps. In some diseases, the primary site of infection is a lymph node; for example, *S. typhi* invades lymphatic tissue in the intestines (Peyer's patches) or throat (tonsils), and then progresses through the various stages described below, including secondary bacteremia or septicemia and possibly fatal typhoid fever.

Bacteremia. A particularly virulent organism passes through the lymph node and enters the bloodstream. This produces a temporary *bacteremia,* a condition in which bacteria circulate in the blood but do not damage it and do not multiply.

The R-E System. The blood passes repeatedly through all organs of the body. The spleen, liver, bone marrow and other tissues with highly phagocytic cells constitute the "reticuloendothelial" (R-E) system (reticular connective tissue cells and capillary or sinus endothelial cells of various organs). Foreign particles of any kind are quickly removed from the blood in its passage through these tissues. Experiments in laboratory animals have demonstrated that millions of bacteria injected into the vein of an animal disappear within a few minutes. Phagocytosis in the R-E system may completely eliminate the pathogen.

Secondary Bacteremia or Septicemia. Occasionally the microorganism is not destroyed and actually multiplies within the R-E system. Reinvasion of the blood then produces secondary bacteremia or even *septicemia,* in which the organisms continue to multiply and damage the blood components. Infection of any or all other organs may occur.

Focal Infection. A local infection from which bacteria continuously or intermittently enter the bloodstream is known as a *focal* infection. Common primary foci of infection are found about the head and throat. Infected teeth or tonsils sometimes provide a constant supply of pathogens, which then infect other organs (e.g., the kidneys). General septicemia with multiple secondary foci of infection is known as *pyemia.*

PHYSICAL AND CHEMICAL SIGNS OF INFECTION

The most obvious physical signs of infection are fever and increased pulse rate. Change in the number or percentage of white blood cells is readily detected by simple laboratory procedures.

Leukocytes. The normal individual has between 5000 and 10,000 white blood cells, or leukocytes, per cubic millimeter of blood. The several kinds of leukocytes are classified according to their size (7μm to 20 μm), presence and type of granules, shape and size of nucleus, and appearance of cytoplasm. The five principal types are usually present in fairly constant percentages:

Per Cent

Polymorphonuclear neutrophils	50–70
Basophils	0.5–1.0
Eosinophils	1–5
Lymphocytes	20–30
Monocytes	2–6

The first three types listed are formed in the marrow of the flat bones. Lymphocytes are believed to be formed in lymph nodes, and monocytes are derived from reticulum cells, particularly in the spleen. Leukocytes are short-lived; the majority are normally replaced within a few hours to four days.

The total number of white blood cells increases in many infectious diseases. *Leukocytosis* is an increase above the normal count; it begins at about 10,000 cells per cubic millimeter, and 30,000 or 40,000 represents marked leukocytosis.

A differential count is made by examining stained blood smears and counting the different types of leukocytes, the results being expressed in percentages. An increase in polymorphonuclear neutrophils, designated *neutrophilia,* is characteristic of acute infections by staphylococci, streptococci, and other pus-producing bacteria. *Neutropenia,* a marked decrease in polymorphonuclear neutrophils, occurs in typhoid fever, undulant fever, and influenza. *Lymphocytosis,* an increase in lymphocytes, is typical of whooping cough and mumps. These examples indicate how the blood picture may change as a result of infectious disease and also show one way in which laboratory examinations may assist in diagnosing disease.

Composition of Body Fluids. The chemical composition of the body fluids may change as a result of infection. The concentrations of sugar, protein, chloride, and other substances in the blood increase or decrease in different diseases. Spinal fluid may contain pus cells or abnormal percentages of protein, glucose, and other chemicals. Urine examinations reveal kidney damage. Hemolytic streptococcus infections affect the kidneys almost specifically and increase the excretion of albumin and solid particles known as casts. Protein catabolism is accelerated, and the urine becomes acid and contains a greater amount of nitrogen than normal. Elimination of ketones and diacetic acid is typical of infectious diseases in children.

These and other changes in the physical and chemical make-up of the individual reflect the profound physiologic disturbance caused by infectious disease. Another type of disturbance, leading in some cases to immunity, accompanies recovery from infectious disease. Active immunization is so important that it will be discussed separately in the next chapter, together with other factors that influence resistance to infectious disease.

EPIDEMIC VS. ENDEMIC DISEASE

A parasite introduced into a completely susceptible population or into a new host species in which it can establish itself is likely to spread widely and produce an *epidemic*. The parasite is passed quickly from one individual to another, and soon a large percentage of the host population is infected.

After a certain proportion of the population has become infected, the rate of transmission decreases as the number of uninfected individuals dwindles. The survivors are probably, on the average, a little more resistant than those who succumb. Moreover, recovered individuals usually possess specific immunity. The parasite persists thereafter in the population at a greatly reduced incidence. Immune individuals sometimes serve as carriers; the organism may produce mild and unrecognized *subclinical* infections in highly resistant individuals or occasional frank cases in the nonresistant. The net result is an *endemic* condition in which the disease is constantly present but in insignificant numbers.

The endemic state is interrupted by epidemics whenever sufficient susceptible and nonimmune individuals accumulate by birth or immigration. Increasing resistance during several generations according to the "survival of the fittest" principle gradually reduces the severity of the disease. At the same time, the parasite may mutate to a form better adapted to the host species and better able to survive. These factors increase the duration of the endemic condition and reduce the frequency of epidemics.

Epidemics of infectious disease, whether in man, lower animals, or plants, are largely the result of interference with the natural distribution of the various species that make up a climax population. A normal tropical rain forest in the East Indies may contain 100 species of trees per acre, together with shrubs and other plants, 30 species of mammals, and other animals. In such a forest there is relatively little opportunity for a parasite to be transmitted from one individual to another of the same species; the climax population consists of many species but few individuals of each. Disease is either absent or endemic; epidemics are rare.

A natural catastrophe such as a forest fire or deliberate deforestation by man changes the ecologic picture completely. Bare areas are soon invaded by rapidly growing plants and animals, and large populations of a few species are quickly produced. Under these conditions, epidemics of infectious disease can readily occur.

In some East Indian forest areas a rickettsial disease, jungle tsutsugamushi or scrub typhus, has long been endemic. It is transmitted by mites among its natural hosts: voles, shrews, field mice, rats, and other small animals. Deforestation of these areas upsets the normal climax state. Grassy scrub grows quickly, and in this type of vegetation the black rat flourishes. This species is one of the hosts of *Rickettsia tsutsugamushi,* and rapid transfer of the parasite soon causes an epidemic among the rat population, which is also transmitted to man, in whom the mortality is 20 to 40 per cent.

Evolution of Epidemic Disease

There is evidence that many infectious diseases of man and domestic animals have appeared within the few thousand years of recorded history. Prehistoric man lived in caves in small family groups and probably did not congregate in large communities. As in the tropical rain forest, major epidemics were not common. Socialization later occurred, the family group became a tribe, agriculture developed, and eventually urbanization took place. Large populations crowded in small areas provided opportunity for epidemic spread of infectious disease; ancient records indicate that this actually occurred. The widespread distribution of infectious disease initiated the process of selection of resistant individuals. It may be noted in passing that the practice of agriculture—cultivation of plants and animals—disturbs the natural balanced condition in any area and favors the spread of plant and animal diseases.

FACTORS THAT DETERMINE THE NATURE OF AN EPIDEMIC

The epidemiology of infectious disease depends upon (1) the degree of adaptation of the parasite and host to each other, (2) the ability of the parasite to survive outside the host, and (3) the mode of transfer of the parasite to new host individuals.

Parasite-Host Adaptation. The first contact between a parasite and a new host species or population is likely to be violent, but mutual adaptation eventually occurs. This process may be rapid.

Myxomatosis is a viral infection of rabbits. It is endemic among the wild rabbits of Brazil, where it is transmitted by the bite of contaminated mosquitoes; the virus pro-vokes only slight swellings under the skin and temporary invasion of the blood, and the rabbit is subsequently immune. However, the disease is highly fatal for the European rabbit. Southern Australia is overrun with European rabbits, the descendants of animals imported from England about 1860. Myxomatosis was deliberately introduced into Australian rabbits in 1950 and at first killed 99.5 per cent of the infected animals. Two years later, however, the virus had become attenuated to such an extent that the mortality was only 90 per cent. A similar change in virulence occurred in Europe when the virus entered the wild rabbit population.

Influenza viruses undergo more or less continuous modification. The virus that caused the pandemic of 1918–19 was a swine influenza strain. It was succeeded by a series of related influenza A viruses and reappeared in man in 1930. A new virus, influenza A (Asian) appeared in 1957–58, followed by the Hong Kong variant in 1968–69. Still another influenza A virus, the Victoria strain, was prevalent in 1975–76. A virus closely related to the 1918–19 swine influenza also appeared in 1976 among United States Army recruits at Fort Dix, New Jersey.

Adaptation of man to the syphilis and tuberculosis organisms has already been mentioned; this was apparently a slower process.

Survival of the Parasite. The ability of each species of parasite to survive outside its host is fairly constant, although some changes are to be expected. A parasite that continues the process of adapting to a given host may suffer further loss of independent survival powers; on the other hand, an organism that encounters strenuous environmental conditions between hosts may become increasingly resistant.

SUPPLEMENTARY READING

Burrows, W.: *Textbook of Microbiology,* 20th ed. Philadelphia, W. B. Saunders Co., 1973.
Davis, B. D., Dulbecco, R., Eisen, H. N., Ginsberg, H. S., and Wood, W. B., Jr.: *Microbiology,* 2nd ed. New York, Harper and Row, Publishers, 1973.
Dubos, R. J., and Hirsch, J. G. (eds.): *Bacterial and Mycotic Infections of Man,* 4th ed. Philadelphia, J. B. Lippincott Co., 1965.

Joklik, W. K., and Smith, D. T.: *Zinsser Microbiology,* 15th ed. New York, Appleton-Century-Crofts, Inc., 1972.

Moulder, J. W.: *The Biochemistry of Intracellular Parasitism.* Chicago, The University of Chicago Press, 1962.

Schlessinger, D. (ed.): *Microbiology–1975.* Washington, D.C., American Society for Microbiology, 1975.

Society for General Microbiology: *Mechanisms of Microbial Pathogenicity.* Cambridge, Cambridge University Press, 1955.

Society for General Microbiology: *Microbial Behaviour "In Vivo" and "In Vitro."* Cambridge, Cambridge University Press, 1964.

Wilson, G. S., and Miles, A. A.: *Topley and Wilson's Principles of Bacteriology, Virology, and Immunity,* 6th ed. Baltimore, The Williams & Wilkins Co., 1975.

19 RESISTANCE AND IMMUNITY

Simultaneously with the evolution of parasitism and the partially related phenomenon of pathogenicity, higher animals may acquire tolerance for their microbial invaders and learn to live with them in reasonable harmony—or they may become completely intolerant and develop means to prevent invasion or to destroy the parasite. In any event, after many generations a balanced state is established. Disease then occurs only when abnormal conditions prevail; for example, when the host is debilitated or encounters an overwhelming number of individual parasites, or when the parasite is unusually virulent.

In this chapter we shall consider the defenses with which the host opposes the invasive and destructive components or products of pathogenic microorganisms. These defenses are sometimes discussed together under a single heading—immunity. However, there are many types of defensive structures and mechanisms, and it seems preferable to subdivide them according to some basic principles, as summarized in the next section. The term *immunity* is properly restricted to a particular kind of defense and will be used hereafter in the sense defined below.

Defenses Against Infection. *Natural resistance* is one of three types of protection against infectious disease. It is associated with physical or physiologic conditions characteristic of an individual. It is nonspecific, that is, directed against no particular disease, and varies from time to time in the same individual and from one individual to another.

Nonsusceptibility is absolute protection against particular diseases and is as-

sociated with species characteristics. It is therefore basically genetic.

Immunity is directed against specific diseases and varies in degree. It is dependent on specifically reactive cells (immunocytes) or on antibodies, which are proteins associated with the globulin fraction of blood serum. It is largely an individual characteristic, although strains or even species of animals possess certain antibodies for reasons that are unknown but are assumed to be genetic.

NATURAL RESISTANCE

Natural resistance is attributed to a variety of anatomic and physiologic factors, which vary from one individual to another and also vary within the same individual at different times or under different environmental conditions. Environmental factors are therefore often included in a discussion of natural resistance.

NORMAL BARRIERS AGAINST INFECTION

Mechanical Barriers. Most living forms are surrounded by a covering layer of cells or a membrane of some sort, which separates the organism from its environment. This integument is usually strong and impermeable, except for openings that permit passage of liquids and gases.

Intact skin is an excellent barrier to infection. It is composed of several layers of cells, which constantly slough off from the outside and are replaced from the inside. Even the minute openings of the sweat and sebaceous glands are protected by chemical substances and are not ordinarily invaded by living microorganisms.

The sticky secretion of mucous membranes entraps microorganisms and keeps them from vulnerable organs such as the lungs. Moreover, many mucous surfaces are covered with cilia, whose active motion constantly removes foreign objects.

Few if any microorganisms are able to penetrate healthy, unbroken skin, but physical abrasion and physiologic disturbances induced by abnormal hormone concentrations, malnutrition, etc., occasionally permit microorganisms to enter hair follicles and sebaceous glands, where they multiply and produce local infections.

Chemical Barriers. Extreme acidity or alkalinity inhibits or kills microorganisms. Gastric juice, which may be as acid as pH 2, undoubtedly kills many organisms in water and food and helps to prevent infection. Some bacteria withstand the acidity or are protected by the food with which they are swallowed and establish themselves as intestinal pathogens. These organisms also survive exposure to the strongly alkaline fluid, bile, which enters the intestine in large quantities from the gallbladder.

A number of substances that are antibacterial *in vitro* have been isolated from animal tissues and body fluids (Table 19–1). Their effectiveness in the body is not known in every case.

Complement is a thermolabile (inactivated at 56° C. in 30 minutes) complex of nine or more components present in the sera of most warm-blooded animals. The various components are designated by numbers, and some of their chemical and physical properties have been determined (see Table 19–2). Some or all components participate in various immunologic reactions *in vivo, in vitro,* or both: lysis of susceptible cell membranes, phagocytosis, intracellular digestion, chemotaxis, liberation of pharmacologically active substances, and immune adherence of bacteria to body cells. All components are necessary for lysis of red blood cells or bacteria (hemolysis or bacteriolysis) sensitized by homologous antibody, and they react in the sequence in which they are listed in Table 19–2.

TABLE 19–1. Antibacterial Substances from Animal Tissue or Fluid*

Name	Common Source	Chemical Nature	Antibacterial Selectivity
Complement	Serum	Euglobulin-carbohydrate-lipoprotein (?)	Gram negative
Properdin	Serum	Euglobulin	Gram negative
Phagocytin	Leukocytes	Globulin	Gram negative
Lysozyme	Ubiquitous	Small basic protein	Gram positive (chiefly)
β-Lysin	Serum	Protein (?)	Gram positive
Histone	Lymphatics	Small basic protein	Gram positive
Protamine	Sperm	Small basic protein	Gram positive
Tissue polypeptides	Lymphatics	Linear basic peptides	Gram positive
Leukin	Leukocytes	Basic peptides	Gram positive
Plakin	Blood platelets	Peptide (?)	Gram positive
Hematin, mesohematin	Red blood cells	Iron porphyrins	Gram positive
Spermine, spermidine	Pancreas, prostate	Basic polyamines	Gram positive

*Modified from Skarnes and Watson, Bact. Rev., *21:*273–294, 1957.

TABLE 19–2.　Some Properties of Human Complement Components

Component	C'1q	C'1r	C'1s	C'4	C'2	C'3	C'5	C'6	C'7	C'8	C'9
Electrophoretic mobility	γ_2	β	α_2	β_1	β_2	β_1	β_1	β_2	β_2	γ_1	α
Molecular weight	400,000	168,000	80,000	240,000	130,000	180,000	185,000	125,000		150,000	70,000
Carbohydrate (%)	17			14		2.7	19				+
Thermolability (56°C., 30 min.)	+	+	+	−	+	−	+	−	−	+	+
Serum concentration (μg/ml)	190		22	430	30	1200	75	60		<10	1–2
Probable source	Small intestine, macrophages			Bone marrow, spleen		Macrophages					

Properdin is a normal serum component of euglobulin nature that possesses antimicrobial activity *in vitro* in the presence of Mg^{++} ions and complement components and is assumed to behave similarly *in vivo*. It is bactericidal against certain gram-negative bacteria such as *S. dysenteriae,* but its action is apparently nonspecific. It also apparently has some antiviral activity. Its action is not well understood.

Tears, nasal secretions, and saliva contain lysozyme, an enzyme that dissolves some staphylococci, intestinal streptococci, and possibly other pathogenic bacteria. Sweat is bactericidal for organisms, such as *Escherichia coli, Salmonella typhi,* and *Streptococcus pyogenes,* that are not normally present on the skin; staphylococci, on the contrary, are resistant. Even the influenza virus is inactivated by skin secretions. The effective substance is probably lactic acid. The reaction of the skin may be as low as pH 5.

Various basic peptides or proteins isolated from lymphatic glands, blood cells, and other tissues appear to kill gram-positive bacteria.

Interferon. Interferon is a relatively nonspecific antiviral agent produced normally in the course of virus infection. It is a cell-coded protein of low molecular weight (about 30,000). It interferes with the synthesis of virus protein by derepressing the synthesis of another cell-coded protein, "translation inhibiting protein (TIP)," whose action is to inhibit the translation of virus mRNA into virus protein. TIP has no effect on translation of host cell mRNA. Synthesis of interferon is induced in animals or in tissue cultures by any foreign nucleic acid, either viral (active or inactivated) or nonviral, and even by double-stranded polynucleotides such as polyinosinic:cytidylic acid (poly I:C), a synthetic polymer of inosinic and cytidylic acids. The inhibitory action of interferon is not virus-specific, but it is host-specific; that is, it inhibits synthesis of many viruses, but only within the species of host that produced the interferon.

INDIVIDUAL FACTORS AFFECTING RESISTANCE

Nutrition. Well nourished individuals tend to resist or recover from infection better than undernourished individuals. Lack of vitamin A decreases resistance to infection, particularly by microorganisms invading the bronchi and upper respiratory passages. Adequate protein in the diet is believed to maintain resistance or immunity or both. Various reports indicate that human populations whose principal dietary staple is animal protein are more resistant to chronic respiratory infection, tuberculosis, etc., than those whose diet is chiefly cereal carbohydrate. The nature of the protein also seems to be important. Rats fed animal protein (casein) are two to three times as resistant to infection by *Salmonella enteritidis* as rats on a plant protein (wheat or soybean) diet.

Paradoxically, virus infections may be enhanced by adequate nutrition. Good health confers no protection against measles, influenza, chickenpox, and smallpox. Undernourished or partially starved mice are more resistant than normal animals to certain encephalitis viruses. Viruses evidently multiply better in healthy cells than in cells damaged by lack of proper nutrients.

Debilitation. General debilitation predisposes to infection by many microbial agents. Chronic disease, infectious or noninfectious, is a debilitating agent. Chronic alcoholism, fatigue, pregnancy, and age also influence the occurrence or severity of infectious disease. Pneumonia is particularly prevalent and severe in infants and in the aged.

Physical Stress. Experimental observations indicate that resistance to infection is correlated with the ability to respond to stress. Rabbits whose body temperatures return to normal promptly after artificial chilling are more resistant than those with a prolonged warming time. Experiments with humans indicate a correlation between the rate of oxygen utilization during exercise and resistance to the common cold.

Temperature and humidity extremes, especially sudden changes of temperature, alter the physiologic state of the nasal mucous membranes and may increase their permeability and susceptibility to invasion by potential pathogens of the normal flora. Fatigue, shock, and oxygen starvation also affect tissue or capillary permeability and increase the likelihood of infection.

Heredity. Heredity is a factor in the response of the individual to infectious agents. Webster, at Rockefeller University, increased and decreased the resistance of mice to *S. enteritidis,* a mouse pathogen, by selective breeding. From normal mice, whose mortality rate was 37 per cent, lines were developed with mortality rates of 85 per cent and 15 per cent.

The resistance of human populations to tuberculosis and syphilis has changed during the last few centuries. Tuberculosis in present civilized populations is usually chronic, but previously unexposed native populations are nearly exterminated by their first contact with the disease. This occurred when tuberculosis was introduced into Tasmania and also in several thousand Kaffirs brought to Ceylon by the Dutch. The few survivors of such an epidemic are more resistant than the majority, and natural selection eventually produces a population in which the disease is chronic and less severe. The physical or physiologic factors responsible for resistance are not known.

ENVIRONMENTAL FACTORS AFFECTING RESISTANCE

Environmental factors affect the resistance of individuals or of populations only indirectly, but should be mentioned because they influence the occurrence of infectious disease.

Climate. Climate determines the distribution of pathogenic microorganisms. Yellow fever and malaria are tropical or subtropical diseases because their mosquito vectors or hosts require a warm climate. Bacillary dysentery is also common in tropical and subtropical areas, but it occurs with some frequency in all climatic regions. It is particularly prevalent in areas of poor sanitation and personal hygiene; climate evidently determines both the socioeconomic status of the populations and their state of civilization.

Urbanization. Population density affects the rapidity of transfer of infectious organisms from one individual to another and hence the occurrence of epidemics or other outbreaks. Respiratory infections tend to spread rapidly in heavily populated areas. However, community sanitation, which is developed best in the industrialized areas of the northern and southern temperate zones, has markedly decreased the incidence of intestinal disease.

NONSUSCEPTIBILITY

Nonsusceptibility is absolute protection against specific disease. It is associated with the species of the host and is inherited in the same sense as other species characteristics. Nonsusceptibility is determined by physiologic and anatomic factors, but in many cases the specific factors are not known.

Man is nonsusceptible to many infectious diseases of animals: chicken cholera, hog cholera, cattle plague, canine distemper. Lower animals are nonsusceptible to many human diseases, including typhoid fever, dysentery, cholera, whooping cough, influenza, measles, mumps, gonorrhea, and syphilis.

Body temperature and diet may contribute to nonsusceptibility. Early experiments demonstrated that frogs and chickens are normally nonsusceptible to anthrax, but frogs inoculated with *Bacillus anthracis* succumbed to the infection when warmed to 35° C., as did chickens artificially cooled from a normal body temperature of about 41° C.

IMMUNITY

Immunity is a specific form of resistance that depends on the presence of immunocytes or antibodies. It varies in degree from one individual to another and may vary during the life of the individual. It occurs in animals but is generally believed not to occur in plants.

THEORIES OF IMMUNITY

The participation of blood components in immunity has long been known, but for several years there were two opposing theories of the mechanism of immunity.

Humoral Theory. According to the humoral theory, soluble substances in the blood serum are responsible for immunity to infectious disease. (Serum is the clear, straw-colored fluid remaining after the cells and fibrin have been removed from blood.) It was observed that bacteria of certain species died when mixed with the serum of animals that had recovered from infection or had been injected with killed bacteria of the same species. Infected animals recovered when injected with immune serum, whereas control animals untreated with serum succumbed. Moreover, bacteria were shown to die and dissolve when injected into an immune animal. The humoral theory was championed by Buchner, Pfeiffer, and others.

Cellular Theory. The cellular theory of immunity was proposed and defended by Metchnikoff. He found by microscopic study of the transparent waterflea, *Daphnia,* that when yeast cells that ordinarily produced a fatal infection were ingested by an ameboid type of body cell, the animal subsequently recovered. When ingestion did not occur, the waterflea succumbed to the infection. Further observations showed that this phenomenon was not limited to *Daphnia,* and that the ameboid cells of other species had the same capacity for ingesting foreign objects, including microorganisms. Metchnikoff postulated that *phagocytosis,* as he called this process, is the mechanism responsible for immunity and recovery from animal disease.

Opsonic Theory. Metchnikoff and the various proponents of the humoral theory debated and defended their respective beliefs for several years. Finally, Wright and Douglas showed in 1903 that both serum and phagocytic cells are important in immunity. Phagocytosis occurs to a limited extent in the absence of serum but is greatly enhanced by a serum component, which they called *opsonin.*

Opsonin is the name applied to antibody as demonstrated by specific phagocytosis. The activity of antibody can be shown in several other ways, and it is often given different names according to the method of demonstration.

Much later it was found that nonphagocytic cells of lymphoid origin may become specifically reactive in an animal that has been sensitized by contact with foreign cells or other substances. Subsequent experience with the same agent produces tissue damage characteristic of transplantation reaction or of certain allergies; under some circumstances the lymphoid cells participate directly in antimicrobial defense.

ANTIGENS

Antigens are substances that incite the appearance of antibodies and that react observably with those antibodies. Most proteins and some polysaccharides are antigenic; lipids are generally not antigenic.

Antigens are either large molecules or portions of large molecules. Few, if any, substances with a molecular weight less than 5000 are antigenic. They possess in their architecture several repeated distinctive molecular configurations equal in size to four or five amino acid or monosaccharide residues; these comprise the *antigenic determinants* or combining sites.

Incomplete antigens, or *haptens,* are too small to incite antibody production and possess too few determinant sites to participate in certain forms of visible reaction with antibody, although they will combine with an antibody of appropriate molecular configuration. Such an antibody can be produced by injecting the hapten joined to a larger, often nonantigenic "carrier." Portions of cell structures, such as the pneumococcus capsule, may also function as haptens.

Antigens are not necessarily pathogenic; in fact, much experimental work on the nature of antigen-antibody reactions has been done with harmless antigenic substances. The sources of antigenic materials are universal; they are found in all animals, plants, bacteria, viruses, and other microorganisms.

In general, it is said that only substances foreign to the injected animal can induce the formation of antibodies. Some exceptional substances, such as thyroglobulin, lens protein of the eye, and spermatozoa, which are normally inaccessible to the blood and lymph, may be antigenic in the same individual when removed from their normal location and injected. Serum proteins can be made antigenic for the donor animal by slight chemical modification (e.g., by addition of iodine, nitrate, or other radicals). Certain human diseases are attributed to the acquisition of antigenicity by normal body components, which undergo chemical modification accompanying aging or other conditions; the altered body constituents induce antibody formation in the same person and provoke a destructive antigen-antibody reaction *in vivo.*

These observations suggest that the immunologic mechanism is of a general rather than a solely defensive function, as was thought in the early days of immunology. It now appears that antibody can be produced against body components that did not in embryonic or neonatal life encounter immunocytes, either because the particular components were sheltered from the blood or lymph or because the immunologic system was not sufficiently developed. This concept will be discussed further on page 368.

Living organisms of any kind are composed of many antigenic materials. Even bacteria and the smaller forms such as viruses contain several antigenic substances, and animals injected with the whole organisms produce numerous antibodies, each corresponding to a single antigenic component. Bacterial cells are considered mosaics of antigens (Fig. 19–1). Some antigens are peculiar to each species of bacterium, a few are also found in other species. Certain antigens are characteristic of flagella, others of capsules, and still others of the cell bodies. Most antigens have not been identified chemically beyond partial determination of their protein or polysaccharide composition, but they can be detected by means of appropriate antibody solutions (antisera).

Bacterial flagella are composed of protein and often possess several serologically distinguishable determinant sites. These have been most intensively investigated in *Salmonella* species, in which their detection and identification are diagnostically useful. Flagellar antigens are commonly called H antigens (German: *Hauch,* veil or film, which describes the swarming growth on agar of certain flagellated bacteria).

The antigens of bacterial cell bodies (i.e., somatic antigens) that are usually detected in serologic reactions are cell wall substances composed of lipopolysaccharide and protein. The distinctive structures are oligosaccharide chains composed of four to six monosaccharide residues, and cells of each species possess several different chains. These were originally identified only by their serologic reactivity with antibodies; this is still the simplest procedure, but it is now possible to determine their chemical composition as well. These somatic antigens are called O antigens (German: *ohne Hauch,*

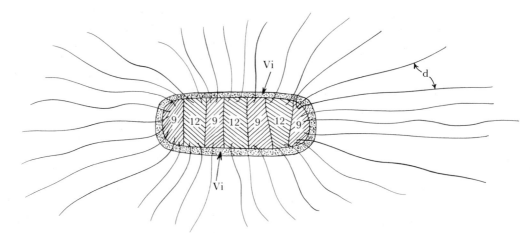

Figure 19–1. *Diagrammatic sketch indicating the mosaic-like antigenic structure of a bacterial cell. This organism,* Salmonella typhi, *possesses a single antigenic component (d) in its flagella. Its cell body may contain an "envelope" antigen (Vi) and two "somatic" antigens (9 and 12), which are actually part of the cell wall structure. Numeral and letter systems are commonly used to label bacterial cellular antigens.*

without veil or film, referring to the failure of nonflagellated bacteria to swarm on agar).

Bacterial capsules are antigenic and differ serologically from other cellular structures. Most bacterial capsules are composed of polysaccharides, and their specificity also is attributed to oligosaccharide residues. Certain gram-positive spore-forming bacilli, such as *B. anthracis,* form polypeptide capsules.

Envelope antigens are produced by many Enterobacteriaceae. The K and Vi antigens are called blocking antigens because they are situated outside the cell wall and prevent somatic antibody from reacting with the O antigens. Removal of the blocking antigen by heat permits the O antigens to react normally and detectably. Antibody made against an envelope antigen reacts observably with cells containing it.

ANTIBODIES

An *antibody* is a protein that has the general properties of an immunoglobulin and reacts specifically with some chemically or biologically defined antigenic determinant. *Immunoglobulins* are defined as proteins of animal origin with known antibody activity, together with certain proteins related to them by chemical structure. The several classes and types of immunoglobulin will be de-

scribed later (page 363). The word "specifically" implies that the antibody can react with only one antigenic determinant. This is not strictly true, because different antigens possessing similar but not identical reaction sites may react with the same antibody if the "fit" between the two substances is sufficiently good (Fig. 19–2).

Normal and Immune Blood Proteins. Blood serum contains 6 to 7 per cent protein. More than 30 proteins can be identified in normal serum by precipitation with sodium or ammonium sulfate or with alcohol under controlled conditions, by their electric charges, and by their molecular weights. Two principal fractions are the albumins and the globulins, each containing about half of the total protein. There are many different globulins, which can be distinguished by their electric charges and molecular weights. Electric charges are indicated by the rate and direction of migration of the protein molecules in an electric field, the process being known as *electrophoresis*. When a spot or band of serum is applied to a strip of filter paper moistened with buffer at pH 8.6 and the paper is placed between buffer reservoirs connected to a source of direct current, the proteins migrate at a speed that depends on their charges. They can then be stained and their concentrations measured photoelectrically. Typical results, shown in Figure 19–3, indicate a large albumin frac-

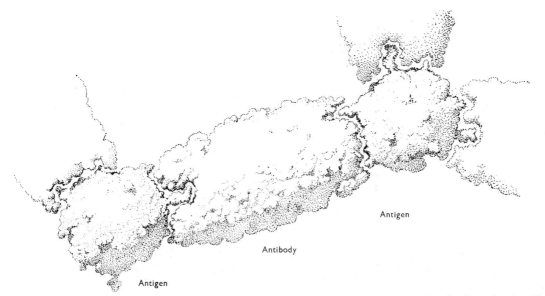

Figure 19–2. *"Lock and key" complementary structures of portions of antigen and antibody molecules. The irregularities of the surface structures of the molecules represent chemical radicals (e.g., a benzene ring or an amino acid). Each antibody can combine with two antigen molecules, and each antigen can react with several antibodies. (L. Pauling, Endeavour, Vol. VII, No. 26, 1948.)*

tion and three major globulins, designated α, β, and γ.

Antibodies are among the serum proteins with the lowest electric charge and are found principally in the gamma globulin fraction. Their molecular weights are between 150,000 and 1,000,000.

Antibody globulin and normal globulin differ in only one *known* way, that is, the ability of antibody to react with its corresponding antigen or hapten. They are chemically and physically identical by all other tests. The difference between them is therefore very subtle, and it is generally believed to consist of a slightly different arrangement of amino acids in the peptide chain, so that when the chain folds into its stable tertiary configuration, it will contain a site complementary in

Figure 19–3. *Electrophoretic separation of major rabbit serum proteins on a filter paper strip. The + electrode was at the right. The curve was plotted from points obtained with a photoelectric densitometer. (From Carpenter: Immunology and Serology, 3rd ed. Philadelphia, W. B. Saunders Co., 1975.)*

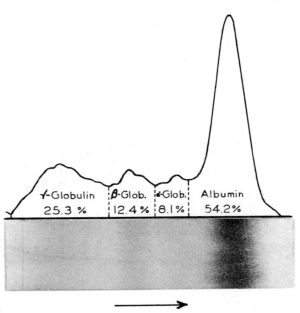

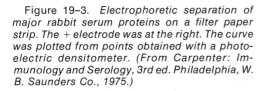

TABLE 19–3. Physical and Biologic Properties of Human Immunoglobulins*

	IgG	IgA (Serum)	IgA (Secretory)	IgM	IgD	IgE
Electrophoretic mobility	$\alpha2$–$\gamma2$	$\beta2$–$\gamma1$		$\beta2$–$\gamma1$	$\beta2$–$\gamma1$	$\gamma1$
Sedimentation coefficient	7S	7S, 9.6S	11S	19S (26S, 32S)	7S	7S
Molecular weight	150,000	160,000, 318,000	370,000–390,000	800,000–950,000	180,000	190,000
Heavy chain	γ	α		μ	δ	ϵ
Molecular weight	53,000	52,000–58,000		65,000–70,000	61,000–65,000	71,000–73,000
Light chains	κ, λ	κ, λ		κ, λ	κ, λ	κ, λ
Molecular weight	22,000 each	22,000 each		22,000 each	22,000 each	22,000 each
Carbohydrate (%)	2.9	7.5		11.8	12	10.7
Half-life (days)	23	5.8		5.1	2.8	2.4
Mean serum concentration (mg./ml.)	12	2.5		0.93	0.03	0.00003
Distribution (% of total in intravascular space)	45	42		76	75	51
Synthetic rate (mg./kg./day)	33	24		6.7	0.4	0.02
Per cent of total immunoglobulin	80	16		4	0.001	0.00003
Serologic properties (in vitro)						
Sensitivity to 2-mercaptoethanol	−			+		
Agglutination	Moderate	+		Strong	?	+
Precipitation	Strong	Weak		Variable	?	?
Complement fixation	IgG1 ++, IgG2 +, IgG3 +++, IgG4 −	Weak		Strong	?	−
Cytolysis	+	−		+	?	?
Opsonization	+	?		+	?	?
Immunologic properties (in vivo)	"Late" response to antigen; antibacterial, antitoxic, antiviral; blood group antibodies	Antibacterial, respiratory virus defense; blood group antibodies	Activity greater in secretions than in serum	"Early" response to antigen; antibacterial, antiviral; blood group antibodies	Unknown; antinuclear antibodies in some SLE and rheumatoid patients	Allergic reagin, possible respiratory tract defense
Cytophilic activity	IgG1 +, IgG2 +					+ in monkey
PCA activity (heterologous skin)	IgG1 +, IgG3 −, IgG4 +					+ in man
Prausnitz-Küstner reaction						
Special biologic properties	Major immunoglobulin in serum; Placental transfer (all subclasses)	Major immunoglobulin in human saliva, colostrum, nasal and bronchial fluids, intestinal tract, urine			Unknown	Fixation to skin

structure to the antigenic determinant. The relationship is like that of a key to a lock or a coin to the die from which it is struck. When the specific determinant sites of antigen and antibody molecules or particles interact, they are held together in a three-dimensional lattice arrangement (see Figure 19–2), which may grow to large dimensions under proper conditions.

Immunoglobulins. Most antibodies are γ-globulins. There are other serum proteins that are chemically and physically indistinguishable from antibody proteins, but for which antibody activity has not been demonstrated. They appear in the course of diseases of lymphoid tissues such as multiple myeloma, Waldenström's macroglobulinemia, and plasma cell tumors. These proteins and antibodies comprise the immunoglobulins. Most of their properties have been determined by study of myeloma and plasma cell tumor proteins, because they can be secured in larger quantities in pure form than can antibodies.

Three major and two minor immunoglobulins are known. Some of their properties are shown in Table 19–3. Ig is a shorthand for immunoglobulin. IgG is the most abundant of the major immunoglobulins. The letter G was originally applied in recognition of the fact that they are principally gamma globulins. IgM was early called a macroglobulin because of its high molecular weight.

Classes of Immunoglobulin. (1) IgG. The chemical and physical structure of IgG has been studied most thoroughly because of its relative abundance and its comparatively small size. IgG consists of four polypeptide chains, two of molecular weight of about 53,000 and two of about 22,000. These are designated the heavy and the light chains, respectively. The four chains are connected by disulfide bridges as shown in Figure 19–4. A minimum of three disulfide bridges is necessary, but there may be as many as seven. Disulfide bonds within the chains help to stabilize their secondary configurations (see also Figure 19–5). There are at least four subclasses, which differ in the number and location of their disulfide bonds and hence in their stereochemical structure. These are designated IgGl, IgG2, IgG3, and IgG4.

When IgG is subjected to mild reduction, the interchain disulfide bonds break and the four chains separate. The heavy chains retain only a fraction of their ability to react with antigen. Enzymatic hydrolysis of IgG with pepsin decreases the molecular weight of the active portion of the molecule to about

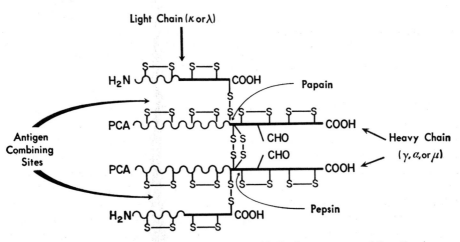

Figure 19–4. *Diagram of the polypeptide chain structure of IgG; the structures of the other immunoglobulin classes are analogous. The two light chains (H_2N ... COOH, top and bottom) are of either κ or λ type; the two heavy chains (PCA ... COOH, center) are of γ type in IgG, α type in IgA, μ type in IgM, δ type in IgD, and ε type in IgE. The chains are joined and their configurations are somewhat stabilized by disulfide bridges (—S—S—). Papain and pepsin cleave the heavy chains at the points indicated, and the intervening area, which includes —S—S— bridges linking the two chains, is a hinge region that provides flexibility to the entire molecule. The wavy portions are the regions of variable amino acid arrangement; this is almost exactly half of the light chains and is probably one-fourth to one-half of the heavy chains. (From Putnam, Science, 163:633, 1969.)*

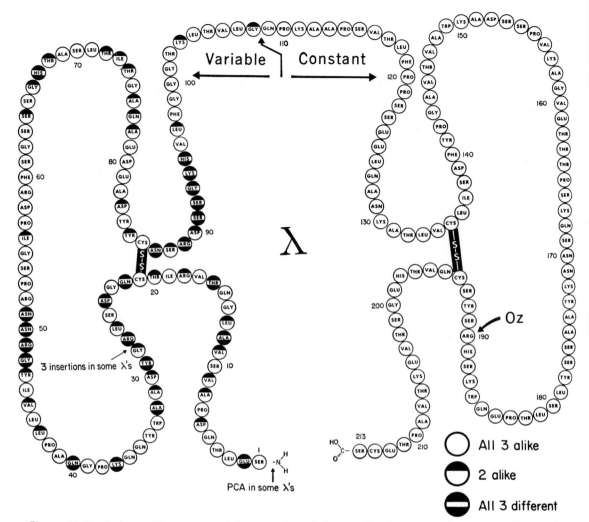

Figure 19–5. *Amino acid sequence of human λ-type IgG—results of a study of three proteins. Amino acids indicated in open circles were alike in all three proteins. There was variation in about half of the 108 amino acids in the NH₂-terminal portion of the molecule, but there was no variation among the amino acids in the COOH-terminal portion. (From Putnam, Science, 163:633, 1969.)*

100,000, and this fragment displays all the antibody function of the original molecule. Careful reduction of this yields fragments of about 25,000 molecular weight, which can combine with antigen but not yield a visible reaction product (i.e., it does not form a lattice structure). When IgG is digested with papain, one fraction retains only combining power, and the remaining fraction is completely inactive.

The structures and amino acid sequences of some light chains have been determined (see Figure 19–5). They are approximately 214 amino acid residues in length and are divisible into two almost equal portions: the amino-terminal half, which ends in a free NH₂ group, and the carboxyl-terminal half, which ends in a free COOH group. The sequence of amino acids in the COOH-terminal half is almost constant, no matter what individual it is secured from, whereas the amino acid sequence in the NH₂-terminal half varies from individual to individual. The light chain consists of two loops of about 60 amino acids each, formed by disulfide bridges joining pairs of cysteine residues.

There are two types of light chains, kappa (κ) and lambda (λ), which differ in the nature of the terminal radicals and in other particulars. The other classes of immunoglobulin

(IgA, IgM, IgD, IgE) also possess light chains of the same types, κ or λ.

The heavy chains of IgG are about twice as long as the light chains and have a similar structure; that is, there is a variable portion, corresponding to the NH$_2$-terminal half of the light chain, and a constant COOH-terminal portion characteristic of the class. The heavy chain of IgG is designated the γ-chain. Heavy chains of IgA, IgM, IgD, and IgE differ from IgG and from each other and are designated α-chain, μ-chain, δ-chain, and ε-chain, respectively.

(2) IgA. About 15 per cent of the γ-globulin in human serum consists of IgA. The molecular weight of the majority of IgA molecules is about 160,000, but 5 to 10 per cent have molecular weights of 300,000 to 400,000. IgA globulins are present in relatively large quantities in secretions such as saliva, tears, and colostrum, and they may also be found in the intestine, where they are known as coproantibodies.

Secreted IgA antibodies are of particular interest because they may play a special role in defense of membranes and exterior body cavities against infectious disease. They contain a "transport piece" produced by glandular epithelium, which assists the secretion process.

(3) IgM. The IgM molecule can be broken by thiol (—SH) reagents into five subunits, each having a molecular weight of about 180,000. Each subunit seems to have the same basic structure as IgG, that is, two light chains and two heavy chains. The molecular weights of the former are about 22,000, and those of the latter are 65,000 to 70,000. The total antibody in serum contains about 4 per cent of IgM. It is the first antibody detected in immunized individuals and is gradually replaced by IgG. Certain human antibodies are usually found in this class: syphilis antibodies and those specific for lipopolysaccharide antigens of gram-negative bacteria.

(4) IgD. This class of immunoglobulin is found in myeloma patients. Its function is not well known.

(5) IgE. IgE is an antibody-like substance, or *reagin,* that occurs in individuals who are hypersensitive or allergic to certain substances such as foods. When the reagin is injected into nonsensitive individuals, they acquire the sensitivity of the donor.

Structural Relationships between Immunoglobulins. Structural relationships between immunoglobulins are illustrated schematically in Figure 19–6. The basic structure consists of the two light and two heavy chains. In IgG and the low molecular

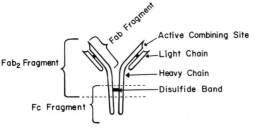

IgG, IgA, IgD, and IgE

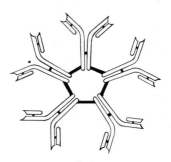

Ig M

IgA with Secretory Piece

Figure 19–6.   *Diagrammatic sketch of immunoglobulin structures. IgA may be either a monomer, dimer, or trimer, and IgM is a pentamer. (From Alexander and Good, Immunobiology for Surgeons. Philadelphia, W. B. Saunders Co., 1970.)*

weight IgA, this composes the entire molecule. The larger IgA molecules appear to be dimers or trimers, and IgM is a pentamer.

Immunoglobulin Active Sites. Inasmuch as a sizable portion of the constant region of the IgG molecule can be digested away without affecting the antibody activity of the residue, and little if any activity remains in the absence of both heavy and light chains, antibody activity must involve the variable NH_2-terminal regions of both chains. As discussed elsewhere (see page 361), antibody activity depends upon attraction between antigen and antibody, and union can occur only at sites of complementary configuration. Since visible evidence of reaction, such as precipitation, requires the formation of a lattice or framework of alternating molecules of the two reagents, antibody molecules must possess at least two combining sites, and

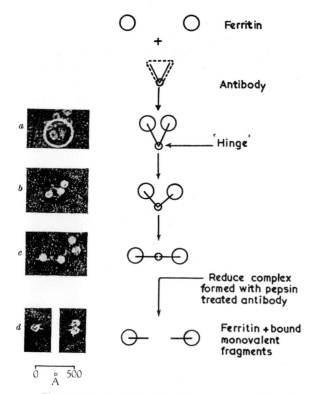

Figure 19–7. Left, *electron micrographs of ferritin-antiferritin complexes.* Right, *sketches representing the corresponding structures.* (d) *shows ferritin + monovalent fragments produced by pepsin digestion of antibody followed by reduction with mercaptoethanol. The photographs illustrate the behavior of the hinge area of the antibody molecule.* (From Feinstein and Rowe, Nature, 205:147, 1965.)

antigens must possess more than two combining sites. The IgG molecule can join two antigen molecules without steric hindrance because the region of the disulfide linkages joining the two heavy chains serves as a hinge and permits the combining areas to separate in a Y configuration (see Figures 19–4 and 19–7).

IgG digested by papain cannot participate in lattice formation because the two antibody-active portions are separated and hence univalent, although each subunit can combine with an antigen determinant site and form a nonprecipitating complex.

Site and Mechanism of Formation of Antibodies. Antibodies are produced by cells of the lymphoid system, particularly the lymph nodes and the spleen. Cells of the spleen and other visceral organs probably produce antibody in response to antigens that cause general infection or have been injected intravenously or intraperitoneally. Regional lymph nodes respond first and perhaps solely to local infections or to antigens injected subcutaneously.

The particular cell type from which antibody is derived appears to be the plasma cell. *Plasma cells* are nucleated, and their extensive cytoplasm stains intensely with basic dyes and therefore presumably contains large amounts of RNA, which helps to direct protein synthesis. They are not usually present in peripheral blood but are found in the spleen, lymph nodes, and other locations, especially following antigenic stimulation.

Humoral and Cell-mediated Immunity. Most bacterial and viral immunity depends upon humoral or circulating antibody. However, clinical observations and laboratory experiments have shown that two immunologic systems participate in various aspects of the response to infectious and other foreign agents. It was noted, for example, that patients with congenital agammaglobulinemia, who are unable to produce antibody, recover uneventfully from measles. Although their sera contain no antibody, they are nevertheless permanently and solidly immune from the disease. This fact points to the existence of a second, nonantibody mechanism of immunity.

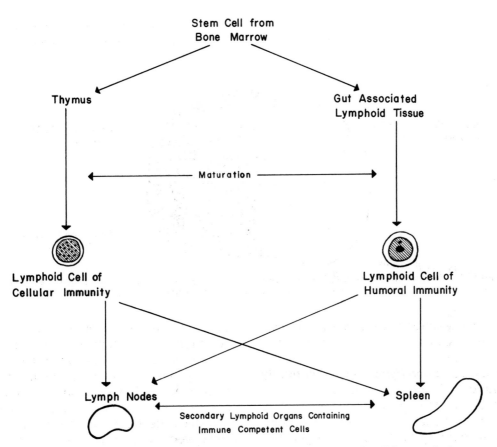

Figure 19-8. *Development of cell-mediated (T cell) and humoral antibody (B cell) immunologic systems from a common bone marrow stem cell. (From Alexander and Good: Immunobiology for Surgeons. Philadelphia, W. B. Saunders Co., 1970.)*

There were other instances of immunologic reactivity in which antibody could not be detected, and it was soon found that specifically sensitive lymphoid cells are responsible for immunologic reactions in chronic bacterial and fungal diseases, such as tuberculosis, brucellosis, and histoplasmosis, in certain forms of allergy, and in rejection of tissues transplanted from other individuals (i.e., the homograft reaction).

Both immunologic mechanisms have the same origin in fetal life (see Figure 19-8). Bone marrow stem cells make their way via the bloodstream to primary lymphoid tissues—either the thymus or the gut-associated lymphoid tissues (GALT), such as the pharyngotonsillar tissues, Peyer's patches, and lymphoid tissue of the appendix. The stem cells develop into lymphoid cells, maturing under the influence of hormones of the respective lymphoid tissues. Those whose maturation is controlled by thymic hormones become the lymphocytes of cell-mediated immunity, and those whose development is regulated by GALT hormones develop into cells that later become plasmacytes and elaborate one of the immunoglobulins (IgG, IgA, etc.). After release from the primary lymphoid organs, the mature lymphocytes circulate and colonize secondary lymphoid organs such as the spleen and lymph nodes. The thymus-dependent and gut-dependent cells become established in particular areas of the spleen and lymph nodes. The former are known as T cells, the latter as B cells (short for Bursa of Fabricius, a lymphoid organ in birds that was found early to be especially significant in antibody production). As shown in Figure 19-9, T lymphocytes populate the deep cor-

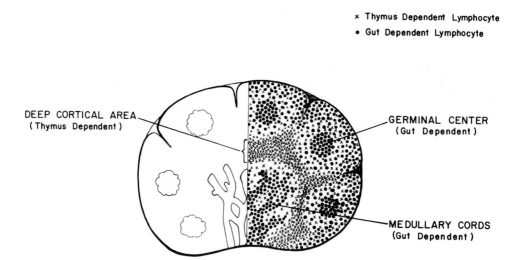

Figure 19–9.   *Diagrammatic sketch of thymus-dependent (T cell) and gut-dependent (B cell) areas of a lymph node. (From Alexander and Good: Immunobiology for Surgeons. Philadelphia, W. B. Saunders Co., 1970.)*

tical areas of lymph nodes, whereas B lymphocytes populate the germinal centers and medullary cords.

Clonal Selection Hypothesis of Antibody Formation. There have been two major types of hypotheses of the mechanism of antibody formation: (1) instructive hypotheses—according to these an antigen behaves like a die, repeatedly stamping out coins from a sheet of metal, and (2) selective hypotheses—these postulate pre-existing patterns from which a selection is made by the antigen. Although the former mechanism at first seemed most attractive, competent investigators were unable to devise reproducible experimental methods for deliberately modifying the physicochemical structure and configuration of proteins to make them complementary to antigenic determinants. Moreover, it did not explain the fact that the immunologic apparatus can distinguish "self" from "not self"; that is, it does not ordinarily destroy an individual's own tissues.

According to the currently favored selective hypothesis, lymphoid cells developing in the fetal primary lymphoid tissues differentiate and bear upon their surface membranes different "recognition sites" (e.g., peptide configurations). Each cell may have a single site, but thousands of different sites are represented among the entire cell population. As blood passes through the primary lymphoid tissues or as the cells leave and travel about the body, some cells encounter substances for which their recognition sites have an affinity and with which they react primarily because their configurations are complementary. The reaction leads to death of the cell, whereupon that particular recognition site is destroyed and can no longer be propagated. Cells whose recognition sites do not encounter complementary structures survive this "censorship" process, settle down in a convenient lymphoid tissue, and start a *clone* of cells in which the same recognition site is replicated. These sites do not react with "self" components, but are available to react with "not self" substances, whenever they come along in the blood or lymph. It has been estimated that 10,000 to 100,000 different recognition sites suffice to encompass the entire range of foreign antigenic determinants that may be encountered.

The recognition sites on B cells consist of immunoglobulin, and it is estimated that 50,000 to 100,000 molecules of immunoglobulin are distributed about the surface of each cell. When a B cell encounters an antigen (Ag) with specific affinity for its immunoglobulin (Ig), an Ig-Ag complex forms and aggregates at a specific locus on the cell surface (the phenomenon of "capping"), and is then taken into the cell. The cell replicates, and its progeny transform into plasma cells, which are especially equipped to manufacture and excrete protein (i.e., Ig). Repeated antigenic stimulation leads to more cell proliferation and to a large population of cells actively synthesizing immunoglobulin.

Immunologic Tolerance. Support for the clonal selection hypothesis of antibody formation is derived from the phenomenon of immunologic tolerance. It was predicted that any antigen—not just "self" antigens—encountered by fetal lymphoid cells would eliminate those lymphocytes bearing the corresponding recognition sites. Animal experiments showed this to be true; in some species, even newborn animals are made immunologically tolerant by injection of antigen, but it may be necessary to continue antigen administration at intervals to maintain the tolerant state. Such animals, when later injected with normally immunogenic doses of the same antigen, do not produce circulating antibody.

The Nature and Activities of T Cells. Lymphocytes that mature under the influence of thymic hormones and become T cells differ from B cells in several respects (see Table 19–4).

They are found particularly in the lymph nodes and hence in the thoracic duct and the blood, whereas B cells are abundant in the spleen but relatively scarce in other lymphatic tissues, and practically absent from the blood. T cells possess distinctive surface antigens designated θ (theta) antigens in the mouse, or T antigens. Where B cells possess a great concentration of immunoglobulin surface receptors, comparable receptors on T cells are somewhat of a mystery. From their behavior, it is evident that T cells have specific receptors, but only a few investigators have been able to detect immunoglobulins. In 1970, Hogg and Greaves reported the presence of a small number of immunoglobulin molecules, and this was confirmed the next year by Marchalonis; the protein was designated IgT. It has been suggested that these immunoglobulin receptors are actually abundant but are nearly submerged in the cell surface and hence difficult to detect and isolate.

T cells are resistant to inactivation by steroids and x-rays, in contrast to B cells. They are stimulated to proliferate by the mitogenic agents concanavallin A and phytohemagglutinin, whereas B cells are stimulated by lipopolysaccharide.

T cells can be made immunologically tolerant more easily than can B cells, and tolerance persists longer.

At first it appeared that the sole function of

TABLE 19–4. Characteristics of T and B Cells

	T Cells	B Cells
Source	Thymus	Bursa of Fabricius (fowl), gut-associated lymphoid tissue (mammals)
Tissue concentration	High in lymph nodes, thoracic duct and blood	High in spleen, low in lymphatic tissue and blood
Surface antigen	θ or T antigen	B antigen
Surface receptors	A postulated immunoglobulin, IgT, in low concentration or submerged	Immunoglobulins in high concentration
Resistance to steroids	High	Low
Resistance to x-rays	High	Low
Susceptibility to mitogens	Concanavallin A, phytohemagglutinin, pokeweed	Lipopolysaccharide, pokeweed
Induction of tolerance	Early and persistent in response to low doses of antigen	Late and less persistent, requires larger doses of antigen
Functional types	Helpers Lymphokine producers Killers Suppressors Stimulators	Antibody producers

T cells is to mediate cellular immunity (that is, immunity against viruses, fungi, and a few other agents) and the delayed hypersensitivity response (e.g., reactions to tuberculin and to poison ivy). Later, it was found that an animal lacking a thymus could not produce antibodies, which was known to be a function of B cells and their progeny, so it appeared that T cells help the process of antibody formation in some way. They were also found to produce and secrete several high molecular weight proteins that mediate various effects: inhibition of macrophage migration (MIF or macrophage inhibitory factor), induction of blast formation of other lymphocytes, destruction of various cultured cell lines, etc. These substances are called *lymphokines*. Certain T cells that appeared to be especially destructive to particular tissues were called "killer" or cytotoxic cells. Other T cells regulate various immunologic processes and are designated *suppressor* and *stimulator* T cells.

In summary, it appears that T cells interact with each other and with macrophages, directly or indirectly, and both of these in turn may affect the activity of B cells. The entire system maintains a delicately controlled balance capable of prompt response to foreign (i.e., "not self") substances but normally unaffected by "self" substances.

Role of Antibodies in Immunity

Types of Antibodies. An antibody is frequently named according to the method in use at the moment to demonstrate its activity; the same antibody may therefore bear several names. Pneumococcal antiserum, for example, contains antibody that can agglutinate (clump) intact pneumococci and that also can precipitate the capsular polysaccharide of the organisms. If an agglutination test is performed, the antibody is called *agglutinin;* in a precipitation test the same antibody is called *precipitin*.

Antitoxins. Antitoxin is antibody that neutralizes toxin, specifically exotoxin. Exotoxins are neutralized according to the law of multiple proportions by homologous (i.e., corresponding) antitoxin; that is, if one unit of

antitoxin neutralizes 100 lethal doses of toxin, 10 units of antitoxin will neutralize 1000 lethal doses. The mechanism of neutralization is not known, but its occurrence is not doubted. Experience during epidemics indicates that individuals possessing a certain small amount of diphtheria antitoxin in their circulating blood are immune from this disease; individuals possessing less antitoxin are likely to acquire diphtheria if exposed.

Cytolysins and Bactericidins. Cytolysins and bactericidal antibodies require the cooperation of the normal serum component, complement, to dissolve or kill bacteria. Complement is not increased by immunization. The effectiveness of cytolytic and bactericidal actions is obvious, but their occurrence is limited to only a few bacterial species.

Opsonins. Opsonins have already been described as antibodies that sensitize certain microorganisms and make them more easily ingested by phagocytic cells. Their full importance is unknown; it is doubtless great.

Precipitins. Precipitins react with and precipitate soluble antigens. This phenomenon occurring *in vivo* probably removes microbial cellular debris and soluble constituents such as capsular substances. The precipitated material is destroyed by phagocytosis. Toxic soluble substances, such as exotoxins, precipitate when mixed in suitable proportions with their corresponding antitoxins *in vitro;* it may be presumed that precipitation also occurs *in vivo*.

Agglutinins. Agglutinins make particulate (cellular) antigens stick together in large masses and settle out like a precipitate. The difference between agglutination and precipitation is that the antigen in agglutination is a particle of microscopic size, whereas a precipitating antigen is in solution.

Agglutination does not necessarily kill the agglutinated organisms, so that the reaction might appear to be of no significance in immunity. However, phagocytosis of clumped bacteria requires little more energy than phagocytosis of a single cell, and the efficiency of phagocytosis is therefore greatly enhanced by previous agglutination, apart from the opsonizing activity of antibody.

Immunizing Effectiveness of Antibodies. Many apparently useless antibodies are produced in the course of infection or artificial immunization with whole bacteria, because the antibody-producing mechanism cannot distinguish between harmful and harmless antigens. Not all the components of a microbial cell are *essential* immunizing antigens. Only antibodies against essential immunizing antigens are necessary for effective immunity, and the nature of these antigens varies according to the microorganism.

Antitoxin appears to be most significant in the case of bacteria that produce exotoxins: diphtheria, tetanus, gas gangrene, scarlet fever, etc. The gas gangrene organism produces at least 12 exotoxins, but only one of the antitoxins is necessary for protection.

Antibodies for the capsular polysaccharides of pneumococci afford much greater protection against pneumonia than antibodies against the rest of the cell bodies. It will be recalled that encapsulated pneumococci resist phagocytosis. Pneumococcal capsules that have combined with antibody no longer resist phagocytosis; the organisms are readily engulfed and destroyed.

The motile intestinal pathogenic bacteria possess flagellar antigens as well as somatic (i.e., cell body) antigens, but only the somatic antibodies are effective in immunity. Some typhoid fever bacteria possess the Vi antigen. Vi antibody is essential for a high degree of immunity against Vi strains but offers no additional protection against strains lacking this antigen.

Importance of Phagocytosis. The phagocytic activity of leukocytes is easily demonstrated, but there are actually many other phagocytic cells in the body, some of which are more important in combating infection. The phagocytic systems of the body, commonly designated the *reticuloendothelial system* (RES), may be outlined briefly as follows:

1. *Cells of reticular and loose connective tissue.* (Reticular cells form the framework of lymph glands, bone marrow, liver, and spleen.)
   A. *Macrophages, including reticular cells.* These cells are actively phagocytic. Most varieties are fixed (i.e., sessile), but histiocytes or clasmatocytes in the loose connective tissue are free or wandering cells.
   B. *Fibroblasts of connective tissue and endothelial cells of blood vessels.* Fibroblasts are structural cells that support and bind tissue together; they are rarely phagocytic. Endothelial cells lining the larger blood vessels and capillaries are also rarely phagocytic.
2. *Free connective tissue and blood cells.* The free connective tissue cells are precursors of macrophages. The blood cells (see page 350), which include many actively phagocytic cells, are the granulocytes (polymorphonuclear neutrophils, eosinophils, and basophils), lymphocytes, and monocytes, in order of presumed decreasing phagocytic activity.

The role of phagocytosis (Fig. 19–10) in prevention of or recovery from infectious disease is more difficult to determine than that of antibodies. Investigators have therefore tended to overemphasize the activities of antibodies and to ignore phagocytosis. Phagocytosis by leukocytes in the absence of opsonizing antibody can be demonstrated upon a "rough" surface, such as filter paper, against which the ameboid cells can trap bacteria or other particles. Phagocytosis occurs *in vivo* in the absence of antibody when the wandering phagocytic cells "corner" invading microorganisms against tissues or other phagocytic cells. Antibody greatly accelerates phagocytosis by leukocytes. The role of antibody in the activity of *fixed* phagocytic cells is unknown, but the dramatic removal of millions of bacteria per milliliter from the blood, as demonstrated repeatedly, indicates that fixed phagocytes are extremely active and are undoubtedly better guardians against infection than has been appreciated.

TYPES OF IMMUNITY

Natural Immunity. Natural immunity is attributed to natural antibodies. These are antibodies that are present in an individual's serum from the time of birth, or shortly thereafter, and persist throughout life. There is no apparent external stimulus for their production. Many animal species possess natural

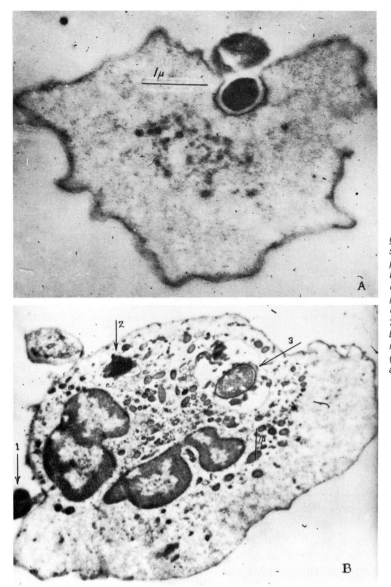

Figure 19–10. *Electron micrographs of two stages in phagocytosis of staphylococci by polymorphonuclear neutrophil cells of the blood. A, A deeply stained bacterial cell nearly surrounded by pseudopodia. B, Arrow 1 points to a coccus about to be ingested, arrow 2 indicates a recently ingested bacterial cell, and arrow 3 designates a coccus within a vacuole. (Photographs by J. R. Goodman et al.; A.S.M. LS-354 and -359.)*

antibodies against human pathogens such as typhoid fever, cholera, and dysentery bacteria. Certain humans possess natural antibodies against the pneumococcus and the typhoid organism, occasionally in considerable amount. Natural antibodies that react with foreign erythrocytes are present in the blood of various species, and reactions attributed to natural antibodies have been reported with bacterial and animal viruses, fungi, protozoa, and metazoa.

Many human sera contain antibodies that react with erythrocytes of other human individuals. These blood group antibodies or *isohemagglutinins* are under genetic control. They appear within a few months after birth, attain their highest concentration by the age of 10 years, and decrease slowly thereafter.

Antibodies that react with erythrocytes apparently have nothing to do with infectious disease. It is not known whether natural antibacterial antibodies of man and lower animals actually prevent infectious disease. Many investigators doubt that the resistance of lower animals to typhoid fever, cholera, dysentery, and other human diseases can be attributed to natural antibodies.

Acquired Immunity. Acquired immunity is attributed to immune antibodies. If the antibodies are produced by the immunized individual, the result is known as *active acquired immunity*. If preformed antibodies are acquired from the mother at birth or by injection from another individual or animal, the result is called *passive acquired immunity*.

Active Acquired Immunity. The stimulus for antibody production may be a natural infection or a series of injections. Recovery from many infectious diseases is followed by specific immunity: typhoid fever, whooping cough, diphtheria, chickenpox, smallpox, etc. Local infections with *Staphylococcus aureus* (e.g., boils and carbuncles) do not usually create immunity; the infectious agent never gains access to the lymph nodes or to the blood, and hence antibodies are never produced. Reasons for the failure of certain other infectious diseases to produce immunity are less clear. Active immunity, particularly that resulting from infection, is often of considerable duration and may persist throughout life.

Artificially acquired active immunity is induced by injecting a variety of agents. Sublethal doses of virulent organisms may be introduced by a route favorable for infection, or larger doses may be given by a route unfavorable for natural infection. The ancient Chinese practice of variolation is an example of the latter. Smallpox virus contained in pustular material from a case of the disease was inoculated into the skin. This virus is normally acquired by the respiratory route; therefore only a mild skin infection was produced, and the individual recovered and was thereafter immune from smallpox. Occasional mishaps, such as frank cases of the disease or concurrent infection with skin bacteria, led to abandonment of this practice, especially when the effectiveness of cowpox vaccine was demonstrated by Jenner in 1796. Immunization with living, virulent microorganisms entails a certain amount of risk and is now usually confined to veterinary practice.

Microorganisms attenuated or weakened by cultivation under unfavorable conditions (e.g., high temperature) or in an unnatural host (e.g., rabies virus transferred through the brains of rabbits and then dried) are often effective immunizing agents and can be used in larger amounts than would be possible with virulent forms. The Sabin poliomyelitis vaccine is composed of living attenuated viruses; it is administered by mouth, the natural infective route.

Killed microorganisms are safer immunizing materials and are used when possible. The standard typhoid-paratyphoid vaccine and pertussis (whooping cough) vaccine are of this type. The Salk vaccine for poliomyelitis consists of virus treated with formalin to inactivate the infective agent.

Bacterial products are used for immunization. Toxoids have already been described. Several laboratories are attempting to isolate the essential immunizing antigens from the typhoid fever, dysentery, and other bacteria. It is hoped that these purified antigens will reduce unpleasant side reactions (sore arm, malaise, etc.) occasionally produced by the routine vaccines.

Artificial immunization or actual infection with one bacterium sometimes induces antibodies and partial immunity against other bacteria possessing some of the same antigens.

The Antibody Response to Artificial Active Immunization. A single injection of antigen is followed by a latent period of several days before antibody appears in detectable amount in the circulating blood. The concentration of antibody rises slowly to a low peak and then diminishes. This is the "primary response" (Fig. 19–11).

A second injection of antigen, several days, weeks, or even months later, induces rapid production of antibody after a shorter latent period; a higher peak is attained and the concentration in the blood decreases more slowly. This is the "secondary response." It can occur even after antibody has completely disappeared from the blood. Additional injections invoke further antibody production within limits: a maximum is eventually reached. The secondary response may also be induced in a previously immunized individual by natural exposure to the homologous infectious agent; antibody is produced rapidly enough to prevent disease.

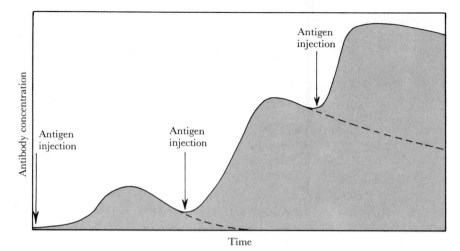

Figure 19–11. *Antibody concentration in the blood following one, two, or more injections of antigen. The responses to second and succeeding injections are quicker, steeper, higher, and more persistent.*

It should be emphasized that immunity is quantitative rather than qualitative and that no immunization procedure, either natural or artificial, can be guaranteed to produce complete immunity against overwhelming infection. Moreover, artificial immunization with killed or attenuated vaccines is often not as effective as natural immunization by actual infection, and periodic reimmunization or "boosters" may be necessary to maintain a satisfactory protective level of immunity.

Experience with immunization against poliomyelitis and whooping cough underlines the statement that artificial immunization does not always assure complete protection; occasionally, individuals who have received the full course of injections become infected, and a few of these patients die. The vast majority are protected, however, and most of those who become ill have mild cases (see Table 19–5).

Passive Acquired Immunity. In this

TABLE 19–5. Duration of Immunity Following Artificial Active Immunization Against Selected Diseases

Disease	Immunogen	Duration of Immunity
Diphtheria	Purified toxoid	4–5 yrs.
Tetanus	Purified toxoid	4–5 yrs.
Smallpox	Attenuated virus	3–5 yrs.
Yellow fever	Attenuated virus	5–10 yrs.
Measles	Attenuated virus	5–6 yrs.
German measles	Attenuated virus	2–3 yrs.*
Poliomyelitis	Attenuated virus (Sabin)	5 yrs.
	Inactivated virus (Salk)	2–3 yrs.
Influenza	Inactivated virus	6 mon.
Typhus fever	Killed rickettsiae	4–6+ mon.
Typhoid fever	Killed bacteria	3–5 yrs.
Whooping cough	Killed bacteria	?†
Cholera	Crude bacterial fraction	1–2 yrs.
Tuberculosis	Attenuated bacteria (BCG)	4–10 yrs.

*Preliminary estimate
†Uncertain; cases usually mild

state, antibodies are acquired naturally from the mother or artificially by injection. Human babies acquire antibodies from the maternal circulation *in utero*. Antibody molecules traverse the single layer of placental cells separating the maternal and fetal circulations and appear in the blood of the newborn baby. Antibodies acquired in this manner are detectable for only three to six months; active immunization is necessary if the child is to retain protection.

Other animals, such as the cow, have a different mechanism for transferring antibodies from the mother to the offspring. The four layers of cells that separate the maternal and fetal circulations inhibit passage of antibodies, but the colostrum, the first milk delivered after birth of the calf, contains a high concentration of antibodies. These are not digested in the stomach of the newborn calf but pass through the wall of the digestive tract intact and enter the blood.

Artificial passive immunization is the injection of whole blood, plasma, serum, or serum fractions. The injected antibodies are detectable for only a few weeks. Antisera from other animal species are usually used, although the antibodies probably do not persist in the recipient as long as antibodies of the same species. Horses are commonly employed because they produce large quantities of antibody.

Small doses of antiserum or antitoxin provide temporary protection of an individual exposed to an infectious agent or toxin-producing organism. Actual cases of the disease are treated with larger doses. Precautions must be taken in the use of antisera of other species lest allergic reactions and even fatal shock result. Some individuals are naturally sensitive to the proteins of foreign species; others acquire sensitivity by repeated injection of the foreign serum.

SEROLOGY

Serology is the study of the nature and behavior of serum antibodies. The antibody concentration attained in human sera as a result of infection or normal immunization is usually not sufficiently great for research or for artificial passive immunization or treatment of disease. Intensive immunization of experimental animals is therefore necessary to produce potent antisera.

PRODUCTION OF ANTISERA

Rabbits are often used for laboratory production of antisera against bacteria, erythrocytes, protein solutions such as foreign serum or egg albumin, and other materials.

Intravenous, intraperitoneal, and subcutaneous injections may be given. Intravenous inoculation ensures prompt distribution of the antigen to all antibody-producing sites. Dissemination of the antigenic material is slightly slower following intraperitoneal inoculation, but it is somewhat safer. Subcutaneous inoculations are usually safest but provide slow distribution of the antigenic material.

The first injection of a toxic antigen is usually small, and succeeding injections are progressively larger as the animal develops antibodies and can tolerate increased doses. Constant and fairly large quantities of nontoxic antigens may be introduced. Doses are spaced one to four days apart; if they are given every day a rest period is allowed after three or four injections, and inoculations are resumed the following week. Trial titrations are performed at intervals, and when the antibody content of the serum is sufficiently great the needed quantity of blood is taken.

Laboratory animals appear quite insensitive to the techniques of injection and bleeding if properly handled. Rabbits will often sit without restraint on the operator's lap or on a laboratory bench while injections are made and small amounts of blood are taken from an ear vein.

ANTIGEN-ANTIBODY REACTIONS

Antigen-antibody reactions as customarily demonstrated in the laboratory are two-stage reactions. The first stage is union of antigen and antibody; this is not directly detectable. The second stage is the visible result of antigen-antibody union: agglutination, precipitation, etc.

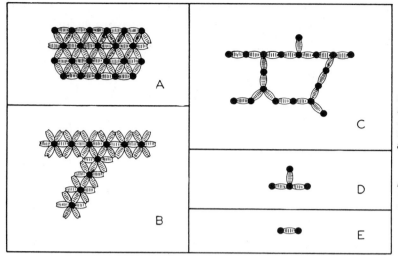

Figure 19–12. *Complexes formed when soluble antigen reacts with its antibody. A, an ideal framework. B, a network formed with excess antibody. C, a network with excess antigen. D–E, soluble complexes formed with excess antigen. (From Carpenter: Immunology and Serology, 3rd ed. Philadelphia, W. B. Saunders Co., 1975.)*

⬭ Antibody ● Antigen

Union of antigen and antibody is specific; that is, an antibody will combine only with the antigen that invoked its formation or with some other antigens closely related chemically. Most antibody molecules are bivalent, that is, capable of combining with two antigen molecules or particles. Antigens are multivalent and hence can combine with several antibody molecules. Union between antigen and antibody therefore produces a lattice or framework in which antibody molecules alternate with antigen molecules or particles. This is the *lattice* or *framework hypothesis* of antigen-antibody reaction (Fig. 19–12). Lattice formation occurs only when antigen and antibody are mixed in optimal proportions. If too little of one reagent or the other is provided, a complete lattice is not formed.

Agglutination.   Agglutination is the visible second stage result of mixing a particulate or cellular antigen with homologous antiserum. The cells clump together and settle to the bottom of the fluid in the form of flocculent masses, compact granules, or thin sheets. Agitation usually does not redisperse the particles completely. The particles behave as though they are sticky and cling tenaciously together.

Tube Agglutination.   Tube agglutination tests are performed in small test tubes. Serial dilutions of serum (1:10, 1:20, 1:40, 1:80, etc.) in saline are prepared, a suspension of the cellular test antigen is added, and the test tubes are incubated. The time and temperature of incubation vary with the antigen-antibody system. After incubation the degree of agglutination in each tube is read (Fig. 19–13). A tube in which all cells are agglutinated is graded + + + +; absence of agglutination is designated −; intermediate reactions are indicated +, + +, or + + +. The titer of the serum is the reciprocal of the greatest dilution causing definite (+) agglutination. This is the *limit dilution* or *extinction dilution* method of estimating antibody content of serum. It is only semiquantitative, and the results can easily vary by a factor of 2 in repeated trials. It is a convenient procedure, especially in clinical applications, and is widely used.

The microscopic appearance of agglutinated bacteria varies according to the cellular antigens that participate. Flagellar antigens become attached to one another and produce a loose aggregate (see Figure 19–14). In tube tests the agglutinate is fluffy or flocculent and can be easily redispersed by shaking. Somatic agglutination consists of polar attachment of the cell bodies (see Figure 19–15) in a crystal-like array, which produces a compact, granular sediment in tube tests.

Adsorption of Agglutinins.   Agglutinin adsorption is a method used to prepare antibody solutions that react with only certain

Figure 19–13. *The results of a tube agglutination test. Tube C is a control; it contains bacterial cells in saline, but no antiserum. The remaining tubes contain bacterial cells and increasing dilutions of antiserum. There is agglutination in tubes 1 through 7, but not in tube 8. (From Burrows: Textbook of Microbiology, 20th ed. Philadelphia, W. B. Saunders Co., 1973.)*

selected antigens. For example, a pure antibody solution that reacts with the H antigen *d* of *S. typhi* (see Figure 19–1), is prepared by adsorbing antityphoid serum with an organism containing antigens 9, 12, and Vi. This may be a nonmotile variant of *S. typhi* or a motile strain in which antigen *d* has been destroyed or denatured by heat or formaldehyde. The antiserum and a heavy suspension of the bacteria are mixed, incubated, and then centrifuged to deposit the cells and their adsorbed antibodies. The supernatant liquid contains only the unadsorbed anti-d. This solution can be kept for future use to detect antigen *d* in any bacterium. Antibodies purified in this manner are used to determine bacterial antigenic structures (that is, their various antigenic determinants) and to help identify the organisms.

Macroscopic Slide Agglutination. Mac-

roscopic slide agglutination is used in rapid identification of bacteria. A drop of concentrated antiserum (e.g., 1:5 or 1:10) is mixed with a drop of a heavy suspension of the cells on a microscope slide or glass plate and rocked for a few seconds to a few minutes. Granulation or flocculation occurs in positive tests. This procedure can be used for immediate identification of colonies from an agar plate.

Hemagglutination and Blood Grouping. Erythrocytes agglutinate when mixed with homologous antiserum, and this is the basis for tests to identify the species origin of blood cells and for grouping and typing of human erythrocytes. Antiserum prepared by injecting a rabbit or other animal with cells of a different species agglutinates the R.B.C. of the donor species and often less strongly those of related species. Rabbit antisera against rhesus monkey cells, for example, agglutinate rhesus erythrocytes and also those of about 85 per cent of humans. The

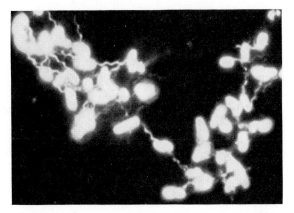

Figure 19–14. *Flagellar agglutination of S. typhi: a loose aggregate with flagella attached to one another by antibody. (Photograph kindly supplied by the late Dr. A. Pijper.)*

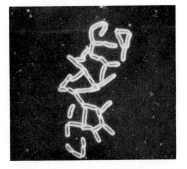

Figure 19–15. *Somatic agglutination of S. typhi: a regular, crystal-like structure in which the cell bodies are joined via antibody molecules. (Photograph kindly supplied by the late Dr. A. Pijper.)*

TABLE 19–6. Agglutination in Mixtures of Serum and Erythrocytes of Individuals of the Four Principal Human Blood Groups

Cells of Individual	Blood Group	Agglutinogens in Cells	Serum of Individual				Per cent in U.S.A.
			1	2	3	4	
			Agglutinins				
			Anti-A Anti-B	Anti-B	Anti-A	None	
1	O	—	—	—	—	—	45
2	A	A	+	—	+	—	41
3	B	B	+	+	—	—	10
4	AB	A, B	+	+	+	—	4

reactive antigenic determinants are known as Rh factors.

Erythrocytes, like bacteria, are mosaics of many antigenic determinants or substances. In addition to the Rh antigens, human cells contain the A and B substances, M, N, S, s, P, and many others. A and B are of particular interest because natural antibodies against them are present in all individuals whose cells lack the respective antigenic substances (see Table 19–6). Thus, an individual whose red blood cells contain A possesses anti-B in his serum, and vice versa. Similarly, the serum of a person with both A and B in his cells contains neither antibody, and that of an individual without either A or B possesses both anti-A and anti-B. Human blood groups are named according to the antigens (A or B) present in the erythrocytes. The red cells can be grouped by use of two sera: anti-B, obtained from a group A individual, and anti-A, from a person of group B.

A macroscopic slide test is commonly used, and the results are obtained within a few minutes.

Blood grouping is important in preparation for transfusion. It is dangerous to introduce incompatible blood. The greatest risk is associated with intravascular agglutination or hemolysis (see page 381) of the transfused cells by the recipient's antibodies; this may cause a fatal reaction. Even when both the prospective donor and recipient possess blood of the same group, "cross match" tests are necessary, because other antigens and natural or immune antibodies may be present and cause a transfusion reaction. The serum of the recipient is mixed with erythrocytes from the prospective donor, and the donor is used only if his cells are not agglutinated by the recipient's serum. The reciprocal test, in which serum of the donor is mixed with cells from the recipient, is also desirable.

TABLE 19–7. Precipitation of Egg Albumin by Antibody in a Rabbit Anti-egg Albumin Serum*

Egg albumin (mg.)	0.0625	0.1250	0.2500	0.5000	1.0000	2.0000
Anti-egg albumin serum (undiluted)	1 ml.	1 ml.	1 ml.	1 ml.	1 ml.	1 ml.
Precipitation:						
Visual inspection†	+	+ +	+ + +	+ + + +	+ + +	±
Protein by analysis (mg.)‡	1.13	1.94	3.32	5.25	3.26	0.56

*From Carpenter: *Immunology and Serology*, 3rd ed. Philadelphia, W. B. Saunders Company, 1975.
†A heavy precipitate with clear supernatant liquid is graded + + + +, smaller precipitates are graded + + +, + +, etc.
‡Nitrogen determined by micro-Kjeldahl method, multiplied by the factor 6.25.

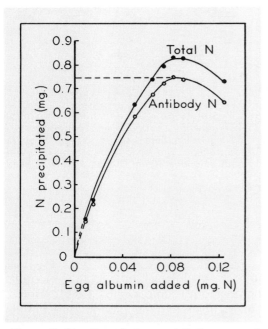

Figure 19-16.   *Quantitative precipitation curve obtained with egg albumin and rabbit anti-egg albumin serum. The antibody nitrogen data were obtained by subtracting egg albumin nitrogen from total nitrogen precipitated.*

Precipitation.   Solutions of proteins or certain polysaccharides are used in precipitation tests with appropriate antisera. The antigen is diluted serially, the antiserum is either undiluted or diluted no more than 1:10, because precipitation requires a very high concentration of antibody.

When simple mixtures of antigen dilutions and antiserum are incubated, maximum precipitate forms in one or two tubes, and the amount of precipitate diminishes in adjacent tubes (see Table 19-7). This *zone phenomenon* demonstrates that optimal pre-

cipitation requires a certain ratio of antigen to antibody. According to the framework hypothesis, precipitation occurs when a complex is formed having a structure similar to that shown in Figure 19-12A; multivalent antigen and bivalent antibody produce an aggregate that grows to large size. With excess of either reagent, complexes form in which all combining sites on only one reagent are occupied (see Figure 19-12B-E); these are more or less soluble.

Quantitative precipitation curves can be obtained by total nitrogen determination in the precipitates formed in simple mixtures of antigen dilutions and antibody. Typical results are shown in Figure 19-16. The amount of antibody in the serum is calculated by subtracting the amount of antigen added from the total amount of precipitate at the peak of the curve.

The *ring* or *interfacial test* is a common method of demonstrating precipitation. A layer of antigen dilution is carefully placed over the antiserum in a very small test tube or capillary. Diffusion of antigen and antibody establishes a layer of optimal proportions at or near the interface between the two liquids within a few minutes, and a fine line of precipitate appears (Fig. 19-17). The reciprocal of the greatest antigen dilution yielding a positive result indicates the sensitivity of the antiserum. This test is used to identify proteins in the medicolegal identification of blood stains and body secretions in murder and other criminal acts. Various foods including meats (horse meat in "hamburger," for example) can be identified if appropriate antisera are available.

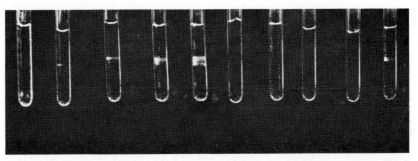

Figure 19-17.   *The results of a medicolegal precipitation test by the interfacial method. The five tubes on the left contain antihuman serum with upper layers of decreasing dilutions of an extract of a blood stain. There is no ring of precipitate in the first tube, but there are increasingly strong zones of precipitate in the next four tubes. The remaining tubes are controls; the last contains known human blood layered over antihuman serum. (From Boyd: Fundamentals of Immunology. New York, Interscience Publishers, Inc., 1956.)*

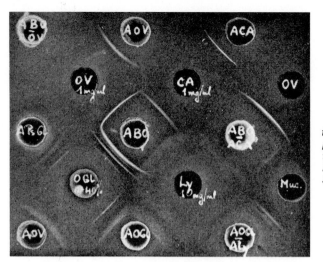

Figure 19–18. *Gel diffusion precipitation test in which antigens and antibodies are placed in alternate wells in an agar film. The top, middle, and bottom rows contained antibody solutions. (From Grabar, Ann. N. Y. Acad. Sci., 69:1957.)*

Gel diffusion precipitation methods utilize agar or other gels as media in which diffusion of antigen and antibody can occur under more controlled conditions than in liquid solutions. In the Ouchterlony technique, alternate wells cut in a layer of agar in a Petri dish or on a microscope slide are filled with antigen and antibody (see Figure 19–18). Lines of precipitate form where antigen and antibody meet in optimal proportions. This technique is useful in studying the antigenic composition of complex mixtures of proteins.

An estimate of the number of different antigenic components in a solution can be obtained when the antiserum used contains antibodies against each component, because separate lines of precipitate usually form owing to the differing rates of diffusion of the various reactants. Additional information about the composition of an antigenic mix-

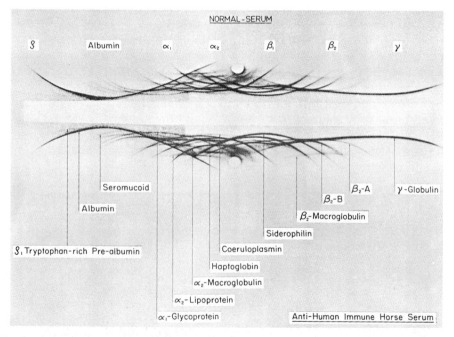

Figure 19–19. *Immunoelectrophoresis of normal human serum. The serum specimen was placed in the two circular wells, and after electrophoresis the horizontal central trough was filled with horse antihuman serum. Lines of precipitate formed where antibodies, diffusing from the trough, reacted with serologically equivalent concentrations of antigens as they diffused from the spots to which they had migrated during electrophoresis. (By courtesy of National Instrument Laboratories, Inc., Washington, D. C.)*

ture such as serum is provided by a combination of gel diffusion and electrophoresis. The results of immunoelectrophoretic analysis of a human serum are shown in Figure 19–19. A well in an agar layer is filled with the serum, which is then subjected to electrophoresis. This separates components of different electric charge. Troughs are cut parallel to the direction of migration and filled with homologous antiserum. Antibodies diffuse from the troughs, and at the same time the various antigens diffuse in all directions. As in the Ouchterlony test, arcs of precipitate form where the two reagents meet in optimal proportions. More than 30 different human proteins have been detected. Immunoelectrophoresis is used to detect pathologic proteins produced during the course of disease.

Toxin-Antitoxin Reactions. Neutralization of toxicity is determined by experiments with animals. Various quantities of exotoxin are mixed with a constant amount of antitoxin and injected after short incubation to permit chemical union. Neutralization is indicated by survival of the test animal (Table 19–8). The potency of the antitoxin is measured by the amount of toxin it neutralizes and is usually expressed in *units*. A unit is an arbitrary quantity contained in a standard preparation at the National Institutes of Health in Bethesda, Maryland, and at other laboratories throughout the world.

Inasmuch as exotoxins are composed of protein, they are also precipitated by homologous antisera. The Ramon flocculation test is set up in a manner similar to that of the simple mixture test, except that varying amounts of the antiserum (antitoxin) are mixed with a constant amount of the toxin. The endpoint of the titration is indicated by the mixture in the series that flocculates most rapidly. This mixture is usually neutral when tested by animal inoculation. Mixtures with less antitoxin are toxic. Flocculation with a known toxin provides a rapid *in vitro* assay of the potency of antitoxin, but animal tests are performed before human use is permitted.

The virulence (i.e., exotoxin producing power) of diphtheria cultures can be ascertained by a form of gel diffusion test. A strip of filter paper soaked in diphtheria antitoxin is placed in a Petri dish, covered with agar medium, and the agar is streaked at a right angle with the cultures under test. After incubation, lines of precipitate bisecting the angles between the bacterial growth and the filter paper indicate the culture that produced toxin (see Figure 19–20).

Lysis and Complement Fixation. Lysis or cytolysis may occur when cells that have been *sensitized* by union with their homologous antibodies react with complement. Bacteriolysis, the dissolving of bacteria, is demonstrated most satisfactorily with *Vibrio cholerae* and *S. typhi.* Hemolysis is the dissolution of erythrocytes; most theoretical studies of lysis have been carried out with hemolysis because the results can be detected readily by macroscopic inspection, whereas bacteriolysis must usually be determined by microscopic examination.

Complement may combine with sensitized antigens and produce no effect detectable by either microscopic or macroscopic examination. If the amount of complement employed in such a preparation is not excessive, the fact that it has combined with the sensitized antigen can be determined by adding another sensitized antigen that does give a visible reaction with complement. Sensitized erythrocytes are usually chosen as the indicator system. Egg albumin and anti-egg albumin serum, for example, mixed in proportions that yield no trace of precipitate, may nevertheless combine with complement to form a three-component complex (Fig. 19–21). Complement is "fixed" in this system and is unable to react with another

TABLE 19–8. Neutralization of Tetanus Toxin by Antitoxin, Tested by Intramuscular Inoculation of Mice

Dose Per Mouse		Death Time (Hours)
Toxin (M.L.D.)*	Antitoxin (Unit)	
200	0.1	Survived
325	0.1	Survived
500	0.1	47
800	0.1	22
1250	0.1	19
2500	0.1	<18

* 1 M.L.D. of tetanus toxin is defined as the least amount that will kill a 20 gm. mouse in 120 hours.

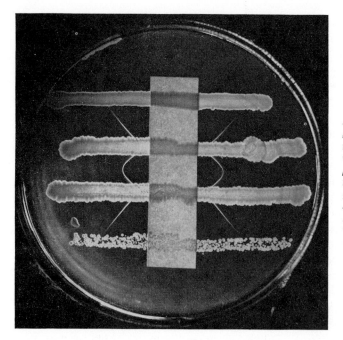

Figure 19–20. *In vitro test for virulence. Serum agar covers a strip of filter paper saturated with diphtheria antitoxin. The agar was then inoculated at a right angle to the paper with strains of Corynebacterium. A thin line of precipitate formed when antitoxin diffusing from the paper met toxin diffusing from the growth in proper concentration. (From specimens prepared by Miss Elizabeth O. King. Photo courtesy U. S. Public Health Service, Communicable Disease Center, Atlanta, Ga.)*

antigen-antibody system subsequently added, such as erythrocytes sensitized by their homologous antibody (hemolysin). Absence of either the egg albumin or its antibody yields a mixture in which complement is free and lyses the sensitized erythrocytes.

The complement fixation test is very sensitive and detects traces of either antigen or antibody, depending on the manner of setting up the test. It is used in the Wassermann test for syphilis, in which an antibodylike substance known as *reagin* is the unknown. The test antigen is a lipid, originally isolated from beef heart. Its chemical composition

and structure have been determined, and the pure compound, cardiolipin, is now employed.

Guinea pig serum is the usual source of complement. Fresh or lyophilized serum is employed. Its potency is determined each day, because complement is unstable and only a known minimal amount should be used in the test. The patient's serum is heated 30 minutes at 56° C. to inactivate normal complement. The Wassermann test is set up by mixing patient's serum, cardiolipin, and complement and incubating to allow cardiolipin to combine with reagin, if pres-

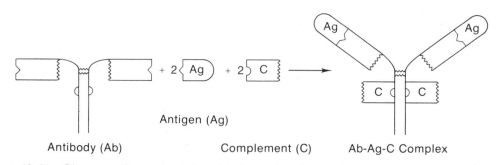

Antibody (Ab)   Antigen (Ag)   Complement (C)   Ab-Ag-C Complex

Figure 19–21. *Diagrammatic representation of the formation of a complex of an antigen such as egg albumin (Ag), its homologous antibody (Ab), and complement (C). Only after union with Ag has caused change in the shape of the Ab molecule can the latter combine with C. If Ag is absent, C is prevented from combining with Ab by steric hindrance. (From Carpenter: Immunology and Serology, 3rd ed. Philadelphia, W. B. Saunders Co., 1975.)*

ent, and for this complex to fix complement. At this stage there is no change in the appearance of the reaction mixture. To determine whether complement has reacted, sheep erythrocytes and antisheep antibody are added. Lack of hemolysis after further incubation indicates that all the complement has reacted with the first reagents; this is a positive test for syphilitic reagin in the patient's serum.

The Wassermann test is nonspecific in the sense that cardiolipin is not a component of the syphilis organism, but appears to be a product formed or liberated from damaged tissue during syphilis and also during certain other diseases. For this reason, false positive Wassermann tests occasionally occur. It would obviously be preferable to use the syphilis organism as the test antigen, but it cannot be cultivated *in vitro.*

In the *Treponema pallidum* immobilization test (TPI), a crude suspension of the bacteria is secured from rabbits inoculated intratesticularly. Tissue emulsion containing the bacteria is mixed with inactivated patient's serum and complement. After overnight incubation under anaerobic conditions, it is examined with the darkfield microscope; loss of motility occurs in the presence of syphilis antibody. The antibody detected in this test is not reagin; it is a true antitreponemal antibody. TPI antibody persists after recovery, whereas syphilitic reagin disappears from the serum when the patient is cured.

Phagocytosis. Phagocytosis *in vitro* can be demonstrated with *S. aureus* and human blood. Equal parts of whole blood, bacterial suspension, and antiserum are mixed and incubated at 37° C. for 30 minutes; a smear is then prepared and stained. The average number of bacteria ingested by the polymorphonuclear neutrophils is determined (Fig. 19–22). A measure of the opsonizing activity of immune serum can be obtained by comparing its behavior with that of normal serum. Opsonin is presumed to assist in recovery from certain infections.

Hypersensitivity Reactions. Anaphylaxis and allergy can be demonstrated only in a living animal or with living animal tissue.

Anaphylaxis. The guinea pig is the most

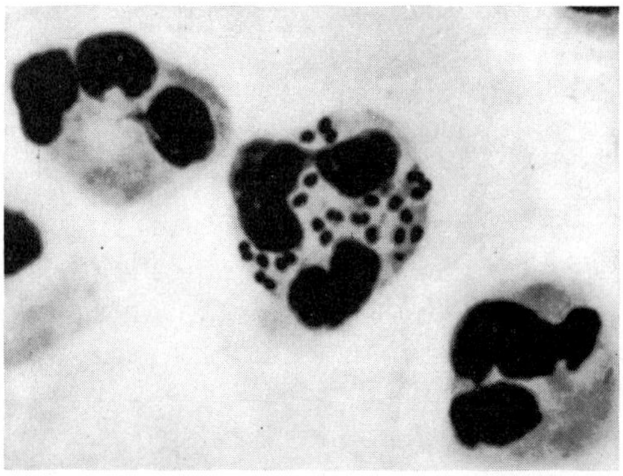

Figure 19–22. *Phagocytosis of* Neisseria *meningitidis. The center polymorphonuclear leukocyte has ingested nearly two dozen of the paired cocci (1200X). (From Ruch and Patton: Medical Physiology and Biophysics, 19th ed. Philadelphia, W. B. Saunders Co., 1965.)*

satisfactory experimental animal for demonstration of anaphylaxis, because it is easily sensitized and shocked. An animal is given a small intraperitoneal injection of a foreign protein such as horse serum or egg albumin. About three weeks later it is injected intravenously with a larger dose of the same antigen. Within a few moments the animal begins to cough, gasps for breath, has convulsions, and dies, often within one or two minutes after injection. A second guinea pig given *repeated* injections of the same antigen becomes "immune"; later reinjection of the antigen is harmless.

Production of anaphylactic shock in the guinea pig requires two injections of antigen separated by a period of about two weeks. The first or sensitizing injection incites production of only a small amount of antibody, and the animal becomes hypersensitive. Most of the antibody is attached to various body cells, and only a low concentration is in the blood.

The second or shocking dose two or three weeks later is larger than the first and is usually injected directly into the bloodstream, where it quickly encounters sensitive cells. The excess antigen reacts with the cellular antibody and damages the cells, which release histamine or a similar substance. The pharmacologic action of histamine in the guinea pig includes the symptoms described above. These symptoms are referable to the contraction of smooth muscle and to capillary dilatation, both induced by histamine.

Passive anaphylactic sensitization is achieved by introduction of serum from a hypersensitive animal or *small amounts* of antiserum from an immune animal. A few hours after injecting the serum, anaphylactic shock can be induced by intravenous injection of the antigen used to prepare the hypersensitive or immune serum.

Anaphylaxis can also be demonstrated *in vitro* by use of smooth muscle from a hypersensitive guinea pig. A strip of uterus or small intestine is suspended from a kymograph needle in a balanced isotonic solution, and antigen is added to the bath. Marked contraction follows within a few seconds. This is the Schultz-Dale test; it has been used to study the nature of antigenicity and the chemical radicals that determine the specificity of antigen-antibody reaction.

Allergy. Allergy is sometimes defined as hypersensitivity occurring naturally in man. There are many familiar manifestations of allergy. Allergy resembles anaphylaxis in many respects. Usually sensitization with the antigenic material precedes establishment of the hypersensitive state.

The antigen is encountered by inhalation, ingestion, or direct contact with the skin, and the type of reaction elicited depends largely on the nature of the antigenic experience. Inhaled substances commonly incite hay fever or asthmatic responses, ingested materials induce gastric distress, contact with poison ivy elicits the typical itching eruption. Urticaria and asthma induced by foods are attributed to antigenic substances in the ingested matter, which reach the skin and bronchi, respectively, via the bloodstream.

Allergy of infection is an example of a thymus-dependent (T cell) immunologic reaction. It is a *delayed hypersensitivity,* in contrast to anaphylaxis and most of the allergies just described, which are *immediate hypersensitivities,* attributable to circulating antibodies and hence to the GALT system. Hypersensitivity of the tuberculin type is of interest in connection with the problem of resistance to certain chronic infectious diseases. An individual infected with *Mycobacterium tuberculosis* becomes hypersensitive

to tuberculoproteins. Sensitivity is detected by the tuberculin test; an extract of *M. tuberculosis* injected into the skin incites local inflammation, and edema appears after several hours. The reaction progresses to a maximum between 15 and 48 hours and then slowly fades. A positive tuberculin test occurs in individuals with present or *past* tuberculous infection; it is therefore not necessarily diagnostic of active infection.

Similar delayed but violent inflammation follows injection of *M. tuberculosis* into a tuberculin positive animal. An extensive ulcer develops but soon becomes walled off and heals completely. The same type of response presumably occurs when a tuberculous human is reexposed to the infection and probably constitutes a protective reaction.

Transplantation Immunity. Transplantation of tissues from one site to another on the same individual has been practiced successfully for many years. The grafted tissue vascularizes and is accepted permanently. Skin grafts, for example, are commonly used in plastic surgery and in repair of burned areas. Attempts to transplant tissues from one animal to another, even of the same species, are usually unsuccessful. Although vascularization may begin, within a few days the blood circulation in the graft decreases, and necrosis and sloughing of the graft take place. This is *graft rejection.*

The fact that an immunologic mechanism is responsible for graft rejection is indicated by the fact that a second graft from the same donor is rejected much more rapidly. This is similar to the secondary immunologic response to antigen injection, and is called the *second set reaction.* However, graft rejection does not involve circulating antibodies. It is caused by T lymphocytes. Transfer of lymphoid cells from an individual who has rejected a transplant to a normal individual confers upon the recipient the ability to reject rapidly a primary transplant of the same kind.

There are four principal types of grafts, defined according to the relationship between the donor and the recipient. (1) An *autograft* is a transplant from one site to another of the same individual. It is readily accepted, as just described. (2) An *isograft* is a transplant from one individual to another genetically identical individual. These are either identical twins or highly inbred strains of animals. An isograft is normally accepted in the same manner as an autograft. (3) An *allograft* (Greek: *allos,* other) is a transplant from one individual to a genetically nonidentical individual of the same species. An allograft often vascularizes and looks healthy for a few days, but then necrosis sets in and the grafted tissue withers and is rejected. (4) A *heterograft* or *xenograft* (Greek: *xenos,* foreign) is a transplant from one species to another. Heterografts are obviously antigenic; they incite an immunologic reaction and are quickly rejected.

The antigens concerned in transplantation immunity are known as histocompatibility antigens; their formation is under genetic control. Histocompatibility genes have been studied most extensively in inbred mice and rats, in which more than a dozen independently segregating gene loci for histocompatibility antigens have been recognized. Multiple alleles are found at each locus. For example, at least 15 alleles have been identified at the H-2 locus, which determines the immunologically most potent histocompatibility antigen in the mouse. There are numerous transplantation antigens, and they occur in an almost infinite number of combinations. The situation in man is more complicated than that in rodents, and there is justification for the statement that each human is antigenically different in some respect from every other (with the exception of identical twins).

One general method used to increase the success of tissue transplantation is to destroy the immunologic mechanism responsible for graft rejection by use of radiations, anti-inflammatory drugs such as corticosteroids, and immunologic suppressants, which include alkylating agents, antimetabolites, and antilymphocyte serum. Most of these agents damage or destroy the immunologic mechanism of the recipient to such an extent that he becomes highly susceptible to microbial infection. He is likely to succumb to disease caused by endogenous microorganisms or normally harmless organisms from the environment.

There are differences in susceptibility of transplanted tissues to allograft rejection. Considerable success has been obtained with kidney transplants, but even in this case identical twins or other closely related individuals are selected as donors. Some authorities believe that donor selection offers the best hope for success in transplantation of organs. The histocompatibility antigens in leukocytes and platelets can be typed in somewhat the same way as erythrocytes, and proper matching of donor and recipient is necessary.

A phenomenon related to graft rejection occurs when immunologically competent adult lymphoid cells are injected into a newborn (i.e., immunologically immature) animal of the same species. If the donor is genetically different from the recipient, a syndrome known as *runting* or the *graft-versus-host* reaction occurs. The inoculated animal fails to gain weight normally and develops diarrhea and skin lesions and dies after a few weeks. This response is attributed to reaction of the transferred cells with histocompatibility antigens of the recipient. It is not graft rejection, because the host animal is immunologically immature. A similar phenomenon can be demonstrated in adult animals whose lymphoid cells have been destroyed by x-irradiation. This is called *secondary* or *homologous disease.*

SEROLOGIC DIAGNOSIS OF INFECTIOUS DISEASE

Serologic reactions are important aids in the diagnosis of infectious disease. Antibodies can be detected in the patient's serum, and microorganisms isolated from clinical specimens can be quickly identified by appropriate procedures.

Detection of Antibodies in Patient's Serum. Antibodies usually appear one to two weeks after an infection begins and increase in concentration during several weeks. Laboratory diagnosis by detection of developing antibody is therefore possible in diseases that run a prolonged course (e.g., typhoid and paratyphoid fevers, brucellosis, syphilis).

An appropriate serologic test is performed with a sample of the patient's serum and test antigens consisting of known organisms. It may be suspected, for example, that a patient has one of the enteric fevers. Agglutination tests are prepared by mixing the patient's serum with antigens consisting of suspensions of the typhoid and various paratyphoid bacteria. If the disease is actually typhoid fever, the typhoid bacteria should be agglutinated by high dilutions of the serum.

A high antibody titer usually indicates active infection if the patient's history reveals no previous infection or artificial immunization with the same organism. A second serum specimen should be secured a week or more later, and if this contains a higher titer of antibodies than the first it is almost certain that the patient is infected with the corresponding organism. Tests with several different antigens are sometimes necessary to determine the cause of disease. Laboratories can prepare their own bacterial test antigens by suspending the organisms from agar slants or broth in saline; certain antigens can be secured from commercial manufacturers.

Agglutination tests are commonly employed to detect bacterial antibodies; phagocytosis tests are also used to a limited extent. Complement fixation is valuable in studying sera of patients with viral disease, and also in the Wassermann test for syphilis.

Identification of Microorganisms. Pure cultures of organisms isolated from patients by the usual methods can be identified by serologic tests with known antisera, usually prepared in rabbits. Serologic procedures are more rapid than most routine cultural and biochemical tests used for identification. They are used for a quick diagnosis to direct the physician toward the treatment most likely to be beneficial, but their results should be confirmed by the other customary tests.

Rapid detection and identification of bacteria in clinical material can be accomplished by use of fluorescent antibodies. This technique was described in Chapter 3 (page 48). It is valuable in laboratory diagnostic work because of its speed and simplicity. Modifications are also used in research to determine the localization of viruses and other antigens in tissues.

SUPPLEMENTARY READING

Abramoff, P., and LaVia, M.: *Biology of the Immune Response.* New York, McGraw-Hill Book Company, 1970.

Alexander, J. W., and Good, R. A.: *Immunobiology for Surgeons.* Philadelphia, W. B. Saunders Company, 1970.

Barrett, J. T.: *Textbook of Immunology.* St. Louis, The C. V. Mosby Company, 1974.

Bigley, N. J.: *Immunologic Fundamentals.* Chicago, Year Book Medical Publishers, Inc., 1975.

Boyd, W. C.: *Fundamentals of Immunology,* 4th ed. New York, John Wiley & Sons, Inc., 1967.

Burnet, F. M.: *Cellular Immunology.* Melbourne, Melbourne University Press, 1969.

Burnet, F. M.: *Immunological Surveillance.* Oxford, Pergamon Press, 1970.

Burrows, W.: *Textbook of Microbiology,* 20th ed. Philadelphia, W. B. Saunders Company, 1973.

Carpenter, P. L.: *Immunology and Serology,* 3rd ed. Philadelphia, W. B. Saunders Company, 1975.

Davis, B. D., Dulbecco, R., Eisen, H. N., Ginsberg, H. S., and Wood, W. B., Jr.: *Microbiology,* 2nd ed. Hagerstown, Md., Harper & Row, 1973.

Glasser, R. J.: *The Body Is the Hero.* New York, Random House, Inc., 1976.

Herbert, W. J., and Wilkinson, P. C.: *A Dictionary of Immunology.* Oxford, Blackwell Scientific Publications, 1971.

Joklik, W. K., and Smith, D. T.: *Zinsser Microbiology,* 15th ed. New York, Appleton-Century-Crofts, Inc., 1972.

Lewin, R.: *In Defense of the Body.* Garden City, N.Y., Doubleday, 1974.

Rose, N. R., and Friedman, H. (eds.): *Manual of Clinical Immunology.* Washington, D.C., American Society for Microbiology, 1975.

Weir, D. M.: *Immunology for Undergraduates,* 3rd ed. Edinburgh, Churchill Livingstone, 1973.

Wilson, G. S., and Miles, A. A.: *Topley and Wilson's Principles of Bacteriology, Virology, and Immunity,* 6th ed. Baltimore, The Williams & Wilkins Co., 1975.

DISEASES 20
TRANSMITTED BY
DIRECT CONTACT

Microorganisms whose multiplication requires nutrients or conditions obtainable only in a living host and whose resistance to environmental hazards is low are usually transmitted from host to host by the shortest possible pathway. The success of physical contact as a mode of transmission is attested by the incidence of venereal disease. Other agents with nutrient requirements that restrict them to a cutaneous habitat are passed readily from hand to hand. The close personal contact that is inescapable in classrooms, theaters, supermarkets, buses, and other features of modern urban life permits transfer of viable pathogens from one respiratory tract to another.

VENEREAL DISEASES

Venereal diseases are the most notorious representatives of those that are transmitted by direct contact. The U.S. Public Health Service estimated that there were 1.5 million cases of gonorrhea in the United States in 1969—about one per 130 persons—and 70,000 cases of syphilis. Most of them were in the 20 to 24 age group, followed by the 15 to 19 and 25 to 29 year age groups. Until the end of 1975, the number of new cases of gonorrhea increased at an annual rate of about 10 per cent. In that year, more than 1 million new cases were reported in the United States.

GONORRHEA

Neisseria gonorrhoeae, the cause of gonorrhea, is transmitted almost solely by sexual contact. This gram-negative diplococcus is cultivable *in vitro* only at 37° C. in a moist atmosphere containing 5 to 10 per cent CO_2 and on a rich medium such as chocolate agar. However, it multiplies rapidly *in vivo* in the subepithelial tissue fluids of man.

Following infection of the mucous membranes of the urogenital tract, the organisms pass between columnar epithelial cells into deeper areas, where their multiplication incites an acute inflammatory response that leads to the characteristic purulent urethral or vaginal discharge of early gonorrhea. In the male, this may extend to the prostate and epididymis, while in the female the infection spreads from the urethra, vagina, and cervix to the fallopian tubes. Unlike the situation in men, the early stages of gonorrhea in women may be symptomless. Chronic infection leading to sterility is not uncommon.

The pathogenicity of *N. gonorrhoeae* is associated with the antiphagocytic properties of the polysaccharide "K" antigen of smooth forms, which is present in fresh isolates and is lost rapidly on laboratory media as the organism changes to the rough phase. This antigen is part of the envelope of gonococci, which lack capsules. Gonococci also possess an endotoxin, but its function is not clearly defined.

Immunity to gonorrhea is weak, if it exists at all, and repeated attacks are common. Even patients with chronic gonorrhea may be superinfected with the acute disease.

Treatment is not difficult since the introduction of antibiotic therapy. Although strains moderately resistant to penicillin arise, they can be successfully combated by vigorous treatment. Therapy is easier in the male than in the female. Prophylactic use of penicillin during the two to five day incubation period usually prevents appearance of disease symptoms.

Strains of *N. gonorrhoeae* highly resistant to penicillin appeared in the United States in 1976. Of 12 cases reported between February and September, 11 were apparently imported from the Far East. In the same year, 40 cases caused by penicillinase-producing strains were detected in Liverpool, England. The worldwide significance of these organisms is not yet clear. Very intensive therapy with a variety of antibiotics is required to cure the patients.

SYPHILIS

Syphilis first came to general attention as a highly dangerous epidemic about 1495 A.D. in Italy, France, and Spain. The disease became modified during the intervening centuries and is not as rapidly fatal as it was formerly, although it is fully as contagious.

Treponema pallidum, like *N. gonorrhoeae,* is transmitted almost solely by sexual contact. The long, very slender, flexuous, motile spirochetes are best detected in syphilitic chancre fluid by darkfield or fluorescent antibody techniques.

It is doubtful that virulent *T. pallidum* has ever been cultivated *in vitro,* and it does not grow in chick embryos or tissue cultures. In man, monkeys, and rabbits the minimum generation time is about 30 hours. The organism is destroyed quickly by drying and is killed at 41.5° C. in one hour. It may retain viability in tissues from infected animals for as long as five days, and suspensions in the most favorable medium devised, containing serum components, pyruvate, and reducing agents, and maintained under anaerobic conditions, survive only one or two days at 37° C. or four to seven days at 25° C.

Following transmission by sexual contact, the organisms rapidly penetrate mucous membranes or minute breaks in the skin. They often remain localized, but invasion of the lymph nodes and bloodstream may occur within five minutes to several hours. The bacteria multiply extensively in the local site for 10 to 90 days (average, 21 days) before the chancre of the primary stage appears.

This painless, eroded sore or ulcer, containing spirochetes, heals slowly in four to six weeks. The secondary stage appears four to eight weeks later and consists of a skin rash and mucous patches on the lips, mouth, and genitalia. These lesions contain the organisms, and there may also be involvement of the eyes, bones, and central nervous system. Symptoms disappear in a few days to several months. A latent period of months or years follows, with no outward signs but with slowly progressive degeneration and chronic inflammation of cardiovascular, central nervous system, and other tissues. Approximately one case in four undergoes spontaneous cure, another remains latent permanently, and the others progress to the tertiary stage, characterized by explosive breakdown of tissues. Any organ in the body may be affected: heart, vascular, bone, brain. Paresis and tabes dorsalis, with generalized paralysis and locomotor ataxia, respectively, are common evidences of tertiary syphilis.

There are two principal types of serologic test for evidence of syphilitic infection: those based on detection of syphilitic reagin, and those based on detection of antitreponemal antibody. The Wassermann and TPI tests were described briefly in Chapter 19 (see page 383).

In view of the fact that the Wassermann test is not absolutely accurate and requires 15 to 18 hours to complete, rapid flocculation procedures were developed. Emulsions of cardiolipin and lecithin or cholesterol in saline flocculate when mixed with serum containing syphilitic reagin. The Kahn, Eagle, and other macroscopic or microscopic methods require only minutes, and are about as accurate as complement fixation methods.

Treponemal antibody can be detected by the *Fluorescent Treponemal Antibody* (FTA) test. A smear of *T. pallidum* cells (i.e., infected rabbit testicular emulsion) is treated with patient's serum and then with fluorescent antihuman gamma globulin. If treponemal antibody is present in the serum, the cells fluoresce when illuminated with ultraviolet light and show distinctly in the dark field. Wassermann reagin disappears upon clinical cure, whether spontaneous or as a result of therapy, but treponemal antibody persists.

Penicillin has replaced the earlier agents —bismuth, mercurials, and arsenicals— for treating syphilis. One to two weeks of treatment is required, because penicillin is effective only against actively multiplying bacteria, and *T. pallidum* multiplies very slowly.

LYMPHOGRANULOMA VENEREUM

Lymphogranuloma venereum is caused by certain strains of *Chlamydia trachomatis,* bacterialike organisms included in the Rickettsias. They are slightly smaller than other rickettsiae; they are filterable, obligate intracellular parasites, possess a cellular rather than a viral type of structure and contain both DNA and RNA. They are classified in a separate order, Chlamydiales (Greek: *chlamys,* cloak). The term chlamydia is often applied to all agents within the family, and other strains cause trachoma, inclusion conjunctivitis, and pneumonitis in numerous animal species; *C. psittaci* is the cause of psittacosis (parrot fever).

The various chlamydiae are morphologically indistinguishable and resemble rickettsiae in being gram-negative, stainable with basic dyes, and barely visible by light microscopy. They go through a fairly complicated multiplication cycle within host cells, as illustrated in Figure 23–2 (see page 423). Dense, spherical particles, 0.2 to 0.3 μm in diameter, penetrate the host cell by an unknown mechanism, and these particles slowly reorganize in internal structure for about 10 hours, enlarging to form "initial bodies," 0.5 to 0.7 μm in diameter, which are less dense and less infectious than the original spherical particles. Between 20 and 25 hours after infection, the particles divide logarithmically, and the large forms gradually differentiate to the small, dense, highly infectious forms, which are ultimately released when the cell bursts.

Chlamydiae live intracellularly, but there is evidence that they can synthesize some amino acids and coenzymes, including arginine, histidine, lysine, glutamine, folic acid, and riboflavin. They possess little, if any, oxidative activity and rely upon the host cell for ATP as a source of energy, coenzyme A, NAD and other coenzymes.

The mechanism of pathogenicity of chlamydiae is unknown. Large intravenous inocula produce fatal shock in mice, with extensive fluid exudation and hemorrhage, which can be prevented by active or passive immunization but not by antibiotics. This suggests an antigenic endotoxin, but the active substance has not been isolated. The protective antibodies that arise during infection or active immunization are actually directed against protein antigens.

Lymphogranuloma venereum occurs naturally only in man, in whom it is transmitted by sexual intercourse. The agent is cultivated in the yolk sac of chick embryos for preparation of diagnostic test antigens.

After an incubation period of 7 to 12 days, the first sign of infection is a painless vesicle on the genitals, which ruptures and heals without scarring. The chlamydiae spread to the regional lymph nodes, which enlarge and become tender and may suppurate one week to two months later. These enlarged, granulomatous nodes are called buboes. As they heal, scar tissue forms and may later obstruct lymph channels and cause edema in the genital area. Perirectal scarring and rectal obstruction sometimes follow in the female.

Clinical diagnosis is aided by a delayed hypersensitivity reaction, the Frei test. Material from infected yolk sac cultures, marketed under the name Lygranum, is infected intradermally, and an erythematous papule greater than 6 mm. in diameter and appearing within 24 to 48 hours indicates present or previous infection. A complement fixation test for antibodies in patients' sera is also available.

Antibiotic therapy with tetracyclines is effective in acute stages but is of little effect when late perirectal scarring has occurred.

CUTANEOUS AND SUPERFICIAL MYCOSES

Fungus infections of the skin, hair, and nails are very common. They are caused by the *dermatophytes*—approximately 30 fungi out of a total of about 55 that are commonly considered pathogenic for man; the others produce deep-seated or systemic infections. Skin lesions produced by dermatophytes are usually more or less circular, with raised borders, and they tend to enlarge equally in all directions. It was early thought that they were caused by worms or lice, so the diseases were called *ringworm* or *tinea* (Latin: worm), further qualified by the area of skin involved (e.g., *tinea capitis*, ringworm of the scalp; *tinea pedis*, athlete's foot).

Two types of hair invasion are seen: (1) in *endothrix* infections the hyphae grow only within the hair, and (2) in *ectothrix* infections the growth is both within and on the surface of the hair shaft.

Most dermatophytes are parasitic, being found principally on or in keratinized tissues of man and animals. Although they have a predilection for keratin—an insoluble, fibrous, highly stable protein—most species hydrolyze it only slowly and grow well *in vitro* in its absence.

There are three pincipal genera of dermatophytes: *Microsporum* species attack hair and skin, *Trichophyton* species infect hair, skin, and nails, and *Epidermophyton* infects skin and nails. The nature of the infection depends on the species of fungus and the location of the infected area.

DERMATOPHYTE INFECTIONS

Tinea capitis, affecting the scalp and hair, is caused by species of *Microsporum* and *Trichophyton.* It varies from small, itching, scaling papules that spread peripherally and form patches in which the hairs break off a few millimeters above the scalp to chronic infections that develop into large, raised, cup-shaped crusts, which may ultimately cover the scalp and cause permanent baldness. Certain species produce a violent inflammation and pustular abscesses around the hair follicles. Some of the fungi are highly contagious, occurring in epidemics spread directly or via hair clippers and other fomites; some are transmitted from pets, such as puppies and kittens (see Plate III*D*).

Tinea corporis, or ringworm of the smooth skin, is also caused by species of *Microsporum* and *Trichophyton.* Lesions range from simple scales to deep granulomata. The typical lesion is ring-shaped with a scaly center that tends to heal, surrounded by an advancing circle of vesicles and papules. The lesions may itch, particularly in warm weather.

Tinea barbae, "barber's itch," involves the bearded area of the face and neck, and is also caused by species of *Trichophyton* and, more rarely, *Microsporum.* Dark red, boggy, pustular abscesses form near the angle of the jaw, and the hairs in this area become brittle and lack luster.

Tinea cruris is an itching, scaly dermatitis of the groin, perineum, and perianal region; it is caused most frequently by *E. floccosum.*

Tinea pedis, or athlete's foot, is the most common fungus infection. *T. mentagrophytes* causes the majority of cases, producing an inflammatory infection with vesicles between the toes and on the soles. The skin peels and fissures form in infected areas. The infection may spread to the hands, thighs, axillae, and other regions, and secondary pyogenic infection sometimes occurs.

Tinea unguium, sometimes called onychomycosis or ringworm of the nails, is caused by various species of *Trichophyton* or by *E. floccosum.* The nails are opaque and brittle, lose luster, become distorted and thickened, and separate from the nail bed.

Diagnosis. Diagnosis of fungus infections is aided by microscopic examination of material from lesions rendered transparent by warming with 10 per cent NaOH or KOH, by cultivation on Sabouraud dextrose agar, followed by macroscopic and microscopic study of the colonies, and by examination of the infected areas with ultraviolet light, which produces characteristic fluorescence with certain fungi.

Treatment. Vigorous treatment is usually necessary to cure fungus infections. Some yield to prolonged local treatment with undecylenic acid or other fatty acids, thymol, salicylic acid, iodine, various ointments, and other antifungal agents. Oral administration of large doses of the antibiotic griseofulvin over periods of weeks or months has given good results. It is deposited in growing keratinized tissues and renders them insusceptible to fungal invasion. It does not kill fungi already within the tissues but prevents their spread to newly formed tissues. This fact explains why prolonged therapy is necessary.

UPPER RESPIRATORY INFECTIONS

Upper respiratory infections caused by highly parasitic or delicate microorganisms whose survival depends on rapid transfer between hosts are included among the diseases transmitted by direct contact. Those caused by organisms of greater resistance and usually disseminated broadly are classified as airborne diseases.

PERTUSSIS

Whooping cough is caused by *Bordetella pertussis,* a small, gram-negative, nonmotile, rod-shaped bacterium with a tendency to be pleomorphic in older cultures. It grows slowly on Bordet-Gengou agar, a rich medium containing glycerol extract of potato and 15 to 25 per cent fresh sheep blood, and produces dome-shaped "bisected pearl" colonies in two to three days. Blood *per se* is not essential, but its proteins combine with toxic fatty acids, which otherwise inhibit growth of the bacteria; charcoal has been used successfully instead of blood. *B. pertussis* survives poorly outside the body because of its sensitivity to drying and to ultraviolet light.

Virulent, fresh isolates are encapsulated and possess several antigens: a thermolabile protein exotoxin of low potency that appears to be dermonecrotic, a thermostable cell wall lipopolysaccharide with the properties of an endotoxin, and a "protective antigen" that immunizes mice against experimental infection; the third antigen is lost during laboratory cultivation.

Following infection by inhalation of infective respiratory mucus or by fomites or hand-to-hand transmission, the organisms localize on the ciliated epithelium of the trachea and bronchi, where they multiply rapidly but do not penetrate the mucosa. Their toxic products irritate the subepithelial tissue, causing necrosis and inflammation, and the mass of capsulated bacteria interferes with ciliary action. After an incubation period of 7 to 16 days, the first, or catarrhal, stage of the disease appears. It starts with a mild cough, which becomes progressively worse during the next 10 to 14 days. This is the most highly contagious period. The irritation and inflammation become more pronounced, and the paroxysmal stage of two to six weeks follows. In severe cases there may be vomiting, cyanosis, and convulsions. The bacteria usually disappear by the fourth to fifth week of disease. Convalescence requires two or more weeks. Complications, which are responsible for a large percentage of the deaths, include bronchopneumonia or lobar pneumonia, otitis media, and malnutrition.

Antibodies detectable by agglutination and complement fixation appear about the third or fourth week. Leukocytosis with marked increase of lymphocytes occurs early. Laboratory diagnosis is aided by use of a "cough plate" (i.e., Bordet-Gengou agar exposed to several "whoops") or preferably by a nasopharyngeal swab, which is passed through a drop of penicillin (to inhibit other bacteria) before being streaked on a plate.

Recovery is usually followed by permanent immunity. Active immunization is also accomplished by use of killed bacteria containing the "protective antigen." Immunity is not absolute, but surveys indicate approximately 90 per cent reduction in incidence and considerable decrease in severity. Inasmuch as the greatest mortality occurs in infants, immunization should be practiced at the age of three months, and booster injections are recommended after one, three, and five years.

INFLUENZA

Epidemics of influenza have occurred for centuries, but the causative agent was not isolated until 1933, when Smith, Andrewes, and Laidlaw found that filtered, bacteria-free nasal washings from patients produced a characteristic febrile illness in ferrets inoculated intranasally.

Three major types of influenza virus—A, B, and C—have been identified, and there are several subtypes of A and B. These viruses vary somewhat in size, but they are generally spherical or ovoid. Their entire surface is covered with radial rods or spikes, about 100 Å in length, beneath which is a lipid-containing envelope 60 to 100 Å thick, surrounding a helical structure, the nucleocapsid. The spikes consist of glycoprotein and serve in the intact virus particle as a hemagglutinin; that is, they attach to specific receptor sites of erythrocytes of many mammalian and avian species, whereupon agglutination occurs. The nucleocapsid contains single-stranded RNA with a molecular weight of 2.5 to 2.8 million daltons.

The differences between the types and subtypes of influenza viruses are based upon the serologic properties of the ribonucleoprotein and surface glycoprotein (hemagglutinin) antigens. Antibodies against the latter also neutralize the infectivity of the virus particle.

After the virus is inhaled in infectious droplets of respiratory secretions, it replicates during the one to two day incubation period in susceptible cells lining the upper respiratory tract, trachea, and bronchi. The invaded cells are destroyed, and epithelial desquamation causes the respiratory symptoms characteristic of the acute disease. The general symptoms—fever, chills, muscular ache, headache, prostration, and loss of appetite—are difficult to explain since there is ordinarily no viremia; that is, the virus remains localized in the respiratory tract. The virus does contain pyrogenic and other mildly toxic components, but it is thought that the constitutional effects are due to tissue breakdown products.

Influenza is normally self-limited and lasts three to seven days. In some cases there are small areas of pulmonary consolidation, and deaths attributed to primary influenzal pneumonia occur mainly in persons whose respiratory function is impaired by other illness (e.g., chronic pulmonary disease) or by pregnancy. Most deaths are associated with secondary bacterial pneumonia caused by *S. aureus, H. influenzae,* hemolytic streptococci, or pneumococci, which behave as opportunists, invading pulmonary tissues damaged by the influenza virus.

Either frank or subclinical influenza confers immunity, but it is only of short duration. The immunity is directed against the surface antigens and hence is subtype specific. It is of interest that whereas the protective function of the immunity is lost within one to two years, antihemagglutinin can be detected for many years, perhaps throughout the remainder of life. By this means the viruses that caused epidemics of influenza prior to 1933 have been identified. For example, when Asian influenza appeared in 1957, sera from numerous persons 80 to 90 years of age were found to contain antibodies that reacted with the virus (A2) isolated. Inasmuch as the greatest incidence of influenza is usually in the 5 to 15 year age group, it was deduced that the 1889–1890 epidemic was caused by a virus similar to A2.

Artificial immunization against influenza is only partially effective. Vaccines prepared by formalin inactivation of viruses propagated in chick embryos induce immunity that persists for about 6 months. The virus must be serologically identical with, or at least closely related to, the prevalent strain. The known history of influenza viruses indicates a constant "antigenic drift," which is of great practical importance. The influenza A virus, isolated in 1933, caused disease until 1943, but in 1947 a different subtype, A1, supplanted the previous A strain. A2 appeared 10 years later and became prevalent. Inasmuch as relatively little cross-immunity is engendered by heterologous vaccine, it is necessary to know the type of virus against which protection is required. National and international health agencies keep close watch over outbreaks and attempt to forecast epidemics in time to permit immunization of those most susceptible to fatal influenza: the aged, pregnant, and those with cardiac disease.

As noted previously (see p. 352), the virus isolated in the spring of 1976 from recruits at Fort Dix, New Jersey, appeared to be closely related to the swine influenza strain implicated in the pandemic of 1918–1919, in which a very high mortality rate prevailed. It is difficult, nearly 60 years later, to say what percentage of deaths was actually caused by the virus and what was due to concurrent infection with other respiratory agents and the stress associated with World War I. Some United States authorities deemed it advisable, however, to immunize as many persons as possible, especially among the group most susceptible to fatal respiratory infection, before the next year. An inactivated virus vaccine was administered to large numbers of the elderly, chronically ill, and others. As this is written (November, 1976), it seems likely that the effectiveness of the immunization program will not be known unless cases of swine influenza occur among non-immunized individuals during the normal 1976–1977 influenza season.

COMMON COLD

The common cold, defined as an acute, afebrile, upper respiratory disease, is the most frequent affliction of man. Although nonfatal, it causes loss of more than 200,000,000 man-days of work and school in the United States alone. The average annual incidence is one to two colds per person, but some individuals have five or six each year.

The cause of the common cold is a virus, and more than 75 different viruses have been isolated from patients. Most of these are assigned on the basis of physical and chemical characteristics to the rhinovirus subgroup (Greek: *rhino,* nose) of the picornaviruses, but symptoms of the common cold are also produced by strains of Coxsackie virus.

Man is the only known host for the rhinoviruses, although chimpanzees in captivity acquire the infection from humans, and gibbons can be infected. No laboratory animal is susceptible, and the viruses have been propagated only in tissue cultures of human and monkey cell lines. Virus can be found in the nose and throat just before symptoms appear and for one to six days thereafter.

The average incubation period following intranasal inoculation of human volunteers is about two days. The nasal membranes become swollen, secrete mucus excessively, and the paranasal sinuses may also be involved. The nasopharynx and the larynx and trachea become inflamed. In a mild cold, there is nasal stuffiness, sneezing, mild headache, slight sore throat, chilliness, and general malaise. Symptoms increase for a day or two and usually subside and disappear in three or four days. The cough, in more severe cases, is usually due to irritating postnasal discharge.

Immunity lasting two years or more may follow recovery. However, repeated attacks are common, and these are attributed in part to the failure of some individuals to develop antibodies to viruses that attack only superficial mucous membranes and in part to the multiplicity of immunologically different viruses that cause colds. The chief problem in artificial immunization of man seems to be the latter aspect.

There are two views of the epidemiology of the common cold. According to the first, the virus is transmitted more or less directly from person to person in discharges from the upper respiratory tract. However, experiments with volunteers indicate a lower transmission rate than would be anticipated, and distribution within family groups is often less than expected. The observation that there are often widespread outbreaks immediately following a sudden cold wave, especially in the Fall, suggests that many individuals carry latent virus, which is activated by climatic changes. These individuals then serve as sources for person-to-person transfer.

MEASLES (RUBEOLA)

Measles is one of the most highly contagious diseases, and almost all children contract it. Since recovery is followed by virtually permanent immunity, measles is almost solely a childhood disease. The cause is a virus, which occurs in only one immunologic form. It is found naturally in man and monkeys but can be cultivated *in vitro* in a variety of mammalian and chick embryo cells.

The virus is transmitted in respiratory secretions. It enters the upper respiratory tract and perhaps the eyes, and multiplies in the epithelium and regional lymphatic tissue. Early symptoms, which appear after an incubation period of 10 to 12 days, include coryza, conjunctivitis, sore throat with dry cough, headache, low-grade fever, and Koplik spots (small red patches with central white specks on the mucosa of the inside of the cheek). Toward the end of the incubation period, the virus enters the bloodstream and is distributed to the skin and to other lymphatic tissues. The characteristic maculopapular (elevated discolored spots) skin rash appears, first on the head and face and then on the trunk and limbs.

Virus is present in the upper respiratory and eye secretions prior to the rash and for two days after it develops. Inasmuch as the disease may not be diagnosed before this time, early excretion of the virus favors its epidemic spread.

Frequent complications include bronchopneumonia and otitis media, with or without secondary bacterial infection. Much more serious, but fortunately much less common, is encephalomyelitis, appearing 5 to 7 days after the rash in about one case in 10,000 in epidemics.

In the United States measles epidemics occur in two to three year cycles as susceptible populations of children accumulate. The highest incidence is in the five to seven year age group, in whom the disease is ordinarily mild. It is more severe in very young children and in adults.

Pooled human γ-globulin contains measles antibodies, and administration to exposed susceptible individuals early in the incubation period prevents the disease or diminishes its severity. Attenuated virus vaccines and formalin inactivated vaccines are used to induce active immunity. The antibody titers produced are not as great as those developed during an actual case, but they protect against moderate exposure.

GERMAN MEASLES (RUBELLA)

German measles resembles measles but is milder and not as serious in the very young. Its principal significance is the fact that women who have German measles during the first three months of pregnancy frequently give birth to babies with severe congenital defects.

German measles is highly contagious and, like measles, is spread by nasal secretions. However, infection is often inapparent, so patients are frequently not isolated and hence transmission of the virus is not interrupted. Following exposure, there is an incubation period of 14 to 25 days before the rash appears, and the virus may be present in the nasopharyngeal secretions for a week before and a week after appearance of the rash. First symptoms include fever, mild coryza, and malaise a day or two preceding the rash. Just before and after the rash appears, the virus is disseminated in the blood.

A fetus infected during the first trimester may be stillborn, or if it survives it may be deaf, have cataracts, heart abnormalities, microcephaly, or other congenital defects. Occurrence of structural abnormality varies between 12 and 83 per cent in different series of observations. At birth and for weeks or months thereafter the nasopharyngeal secretions of these infants contain the virus.

Fortunately there is only a single immunologic type of rubella virus, and a vaccine has been developed that induces satisfactory active immunity. Health authorities recommend immunization of all school children in order to reduce the likelihood that pregnant women will become infected.

PYOGENIC COCCAL INFECTIONS

The pyogenic or pus producing cocci include three major genera: *Staphylococcus*, *Streptococcus*, and *Neisseria*. They characteristically incite a reaction in which leukocytes and other phagocytic cells accumulate, along with plasma proteins, fluid, red blood cells, and other debris. This is known as pus. Although, as pointed out in Chapter 17, representatives of these genera comprise part of the natural flora of most individuals and are harmless in their normal location or under customary conditions, they may produce disease in other locations or under abnormal conditions or in a recipient of lowered resistance.

STAPHYLOCOCCI

The principal pathogen among the staphylococci is *S. aureus*. Pathogenic strains are usually coagulase-positive, hemolytic, and ferment mannitol. Staphylococci are catalase-positive.

Coagulase production is detected by rapid slide or tube tests. The organisms are mixed with undiluted oxalated human or rabbit plasma on a slide, or with diluted plasma in a small test tube. Positive results consist of rapid clumping on the slide or any degree of clotting in the tube within three hours. A positive slide test is attributed to coagulase bound to the cells, whereas a positive tube test is due to either bound or free (extracellular) coagulase. The role of coagulase in pathogenicity is not clear.

The virulence of *S. aureus* is attributed in large measure to its ability to resist or survive phagocytosis. Virulent strains possess one or more antiphagocytic surface antigens, and some produce capsules that inhibit phagocytosis. The α-toxin of *S. aureus* has several properties that enhance virulence. This protein kills phagocytic cells, constricts smooth muscles and paralyzes blood vessel walls, is dermonecrotic, and in sufficient dosage it is lethal to experimental animals. Another virulence factor, especially in chronic or recurrent infections, is the delayed hypersensitivity response of the host, which enhances tissue necrosis.

The most significant factor in disease production by *S. aureus* is probably reduced host defense. The newborn baby is promptly colonized by this organism, and staphylococcal infections are common during childhood and puberty. Adults therefore have a considerable degree of immunity. *S. aureus* infections in adults are usually mild, except when the normal antibacterial defenses have been reduced by accidental or surgical trauma, burns, and chronic debilitating diseases such as diabetes, cancer, and cirrhosis of the liver.

Formation of a fibrin barrier is a typical feature of local staphylococcal infections such as boils. *S. aureus* multiplies in the subcutaneous tissues (e.g., in or near a hair follicle or sebaceous gland) and destroys body cells in an area several millimeters in diameter and bounded by a fibrin clot and filled with pus, including phagocytized and unphagocytized bacteria. Under favorable circumstances, phagocytosis continues and the infection terminates and heals spontaneously. The pus is replaced by granulation tissue and a scar remains. If, through accident or premature amateur surgery, the fibrin wall is broken, the organisms enter the lymphatic channels or blood vessels and spread to other parts of the body. The general bloodstream infection or septicemia that follows is frequently fatal.

The organisms may also reach the kidneys, liver, or other organs and produce abscesses. Inflammation of the heart lining is known as endocarditis. If the membranes surrounding the bones become infected and inflamed, periostitis occurs, and infection of the bone marrow causes osteomyelitis. Other staphylococcal infections include conjunctivitis, sinusitis, pneumonia, meningitis, and mastoiditis. Nearly every part of the body may be infected by this organism.

Before antibiotic therapy, the mortality rate due to a staphylococcal septicemia was 80 to 90 per cent. Immediately after introduction of penicillin, the mortality rate dropped drastically, but with the emergence of penicillin resistant mutants and their selection as a result of widespread use of the drug, resistant strains became so prevalent that the mortality rate once more increased to an alarming figure. This has necessitated continuous search for new antibiotics that are effective against the penicillin resistant strains.

STREPTOCOCCI

Most pathogenic streptococci are hemolytic, although viridans streptococci of the throat and upper respiratory tract behave opportunistically and produce disease under special circumstances. The normal enterococci of the intestinal tract are rarely pathogenic.

The morphology of streptococci is affected by the medium. Chains are generally shorter *in vivo* than in laboratory culture, and longer in liquid than on solid media. Smears from the latter often contain clusters that resemble staphylococci. They can be distinguished by the catalase test, in which streptococci are negative.

Streptococci produce three principal colony types. Mucoid colonies are large, and the cells possess large capsules of hyaluronic acid. Matt colonies are flatter, rougher, and dull in appearance. Glossy colonies are small and shiny; the cells do not produce hyaluronic acid or else it dissolves into the culture medium.

Figure 20–1 shows diagrammatically the various layers of capsular and cellular material that comprise the surface of a streptococcal cell. Some of these substances are significant factors in pathogenicity, and many are antigenic.

Streptococci are more highly specialized than staphylococci and require about half a dozen vitamins and more than a dozen amino acids. Blood is often included in culture media as a source of nutrients and also as an indicator of hemolysin production. Incubation temperature requirements are fairly close to that of the human and animal body.

Hemolytic streptococci are typed serologically by the Lancefield precipitation method. The test antigen is the "C" carbohydrate, which is extracted from cells sedimented from broth cultures by boiling with dilute HCl at pH 2 for 10 minutes. After neutralization with NaOH and centrifugation to remove the cells, ring tests are set up in capillary tubes with the C substance extract as one layer and rabbit antiserum as the other. Most human pathogenic hemolytic

streptococci have the same C antigen and are designated group A. Streptococci with different C substances comprise groups B through O. Group A organisms can also be tentatively identified by their sensitivity to bacitracin in a filter paper disk test.

The strains in group A are further subdivided into about 50 types by precipitation or agglutination tests in which the significant antigen is a cell wall protein, the M substance. Anti-M antibody is also protective, which indicates that the M protein is a factor in virulence; both it and hyaluronic acid capsules are antiphagocytic.

Hemolytic streptococci cause a greater variety of diseases than any other bacteria and possess the ability to invade practically all organs and tissues. It is therefore not surprising to find that they liberate at least a score of different antigenic substances during human infection; some are important in virulence, but the significance of others is not certain.

The *erythrogenic toxin* (Dick toxin) causes the rash of scarlet fever. It is produced by lysogenic strains. Introduction of a small amount of toxin into the skin incites local erythema (redness), which reaches a maximum at about 24 hours in susceptible individuals. This is the Dick test. During convalescence, scarlet fever patients become Dick-negative, an indication that antibody neutralizes the erythrogenic effect of the toxin. A positive Dick test indicates lack of antitoxin.

Two hemolysins, known as *streptolysins,* are liberated by streptococci. Streptolysin S, extractable from intact cells with serum, is stable in oxygen and produces hemolysis around surface colonies on blood agar. Streptolysin O is reversibly inactivated by atmospheric oxygen and lyses red blood cells only around subsurface colonies. It is antigenic, and the antibodies neutralize the hemolytic action of the toxin. The sera of patients recovering from streptococcal infection contain antistreptolysin O antibodies, which are detected by the ASTO test. Streptolysins S and O are toxic for leukocytes, and streptolysin O produces fatal cardiac arrest when injected intravenously into experimental animals.

NADase, an enzyme that liberates nico-

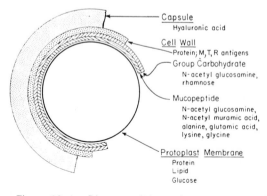

Figure 20–1. *Diagram of the surface structures of a group A hemolytic streptococcus. (From Krause, Bacteriol. Rev., 27:369, 1963.)*

tinamide from NAD, is produced by many strains of group A streptococci of type 12, which are frequently found in glomerulonephritis; its significance in pathogenesis of this disease is not known.

Streptokinase, originally called fibrinolysin, catalyzes the conversion of plasminogen to plasmin, a protease that hydrolyzes fibrin clots. It is produced by most strains of type A hemolytic streptococci and has been assumed to lyse or prevent the formation of fibrin barriers around local streptococcal lesions, thus permitting the organisms to spread about the body. However, there is no conclusive evidence to support this hypothesis.

Hyaluronidase hydrolyzes the antiphagocytic hyaluronic acid gel comprising the streptococcal capsule and therefore contributes to its transient nature. The sera of most patients who are recovering from streptococcal disease contain antihyaluronidase antibodies. Hyaluronidase has long been considered of possible significance in promoting the dissemination of streptococci throughout mammalian tissues.

Many strains of streptococci produce a potent *proteinase,* which damages the heart muscle of laboratory animals injected intravenously.

With the exception of the erythrogenic toxin, streptococci owe their pathogenicity to factors that determine invasiveness, principally antiphagocytic properties, since nearly all pathogenic streptococci are killed within a matter of minutes after they are ingested. As mentioned previously, the hyaluronic acid capsule and M protein are the principal antiphagocytic agents.

Hemolytic streptococci are notorious for the great variety of diseases that they cause. The infections may be *primary,* such as septic sore throat or pharyngitis, *secondary,* as in the case of streptococcal pneumonia, or *mixed,* such as wound infections. They may be *focal,* such as tonsillitis, in which the organisms are continuously or intermittently released to establish secondary sites of infection elsewhere in the body (e.g., kidneys, heart, joints), and *septicemia* is a general body infection.

The majority of streptococcal infections are suppurative: infections of the skin and subcutaneous tissue, such as erysipelas and cellulitis, and infections of the respiratory tract, including pharyngitis, with or without scarlet fever, tonsillitis, bronchopneumonia, otitis media, or mastoiditis. The scarlet fever rash appears in susceptible (i.e., nonimmune) individuals with pharyngitis caused by a toxigenic strain of streptococcus; these individuals are usually children over two years of age.

Streptococcal infections are likely to progress to nonsuppurative complications or sequelae, among which are carditis, arthritis, and nephritis. Acute glomerulonephritis symptoms, which appear about one week after an acute pyogenic infection, usually of the pharynx, include hematuria, edema, and hypertension; these symptoms indicate kidney damage. Rheumatic fever also often starts as an acute streptococcal infection of the nasopharynx, followed by a quiescent period of one to five weeks in which streptococci are still present in the throat. Almost any group A streptococcus may be the cause, and about 3 per cent of epidemic pharyngitis cases progress to rheumatic fever.

Puerperal sepsis, or childbed fever, is caused by hemolytic streptococci that infect tissues damaged at the time of childbirth. The source of the organisms is sometimes endogenous, but it may be the pharynx of the attending physician or someone else in the delivery room.

The alpha-streptococci or viridans group are almost universal inhabitants of the upper respiratory tract. They are generally of very low pathogenicity. Their principal significance in disease is as opportunists that cause subacute bacterial endocarditis (see Chapter 17, page 341).

Hemolytic streptococci are susceptible to a variety of antimicrobial drugs, including sulfonamides, which are bacteriostatic, and penicillin, which is bactericidal. Group A streptococci do not produce mutants resistant to penicillin, so this antibiotic can be used in small doses to prevent streptococcal infections in rheumatic patients and thus reduce the probability of recurrences of rheumatic fever.

Bacterial Pneumonia. *Pneumonia* is a clinical term applied to any acute inflammation of the lung. Prior to the Second World War, *Streptococcus pneumoniae* was considered to be the major etiologic agent, and a small percentage of cases was attributed to hemolytic streptococci, staphylococci, and the gram-negative rod, *Klebsiella pneumoniae*. During the war there were many cases of pneumonia for which a bacterial cause could not be determined, and it was assumed that they were of viral origin. Later it was found that some of these were actually due to *Mycoplasma pneumoniae*.

S. pneumoniae is gram-positive in young cultures, and the cells are elongate or lanceolate cocci arranged in pairs; they are also seen frequently as single cells and chains of paired cells. Virulent organisms are encapsulated, especially in clinical material (sputum, blood, and exudates) or infusion media containing blood constituents, sugar, or both. The organisms are nutritionally fastidious, and grow only between 25° and 42° C. They produce neither catalase nor peroxidase and hence accumulate H_2O_2 when growing aerobically. They therefore die quickly in culture unless a source of catalase, such as red blood cells, is provided. Growth is rapid on blood agar. The colonies are α-hemolytic and are difficult to distinguish from those of other alpha streptococci.

In older cultures (i.e., 24 hours) the cells become progressively gram-negative and the viable count decreases. These changes are due to autolytic enzymes, and in one or two days a turbid broth culture becomes almost completely clear. Autolysis is stimulated by surface tension depressants such as bile or bile salts, and bile solubility differentiates pneumococci from other streptococci. Most strains of pneumococcus also produce acid from the polysaccharide inulin and are sensitive to optochin (ethylhydrocupreine), which does not inhibit other alpha streptococci.

There are more than 75 serologic types of *S. pneumoniae*. Serologic differences were first detected by agglutination, but Neufeld later discovered that homologous antisera caused the polysaccharide capsules to enlarge, whereas heterologous antisera produced no change. The Neufeld "quellung-reaktion" (swelling test) can be performed in a few minutes by mixing sputum or cultured bacteria with antisera on microscope slides. Serologic typing was formerly important, because only homologus antiserum is effective in treating the disease, but since introduction of chemotherapy, this is of no significance.

Although pneumococci release a hemolysin and a necrotizing agent, disease production is attributed almost entirely to their invasiveness. Only encapsulated strains are virulent, and the degree of virulence is correlated with the amount of polysaccharide or specific soluble substance (S.S.S.). The type 3 pneumococcus, which possesses the largest capsule, is generally considered to be most virulent. Any agent or procedure that removes the capsule or counteracts its antiphagocytic power lowers the virulence of a pneumococcus. Most strains lose their capsules and become avirulent when cultivated on laboratory media, and encapsulated bacteria that have reacted with specific antiserum are readily ingested and digested by phagocytes.

Invasion and disease production depend upon decreased host resistance. Factors affecting resistance include (1) the epiglottal reflex, which normally prevents aspiration of infected secretions into the bronchial tree, (2) secretion of bronchial mucus, which traps microorganisms before they reach the more vulnerable alveolar membranes, (3) cilia of the respiratory endothelium, whose activity moves debris upward at the rate of 1 to 3 cm. per hour, (4) the cough reflex, which also removes foreign objects from the lower membranes, (5) lymphatics that drain the bronchi and bronchioles, and (6) phagocytic alveolar mononuclear cells. These defenses are weakened by various predisposing conditions: (1) exposure to extreme or sudden cold; this increases the permeability of mucous membranes and stimulates the secretion of mucus, which inhibits the epiglottal reflex, ciliary action, and phagocytosis; (2) bacterial and viral diseases, including typhoid fever, whooping cough, scarlet fever, influenza, and the common cold; and (3) pulmonary edema resulting from anesthetics, irritating chemicals, cardiac failure, and prolonged bed rest.

At the outer edge of a spreading pneumonic lesion is a zone of edema, in which serous fluid that has passed from the vascular system into the alveoli serves as culture medium for the bacteria and also transports them to adjacent alveoli directly and via the bronchioles. Leukocytes and some erythrocytes accumulate in the edema fluid, producing a thick, purulent exudate. This is the *stage of consolidation,* during which the affected area is useless as a medium for exchange of O_2 and CO_2. Toward the center of the lesion, phagocytosis occurs in favorable cases, resulting in destruction of the pneumococci, and ultimately macrophages replace the granulocytes and the lesion undergoes *resolution.* Otherwise, the lesions spread and involve other lobes of the lung or the alveoli beneath the pleura, producing pleurisy. Progressing pleural infection with abscess formation may develop into empyema, and pericarditis follows spread to the adjacent pericardium. In any of these situations the bloodstream may be invaded, and the resulting bacteremia distributes the organisms to such locations as the meninges or the heart valves. The mortality due to pneumococcal pneumonia varies directly with the number of lobes involved, the occurrence of bacteremia, and other complications. It also varies with age, being highest in infants and the aged, and with the type of pneumococcus.

Serum antibodies appear after 7 to 10 days, and if their concentration is adequate, spontaneous crisis follows with resolution and cure. This occurs in about 70 per cent of the untreated cases. Antibodies persist only a few months, so second attacks of pneumonia are not uncommon. While they may be caused by the same serologic type, they are usually precipitated by other types.

Penicillin is effective in treating pneumococcal pneumonia because the organisms do not become significantly resistant. Broad spectrum antibiotics, such as tetracyclines and erythromycin, are also used.

NEISSERIA

Neisseria meningitidis is one of the two important pathogenic gram-negative diplococci. They require special media such as blood or chocolate (i.e., heated blood) agar, cystine-trypticase semisolid agar, or media containing starch. One function of protein or starch is to absorb toxic fatty acids and metals present in peptone, agar, and other culture medium ingredients. The meningococcus and gonococcus are more sensitive to these substances than are the "saprophytic" gram-negative diplococci, which will grow on ordinary nutrient agar.

N. meningitidis is normally present in the nasopharynx of 4 to 5 per cent of the population. The carrier rate increases with the population density and the incidence of upper respiratory disease, at times reaching 30 per cent.

Children under the age of 10 years possess little immunity, but mild nasopharyngeal infections increase their immunity to such a point that in a normal population exposed to epidemic meningitis only one to 30 out of 10,000 persons develop the disease. The first and often only sign of infection is nasopharyngitis, but if the organisms enter the bloodstream and pass to the meninges (the membranes that cover the brain and spinal cord), they produce painful inflammation. They also infect localized areas in the skin, which appear as reddish spots or petechiae and give rise to the name *spotted fever.* Spinal fluid from patients usually contains many white blood cells and gram-negative diplococci, some of the latter being ingested by the phagocytes and appearing in their cytoplasm. The mortality rate in cerebrospinal meningitis epidemics varies from 35 to 80 per cent.

The virulence of meningococci is correlated with the antiphagocytic nature of their capsules and with their content of endotoxin, which damages the vascular system and causes localized subcutaneous hemorrhage. Phagocytized meningococci do not multiply, so phagocytosis is an effective defense. It is aided by antibody.

SUPPLEMENTARY READING

Burrows, W.: *Textbook of Microbiology,* 20th ed. Philadelphia, W. B. Saunders Company, 1973.

Davis, B. D., Dulbecco, R., Eisen, H. N., Ginsberg, H. S., and Wood, W. B., Jr.: *Microbiology,* 2nd ed. New York, Hoeber Medical Division, 1973.

Dubos, R. J., and Hirsch, J. G. (eds.): *Bacterial and Mycotic Infections of Man,* 4th ed. Philadelphia, J. B. Lippincott Company, 1965.

Fenner, F. J., and White, D. O.: *Medical Virology,* 2nd ed. New York, Academic Press, 1976.

Gardner, P., and Provine, H. T.: *Manual of Acute Bacterial Infections.* Boston, Little, Brown and Company, 1976.

Horsfall, F. L., and Tamm, I. (eds.): *Viral and Rickettsial Infections of Man,* 4th ed. Philadelphia, J. B. Lippincott Company, 1965.

Joklik, W. K., and Smith, D. T.: *Zinsser Microbiology,* 15th ed. New York, Appleton-Century-Crofts, Inc., 1972.

Stewart, F. S.: *Bacteriology and Immunology for Students of Medicine,* 9th ed. London, Bailliere Tindall & Cassell, 1968.

Wilson, G. S., and Miles, A. A.: *Topley and Wilson's Principles of Bacteriology, Virology, and Immunity,* 6th ed. London, Edward Arnold (Publishers) Ltd., 1972.

FOODBORNE AND 21
WATERBORNE
DISEASES

Microorganisms that are less demanding than the groups discussed in the preceding chapter can be transmitted to new hosts by indirect routes. Those that leave the body in the excretions often find their way to food or water, in which they may even multiply, and are there assured of passage to the digestive tract of another host. Inasmuch as the mouth is the only portal of entry for these organisms, the unpleasant but inescapable fact is that a case of intestinal disease is the direct consequence of some fault in sanitation or personal hygiene.

Among diseases transmitted in this way are the enteric fevers, gastroenteritis, and bacillary dysentery (all caused by members of the Enterobacteriaceae), amebic dysentery (caused by a protozoon), brucellosis, food intoxications, and certain viral infections, including poliomyelitis.

ENTEROBACTERIACEAE

This large family, whose principal habitat is the intestinal tract of man and animals, includes many bacteria that are normally considered harmless and two genera whose members are pathogenic: *Salmonella* and *Shigella*. All are gram-negative nonspore-forming rods; all attack carbohydrates fermentatively, reduce nitrates to nitrites, and are oxidase-negative. Table 21–1 lists characteristics of some of the genera and species.

Escherichia and *Klebsiella pneumoniae* (formerly *Aerobacter aerogenes*) comprise the *normal coliforms,* and these, together with *Proteus,* are found in the intestinal tract of most individuals. They are usually nonpathogenic, but they are opportunists and produce disease when they gain access

TABLE 21–1. **Characteristics of Some Enterobacteriaceae***

	Glucose	Lactose	Sucrose	Salicin	Maltose	Mannitol	Indole	M.R.	V.P.	Citrate	Urease	H$_2$S	Motility
Escherichia	AG	AG	AG	AG	AG	AG	+	+	−	−	−	−	±
Klebsiella pneumoniae	AG	AG	AG	AG	AG	AG	−	−	+	+	+	−	−
Proteus	AG	−	AG	−	AG	AG	±	+	−	±	+	−	+
Salmonella (most species)	AG	−	−	−	AG	AG	−	+	−	±	−	±	+
Salmonella typhi	A	−	−	−	A	A	−	+	−	−	−	+	+
Shigella dysenteriae	A	−	−	−	−	−	−	+	−	−	−	−	−
Shigella (other species)	A	−	A	−	A	A	±	+	−	−	−	−	−

*A, acid   G, gas   ±, some species or strains + and some −

to a susceptible tissue or organ, such as the bladder, and some strains of *Escherichia* produce intestinal disease in infants and debilitated individuals. The normal coliforms are distinguished from the rest of the Enterobacteriaceae by their ability to ferment lactose. *Proteus* species are readily recognized by hydrolysis of urea. Nearly all *Salmonella* species, with the exception of *S. typhi,* produce gas from sugars that they ferment. *S. typhi* is differentiated from *Shigella* by its motility.

ISOLATION

Salmonella and *Shigella* are isolated from feces or urine by plating on differential or selective media. Differential media contain lactose and an indicator that helps to distinguish colonies of lactose fermenting organisms from those of nonfermenters. Eosin methylene blue (EMB) agar, for example, is a buffered peptone lactose medium with eosin and methylene blue dyes, on which *Escherichia* colonies are black with a greenish metallic sheen, and *K. pneumoniae* colonies are larger, mucoid, and pink with a dark center. *Salmonella* and *Shigella* (and *Proteus*) colonies are colorless to grayish-white and often translucent, and can readily be distinguished from the normal coliforms. Selective media, such as *Shigella-Salmonella* (SS) and desoxycholate citrate (DC) agar, contain lactose and peptone, inhibitory substances (bile salts, citrate, and/or thiosulfate in high concentrations), the bacteriostatic dye brilliant green, and neutral red indicator. These media inhibit most bacteria except *Shigella* and *Salmonella,* which produce translucent colonies. Some coliforms multiply, especially with very large inocula, but the colonies are distinctive in color and appearance (see Plate IV).

IDENTIFICATION

Tentative identification of isolates is made by glucose, lactose, and mannitol fermentation, urea hydrolysis, and motility determination. Final identification depends upon serologic tests, because there are approxi-

mately 2000 different serologic types of *Salmonella* and two score or more of *Shigella.*

Serologic identification of *Salmonella* is based upon detection of somatic (O) and flagellar (H) antigens. The somatic antigens are cell wall lipopolysaccharides, and their specificity is determined by the nature and arrangement of the sugars of which they are composed. Although the structure of some of them has been determined, they are designated by a numerical system with reference to specific antisera, which are used to identify them by agglutination tests. Flagellar antigens are protein.

The flagella of most species of *Salmonella* undergo a reversible change in antigenic composition known as phase variation. A fresh isolate of *S. schottmuelleri,* for example, contains four somatic antigens, 1, 4, 5, and 12, and either flagellar antigen b or flagellar antigens 1 and 2 (see Table 21–2). No matter which form is isolated, after several laboratory transfers the culture will contain a mixture of bacteria with both antigenic formulae: 1,4,5,12;b and 1,4,5,12;1,2. When this mixture is plated and sufficient colonies are selected, both forms can be reisolated, but after further transfers each culture will again contain a mixture of the two forms.

Monospecific antisera, that is, solutions containing only one antibody, are prepared by adsorption, as described in Chapter 19 (page 376). For example, *Salmonella* somatic antibody 9 can be secured by immunizing a rabbit with an old culture of *S. typhi* (lacking Vi antigen) that has been heated to destroy H antigen d, followed by adsorption of

TABLE 21–2. Partial Antigenic Compositions of Some Salmonella Species

Species	Somatic Antigens	Flagellar Antigens	
		Phase 1	Phase 2
S. paratyphi A	1 2 12	a	—
S. schottmuelleri	1 4 5 12	b	1 2
S. typhimurium	1 4 5 12	i	1 2
S. enteritidis	1 9 12	g m	—
S. typhi	9 12 [Vi]*	d	—

*Vi is a special envelope antigen present in fresh isolates but lost upon subculture.

the serum with *S. typhimurium*. This can be used to detect the corresponding antigen in an unknown bacterium by an agglutination test.

It can be seen that, of the five organisms listed in Table 21–2, *S. paratyphi A* can be identified by its reaction with somatic antibody 2 or flagellar antibody a. Appropriate antibodies to distinguish the other bacteria are readily apparent. This is the principle used in serologic identification of *Salmonella* and *Shigella* species.

GASTROINTESTINAL DISEASES

Two principal types of gastrointestinal disease are attributed to *Salmonella*: enteric fever and gastroenteritis. Enteric fevers are characterized by an initial bacteremia, more or less continuous fever, and symptoms of a general nature. Gastroenteritis is acute, with vomiting and diarrhea and only occasional bacteremia.

Enteric Fevers. Typhoid Fever. Following oral infection with *S. typhi,* the organism passes through the lymphoid tissues of the pharynx (tonsils) or the intestine (Peyer's patches) into the bloodstream, from which it is removed by the cells of the reticuloendothelial system: liver, spleen, mesenteric lymph nodes, bone marrow. The organisms multiply, particularly in the spleen and liver, and by the tenth to fourteenth day they spill over once more into the bloodstream. This is the end of the incubation period and marks the beginning of clinical disease. Liberation of endotoxins from the bacteria causes fever, and other symptoms include headache, loss of appetite, weakness, diarrhea, and rose spots on the abdomen. The disease progresses for several weeks, and the fever drops about the time that antibodies appear in the circulation. Recovery occurs in nine out of ten cases.

Before the advent of modern sanitary practices, about 50 per cent of typhoid fever outbreaks were waterborne and 30 per cent were milkborne. With modern emphasis on sanitation, the distribution of cases has shifted to food. *S. typhi* enters foods on the hands of carriers or ambulatory cases. A carrier is an individual who does not have an active infection, but who harbors and excretes the organisms. Excreted bacteria frequently contaminate the hands of both patients and carriers. *S. typhi* survives well in nonacid foods. The greatest hazard in the United States is undetected carriers working in food handling positions. In countries where human excrement is used as fertilizer, typhoid is disseminated on salad vegetables such as lettuce, watercress, radishes, and cabbages. There is also danger that wells used for drinking water will be contaminated by nearby use of human fertilizer.

Paratyphoid Fevers. Paratyphoid fevers resemble typhoid fever but run a milder course. They are characterized by sudden onset, after an incubation period of one to ten days, with chills and most of the same symptoms as typhoid fever but to a lesser degree. The course is usually shorter and the average mortality rate is lower. The only certain means of distinguishing typhoid fever from paratyphoid fever is by isolating and identifying the causative organism.

Recovery from typhoid and paratyphoid fevers is accompanied by a good grade of immunity, and artificial immunization by a killed bacterin is effective. The usual immunogen is a mixture of typhoid bacteria and *S. paratyphi A* and *S. schottmuelleri,* formerly known as paratyphoid A and B, respectively. This so-called TAB vaccine, administered in three weekly doses, protects against ordinary exposure. It should be reinforced by booster immunizations at frequent intervals if there is likelihood of contact with the infecting organisms.

Gastroenteritis. Gastroenteritis attributed to *Salmonella* species follows ingestion of contaminated food, often after the organisms have multiplied and attained large numbers. The incubation period is 8 to 48 hours, and symptoms include nausea, acute vomiting, diarrhea, prostration, and a slight rise in temperature; recovery usually occurs within a few days. Serious septicemia occasionally develops; the disease is more severe in children than in adults. Mortality is less than 1 per cent.

Gastroenteritis is a food *infection,* although from its severity and sudden onset it is often erroneously referred to as food poisoning. The latter term should be reserved for botulism and *Staphylococcus aureus* intoxication, which will be discussed later.

Many species of *Salmonella* have been isolated from foodborne gastroenteritis; *S. typhimurium* and *S. enteritidis* head the list. Foods by which these organisms are transmitted include meats and fish, milk, and eggs or egg products (e.g., spray-dried egg powder). Partially cooked meat products such as sausage are frequently responsible. Vegetables, cereals, and fruits are rarely involved.

Foods are infected by two principal methods. Some *Salmonella* species are natural pathogens of domestic animals, and their meat or products may contain the organisms. Eggs from infected chickens may be contaminated, and the bacteria are present in the yolks. Between 2 and 10 per cent of hogs and cattle have been reported to be infected with *Salmonella*. Foods also become infected during storage and preparation for eating. Diseased rats and mice that inhabit storage warehouses and human carriers are sources of contamination.

Bacillary Dysentery. Dysentery is an acute infectious disease of the lower ileum and colon. The incubation period following ingestion of *Shigella* species is usually about 48 hours, and the disease begins suddenly with fever, abdominal pain, vomiting, and diarrhea. There is often ulceration and marked inflammation of the intestinal mucosa, and the watery excreta frequently contain blood and mucus. Prostration is due largely to loss of fluid and electrolytes resulting from the purging diarrhea.

S. dysenteriae produces a very potent, thermolabile, toxic protein, which has long been described as a neurotropic exotoxin because injected rabbits develop paralysis of the limbs. However, it seems to affect primarily the blood vessels, so the mechanism by which it causes paralysis is uncertain. Although its toxicity is approximately the same as that of the tetanus and botulism toxins, and it has other properties of an exotoxin, it is released chiefly by autolysis of dead cells, rather than by excretion from living organisms.

Most *Shigella* species produce only endotoxin, which is responsible for the intestinal hemorrhage, loss of weight, and other general symptoms typical of endotoxins. *Shigella* species do not pass through the intestine wall but remain localized within the intestine or in cells lining the intestinal mucosa. Blood and urine cultures are therefore negative; the stools contain essentially a pure culture of the *Shigella*.

Mortality due to bacillary dysentery varies greatly according to the species of organism and the population affected. *S. dysenteriae* mortality is sometimes as high as 20 per cent, but most species produce a lower mortality. Many cases are mild and are not recognized as bacillary dysentery.

The immunity engendered by a case is poor and temporary. One reason is that the infection remains localized within the intestine and the organisms do not reach efficient antibody producing sites. Moreover, antibodies that are formed frequently do not reach the lumen of the intestine in sufficient concentration to be effective. Artificial active immunization is not available.

Bacillary dysentery is a particular problem in institutions, jails, and military installations. Epidemic dysentery is common in tropical and subtropical areas and in the warm months of the year, but dysentery is endemic in temperate zones and even in countries with improved sanitary conditions. Personal hygiene is a factor, and convalescent carriers may discharge the organisms for four or five weeks. Infected food is the principal vehicle by which the organisms are transmitted.

AMEBIC DYSENTERY

Entamoeba histolytica is one of less than 35 species of protozoa known to parasitize man. In the intestinal ulcers of dysentery it appears as a granular, colorless or pale green cytoplasmic mass, 15 to 50 μm in diameter, with no definite shape. It is motile by means of pseudopodia. The granular cytoplasm often contains erythrocytes and cell debris in various stages of digestion. Persons with active cases of amebic dysentery discharge the active ameboid parasites in their feces, but encystment occurs in the intestine of a carrier, so that his stools contain cysts in large numbers. A cyst is only 10 to 20 μm in diameter, round, nonmotile, surrounded by a clear wall, and contains four nuclei. The cyst is the infective stage, because ameboid forms are highly susceptible to drying and to changes in temperature and salt concentration and are also quickly destroyed by gastric juice. They therefore fail to survive passage from the intestine of a patient to that of another individual. Cysts, on the contrary, are considerably more resistant to temperature changes, drying, and to gastric acidity. They can even survive for several months at 0° C.

After ingestion, cysts of *E. histolytica* traverse the intestinal tract, and excystment, followed by division, yields eight small amebae per cyst. These establish themselves in the large intestine. Many individuals harbor the amebae without disease symptoms. In a highly susceptible host or when the parasite is unusually virulent, the organisms penetrate the intestine wall, where they live on blood and other tissue cells. An abscess forms which breaks into the lumen of the intestine and produces an ulcer. Initial diarrhea is followed by dysentery. The host may recover spontaneously; otherwise the invasion progresses and symptoms become more severe, death resulting from exhaustion, intestinal hemorrhage, or peritonitis following perforation of the intestinal wall. It is not uncommon for the amebae to enter the bloodstream and be carried to the liver, spleen, lungs, and brain, where secondary abscesses form. Brain abscesses are universally fatal.

Little immunity results from recovery, but antibodies detectable by complement fixation and by various other techniques are produced.

A number of drugs are available for treating amebic dysentery: emetine, iodine, arsenic compounds, and several antibiotics.

The distribution of *E. histolytica* resembles that of other intestinal pathogens, that is, any vehicle and route by which the excreta of a patient reach the mouth of another individual. Drinking water, food handlers, and houseflies are the most important. Active cases are not significant sources of the infectious agent because they excrete only the ameboid form, whereas carriers excrete infective cysts.

BRUCELLOSIS

Brucellosis in man is a protracted, febrile illness of low mortality; it is very debilitating and occasions great loss of time and productivity. It is caused by three species of *Brucella: B. melitensis, B. abortus,* and *B. suis.* These are gram-negative, minute, cocco-bacillary organisms with a tendency to irregular staining. They are usually cultivated in enriched media such as liver infusion, tryptose, or trypticase, because a variety of amino acids and vitamins must be supplied in synthetic media. Growth is slow, particularly upon primary isolation, and *B. abortus* requires an atmosphere containing 10 per cent CO_2.

The three species can produce disease in a variety of mammals, including man, although each is normally found in a particular host. *B. melitensis* occurs naturally in goats, *B. abortus* in cattle, and *B. suis* in swine. The organisms gain entrance to the animal or human body via the alimentary tract or through broken skin or the conjunctivae. Their first target seems to be the local lymphatics, through which they pass to regional lymph nodes and eventually via lymphatic channels to the bloodstream. They are removed from the blood by the reticuloendothelial system and finally localize in the liver, spleen, bone marrow, and kidneys. In animals other than man they accumulate in the mammary glands and the genital organs, particularly the pregnant uterus, and often produce abortion. The predilection of *Brucella* for the pregnant uterus is associated with the four-carbon alcohol, erythritol ($CH_2OH \cdot CHOH \cdot CHOH \cdot CH_2OH$), which is present in significant quantities only in this organ and associated fluids. The absence of erythritol from the human placenta appears to explain why abortion is not a customary feature of human brucellosis.

The lesions of brucellosis are granulomata, which consist of local accumulations of lymphocytes, epithelioid cells, and giant cells, and in severe cases abscesses develop. The organisms are present in the lesions and may be ingested by mononuclear phagocytic cells, where they survive and multiply intracellularly. In this situation they are protected from antibodies and from antibacterial drugs.

Brucella organisms are acquired by man in raw milk from infected cattle or goats and by handling infected tissues. Brucellosis is particularly common among farmers, slaughter house workers, and veterinarians. The incubation period is from one week to four months. Symptoms include weakness, chilliness, headache and backache, insomnia, pain in the joints, and fever; the patient may become irritable and nervous in severe cases or suffer fits of depression. In some patients the fever rises during the day to 101° to 104° F., and in others the temperature rises stepwise at daily intervals to a peak, from which it gradually returns to normal. This cycle may be repeated several times. Atypical chronic brucellosis, in which there is muscular stiffness, gastric disturbance, and neurologic symptoms, lasts from 1 to 20 years. The course of acute brucellosis is one to four months, but relapses are not uncommon. The mortality rate is between one and three per cent.

Antibodies appear three to six weeks after the first symptoms and can be detected by agglutination, complement fixation, or by the opsonocytophagic test (enhancement of phagocytosis *in vitro*). These antibodies have no protective action and their presence in the blood does not even prevent bacteremia. Delayed hypersensitivity detectable by a skin test with a brucella extract, brucellergin, is often a feature of the disease. As in the case of the tuberculin test, a positive result indicates past as well as present infection.

Chemotherapy with tetracyclines and streptomycin, preferably used in combination, is often effective. Therapy should continue for three or four weeks because the bacteria are protected by their intracellular location.

FOOD INTOXICATIONS

There are two major food poisoning agents: (1) the enterotoxin produced by certain strains of staphylococci and (2) the exotoxin of *Clostridium botulinum*. In each case, intoxication depends upon ingestion of food in which the toxin has been formed as a result of incubation of the contaminated product for a suitable interval under appropriate conditions of temperature, aerobiosis, pH, and other factors. Both toxins are relatively resistant to proteolytic enzymes, which permits their absorption from the alimentary tract.

STAPHYLOCOCCAL FOOD INTOXICATION

The incubation period following ingestion of food in which staphylococcal enterotoxin has been produced is only one to three hours, and the onset of symptoms is sudden and violent, with nausea, vomiting, diarrhea, and sometimes prostration. Fever may be absent. Recovery is rapid, usually within 24 to 48 hours. Fatalities are rare.

The widespread distribution of staphylococci upon the skin and mucous membranes of the human body, both in healthy individuals and in those with upper respiratory disease, has been noted. A few of these organisms are capable of producing enterotoxin. There is ample opportunity for the production of adequate amounts of the toxin in food inoculated with these bacteria and stored for a few hours without proper refrigeration.

Foods most commonly involved include salads, bakery products containing cream or custard fillings, and creamed potatoes; canned or potted meat or fish, meat pies, pressed beef, and tongue have also been implicated. Staphylococcal enterotoxin is unusual because it is thermostable; that is, it withstands boiling for 30 minutes or longer, and therefore is not destroyed by cooking.

At least four immunologically different enterotoxins have been isolated. They are low molecular weight, basic proteins. They are not highly antigenic, but antisera against them can be used for serologic identification by immunodiffusion.

BOTULISM

Botulism is caused by the exotoxin of *C. botulinum,* one of the most powerful poisons known. The toxin is produced in food before it is consumed and is absorbed by the mucous membranes of the stomach and the upper intestines. Animal experiments indicate that a fatal dose for an adult may be as small as 0.01 mg. or even less, an amount that might be contained within a single infected bean.

The incubation period is usually less than 24 hours. Symptoms include vomiting, constipation, double vision, thirst, paralysis of the pharynx, and secretion of thick, viscid saliva. Consciousness remains unaffected until the patient is near death; the temperature is usually subnormal. Death may occur within 24 hours after onset or may be delayed for a week. Complete recovery may require as long as six or eight months. Sixty to 70 per cent of cases are fatal.

Nearly all outbreaks of botulism are attributed to foods that have been smoked, pickled, or canned, allowed to stand for a time, and then eaten without cooking or with insufficient cooking. Most cases in the United States are associated with home canned vegetables, such as corn, string beans, spinach, and peas. Outbreaks in Europe are usually due to infected sausages, ham, preserved meats, fowl, or fish. Frequently the foods have been obviously spoiled, but this is not always true.

The danger from home canning lies in the fact that *C. botulinum* is widely distributed in garden soil, and its spores are frequently very resistant to heat. Boiling for several hours is necessary to ensure their destruction, and even prolonged heating under steam pressure is required to kill them in nonacid foods.

Fortunately, the toxin of *C. botulinum* is destroyed in a few minutes at 65° C. and very quickly by boiling. Home canned food should always be heated at the boiling point for several minutes before use.

There are at least seven immunologic types of botulinum toxin, designated by the letters A through G. Botulism in man is caused by types A, B, E, and F. Type C produces disease in poultry, and type D in cattle and horses. Types E and F are found in fish and other marine products. Type G is derived from the soil. The various toxins produce the same symptoms but have different im-munologic specificity, that is, each toxin is neutralized only by its homologous antitoxin.

ENTERIC VIRUSES

More than 50 viruses have been isolated from the intestinal contents of man, and some are known to be transmitted by milk, food, and water. It has been postulated that many cause intestinal disease, particularly in infants, and some produce symptoms of other kinds. Many intestinal viruses have not yet been demonstrated to have pathogenic significance, aside from their cytopathic effects in tissue cultures.

POLIOMYELITIS

The poliomyelitis virus is one of the smallest—25 to 30 nm in diameter, largely nucleoprotein, and containing single-stranded RNA. It is inactivated by heat at 50° to 55° C. for 30 minutes, but resists chlorine in concentrations that kill many waterborne bacteria. There are three immunologic types.

Poliomyelitis occurs naturally only in man. Infection by way of the respiratory or alimentary tract is followed by an incubation period that averages between 7 and 14 days but it may be as long as 35 days. Early symptoms consist of fever with a mild pharyngitis and headache, or gastroenteritis with nausea, vomiting, and constipation. In many cases these are the only symptoms, and they disappear after three or four days. This is known as abortive poliomyelitis and can ordinarily be diagnosed only by isolation of the virus or by detection of a rise in serum antibody when two or more weekly specimens are examined. Cases that do not cure spontaneously may progress to central nervous system involvement with pain and stiffness in the muscles of the neck and back. This is non-paralytic poliomyelitis. Flaccid paralysis resulting from destruction of motor nerve cells in the anterior horn of the spinal cord is characteristic of paralytic poliomyelitis.

There is a high incidence of symptomless infections—probably 99 per cent, according to surveys of antibody titers in large populations. Symptomless infections as well as abortive cases may engender protective antibodies that immunize a large percentage of the population. The situation in poliomyelitis resembles that in cerebrospinal meningitis, that is, frank cases of disease are rare in a population at risk. The average case fatality rate is 10 to 15 per cent, but it may be as high as 40 per cent in some epidemics. Fifteen to 40 per cent of survivors have residual paralysis.

Serum from the great majority of adults contains protective antibodies against poliovirus, and passive immunization of ex- posed individuals with γ-globulin from adult sera provides effective protection. Active immunization is afforded by two varieties of vaccine: Formalin-inactivated (Salk) vaccine induces a protective level of serum antibody that prevents bloodstream distribution of the virus. Attenuated live virus vaccine (Sabin) is administered orally and produces a mild alimentary tract infection that leads to immunity in which IgA coproantibody present in the intestine suppresses establishment of virulent virus. Both types of vaccine must contain the three immunologic types of virus in order to afford full protection.

SUPPLEMENTARY READING

Burrows, W.: *Textbook of Microbiology*, 20th ed. Philadelphia, W. B. Saunders Company, 1973.

Davis, B. D., Dulbecco, R., Eisen, H. N., Ginsberg, H. S., and Wood, W. B., Jr.: *Microbiology,* 2nd ed. New York, Hoeber Medical Division, 1973.

Dubos, R. J., and Hirsch, J. G. (eds.): *Bacterial and Mycotic Infections of Man,* 4th ed. Philadelphia, J. B. Lippincott Company, 1965.

Fenner, F. J., and White, D. O.: *Medical Virology,* 2nd ed. New York, Academic Press, Inc., 1976.

Horsfall, F. L., and Tamm, I. (eds.): *Viral and Rickettsial Infections of Man,* 4th ed. Philadelphia, J. B. Lippincott Company, 1965.

Joklik, W. K., and Smith, D. T.: *Zinsser Microbiology,* 15th ed. New York, Appleton-Century-Crofts Inc., 1972.

Stewart, F. S.: *Bacteriology and Immunology for Students of Medicine,* 9th ed. London, Bailliere Tindall & Cassell, 1968.

Wilson, G. S., and Miles, A. A.: *Topley and Wilson's Principles of Bacteriology, Virology, and Immunity,* 6th ed. London, Edward Arnold (Publishers) Ltd., 1972.

DISEASES 22
TRANSMITTED
BY INOCULATION

Diseases transmitted by direct inoculation are usually caused by microorganisms that are obligately parasitic and survive poorly outside a host, or by organisms with a particular growth requirement that is satisfied only within damaged tissues (e.g., low oxidation-reduction potential). There are three principal groups of diseases in this category: (1) arthropod-borne diseases, in which the arthropod may be only a vector or may be a necessary component in a complicated cycle; (2) diseases transmitted by animal bites; and (3) diseases in which the infectious agent is introduced mechanically into deep, injured tissues.

ARTHROPOD-BORNE DISEASES

ROCKY MOUNTAIN SPOTTED FEVER

The etiologic agent of Rocky Mountain spotted fever is *Rickettsia rickettsii,* an obligate, intracellular parasite, about 0.6×1.2 μm. in size. The natural host of *R. rickettsii* is one of four species of tick: *Dermacentor andersoni,* the Rocky Mountain wood tick; *D. variabilis,* the American dog tick; *Amblyoma americanum,* the lone star tick; and *Haemaphysalis leporis-palustris,* the rabbit tick. The rickettsiae are found in all developmental stages of the ticks, from egg to adult. They may be transmitted during copulation and are then passed to the progeny of infected females. The various species of ticks normally inhabit and feed on certain animals, often rodents and rabbits, but larger wild and domestic animals also serve as hosts. Ticks infest susceptible animals, but

this does not seem to be a necessary stage in the life cycle of the parasite; tick-to-tick transfer is all that is required. Similarly, man is an incidental victim and plays no part in maintenance of the infection in Nature.

D. andersoni is found particularly in the Rocky Mountain region where its chief hosts are rodents and rabbits. *D. variabilis* is widespread throughout the United States, Canada, and Mexico. It is parasitic upon species of wild mice and dogs, and it also lives on cattle, horses, deer, and other animals. *D. variabilis* is important because of its predilection for dogs, which brings it and the rickettsiae into close association with human habitations.

The incubation period in man following the bite of an infected tick is between 3 and 12 days. Onset of illness is abrupt, with severe headache, prostration, muscle pains particularly in the back and legs, nausea and occasional vomiting, and fever reaching 103° or 104° F. within the first two days. The fever continues for 15 to 20 days in severe cases and usually terminates gradually over a period of several days. The headache is generalized but is very intense in the frontal area. The red rash appears on the fourth day of fever and is a characteristic sign. It begins on the wrists, ankles, palms, soles, and forearms, and spreads in a few hours to the axillae, buttocks, trunk, neck, and face. The rash becomes deeper red in two or three days, and after another day or two the minute, hemorrhagic lesions coalesce to form large blemishes. The rash does not heal rapidly, and brownish pigmented areas remain for several weeks during convalescence.

Many of the symptoms are explained by

the fact that the rickettsiae have a predilection for endothelial cells of small blood vessels. Local clot formation and multiplication of the endothelial cells hinder the flow of blood and permit escape of red cells into surrounding tissues. Inflammation of the blood vessels of the skin and brain is responsible for the rash and for the stupor and intense headache. Leakage of plasma through the damaged blood vessel walls decreases blood volume and leads to shock.

The mortality rate of untreated cases is as high as 90 per cent in the Bitterroot Valley of Montana, and as low as 5 per cent in the Snake River Valley of Idaho. Over the rest of the country it averages 20 to 25 per cent. Chloramphenicol and tetracyclines are effective chemotherapeutic agents, and present mortality rates are only about 5 per cent.

PLAGUE

Bubonic plague is caused by *Yersinia pestis*. This gram-negative organism is coccobacillary in young cultures on laboratory media at 25° to 30° C., but produces large involution forms in older cultures. It is readily killed by sunlight, heat, and chemicals, and soon disappears from soil and dead animal carcasses, but remains viable several weeks when dried in flea feces or in sputum.

Human infections of classic bubonic plague are acquired from rats, which are the natural reservoir. The bacteria are present in rat blood, often to the extent of 10 million per milliliter. They are transferred from rat to rat by blood-sucking fleas. After a flea feeds upon an infected rat, the bacteria multiply prolifically in the digestive tract of the flea, and soon the mass of bacteria completely blocks the proventriculus. When the flea feeds on another animal, it regurgitates some of the bacteria, thus infecting the second animal. This cycle continues, with passage of *Y. pestis* from rat to flea to rat. When the rat population diminishes, a hungry flea will feed on man, and human infection takes place in the same way, or else flea feces are scratched into the bite.

Y. pestis and its toxins have a predilection for lymphatic and vascular systems. The bacteria multiply rapidly *in vivo,* and they and their toxins cause inflammation, coagulation, and necrosis. There may be only a local vesicle or pustule, or primary lymph nodes may be invaded and develop gelatinous edema. If the initial site of infection is in the leg, the inguinal lymph glands become swollen; they are called *buboes*. Secondary lymph nodes may become infected, and finally the organisms enter the blood and invade the liver, spleen, lungs, and other organs. In the terminal stages the patient is often cyanotic, which is probably the origin of the name *black death,* often applied to the disease.

The incubation period following a flea bite is 3 to 7 days. Sudden onset is followed by high fever, rapid pulse, nervous symptoms, and pain where the bubo appears. The mortality rate is 60 to 95 per cent. At autopsy lymph nodes appear enlarged and engorged with blood and hemorrhagic, and the lungs are congested.

If the lungs become infected, the pneumonic form of the disease ensues. This is spread by droplet infection as any respiratory disease. Pneumonic plague is highly contagious. Its mortality rate is 100 per cent.

Artificial immunization with killed or attenuated vaccines is only partially effective. Prophylactic use of sulfonamides markedly reduces the incidence of disease in epidemics, and streptomycin, chloramphenicol, and the tetracyclines are valuable if given early.

YELLOW FEVER

Yellow fever is caused by one of a large group of viruses—200 or more—that are transmitted from animal to animal by arthropods. Many of these arthropod-borne viruses, known as *togaviruses,* are transmitted to man, but human infection is often an incidental event and not essential in the life history of the virus.

The yellow fever virus has viscerotropic properties; that is, it has a particular affinity for the internal organs, including the liver, kidney, and heart, in which it produces degeneration, necrosis, and hemorrhage. Man is infected by the bite of infected mosquitoes.

After an incubation period of one to three days the onset is sudden with headache, backache, rigor, and with fever rising rapidly to a maximum within 48 hours. The face is flushed, and there is nausea and vomiting. The temperature falls after three or four days, then rises again, and in the later stage of the disease the venous circulation is impaired and there is marked hemorrhage, prostration, jaundice, and kidney involvement with albuminuria. In fatal cases death usually occurs by the sixth or seventh day; otherwise recovery is rapid and uneventful. The mortality rate is about 5 per cent, and many infections are mild.

There are two epidemiologic varieties of the disease: urban yellow fever and jungle yellow fever. In urban yellow fever the virus is transmitted to man by the mosquito *Aedes aegypti,* and the life history of the virus includes a man-mosquito-man cycle, in which the mosquito feeds on an infected human and in turn becomes infectious after 10 to 15 days, remaining infectious throughout life. *A. aegypti* is a household mosquito, and since man-to-man infection does not normally occur, control of the mosquito is an effective means of eliminating the disease.

Jungle yellow fever is a natural infection of monkeys and certain other lower animals and is usually transmitted by mosquitoes other than *A. aegypti,* notably species of *Haemogogus.* Human infection with jungle yellow fever is peripheral to the life history of the virus. It occurs principally in men who work in or at the border of the forest. Infected humans may initiate an outbreak of urban yellow fever, because if they are bitten by *A. aegypti,* the mosquito becomes infected and can transmit the virus to susceptible humans after the usual 10 to 15 day interval.

In man, viremia begins a day or two before the onset of clinical illness and persists for two to four days after. It is during this period that feeding mosquitoes are particularly apt to become infected. The 10 to 15 day period of virus replication in cells lining the intestinal tract of the mosquito permits sufficient virus to accumulate in the salivary glands to infect a human host.

Recovery from yellow fever is usually accompanied by permanent immunity. Artificial active immunization is possible, and at least two attenuated virus vaccines are available: (1) strain *17D* was attenuated by passage through mouse embryo and chick embryo tissue cultures; (2) the *neurotropic virus* was attenuated by serial mouse brain passage. Virus 17D produces fewer and less severe toxic reactions, but the neurotropic strain gives a better antibody response. In either case protection for upwards of six years is usually engendered.

MALARIA

The life cycle of the malaria parasites, species of the protozoon *Plasmodium,* has already been described (see page 289). It will be recalled that man is the host for part of their life cycle, and various species of *Anopheles* mosquitoes serve as host for the remainder.

In general, the symptoms of malaria begin several days or weeks after the bite of an infected mosquito. A shaking chill is followed by fever, which suddenly reaches 104° to 105° F. After several hours the fever abruptly drops to near normal, and the patient sweats profusely owing to the rapid decrease in body temperature. The paroxysm of chills and fever accompanies release of merozoites from the red blood cells and is thought to be caused by parasite products or blood cell contents. Between paroxysms the patient appears to be normal.

Three principal species of plasmodium cause malaria in man. *Plasmodium vivax* is the etiologic agent of *benign tertian malaria.* The chills and fever occur every 48 hours, coincident with the release of merozoites.

Plasmodium falciparum is the cause of *malignant tertian* or *estivo-autumnal* malaria. The chills are less pronounced than in benign tertian malaria, fever is more intense and prolonged and tends to be continuous, and since it does not drop sharply the sweating stage is usually absent. This infection is most dangerous because it is often accompanied by coma, convulsions, and cardiac failure. The parasites localize in any organ, and the symptoms depend upon the organ affected.

Most malaria deaths are caused by *P. falciparum*.

Plasmodium malariae causes *quartan malaria*. Paroxysms occur every 72 hours, when merozoites are released. Quartan malaria otherwise resembles benign tertian malaria except that the paroxysms are often more severe and clinical activity lasts for several months, whereas the duration of benign tertian malaria is only three to six weeks.

Partial immunity is acquired as a result of malaria infection, but it is limited to only a few months. Immunity appears to depend upon phagocytosis and digestion of parasitized erythrocytes by macrophages of the spleen, liver, bone marrow, and other organs. Opsonization by antibodies accelerates and increases phagocytosis.

Chemotherapy of malaria has been practiced for centuries. The early treatment utilized an extract of the bark of cinchona trees containing the alkaloid quinine. Quinine does not prevent or cure the natural infection but reduces the population of parasites to a number below the minimum required to produce symptoms. Atabrine, a synthetic drug, behaves in similar fashion. Many drugs developed during and after World War II are effective both prophylactically and therapeutically, so it is now possible to prevent or cure the disease. This is a valuable advance, because malaria is the most common infectious disease of man.

DISEASES TRANSMITTED BY ANIMAL BITES

RABIES

Rabies is a viral disease of wild and domestic animals. It is transmitted from one animal to another or from animal to man by biting, because the virus is present in the salivary glands as well as in other organs and tissues. All mammals tested to date can be infected, and birds are also susceptible. Domestic animals that are particularly susceptible include dogs, cats, and cattle. Since they are usually household pets, dogs and cats are the most frequent source of human infection. Among wild animals, wolves, foxes, skunks, squirrels, and mongooses are important hosts, and are subject to highly fatal disease. Bat infection is very common but is asymptomatic. The virus is present in the saliva of infected bats and may be transmitted by inhalation of their secretions (e.g., in a cave where large numbers of bats congregate). Rabies virus is readily cultivated in laboratory animals inoculated intracerebrally, and in chick embryos and mammalian cell cultures.

The usual portal of entry is a wound or abrasion, through which the infected saliva enters. The virus remains localized for as little as six days or as long as a year, depending upon the size of the inoculum and apparently also the length of the nerve path from the wound to the brain. The virus travels along nerves to the central nervous system, where it multiplies and causes severe encephalitis (brain inflammation). The incubation period is usually the least following bites on the face and head; not only is the path to the brain short, but facial injuries are also often severe.

Prior to clinical disease, the patient is irritable, complains of unusual sensations about the site of injury, and the skin is highly sensitive. Clinical signs include increased muscle tone and difficulty in swallowing, due to spasmodic contraction of throat muscles whenever liquid enters the back of the mouth. Even the sight of fluids often causes the muscles to contract. This is the origin of the common name *hydrophobia* (Greek: fear of water). The mortality rate in untreated rabies is about 50 per cent. The central nervous system shows evidence of damage during the later stages of the disease. Cytoplasmic inclusions known as Negri bodies appear in infected nerve cells. These stain typically and are readily recognized by microscopic examination. They contain virus particles and virus antigens. Detection of Negri bodies in animal brain tissue is indicative of rabies infection.

Rabies was one of the early diseases for which an active immunization procedure was developed. Pasteur discovered that the causative agent (he did not know it was a virus, although he realized it was not a bacterium) could be attenuated by cerebral passage through rabbits. After 50 transfers the original virulent "street virus" lost its virulence and did not produce disease when tested in susceptible animals. This attenuated virus was called "fixed virus." Pasteur dried homogenized spinal cords from rabbits infected with fixed virus in order to control the quantity of infectious agent. Immunization of man was initiated with virus dried for 14 days, and successive daily injections for 15 to 20 days were made with virus dried for progressively shorter intervals. Semple vaccine, commonly used in the United States, is a 4 to 5 per cent suspension of rabbit nerve tissue infected with attenuated virus and treated with phenol. It can be preserved for six months without loss of immunizing capacity. The use of attenuated virus vaccine to treat an individual already bitten by a rabid animal is based upon the fact that the incubation period is frequently very long. Inasmuch as protective levels of antibody arise within a few weeks, 7 to 14 daily subcutaneous injections of Semple vaccine often prevent active disease. Combined inoculation of vaccine and immune serum is also used, the latter to provide a temporary barrier to passage of virus. This method of preventing the disease in infected individuals is not completely effective. In one series of observations, 8 per cent of those who received the full treatment died, whereas 50 per cent of those who refused treatment succumbed.

WOUND INFECTIONS

Any environmental microorganism introduced into a wound at the time of injury may participate directly or indirectly in the infectious process. Because of their special characteristics and frequent high mortality rates, two anaerobic wound infections will be discussed: tetanus and gas gangrene. The organisms principally concerned, members of the genus *Clostridium,* are anaerobic, spore-forming rods, usually found in the intestinal tract of man and animals, and considered facultative parasites. Since they are spore formers and are discharged in the

feces they are also present in the soil, particularly in cultivated soil treated with animal fertilizer.

TETANUS

Tetanus is a toxemia, caused by the potent exotoxin of *C. tetani* in a deep, dirty wound or in a burn. The organism is not invasive, and must be placed mechanically in the site of infection, where it remains localized. As an anaerobe, it requires a lower oxidation-reduction potential than that provided by healthy, viable tissue. Damage resulting from trauma and the consequent impaired blood supply reduces the O-R potential, aided by the metabolic activities of aerobic and facultative organisms also introduced into the wound.

The incubation period following injury may be as short as two days or as long as 50 days. Tetanus appearing within 10 days is called acute tetanus, whereas that occurring after 10 days is chronic tetanus. The mortality varies from 80 per cent in acute tetanus to 15 per cent in chronic tetanus. Disease symptoms include convulsive, tonic contractions of voluntary muscles, beginning at the site of infection and then affecting the neck, arms, trunk, and legs. Death is attributed to asphyxia when the respiratory muscles become involved.

C. tetani produces two toxins: *tetanospasmin,* a neurotropic toxin, whose main site of action is the anterior horn cells of the spinal cord, and *tetanolysin,* which is hemolytic. Tetanospasmin is most significant as the cause of nerve damage. The toxin is transported from the site of infection to the central nervous system, where it causes motor reflex convulsions. It is well established that the principal transport route is the peripheral nerves, and the toxin is propelled by muscular contractions in the fluid between the neurons. Stimulation of the motor end organs produces unremitting rigidity of the voluntary muscles, and if large amounts of toxin are formed it may reach the blood via the local lymphatics and cause general motor neuron involvement, even prior to symptoms of local tetanus.

Antitoxin is administered in the attempt to prevent tetanus in individuals exposed to possible infection of wounds or burns. If sufficient antitoxin is given soon enough, it neutralizes the toxin as it is produced, thus preventing symptoms. Therapeutic use of antitoxin is of limited value, because tetanus toxin has a very high affinity for nerve tissue, and damage to the latter is essentially irreversible. Antitoxin is therefore of little effect in acute tetanus.

Recovery from tetanus is not usually accompanied by permanent immunity, because the toxin is so potent that the amount necessary to produce disease is insufficient to incite antibody formation. The only successful method of active immunization is injection of toxoid, which is nontoxic but highly immunogenic. It can be administered in doses large enough to incite effective immunity. Active immunization of infants during their first year is a common practice.

GAS GANGRENE

Gas gangrene is usually a mixed infection caused by more than one species of anaerobe and several aerobic or facultative bacteria. Among the anaerobes are *Clostridium perfringens, C. oedematiens, C. histolyticum, C. septicum,* and *C. sporogenes.* None of these is as toxic as *C. tetani,* and some are completely nontoxic. *C. sporogenes* and *C. histolyticum* are proteolytic and promote the dissemination of the more highly toxigenic *C. perfringens* and *C. septicum* throughout the tissues. One form of gas gangrene, *clostridial myositis* or inflammation of the muscle, is an invasive infection of the muscle. After an incubation period of six hours to three days there is severe local pain, high pulse rate, relatively low fever, listlessness, apathy, anemia, and either tissue edema or gas formation. The wound often gives off a foul, serous discharge. Death occurs in about 30 per cent of patients in two to three days.

Anaerobic cellulitis is a diffuse inflammation of connective tissues, characterized by gradual onset without pain or systemic symptoms. There are no mental changes and the prognosis is good.

C. perfringens, the most commonly encountered organism in gas gangrene, produces a dozen or more toxins. Alpha-toxin, a lecithinase, hydrolyzes lecithin to a diglyceride and phosphorylcholine. Since lecithin is an essential component of the membranes of all body cells, α-toxin is very important in the pathogenesis of gas gangrene. It is one of the few exotoxins whose chemical action is known. Other toxins of this organism have lethal, hemolytic, necrotizing, and hydrolytic activities.

Treatment of gas gangrene consists of débridement to remove dead tissue in which the anaerobic bacteria grow, use of polyvalent antiserum, and intensive antibiotic therapy.

SUPPLEMENTARY READING

Burrows, W.: *Textbook of Microbiology,* 20th ed. Philadelphia, W. B. Saunders Company, 1973.

Davis, B. D., Dulbecco, R., Eisen, H. N., Ginsberg, H. S., and Wood, W. B., Jr.: *Microbiology,* 2nd ed. New York, Hoeber Medical Division, 1973.

Dubos, R. J., and Hirsch, J. G. (eds.): *Bacterial and Mycotic Infections of Man,* 4th ed. Philadelphia, J. B. Lippincott Company, 1965.

Fenner, F. J., and White, D. O.: *Medical Virology,* 2nd ed. New York, Academic Press, Inc., 1975.

Horsfall, F. L., and Tamm, I. (eds.): *Viral and Rickettsial Infections of Man,* 4th ed. Philadelphia, J. B. Lippincott Company, 1965.

Joklik, W. K., and Smith, D. T.: *Zinsser Microbiology,* 15th ed. New York, Appleton-Century-Crofts, Inc., 1972.

Stewart, F. S.: *Bacteriology and Immunology for Students of Medicine,* 9th ed. London, Bailliere Tindall & Cassell, 1968.

Wilson, G. S., and Miles, A. A.: *Topley and Wilson's Principles of Bacteriology, Virology, and Immunity,* 6th ed. London, Edward Arnold (Publishers) Ltd., 1972.

AIRBORNE DISEASES 23

Airborne diseases spread widely within a short time. Moreover, transmission of microorganisms through the air is not easily interrupted. Despite the fact that many parasitic organisms die quickly after their discharge from a natural host, they are usually so numerous that enough survivors remain to infect fresh hosts.

There are two chief vehicles for distribution of microorganisms through the air: (1) droplets of sputum, saliva, or other respiratory secretions, and (2) dust.

Infectious Droplets

Nasal and oral secretions are expelled during sneezing, coughing, and talking (see Figure 23–1). Sneezing is a particularly effective means of broadcasting infectious agents because the explosive discharge of a considerable volume of air atomizes the secretions and ejects them to a distance of several feet. Droplets larger than 0.1 mm. in diameter quickly fall to the ground (see Table 23–1). Smaller droplets evaporate and leave suspended nuclei containing microorganisms. These minute particles drift about for hours, and the microorganisms they contain remain viable, protected by the residue of evaporated saliva or mucus.

The nostrils and upper respiratory tract contain hairs and sticky mucous surfaces that entrap particulate matter. This mechanism protects the lungs against inhalation of many harmful agents. Large particles (5 μm or more in diameter) are usually removed from inhaled air in this manner. Smaller and lighter particles often pass the nasal area and reach the terminal bronchioles and alveoli of the lungs. Some very small particles are apparently reexhaled.

Figure 23–1. *Droplets of saliva and mucus discharged during a violent, unstifled sneeze. (Highspeed photograph by M. W. Jennison; A.S.M. LS-4.)*

TABLE 23–1. Approximate Rate of Fall of Water Droplets Through Air

Diameter of Droplet	Rate of Fall (Ft./Hr.)
1 μm	0.36
2 μm	1.44
5 μm	9
10 μm	36
50 μm	900
100 μm	3600

Human epidemics of respiratory or upper respiratory infection are common during cold seasons, when people congregate indoors. Distribution of respiratory secretions is enhanced by modern transportation; an individual in the incubation period of influenza, for example, may travel to the other side of the Earth before his symptoms appear. Shortly after Asian influenza was recognized in the Far East in 1957, it was reported in Newport, Rhode Island, and within a few months was widespread across the United States, despite a vigorous program of artificial immunization.

The so-called childhood diseases—chickenpox, measles, mumps, scarlet fever, diphtheria—are readily transmitted from person to person by infective droplets of nasal mucus or saliva. The causative organisms are usually present in a population, either harbored by carriers or as causes of subclinical infections. These diseases engender lasting immunity and occur in epidemic form only at intervals of a few years, after a sufficient population of nonimmune individuals accumulates. The tuberculosis organism is also present in sputum and, when inhaled in sufficient numbers by individuals of low resistance, produces disease.

A few bacterial, mycotic, and rickettsial diseases will be discussed as examples of airborne diseases.

BACTERIAL DISEASES

DIPHTHERIA

Diphtheria could be completely eradicated because all the necessary information is at hand: cause, methods of laboratory diagnosis, specific treatment, detection of carriers, mode of transmission, detection of susceptibility, and methods of immunization. The fact that 285 cases occurred in the United States as recently as 1975 indicates laxity in application of some of the available knowledge.

Corynebacterium diphtheriae. *C. diphtheriae* is a gram-positive, nonspore-forming, club-shaped, granular, rod-shaped bacterium. While it is usually cultivated on blood agar or Loeffler's serum (infusion broth solidified by means of heat-coagulated serum), its nutritional requirements are not as strict as those of hemolytic streptococci, and it grows on tryptose or trypticase soy media (see Plate II *E*).

Pathogenesis. *C. diphtheriae* is an obligate, extracellular parasite. It is not invasive, and usually grows only in the human upper respiratory tract, where it produces a potent exotoxin. This, when absorbed and distributed around the body by the vascular system, is responsible for symptoms of the disease. The organs principally affected are the kidneys, heart, and nerves. Acute interstitial nephritis is the most common kidney lesion. Fatty degeneration of the heart muscle fibers and of the myelin sheath of peripheral nerves and the white matter of the brain and spinal cord account for the cardiac weakness and paralysis that often occur.

Toxin production is limited to lysogenic strains of *C. diphtheriae*, that is, strains that harbor phage β or a closely related bacteriophage. An important factor in toxin production is the concentration of iron in the medium; approximately 100 μg of iron per liter is required. The reaction of the medium must be above pH 6.

Diphtheria is usually considered an airborne infection in man, but the organisms may be transmitted by food or by fomites. In any event, they become established on the superficial layers of the mucous membranes of the nasopharynx or the larynx. The toxin causes necrosis, and the resulting inflammation permits passage of plasma and blood cells into the air passages, where coagulation occurs and a pseudomembrane forms. The bacteria continue to multiply in the pseudomembrane, and the toxin is absorbed and passes into the vascular system, which transports it to susceptible tissues. Fever is not severe—100° to 102° F.—but the pulse rate is rapid owing to the effect of the toxin on the heart. Cardiac damage is usually not obvious before the ninth day, but the patient may die suddenly from heart failure at any time during the next 6 to 10 weeks. Paralysis of muscle groups such as those involved in swallowing and respiration and muscles of the extremities may occur at any time between the third and tenth week of disease.

Laryngeal diphtheria is less common than pharyngeal diphtheria but is more serious because the pseudomembrane obstructs the air passage and the patient suffocates unless tracheotomy is performed. The mortality rate varies, probably with the strain of the infecting organism and with the population infected. During the nineteenth century, there were epidemics with mortality rates of 35 to 40 per cent. Between 1920 and 1960, the average mortality rates in the United States were between 6 and 10 per cent, but beginning about 1960 there was a rise to about 15 per cent. At the same time the incidence of diphtheria dropped steadily.

Laboratory Diagnosis. Laboratory diagnosis of diphtheria is based upon detection of *C. diphtheriae* in the throat and nasal passages of patients. Stained smears of throat swabs are examined directly, and cultures are prepared for isolation of the bacteria. Inasmuch as many strains of the organism are not virulent and a number of diphtheria-like organisms, or diphtheroids, occur in normal throat and nasal passages, virulence tests are performed to determine whether the isolated strain produces toxin. These are carried out by inoculation of guinea pigs, with and without antitoxin. If an animal inoculated subcutaneously with the organism alone dies and a parallel animal that received both the organisms and antitoxin survives, the strain of bacteria is considered toxigenic. Several virulence tests can be performed on a pair of guinea pigs by injecting very small doses intracutaneously into the shaved skin of each of two guinea pigs, one of which has been protected by antitoxin injected intraperitoneally

four hours previously. Toxigenicity is indicated by a local lesion with superficial necrosis in two to three days in the animal that did not receive antitoxin.

Specific Treatment. Diphtheria is treated specifically by injecting antitoxin. Most antitoxin is a purified preparation of serum from a horse that has been injected repeatedly with diphtheria toxoid and later with toxin. Inasmuch as toxin that has been absorbed by susceptible tissue cells can no longer be neutralized by antitoxin, therapy must be started as quickly as possible. In one study of more than 6000 cases, none of the patients treated with antitoxin on the first day died, whereas nearly 19 per cent of those treated on or after the fifth day died. Some patients are hypersensitive to horse serum, so a test for sensitivity is necessary before administering antitoxin, and epinephrine should be ready for immediate injection in case the patient has an anaphylactic reaction. Sensitive patients can often be given antitoxin, but only in small doses, which are increased gradually until the entire amount has been injected.

Carriers. Convalescent patients or contacts may carry the diphtheria organism for 10 months or longer. Carriers can be detected by examination of throat and nasal swabs in the same manner as used in diagnosis, and virulence tests are required to determine the toxigenicity of the cultures isolated.

Immunity. Susceptibility to diphtheria can be detected by the Schick test. Intradermal injection of a minute amount of diphtheria toxin incites local inflammation, which reaches a maximum at about five days in individuals who are susceptible to diphtheria, that is, who do not have circulating antitoxin. These individuals can be immunized actively by injection of toxoid. Two or three doses at monthly intervals usually suffice for primary immunization. A booster injection about a year later induces a secondary response. Primary immunization should be started at the age of three to four months. Booster injections are advisable to maintain maximum protection. Alum precipitated toxoid is somewhat more effective and is most widely used.

TUBERCULOSIS

Despite the fact that the incidence of tuberculosis has decreased in the past century to a small fraction of its incidence in the mid-nineteenth century, there are still over a million persons in the United States with active cases, and tuberculosis is one of the most important communicable diseases in the world, affecting more than 50,000,000 people. In the United States, 50,000 new cases and about 10,000 deaths are reported each year.

Mycobacterium tuberculosis. There are three major types of M. tuberculosis: human, bovine, and avian. Only the human and bovine types affect man. M. tuberculosis is a slender rod, often slightly curved, frequently seen with granules that stain deeply in specimens from a patient or animal. The cells may appear as bundles or random aggregates in sputum specimens. Their acidfast staining property is distinctive, because few other bacteria are acidfast (see Plate II C).

The nutritional requirements of M. tuberculosis are very simple, but this was not realized for many years because the usual isolation media contain serum, egg, or potato extract, and even on these media growth is slow—the first sign of growth appears only after two weeks or longer. The difficulty in cultivating these organisms is that, as is true of various bacteria already cited, long-chain fatty acids and other impurities are toxic. They are neutralized by serum or egg proteins in the traditional complex organic media. With proper precautions to eliminate these inhibitory substances, simple synthetic media suffice: glycerol provides carbon and energy, and ammonia or amino acids provide nitrogen; minerals are necessary, but known vitamins or other growth factors are not. The organisms are aerobic. The human variety grows at 30° to 42° C., but the minimum generation time is about 18 hours. Growth is luxuriant after several weeks; it is heaped up on solid media and forms a thick, granular pellicle on liquid media.

The chemical composition of mycobacteria is unusual; they contain a high percentage of lipid, ranging from 20 to 40 per cent of the dry weight. This is largely concentrated in the cell walls, where it may amount to 60 per cent. The high lipid content is partly, if not entirely, responsible for the unusual resistance and staining properties of the organisms. Mycobacteria can survive considerable treatment with strong mineral acids and alkalies and resist the bacteriostatic action of dyes such as malachite green and gentian violet. Exposure to 5 per cent phenol or 2 per cent cresol for 12 hours is required to kill them. They are resistant to drying, especially in sputum. They are killed at 60° C. in 20 to 30 minutes. M. tuberculosis is the most resistant pathogenic bacterium ordinarily spread by milk, and the pasteurization process was devised to destroy it specifically. The lipid cell wall resists penetration of stains, but incorporation of aniline or phenol into a dye solution and use of heat, as in the Ziehl-Neelsen procedure, permits penetration and staining to take place. The stained cells do not decolorize when treated with acid or acid-alcohol, and this property seems to depend upon certain cell wall lipids and the integrity of the cell wall itself.

Pathogenesis. Tuberculous disease takes many forms, and nearly any part of the body may be infected. In addition to pulmonary tuberculosis, there is also osseous (bone and joint) tuberculosis, enteritis and peritonitis, meningeal tuberculosis, and infection of the cervical glands (adenitis, formerly known as scrofula).

Experimental infections in guinea pigs, which are assumed to represent a reasonable model of human infection, indicate that inhalation of only one or a very few tuberculosis bacteria, deposited deep in the airways, will produce a single primary focus of infection. Prompt ingestion by fixed mononuclear phagocytes is followed by intracellular multiplication of the bacteria. The phagocytes burst and release a great many bacteria, which are then ingested by other phagocytes. The bacteria are relatively nontoxic per se, but after about 21 days in the guinea pig or 40 days in man, delayed hypersensitivity develops. When this happens, acute inflammation and central necrosis take place at every focus of infection, and each lesion contains approximately 1,000,000 bacteria.

The primary lesion, usually encountered in infants, children, and in some adults, is an exudative lesion with rapid inflammation and accumulation of polymorphonuclear cells. If necrosis progresses unchecked, extensive cavity formation occurs and death follows. In the most favorable situation, the infected area heals by resolution, and the disease is arrested. Arrest may be permanent.

In other patients, particularly adults, the exudative lesion evolves into a productive or proliferative lesion, characterized by accumulation of macrophages, which modify into elongated epithelioid cells and become arranged concentrically to form the tubercles. Some of these cells fuse in the centers of tubercles and become "giant" cells, containing many nuclei and viable tuberculosis bacteria in their cytoplasm. On the outside of the tubercle is a layer of lymphocytes and proliferating fibroblasts. If the tubercle calcifies, the progress of the disease is arrested. On the other hand, continued growth of the tubercle may be followed by caseation (cheesy) necrosis, cavitation, and death.

Tuberculosis has been called a disease of civilization, industrialization, and urbanization; it is prevalent in poor houses, monasteries, and prisons. Socioeconomic and occupational factors are significant, as indicated by the high incidence of the disease in crowded, low income areas and in the dusty trades. Inadequate diet, alcoholism, continual fatigue, and pregnancy predispose to disease. The incidence of tuberculosis within families has been attributed to prolonged and intimate contact, but one interesting set of observations involving identical twins suggests that heredity is a factor. In this study, 2534 persons, comprising 308 families, were surveyed; this included 78 pairs of identical twins and 230 pairs of fraternal twins. At least one case of tuberculosis had occurred in each family, and the families were studied to determine the incidence of tuberculosis in other members of the families. From the data in Table 23–2, it is apparent that if one twin had tuberculosis, the chance that a monozygotic twin would also have tuberculosis was more than three

TABLE 23–2. Effect of Heredity on Incidence of Tuberculosis

Relationship	Incidence of Tuberculosis
Identical twins	87.3%
Fraternal twins	25.6%
Full siblings	25.5%
Half siblings	11.9%
Wife or husband	7.1%

times greater than the chance that a dizygotic twin or a nontwin sibling would have tuberculosis.

Immunity. The immunologic mechanism operating in tuberculosis is not clear. Koch early discovered that a guinea pig infected three weeks previously by subcutaneous inoculation of M. tuberculosis could be reinfected by inoculation into another location, but that there was an accelerated, violent reaction at the second site of infection. A lesion appeared in two or three days and soon ulcerated, but it remained circumscribed and promptly healed. The primary infection progressed normally, and the animal died as though it had not received the second inoculation. In the diseased animal, dissemination and multiplication of the second infecting inoculum were restricted, the bacteria were destroyed, and the infected area healed. Koch later found that injection of a sterile culture filtrate (tuberculin) induced the same type of response in a tuberculous guinea pig as living bacteria: local ulceration in two or three days and prompt healing. The Koch phenomenon is an example of delayed hypersensitivity and as such is cell-mediated rather than antibody-mediated.

Immunization is practiced in some countries by injection of an attenuated bovine strain developed by Calmette and Guérin in 1923 in France. The vaccine, known as BCG (Bacille Calmette-Guérin), increases resistance to tuberculosis but does not afford complete protection. It reduces the incidence and case fatality rate of the disease in man.

Diagnosis. Diagnosis of tuberculosis includes use of the tuberculin test, chest x-ray examination, and detection of *M. tuberculosis* in clinical specimens. The tuberculin test, often used in surveys, is useful particularly to exclude tuberculosis, since a positive result indicates past as well as possible present infection. X-ray examination is usually confined to detection of active or calcified lesions of pulmonary tuberculosis, and does not indicate infection elsewhere in the body, as does the tuberculin test. Sputum samples from pulmonary patients are examined by direct staining, using the acidfast procedure, fluorescent staining, and also by attempted cultivation of *M. tuberculosis*. Gastric washings and urine in suspected urinary tract infections are also often examined. Inasmuch as the tuberculosis organism multiplies slowly and contaminants are usually present, the specimens are treated briefly with acid or alkali to kill other bacteria, neutralized, and planted onto media containing dyes to inhibit contaminants that survive or that enter later. Cultures are incubated for six to eight weeks before being discarded as negative. Indication of even a single acidfast bacterium in a sputum sample is evidence of active infection, because such bacteria are not found in the sputum of nontuberculous individuals.

Treatment. Treatment of tuberculosis has improved greatly since 1940. Whereas formerly rest and diet and chest operations were the only treatments available, antibiotic therapy now permits many patients to resume active life in a short time, often without hospitalization. Streptomycin, *p*-aminosalicylic acid, isonicotinylhydrazide, and other drugs are employed, usually in combination, because *M. tuberculosis* often becomes drug-resistant during the prolonged course of therapy that is required.

SYSTEMIC MYCOSES

The systemic mycoses are fungus infections of any or all internal organs of the body as well as of the skin and of subcutaneous and skeletal structures. The customary portal of entry is the lungs, but the organisms find their way to any organ or tissue for which they have affinity. For example, *Blastomyces dermatitidis* usually infects cutaneous and mucocutaneous tissue, *Cryptococcus neoformans* has a predilection for the central nervous system, and *Histoplasma capsulatum* parasitizes the reticuloendothelial system.

Systemic fungus infections are not transmitted from one individual to another. The fungi that cause them are often soil saprophytes; when inhaled, they produce a primary pulmonary infection, an acute but self-limited pneumonitis, easily overlooked or ascribed to bacteria or viruses. The subsequent chronic infection begins insidiously and progresses slowly, with formation of suppurative or granulomatous lesions. The organism spreads, either by direct extension or in the bloodstream, and produces metastatic abscesses or granulomas in almost any organ. Prior to the antibiotic era these infections were nearly always fatal.

Little or no immunity follows recovery from many systemic mycoses, although complement fixing and precipitating antibodies can frequently be detected. Hypersensitivity of the delayed type is usually demonstrable by skin tests similar to the tuberculin test.

NORTH AMERICAN BLASTOMYCOSIS

North American blastomycosis is caused by *Blastomyces dermatitidis* (see page 316). Primary lung infection closely resembles tuberculosis, with cough, chest pains, weakness, and productive sputum, sometimes tinged with blood. There may be consolidation of the lungs, and small abscesses or larger nodules. Pulmonary infections sometimes generalize, and multiple abscesses appear throughout the body, most commonly in the subcutaneous tissues, but also in the viscera, in muscle, under the periosteum, and elsewhere. *B. dermatitidis* also produces a primary cutaneous infection. This is usually slow and chronic; lesions on the exposed skin appear as reddish papules or pustulopapules and often persist for months or years.

North American blastomycosis occurs in Canada and the United States, particularly in the north central and southeastern states.

COCCIDIOIDOMYCOSIS

Previously considered to be relatively uncommon but highly fatal, coccidioidomycosis now appears to be very prevalent in certain areas (e.g., southwestern United States) and often with much less severity than had been supposed. Many benign, acute, self-limited, respiratory infections occur. Nevertheless, this is considered to be one of the most virulent and dangerous of the fungus diseases. Numerous laboratory infections have occurred, some of them fatal.

Primary coccidioidomycosis is acquired by inhalation of the spores of *Coccidioides immitis* (see page 316), and the severity of the disease varies from that of a common cold or influenza to pneumonia, with cavity formation and high fever. Frequently there are no symptoms. Spontaneous recovery usually occurs, and only about 1 per cent of clinically recognized infections progress to the more serious secondary coccidioidomycosis. In the latter, there may be cutaneous, subcutaneous, visceral, bone, and central nervous system lesions, as well as extension of the pulmonary lesions. Lung lesions resemble those of tuberculosis. Infection of the meninges is almost always fatal.

A considerable degree of effective immunity follows recovery from active infection. Humoral antibodies appear, and can be detected by complement fixation and precipitation. Delayed type hypersensitivity is demonstrated by the coccidioidin test, which resembles the tuberculin test and has the same significance.

The reservoir of infection of *C. immitis* appears to be the soil. The organism is a saprophyte, and infection of man and other animals is a chance occurrence and of no significance in the life history of the fungus.

HISTOPLASMOSIS

Histoplasmosis is caused by *Histoplasma capsulatum* (see page 316), a fungus of world-wide distribution but concentrated in certain areas such as the Ohio and Mississippi valleys of the United States, where it is believed that 30,000,000 people have or have had the disease.

Acute pulmonary histoplasmosis begins suddenly with malaise, fever, chills and sweats, cough, chest pains, and difficult breathing. It ranges in severity from mild, subclinical cases to those resembling severe influenza. Recovery is usually rapid and is accompanied by development of fairly strong immunity. In 0.1 to 0.2 per cent of cases the infection progresses to the hilar lymph nodes, which later calcify, producing a lesion resembling that seen in x-ray examination of tuberculosis patients.

Chronic progressive pulmonary histoplasmosis resembles the preceding acute pulmonary type, but the symptoms are more exaggerated and the disease progresses more slowly, like tuberculosis, to necrosis, caseation, and cavitation. This form of disease is usually found in middle aged males.

Disseminated histoplasmosis occurs in infants and children and also in aged persons, particularly those with lymphoma, Hodgkin's disease, diabetes, and certain other ailments. It is rapidly fulminating and fatal, and the infection may involve any organ, but especially the spleen and liver.

Past or present infection is detected by a delayed type hypersensitivity reaction with histoplasmin, a material derived from liquid cultures of the fungus. Evidence from the histoplasmin test indicates that the incidence of active infection is high in areas where the fungus is endemic. Positive reactors range from 50 to 90 per cent of the population of Missouri, Tennessee, Kentucky, Arkansas, and adjacent areas, whereas in regions west and north the proportion of reactors is less than 2 per cent.

Various animals are naturally infected, including dogs, cats, wild rodents, and bats. The fungus is found in the soil, and appears to spread to both man and animals by inhalation. Another source of infection is chicken coops, pigeon roosts, bat-infested caves, and old decaying farm outbuildings where the fungi multiply on excreta and other decomposing organic matter under highly humid conditions.

CRYPTOCOCCOSIS

Cryptococcus neoformans (see page 316) causes two general types of disease: (1)

European blastomycosis, consisting of deep-seated cutaneous or subcutaneous infections, which tend to become generalized, and (2) central nervous system infections that usually begin as lung infections and are known as cryptococcus meningitis. Contaminated soil is the usual source of the organisms. Primary cutaneous infection is rarely established by direct inoculation of the skin. It is usually acquired by inhalation of dust and dirt contaminated with *C. neoformans*. Pulmonary disease is followed by lymphatic or vascular distribution to cutaneous and subcutaneous tissues or to the visceral organs or the central nervous system.

In European blastomycosis, the primary ulcers appear on the face and neck, and infection is later seen in the cervical lymph nodes. Deep ulceration of the skin is common, and the spleen, liver, kidneys, and mesenteric lymphatics may also be infected. Shortly before death the organisms appear in the bloodstream. In cryptococcus meningitis, the central nervous system is the predominant site of infection, with brain involvement. The disease develops slowly, usually without fever or other signs of infection. The yeast is often present in the spinal fluid, and the lungs and kidneys are frequently infected, but any tissue may be attacked as the infection generalizes. Untreated cryptococcosis is almost always fatal.

RICKETTSIOSES

PSITTACOSIS

Chlamydia psittaci, the organism that causes psittacosis, is a member of the Rickettsias, in the order Chlamydiales. Like the causative agent of lymphogranuloma venereum, it is gram-negative, coccoid, obligately parasitic, and is found in the cytoplasm of host cells.

The mature particle or elementary body is 0.2 to 0.3 μm in diameter. Following entrance into a host cell, elementary bodies enlarge to as much as 0.8 μm in diameter. These "initial bodies" then divide and form a cluster of the small, elementary bodies, which are contained within a vesicle (see Figure 23–2).

This subsequently disintegrates and liberates the elementary bodies. The entire growth cycle requires 24 to 48 hours.

Among the natural hosts of these microorganisms are psittacine birds, including parrots, parakeets, and budgerigars. Canaries, finches, and sparrows contract the disease when exposed to infected psittacines. A diseased parrot shivers, is weak and apathetic, and has diarrhea and respiratory disturbance. The liver, spleen, and occasionally the lungs are infected, and the agent is present in the blood, nasal secretions, various tissues, and feces. Dried fecal material is an important source of airborne infection.

Infection of man is most common in professional handlers and breeders of birds, and in women who keep birds as pets. Following inhalation of dried infectious material such as cage dust, the incubation period in man is one to two weeks. Onset of psittacosis is sudden or insidious, and symptoms include chills and fever, headache, intolerance of light, loss of appetite, sore throat, nausea, and vomiting. There may be a persistent dry cough, cyanosis, and low blood pressure, apathy, insomnia, occasional delirium. X-ray examination shows patchy areas of consolidation of the lungs. Autopsy indicates brain and spinal cord damage, necrosis of the liver, spleen enlargement, and damage to the cardiac muscle and kidneys. These findings correlate with the generalized nature of the infection. The case mortality rate is between 10 and 20 per cent.

Antibody detectable by complement fixation appears in 10 to 14 days, and hypersensitivity can be demonstrated by the third to fourth week. There is some degree of protective immunity, which appears to depend upon the continued presence of the infectious agent. Active immunization by means of vaccines gives uncertain results.

Tetracyclines are effective chemotherapeutic agents, and penicillin in large doses is also useful.

Q FEVER

Q fever is primarily a disease of wild and domesticated animals that is transmissible to man, in whom it causes an acute systemic

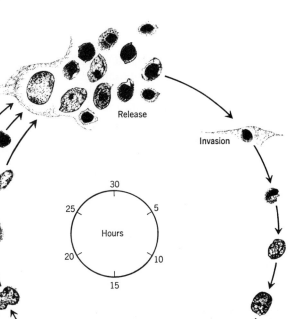

Figure 23-2. *Idealized representation of the growth cycle of* Chlamydia psittaci. *(From Moulder, J. W., The Psittacosis Group as Bacteria. New York, John Wiley & Sons, Inc., 1964.)*

disease, often a form of pneumonia. The infection in animals is usually inapparent, and although the human disease may persist for weeks or months, the mortality rate is only about 1 per cent.

Q fever is caused by the rickettsia, *Coxiella burnetii*. This obligate, intracellular parasite is highly pleomorphic, appearing typically as a gram-negative bipolar rod, 0.25 by 1.0 μm, and also as a diplobacillus, a lance-shaped rod, and a coccus, 0.25 μm in diameter. *C. burnetii* is more resistant than other rickettsiae to physical and chemical agents. It can withstand drying for months or years in feces, blood, urine, or in tap-water. Treatment required for disinfection includes heating at 60° C. for one hour, or exposure to 2 per cent formaldehyde or 1 per cent Lysol.

C. burnetii multiplies in the cytoplasm, not in the nuclei, of cells lining the vascular system, where it appears in large, packed masses, and it is also found in the peritoneal fluid of infected laboratory animals. It grows in chick embryos, and can be cultivated in tissue cultures. It can carry out a number of independent biochemical reactions associated with carbohydrate metabolism, amino acid synthesis, and growth factor incorporation. The evidence suggests that this organism is a bacterium, highly adapted to intracellular life, but the basis of its intracellular parasitism is not known.

Although Q fever was recognized only in 1937, its distribution has been found to be essentially world-wide. The host range is very broad. *C. burnetii* infects a large variety of arthropods, including ticks, lice, mites, and parasitic flies. Naturally infected animals include bandicoots, rats, mice, rabbits, porcupines, hedgehogs, tortoises, and domestic animals such as cattle, sheep, goats, horses, swine, and dogs. Serologic evidence indicates infection in domestic birds, including chickens, ducks, geese, turkeys, and pigeons.

Infection among animals is maintained by arthropods and also by direct animal-to-animal transfer among domestic livestock.

The parasite is transmitted transovarially in several species of ticks.

Human infection is characteristically acquired by inhalation of infectious material from animals, although the organism may be present in minimally pasteurized milk from infected cattle or goats. Slaughterhouse and dairy workers, herders, woolsorters, and tanners are particularly exposed to infection.

The incubation period averages 18 to 21 days. The onset of disease is sudden with fever, chills, general malaise, frontal headache, loss of appetite, muscle pains and often chest pains. During the first 9 to 14 days the fever ranges between 101° and 104° F., but milder fever of less than 102° F. may persist as long as three months. Symptoms are those of an atypical pneumonia, demonstrable by x-ray examination in even mild cases. There may be evidence of liver involvement, and headache is a prominent symptom. Unlike other rickettsial diseases there is no rash. Convalescence is often protracted.

Chronic Q fever may appear as a cardiovascular disease, including subacute endocarditis. This is not a common form of disease but is uniformly fatal.

Diagnosis of Q fever in man is aided by isolation of the causative organism and demonstration of a significant rise in serum antibody titer. The organism is detected in clinical material (blood, sputum, urine, spinal fluid) by intraperitoneal inoculation into guinea pigs; the animals are bled 30 to 40 days later and their sera are examined for specific antibodies. The high incidence of laboratory infection with *C. burnetii* indicates the need for very careful technique in handling animals and cultures. Serologic indications of infection are simpler and safer. Two or more serum samples taken during the acute and convalescent phases are examined by agglutination or complement fixation. A rising titer indicates active infection.

Antibiotic therapy is somewhat effective, particularly if treatment is initiated within the first few days after onset of symptoms. Tetracyclines are frequently employed, and vigorous and prolonged treatment is advisable to avoid relapse.

Since Q fever infection in livestock is of no great economic significance, and man is the principal species in whom disease occurs, there is not likely to be much pressure for application of control measures to animal populations. Control of the human disease is based especially on immunization. Vaccines made from killed organisms elicit humoral antibodies in man and induce some degree of protection. Only those individuals whose daily work involves livestock or their products generally require this form of protection. In areas where the disease is prevalent in domestic animals, milk should be thoroughly pasteurized at the upper limits of flash or holding processes, because any lesser treatment is likely to be inadequate.

SUPPLEMENTARY READING

Burrows, W.: *Textbook of Microbiology*, 20th ed. Philadelphia, W. B. Saunders Company, 1973.

Davis, B. D., Dulbecco, R., Eisen, H. N., Ginsberg, H. S., and Wood, W. B., Jr.: *Microbiology,* 2nd ed. New York, Hoeber Medical Division, 1973.

Dubos, R. J., and Hirsch, J. G. (eds.): *Bacterial and Mycotic Infections of Man,* 4th ed. Philadelphia, J. B. Lippincott Company, 1965.

Fenner, F. J., and White, D. O.: *Medical Virology,* 2nd ed. New York, Academic Press, Inc., 1975.

Horsfall, F. L., and Tamm, I. (eds.): *Viral and Rickettsial Infections of Man,* 4th ed. Philadelphia, J. B. Lippincott Company, 1965.

Joklik, W. K., and Smith, D. T.: *Zinsser Microbiology,* 15th ed. New York, Appleton-Century-Crofts, Inc., 1972.

Stewart, F. S.: *Bacteriology and Immunology for Students of Medicine,* 9th ed. London, Bailliere Tindall & Cassell, 1968.

Wilson, G. S., and Miles, A. A.: *Topley and Wilson's Principles of Bacteriology, Virology, and Immunity,* 6th ed. London, Edward Arnold (Publishers) Ltd., 1972.

SECTION FIVE

ENVIRONMENTAL AND APPLIED MICROBIOLOGY

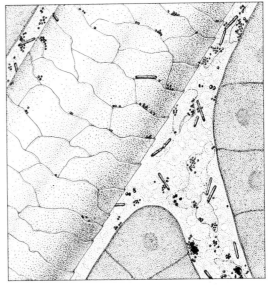

MICROORGANISMS 24 IN THEIR NATURAL HABITATS

NATURAL HABITATS

The natural habitat of early microorganisms was undoubtedly aquatic. Some of the descendants of these organisms became adapted to terrestrial life; that is, they gradually acquired the ability to secure moisture and nutrients from comparatively dry soil particles.

SOIL, WATER, AND AIR

Soil and water are today the principal habitats of microorganisms. Most microorganisms in soil are found in a thin layer within a few feet of the surface. Microorganisms in bodies of water such as lakes or oceans are concentrated at the surface and at the very bottom, in the few inches just above the mud. Air is a reservoir of microorganisms, but little if any multiplication occurs here, as it does in soil and water. Airborne microorganisms are accidental wanderers blown from the soil or water; powerful updrafts sometimes carry them to heights of 20,000 feet or more.

Conditions for Survival and Growth of Microorganisms. Certain conditions must be met if an environment is to serve as a habitat or a reservoir for living organisms. Growth depends on adequate supplies of water and food materials, proper pH, suitable oxygen tension, moderate temperature, and other factors. Requirements for survival are not as rigid as those for growth. Some organisms can resist severe desiccation; nutrients may be in very short supply, and temperatures may vary more widely. It is obvious that some organisms will survive better than others; spore-forming bacteria and cocci are particularly well adapted for survival under adverse conditions.

The size and nature of the population in a given habitat depend on and are limited by these various factors. A body of water containing little organic matter can support only a limited microbial population, even though all other conditions are favorable. Another body of water with a higher concentration of organic matter can support a higher population, but there may be too little dissolved oxygen to permit growth of aerobic organisms. Soil often provides abundant food supplies but may be too dry to permit extensive microbial multiplication. Air lacks moisture and nutrients and is an extremely poor culture medium.

METHODS OF STUDYING ENVIRONMENTAL MICROORGANISMS

DIRECT MICROSCOPIC EXAMINATION

Direct microscopic examination indicates the morphologic types of organisms in a given location but provides only comparative estimates of their numbers.

A survey of soil microorganisms can be made by burying glass slides for an appropriate period, carefully removing them from the soil and staining (i.e., by the Cholodny method). Bacteria that stick to the slides are classified according to their general morphology; many fungi, algae, and protozoa can be identified completely.

Microscope slides immersed in lakes or oceans are used to collect bacteria and other microorganisms that attach to solid surfaces. Algae and protozoa are routinely identified by direct examination of a drop of water on a microscope slide, but frequently the specimen must be concentrated by filtration (e.g., through sand, paper, or fine cloth).

Airborne plant pollen and mold spores and other microorganisms are collected on slides coated with glycerin. This technique was used many years ago in airplanes to determine the types of bacteria at high altitudes.

CULTIVATION

Cultural methods are used to count and identify viable bacteria. The plate count and the membrane filter count (see page 450) are the most common procedures. The number and kinds of bacteria determined by these methods depend on the composition of the culture medium and the incubation conditions. Autotrophic bacteria, for example, will not grow on media containing organic matter, nor will heterotrophic bacteria grow on completely inorganic media. Oxygen supply, temperature, pH, and other factors also affect the ability of bacteria to produce colonies. Numerous media have been devised for estimating bacterial populations in soil, water, and air, but no single medium or set of cultural conditions provides an accurate total bacterial count.

The foregoing limitation of colony counts can sometimes be used to advantage. Certain specific groups of bacteria can be counted in mixed natural populations by judicious selection of media and cultural conditions. For example, bacteria that attack urea are counted on a simple medium containing only this compound and appropriate minerals, and in which urea provides the sole source of carbon, nitrogen, and energy.

Certain types of bacteria cannot be counted satisfactorily on solid culture media, but an indication of their density in a mixed population can be obtained by the dilution count in suitable liquid media. This method is appropriate for estimating the number of organisms that ferment lactose, reduce nitrates, digest cellulose, produce hydrogen sulfide, and so forth. It is also used to determine the number of algae, protozoa, and certain other microorganisms in soil and water.

CHEMICAL ANALYSES

The results of the physiologic activities of certain groups of organisms can sometimes be assayed by chemical analysis. The activity of nitrate-reducing bacteria in sewage, for example, is indicated by quantitative determination of nitrites and ammonia. Conversely, the action of soil bacteria that oxidize ammonia or nitrites is demonstrated by analyses for nitrite and nitrate. Chemical analyses do not reveal the actual number of bacteria but do indicate their total activity, which is often of more importance.

MICROORGANISMS IN SOIL, WATER, AND AIR

THE BIOSPHERE

Those regions of the Earth that support or contain life include the upper few feet of soil, bodies of water, and the lower part of the atmosphere. Microorganisms from each habitat are found in both of the others, but their persistence is subject to the limitations on growth and survival previously mentioned. Certain organisms therefore predominate in the soil, others in water, and some in air.

MICROORGANISMS OF THE SOIL

The Nature of Soils. The numbers and kinds of microorganisms in soil depend upon its physical and chemical characteristics. Most soils are derived from rock and other inorganic matter by weathering, a combination of physical, chemical, and biologic processes. Growth of photosynthetic plants such as lichens and algae on rocks provides organic matter, which supports a population of heterotrophic organisms. Their metabolism yields various organic acids and carbon dioxide, and the latter reacts with water to form carbonic acid. These acids dissolve some of the rocks and minerals. Rocks are broken by freezing and by growth of plant roots in cracks and crevices. Excretions from roots and the accumulated dead plant material stimulate further development of microorganisms and continued dissolution of minerals. Water carries these substances downward, larger plants develop, and the soil increases in depth. Earthworms and other animals invade the area and keep the soil mixed and aerated.

Movement of materials gradually causes layering, and a typical soil profile is established. At the top is undecomposed plant material. Beneath that is a layer of partly decomposed matter, rich in microbial life. The subsoil contains humus and minerals leached from the surface, but there is little usable organic matter. Humus is the organic fraction of soil that is resistant to decomposition. It contributes greatly to good texture and to water holding capacity. The soil base is composed of stones and coarse particles, derived from the underlying bedrock.

The chemical and biologic nature of a soil is determined by the vegetation it supports, and this depends largely upon climate. (1) *Podzol* soils are found under oaks and conifers, such as pine, spruce, and fir. Their leaves and litter decompose slowly and yield acid products; the reaction may be as low as pH 3. Minerals rapidly leach away, and this soil type is relatively infertile. The predominant microbial life consists of fungi instead of bacteria and animal forms, most of which are sensitive to acidity. (2) *Brown* soils develop in forests of deciduous trees such as birch and maple. The reaction is only slightly acid; the topsoil contains considerable organic matter derived from leaf litter, together with many bacteria and animals, and is relatively fertile. (3) *Grassland* soil is rich in organic matter derived from the deep root systems of grasses. As these gradually decompose, a layer of black humus accumulates. (4) *Tropical* soils are found under rain forests. Although the vegetation is lush, the soil contains comparatively little organic matter because the high temperature and rainfall stimulate rapid decomposition of leaf litter. Nutrients are rapidly depleted and fertility is lost within a few years after the forest is cleared.

Water is an essential ingredient of soil, both for plant growth and for microbial activity. The amount present varies with rainfall, drainage, and plant cover. A well drained soil consists of approximately 50 per cent particles, 40 per cent water, and 10 per cent air. Some of the water is free and the remainder is adsorbed to particles. The relative proportion of each varies greatly with the composition of the soil and other factors. Usable nutrients are concentrated in water adsorbed to particle surfaces, and hence microorganisms are most abundant and active in close association with particles. Inasmuch as the total volume of water and air together is constant, the amount of water in the soil determines the amount of oxygen that can be present. This in turn affects microbiologic activity and soil fertility. The only oxygen in a waterlogged soil is that dissolved in the water, and this is soon exhausted by microorganisms. Desirable minerals such as nitrates and sulfates become reduced in anaerobic soils. The reduced compounds are often less satisfactory plant nutrients, so fertility is diminished.

Soil Populations

There is no way to determine precisely how many microorganisms a gram of soil contains, and quoted figures are always estimates. Soil populations are often expressed as ranges. Table 24–1 indicates very roughly the relative abundance of various microorganisms per gram of surface soil. Microbial counts decrease markedly at greater depths.

Plate counts of soil bacteria and molds are much lower than microscopic counts. There are two major reasons for the great difference noted in Table 24–1: (1) the plating media permit growth of only a small segment of the mixed flora, and (2) many dead organisms are detectable with the microscope.

Soil Bacteria. Bacteria are the most abundant microorganisms of the soil. Soil bacteria are of many types and physiologic activities: aerobic, facultative, and anaer-

TABLE 24–1. Relative Abundance of Microorganisms in Surface Soil

Group	Method	Microorganisms Per Gram
Bacteria	Microscopic count	1,000,000,000–22,000,000,000
	Plate count	100,000– 100,000,000
Actinomycetes	Plate count	500,000– 14,000,000
Molds	Microscopic count (mycelium pieces)	3,000,000– 50,000,000
	Plate count (mycelium pieces and spores)	30,000– 1,000,000
Algae	Dilution count	10,000– 30,000
Protozoa	Dilution count	
Ciliates		50– 200
Flagellates		1,000– 10,000
Amebae		500– 2,000

obic; autotrophic and heterotrophic; carbo-hydrate- and protein-decomposing.

Of the 19 Parts into which *Bergey's Manual of Determinative Bacteriology* divides the bacteria, soil bacteria are found mostly in five; however, many of the other Parts include some organisms often present in soil.

Part 7 (gram-negative aerobic rods and cocci) includes *Pseudomonas* as one of its important genera. These rod-shaped organisms frequently produce green or blue fluorescent, water-soluble pigments. Part 7 also includes unpigmented rods of the genera *Rhizobium* (associated with roots of legumes), *Agrobacterium,* and *Achromobacter.* In Part 8 (gram-negative, facultatively anaerobic rods) are found the pigmented forms *Chromobacterium* (violet) and *Flavobacterium* (yellow to orange). Among the gram-positive cocci of Part 14 are *Micrococcus* species that form either irregular groups or regular packets of cells. Part 15 consists of endospore-forming rods and cocci, notably *Bacillus* and *Clostridium.* Gram-positive nonspore-forming rods are found in Part 17 (actinomycetes and related organisms). Soil forms include *Arthrobacter*, certain species of *Corynebacterium,* and *Streptomyces.* The latter produce filamentous, often hard and chalky colonies on media used to cultivate other bacteria and frequently give off an odor like that of decaying leaves (see Figure 24–1). They are particu-

larly abundant in forest soil and in other soils containing large amounts of organic matter, and are active in its decomposition.

Molds in Soil. There is no good method of determining the amount of mold in a specimen. Fragments of mycelium appear in microscopic preparations, and fragments and spores give rise to colonies upon agar media. The total mass of mycelium appears to make up an appreciable proportion of the soil microflora.

Molds actively decompose cellulose, plant and animal proteins, and other complex organic substances. The carbon and nitrogen of these substances therefore become available to other microbial life and eventually to plants.

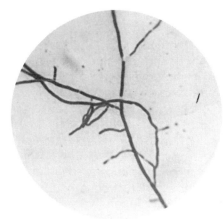

Figure 24–1. *A branching, filamentous soil actinomycete (approx. 1000X).*

Algae in Soil. Algae carry on photosynthesis at or near the soil surface and in so doing accumulate organic matter and energy. Their activity may temporarily decrease fertility because they need nitrogen and compete for this element with higher plants. However, the utilization of soluble nitrogen compounds may prevent the loss of these substances from the soil by leaching, and the nitrogen incorporated in algal protoplasm is gradually returned as dead algae decompose. The net result is beneficial.

Soil Protozoa. Protozoa are the most numerous animal forms in soil, and, by virtue of their relatively large size, they occupy a considerable percentage of its mass. Many protozoa, especially ciliates and amebae, ingest bacteria; some species feed solely on bacteria. Other species are saprophytic and utilize decaying organic matter. Destruction of soil protozoa is usually followed by a marked increase in bacterial population. It might be anticipated that the apparent battle between protozoa and bacteria is an important factor in soil fertility, but this does not seem to be the case. It has even been proposed that destruction of bacteria by protozoa permits more active and extensive multiplication and metabolism of the remaining bacteria, with the result that decomposition proceeds more rapidly. In general, protozoa appear to exert little direct effect on microbial processes within the soil.

Microorganisms in the Rhizosphere. The immediate vicinity of plant roots—*the rhizosphere*—is a region of great microbial activity and hence of large numbers of organisms. Bacteria appear to be most greatly stimulated in this area, and counts of a billion or more per gram are common. Gram-negative rods of the genera *Pseudomonas, Achromobacter,* and *Agrobacterium* are especially abundant, and there is also a marked increase in anaerobes, doubtless due to the utilization of soil oxygen and the formation of carbon dioxide by other organisms. Fungi and algae do not increase greatly, but small flagellate or ciliate protozoa are found in the water films on root hairs and plant epidermis.

The relationship between microorganisms and plants is often mutually beneficial. The roots excrete amino acids, simple sugars and nucleic acid derivatives, which are used as nutrients by bacteria. Some bacteria in turn produce plant stimulants known as auxins, such as indoleacetic acid. It has also been found that a flourishing rhizosphere microbial population provides a "buffer zone" in which many plant pathogens are suppressed, perhaps in part by antibiotics formed by the normal population.

MICROORGANISMS OF WATER

Some microorganisms are indigenous to water; many gain access to it from the soil, air, or animals.

Rain Water. Rain water is practically free from microorganisms except during the early period of a storm, when it washes floating microorganisms from the air. Its flora is therefore that of the air, and usually includes soil organisms blown upward by the wind: bacteria and cysts or spores of protozoa, algae, and fungi. The number of these organisms during the first part of a storm is rarely more than a few score per milliliter.

Surface Water. Rain water that has collected in pools, brooks, and rivers is surface water. It contains many microorganisms derived from the soil. Bacterial counts may be as high as several hundred thousand per milliliter, depending on the extent of soil contamination. Rivers and streams are frequently contaminated by sewage, which greatly increases the number of bacteria. Algae, fungi, and a few protozoa are present in most surface waters.

Stored Water. Stored water is that in reservoirs, ponds, and lakes. Bacterial counts are usually no more than a few hundred per milliliter. The gradual settling of microorganisms tends to reduce the population in the main part of a large body of water. Many algae and protozoa are often present, particularly during warm seasons when the surface water temperature is high.

Ground Water. Ground water is subterranean water that has been filtered through successive layers of soil. It appears in springs and wells. Water from springs and deep wells is usually nearly sterile; bacterial counts range from zero to a score or so per milliliter. Chromogenic (pigment-producing) bacteria are prominent. Ground water is usually free from other microorganisms, which are larger and more effectively removed by filtration through the soil.

Salt Water. Most water eventually reaches the sea, carrying with it dissolved minerals and some microorganisms. Near the coast, rivers maintain a concentration of mineral nutrients sufficient to support multiplication of algae. Rivers that drain a heavily populated area are likely to carry large quantities of sewage, either raw or only partially treated. This means that not only are intestinal and other bacteria discharged into the coastal waters, but also sufficient organic matter to permit their survival. In areas near large cities such as New York City these organisms may be a health hazard to those using the beaches for recreation. Aesthetic as well as health problems are also created by dumping garbage from large coastal cities at sea. In the vicinity of New York, garbage has been detected slowly drifting many miles toward beach and residential areas.

Algae, both freely floating (*phytoplankton*) and attached to the bottom (*benthic algae*), utilize light energy in the initial production of organic matter and are called *primary producers*. The biologic activity of the marine environment depends on the rate of primary production. Algae are either eaten by protozoa and crustaceans (*zooplankton*) or die and decompose owing to bacterial action. In the latter case, protozoa may ingest the bacteria. In any event, the zooplankton are themselves consumed by larger invertebrates, which in turn are eaten by fish. The food chain in an aqueous environment therefore begins with algae as primary producers and continues through zooplankton to larger animals and fish. Primary productivity, and hence fish populations, are especially high in coastal areas.

Recent observations indicate that there are five principal populations of microorganisms in the sea. (1) In the open sea, especially in the upper 300 meters, where phytoplankton release an average of about 25 per cent of their photosynthesate in the form of fatty acids, amino acids, and sugars, bacterial populations of 10,000 to 100,000 per ml. may be found. (2) At the sea-air interface there is a very thin film (approximately 0.1 μm) containing a high concentration (e.g., 0.2 per cent) of dissolved organic carbon, together with nitrogenous and other nutrients sufficient to support a population of 100,000 to 100,000,000 bacteria per ml. (3) Most surfaces, whether inanimate or living, are covered with considerable numbers of attached microorganisms. Bacteria colonize surfaces reversibly by adsorption because of van der Waals forces or electric attraction, or irreversibly by chemotactic attraction to organic substances such as protein-polysaccharide complexes. In the latter case, they attach by means of polysaccharide fibrils and multiply, forming a dense microbial film. Diatoms also accumulate, and protozoa then feed on the smaller microorganisms. (4) A large population of microorganisms is associated with the waste products and remains of larger organisms. Invertebrates and vertebrates convert about one fourth of their feed into excrement, which provides nutrient for heterotrophic microorganisms, both within the gut and in the environment after excretion. In addition, the animal bodies themselves, after death due to normal, seasonal, or catastrophic causes, serve as microbial nutrients. (5) These materials, often macroscopic or particulate in nature, settle to the sea bottom, where a large and varied population of organisms can be found. The bottom of the deep sea is dark, cold (2° C.), and under high pressure (200 to 1000 atm.), so microbial activity is limited to forms adapted to these conditions. Bacteria are the primary decomposers in the upper, aerobic zone, succeeded by flagellates and ciliates. In deeper, anaerobic layers, such bacteria as *Desulfovibrio* reduce sulfates and release hydrogen sulfide.

Kinds of Microorganisms in Water. Bacteria found in water include organisms of the *Pseudomonas fluorescens* type and other chromogenic (violet, red, yellow) rods, coliform bacteria (organisms that resemble *Escherichia coli*), *Proteus* species, sporeformers, and cocci (white, yellow, pink). The significance of coliform bacteria is discussed in considerable detail on page 448.

The number of microorganisms in water

depends on the initial contamination and the ability of the organisms to survive or to multiply. Multiplication is a function of food supply, oxygen, pH, temperature, and other factors. Food supply is determined by the surrounding terrain (e.g., gravel vs. rich, agricultural loam) and by waste materials introduced by man. Water containing organic matter supports a more luxuriant microbial population than pure water.

Microbial, Plant, and Animal Cycles. The interrelationships of microbial and higher organisms are illustrated diagrammatically in Figures 24–2 and 24–3. Algae flourish in bodies of water when the surface layers are warmed by the summer sun. They constitute the principal food for many fish, but in a hot season algae multiply faster than fish can eat them. The excess die and are decomposed by bacteria; this frequently causes the liberation of unpleasant odors. Bacteria in turn are succeeded by protozoa, and their death again provides food for bacteria.

MICROORGANISMS OF AIR

Most microorganisms in the air are in a state of suspended animation. Many are soon killed by desiccation, starvation, ultraviolet irradiation, or other unfavorable conditions. Living airborne organisms are either unusually resistant or else have been in the air only a short time. Resistant forms may be capable of producing spores or cysts; some are spherical (spherical cells offer the least surface area and appear to resist harmful agents better than cells of other shapes); and some resistant organisms possess pigments that protect against injurious radiations.

Sources of Airborne Microorganisms. Extramural (i.e., outdoor) microorganisms are derived from the soil and from bodies of water. Soil organisms on dust particles are carried into the atmosphere by the wind and rising currents of warm air; aquatic organisms enter the atmosphere in water droplets through the combined action of waves and wind.

Bacteria and fungi predominate in outdoor air. Terrestrial bacteria that survive in the air include gram-positive rods, spore-formers, and cocci. Marine bacteria found in air are principally gram-negative rods; fewer than half are spore-forming bacteria. Species of *Vibrio* or *Spirillum* have not been found, al-

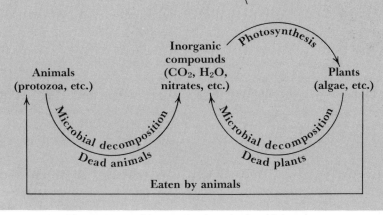

Figure 24–2. *Plant, animal, and microbial cycles.*

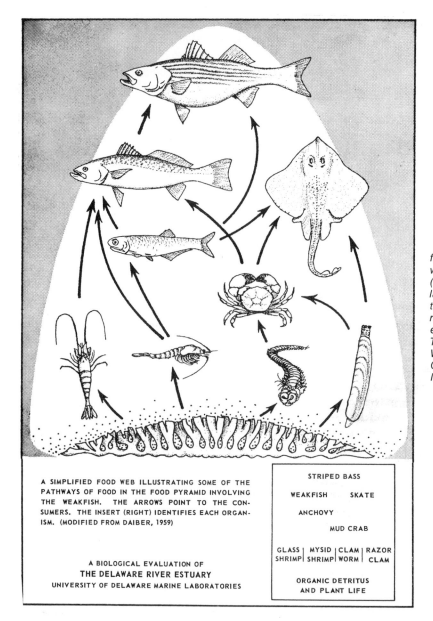

Figure 24–3. *An aquatic food web, showing the pathway from primary producers (plants) at the bottom to a large fish (striped bass) at the top, in a circumscribed environment (the Delaware River estuary). (From Wilber, C. G.: The Biological Aspects of Water Pollution. Courtesy of Charles C Thomas, Publisher, Springfield, Illinois.)*

A SIMPLIFIED FOOD WEB ILLUSTRATING SOME OF THE PATHWAYS OF FOOD IN THE FOOD PYRAMID INVOLVING THE WEAKFISH. THE ARROWS POINT TO THE CONSUMERS. THE INSERT (RIGHT) IDENTIFIES EACH ORGANISM. (MODIFIED FROM DAIBER, 1959)

A BIOLOGICAL EVALUATION OF
THE DELAWARE RIVER ESTUARY
UNIVERSITY OF DELAWARE MARINE LABORATORIES

STRIPED BASS

WEAKFISH   SKATE

ANCHOVY

MUD CRAB

| GLASS SHRIMP | MYSID SHRIMP | CLAM WORM | RAZOR CLAM |

ORGANIC DETRITUS
AND PLANT LIFE

though they constitute 2 to 3 per cent of the bacterial flora of the oceans. Mold spores are always of terrestrial origin. Tremendous numbers are sometimes encountered as "clouds" or "storms," and they may travel thousands of miles from their point of origin. Spores or cysts of yeasts, algae, and protozoa are also present in the air. Some extramural microorganisms are important causes of plant disease (e.g., wheat rust).

Intramural microorganisms include not only the outdoor organisms characteristic of a geographic location but also those of man and the other inhabitants of the area. As was illustrated in Figure 23–1, man expels large numbers of microorganisms in saliva and mucus droplets when sneezing or—to a lesser extent—when coughing or even talking. Human pathogens are frequently found indoors, where ready access to susceptible hosts permits transfer despite their limited survival powers. Few if any human pathogens are transmitted long distances out-of-doors.

Number of Airborne Microorganisms. The number of microorganisms in the air is difficult to determine with accuracy. Simple comparisons are made by gravity plates:

a nutrient medium in Petri dishes is exposed to the air for a definite period (e.g., 15 or 60 minutes), and the colonies that develop after incubation are counted.

Various devices are used to sample a definite volume of air: the air is drawn through moist sand or cotton or bubbled through sterile water; plate counts are then made of the entrapped organisms. Such methods rarely recover 100 per cent of the air flora. Cellulose acetate membrane filters have also been used for counting air bacteria. A known volume of air is drawn through the filter, and the filter is placed in a Petri dish on an absorbent pad moistened with a nutrient broth; visible colonies appear after short incubation.

Counts of airborne microorganisms vary widely with the time of day and season of the year, weather conditions, location, and other factors. The numbers determined in a city street by bubbling air through sterile liquid vary from a few hundred to many thousands per liter.

BIOGEOCHEMICAL ACTIVITIES OF MICROORGANISMS

ENERGY FLOW AND TRANSFORMATION OF MATTER

As previously defined, an ecosystem is usually an open system in a steady state, that is, formation of each component is balanced by its disappearance. The adult human body was cited as an example. Another familiar illustration is an aquarium, containing both plant and animal life and completely sealed so that nothing escapes and nothing enters except light. This is a small model of larger ecosystems such as a lake, an isolated, self-sufficient farm, or Earth itself. Any community of organisms, with its physical and chemical environment, can be said to comprise an ecosystem.

The steady state of an aquarium or a lake exists only because plant and animal nutrient and energy requirements complement each other. Plants utilize inorganic nutrients and CO_2, whereas most animal food is organic. The energy requirement of plants can be satisfied by light, but animals must secure energy by oxidation.

The energy that plants store in carbohydrates is released and utilized by herbivorous animals, and their energy in turn finds its way to carnivores. Dead plants and animals are decomposed by microorganisms, which also release and utilize some energy. At each step in the food chain there is some loss of energy in the form of heat, and the net result is a constant flow of energy through the system (see Figure 24–4).

The various chemical constituents of an ecosystem are converted from one form into another in a cyclic pattern, appearing first in plants, then in animals, and again in plants. The basic changes in elements such as carbon, nitrogen, and sulfur are from the reduced state to the oxidized state and back to the reduced condition. Respiration and fermentation release energy, and these processes are characteristic of animals, non-photosynthetic organisms, and photosynthetic organisms in the dark. The CO_2 produced by animal respiration is then reduced by plants, and the cycle is completed. Another cyclic pattern is illustrated by the synthesis of protein, its breakdown to organic acids and ammonia, and the oxidation of ammonia to nitrate, which is utilized again in the synthesis of protein.

Microorganisms play vital roles in the transformation of carbon, nitrogen, and sulfur. The cycles of these elements in nature will therefore be discussed in some detail. In each it will be noted that there are two principal phases: (1) *immobilization* of elements by the formation of organic substances, and (2) *mineralization* or return of the elements to the inorganic form.

The Carbon Cycle. The main outlines of the carbon cycle are familiar: photosynthetic reduction of atmospheric CO_2 by plants, assimilation of plants by animals, and release of CO_2 by animal and plant respiration. If this were all there was to the cycle it would soon come to a halt, because some important plant constituents cannot be digested by animals, nor can animals return all their carbon to CO_2 by respiration.

Role of Microorganisms. Microorganisms play the essential role of scaven-

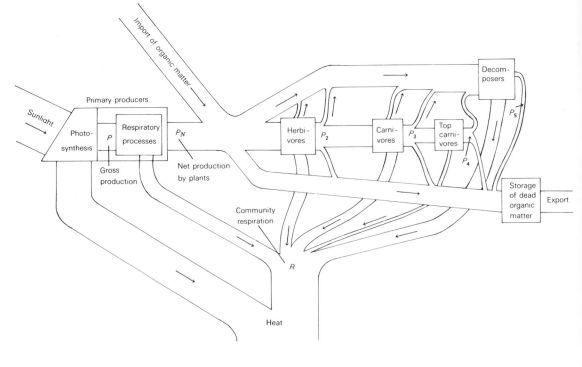

Geochemical activities of microorganisms

Figure 24–4. *Energy flow through the food chain of an ecosystem. (After Odum, E. P.: Fundamentals of Ecology, 3rd ed. Philadelphia, W. B. Saunders Co., 1971.)*

gers, digesting the residues of plant and animal materials. Dissimilation and fermentation (see page 178 ff.) yield acids, alcohols, and other intermediate waste products; respiration oxidizes these to CO_2 (Fig. 24–5). Microbial transformations of carbon are carried out largely in the soil but also occur in water and in any other situation where organic matter decomposes. Under anaerobic conditions (e.g., some sewage disposal processes) methane (CH_4) is formed.

The production of compost is a practical illustration of the decomposition of plant residues. Leaves, lawn clippings, garbage, and other organic materials are piled in a heap, usually with the addition of inorganic fertilizer rich in nitrogen, which is often in short supply. The piled materials have good insulating properties, and the temperature in the center rises rapidly within two or three days, reaches a peak of 65° to 70° C., and gradually returns to ambient temperature as the easily oxidized material is utilized. Mesophilic microorganisms are quickly re-

placed by thermophilic bacteria, actinomycetes, and fungi. They decompose cellulose and other complex carbohydrates, and in the course of one or two years convert the plant material into friable humus. Addition of this to garden soil greatly improves its texture.

Energy Transformations. Plants and animals are placed at the top of the carbon cycle in Figure 24–5 to indicate that they represent the highest form of organization of carbon compounds; CO_2, the simplest carbon compound, is at the bottom of the cycle. The carbon cycle is essentially a cycle of energy changes. Energy is acquired in the transformation from CO_2 to plant material. The dissimilation and respiration processes by which plant and animal substances are reconverted to CO_2 are energy-yielding reactions.

Microorganisms must not be considered external agents in this cycle or in any of the cycles; the cycles should actually be drawn to indicate incorporation of carbon and other elements within the microbial cells as pro-

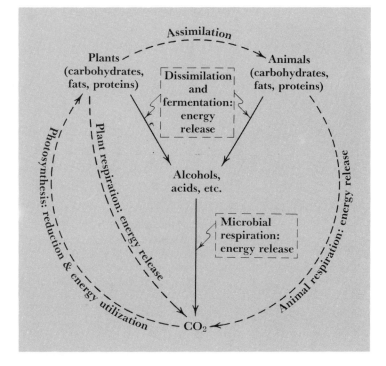

Figure 24–5. *The carbon cycle. Microbial activities are indicated by the solid arrows.*

toplasmic constituents and their subsequent release by dissimilation and respiration. There are therefore cycles within cycles, but it is simplest to present the over-all picture, as in Figure 24–5.

The Nitrogen Cycle. The nitrogen cycle looks more complicated than the carbon cycle, but this is because it is easier to represent some of the intermediate stages in a diagram (Fig. 24–6).

The nitrates assimilated by plants are reduced, and most of the nitrogen in protoplasm is in the amino radicals of protein molecules. Plant proteins, when eaten by animals, are converted principally into animal proteins. Animal metabolism yields excretions containing nitrogenous compounds, such as urea or uric acid, from which the nitrogen is set free as ammonia by appropriate organisms. These excretions, however, account for only a small proportion of the nitrogen within the animal body.

Peptonization and Ammonification. The nitrogen of dead animal and plant tissues returns to the inorganic state by hydrolytic decomposition through the familiar stages—proteoses, peptones, peptides, amino acids—and deamination of amino acids, which releases ammonia. The early steps in hydrolysis are called *peptonization*. Many, but not all, microorganisms can attack native proteins. This hydrolysis is usually performed extracellularly, and other microorganisms may utilize the proteoses or peptones and cause *ammonification*. Ammonification converts the nitrogen of organic compounds to its most reduced form and is logically indicated at the bottom of the cycle.

Oxidation of Ammonia. It would seem appropriate for nature to have provided a mechanism whereby plants could assimilate ammonia directly but, as mentioned previously, most plants secure nitrogen from nitrates. There are, however, a few bacteria that can perform the necessary oxidations, but no one species oxidizes ammonia completely to nitrate. The first step, *nitrosification*, is the oxidation of ammonia to nitrous acid or nitrite. This is accomplished by species of *Nitrosomonas*. The second step is *nitrification,* the oxidation of nitrite to nitrate by *Nitrobacter* species. These and a few related species are apparently the only microorganisms that can convert ammonia into the form plants utilize. Destruction of either the *Nitrosomonas* or *Nitrobacter* type of organism

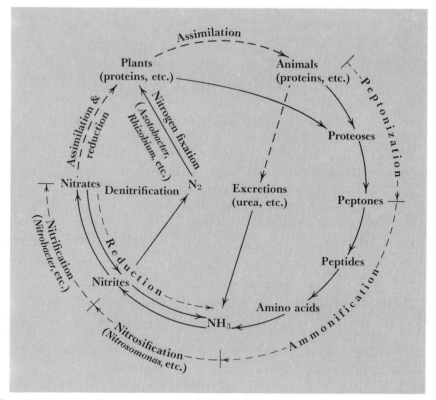

Figure 24–6. *The nitrogen cycle. Microbial activities are indicated by the solid arrows.*

quickly causes serious loss of soil fertility. Fortunately these bacteria are widely distributed.

Reduction. Nitrates that are not promptly assimilated by plants are likely to be lost, either by leaching or by reduction. It was mentioned on page 183 that many microorganisms utilize nitrate as a hydrogen and electron acceptor in the course of anaerobic respiration. Reduction of nitrate yields nitrite, ammonia, or free nitrogen.

Nitrate reduction provides an important route of microbial assimilation of inorganic nitrogen. Ammonia formed intracellularly can react with organic acids derived from carbohydrate dissimilation to produce amino acids, from which cellular proteins may be built. *Denitrification* can be considered a special form of nitrate reduction or respiration in which the end product is gaseous N_2. This is given off to the atmosphere and constitutes a leak in the nitrogen cycle.

From the viewpoint of soil fertility, denitrification is definitely harmful, whereas nitrate reduction with the liberation of nitrite or

ammonia may be harmful. Nitrite is toxic to some plants, and ammonia is not assimilated by many plants as readily as is nitrate. Microbial assimilation of nitrogen temporarily reduces fertility, but the nitrogen, in combined form, is still present in the soil and eventually becomes available for plant growth.

Nitrate reduction, including denitrification, is particularly apt to occur under anaerobic conditions, for example, in swampy or water-logged soil.

Nitrogen Fixation. The inability of higher plants and animals to utilize atmospheric nitrogen would be a catastrophe were it not for the fact that a considerable number of microorganisms can do so, because only a limited amount of nitrogen is converted into combined form by lightning and other natural means. Artificial methods of nitrogen fixation are expensive. Microorganisms that fix nitrogen help to maintain a balanced condition in the soil and make up for loss of nitrogen by denitrification.

The mechanism of biologic nitrogen fixa-

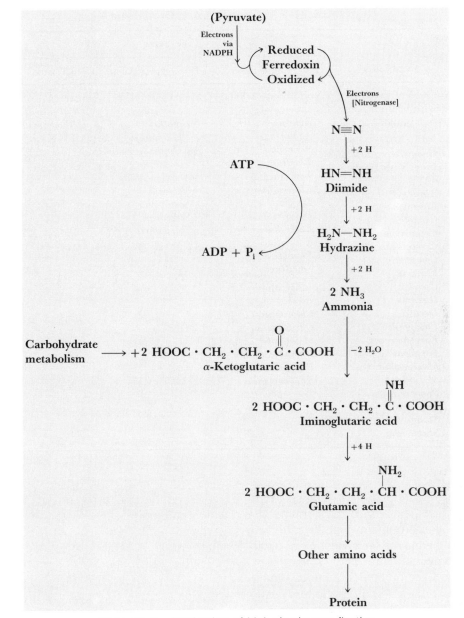

Figure 24-7. *Mechanism of biologic nitrogen fixation.*

tion varies somewhat with the organism. In general, however, free nitrogen is reduced to ammonia, from which glutamic acid and other amino acids can be produced (see Figure 24-7). Reduction of nitrogen consumes a large amount of energy, and this is derived from ATP. The electrons required for the process come from NADPH via the iron-protein ferredoxin. Stepwise reduction in two-electron stages is catalyzed by the en-

zyme nitrogenase, through the intermediate compounds diimide and hydrazine.

The nitrogen fixing microorganisms are usually separated into two groups (see Table 24-2): (1) symbiotic nitrogen-fixing organisms, which fix nitrogen in nodules on the roots or leaves of plants, and (2) free-living nitrogen-fixing organisms, which live independently in soil or water.

Symbiotic nitrogen-fixing organisms in-

TABLE 24–2.   Nitrogen-fixing Organisms

| Symbiotic | Free-living | | |
	Bacteria	Blue-green Algae	Yeasts
Rhizobium species with legume plants	Non-photosynthetic species of:	Species of:	Species of:
Klebsiella-like species with *Psychotria*	*Achromobacter*	*Anabaena*	*Pullularia*
Actinomycetes and/or fungi with species of:	*Azotobacter*	*Calothrix*	*Rhodotorula*
Alnus	*Azomonas*	*Chlorogloea*	
Casuarina	*Bacillus*	*Cylindrospermum*	
Ceanothus	*Beijerinckia*	*Fischerella*	
Ceratozamia	*Clostridium*	*Mastigocladus*	
Cercocarpus	*Derxia*	*Nostoc*	
Comptonia	*Desulfovibrio*	*Scytonema*	
Coriaria	*Enterobacter*	*Stigonema*	
Cycas	*Nocardia*	*Tolypothrix*	
Discaria	*Pseudomonas*		
Dryas	*Spirillum*		
Elaeagnus	Photosynthetic species of:		
Encephalartos	*Chlorobium*		
Hippophae	*Chromatium*		
Macrozamia	*Methanobacterium*		
Myrica	*Rhodomicrobium*		
Podocarpus	*Rhodopseudomonas*		
Purschia	*Rhodospirillum*		
Stangeria			
Shepherdia			

clude the genus *Rhizobium,* whose members are associated with the legumes; a *Klebsiella*-like organism that is found in nodules on the leaves of *Psychotria;* and various actinomycetes and fungi, which fix nitrogen in nodules on nonleguminous plants. The plant supplies a source of energy, and the symbiont returns nitrogen in a combined form that the plant can eventually utilize.

Free-living nitrogen-fixing organisms include some nonphotosynthetic and photosynthetic bacteria, more than 40 species of blue-green algae, and a few yeasts. Most of these organisms reside in the soil or water where they utilize plant residues and other organic matter or CO_2 as a source of carbon, secure energy by oxidation or from sunlight, and fix nitrogen from the atmosphere, if the environment does not contain sufficient combined nitrogen to support growth. Some of the bacteria, such as *Azotobacter, Pseudomonas,* and *Spirillum,* are aerobic. With

the exception of *Clostridium pasteurianum,* which has long been known to fix nitrogen, it was not realized until about 1950 that most nitrogen-fixing bacteria are anaerobic. *Desulfovibrio desulfuricans* is nonphotosynthetic, and among photosynthetic bacteria are *Chlorobium, Chromatium, Rhodospirillum rubrum,* and *Rhodomicrobium vanniellii.*

Because of its significance in agriculture, the symbiosis between leguminous plants and *Rhizobium* species has been investigated more thoroughly than any other such relationship. For many years the rhizobia were known as the *root nodule bacteria,* because they infect the plant roots and produce nodules or tubercles containing millions of the bacterial cells (see Figure 24–8). Infection can occur only when the proper species of bacterium is present on the root hairs of the young plant. Multiplication of the bacteria (and also of other bacteria in the rhizo-

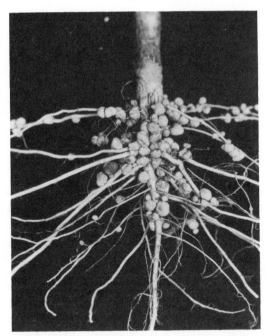

Figure 24–8. *Photograph of roots of a legume (soybean) showing nodules. (Photo courtesy The Nitragin Company. From Minicourses in Biology: Environmental Biology, Interactions, p. 4. Philadelphia, W. B. Saunders Co., 1975.)*

sphere) is stimulated by indoleacetic acid, which is formed from tryptophan secreted by the roots. A bacterial exudate then causes the roots to secrete polygalacturonase, an enzyme that loosens the fibrils of the root hair wall and allows the small, rodlike "swarmer" rhizobia to enter and form "infection threads," by which they eventually make their way into cells of the inner cortex of the root. The bacteria then multiply within the root cells and enlarge, assuming bizarre, irregular shapes called *bacteroids* (see Figure 24–9). The masses of root cells, swollen with bacteroids, comprise the nodules. Physical and physiological communication between the nodules and the vascular system of the root allows exchange of materials. Plant carbohydrates provide carbon and energy, and the bacteroids secure nitrogen from the soil atmosphere. Bacterial growth yields a store of combined nitrogen, and this ultimately passes in some form to the plant tissues. Plant growth is therefore improved, especially on a soil that contains little combined nitrogen.

Legumes comprise one of the great groups of seed plants; they include such common plants as alfalfa, clover, peas, beans, and soybeans. Legumes contain more nitrogen than nonlegumes; alfalfa, for example, contains 300 to 350 pounds of protein per ton, whereas timothy contains only 115 to 150 pounds of protein per ton. Much of the nitrogen in alfalfa protein can be derived from the atmosphere through the action of root nodule bacteria. Legumes can therefore be grown in soil that would not support nonlegumes. Moreover, if the entire legume crop is plowed into the soil the nitrogen content of the soil increases markedly. Even if only the roots are left, fertility is not depleted as much as by comparable harvesting of a nonlegume.

Rhizobium species are differentiated mainly according to the legumes they can

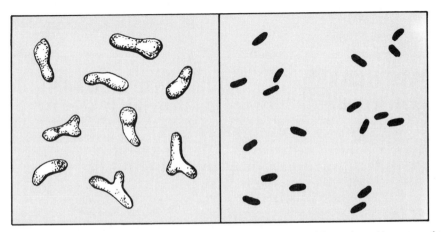

Figure 24–9. *Bacteroids* (left) *and rod forms* (right) *of rhizobia. Bacteroids are found in smears from crushed nodules, rod forms in smears from agar slant cultures.*

infect. *Rhizobium meliloti,* for example, infects and produces nodules upon alfalfa, sweet clover, and bur clover. *R. leguminosarum* produces nodules upon the garden pea, field pea, sweet pea, vetch, broadbean, and lentil. The species of plants upon which the same species of *Rhizobium* fixes nitrogen are called *bacterial-plant groups* or *cross-inoculation groups*. There are many bacterial-plant groups.

Root nodule bacteria are reasonably well distributed in soil all over the Earth, but often not enough are present to ensure complete nodulation of all susceptible plants. The number of rhizobia decreases during periods of drought, unfavorable temperatures, and acidity, and as a result of the activity of bacteriophage and other antagonistic microorganisms. Moreover, effective nodulation occurs only when species of *Rhizobium* of the proper bacterial-plant group are present, and these are not likely to be abundant in a soil upon which plants of another group have been raised for several years.

Inoculation of the seed is recommended to ensure an adequate number of bacteria for a given crop. A common practice is to mix the seed with the bacteria just before planting. After germination of the seed, the minute hairs covering the main root and its branches can be infected by rhizobia in the immediate vicinity, and the nodules that develop are the site of nitrogen fixation. Direct inoculation of the seed places the bacteria in the most advantageous position for infection.

Leguminous plants without nodules grow slowly in nitrogen-poor soils; their leaves turn yellow, and they may die of nitrogen starvation. Nodulated plants grow normally under the same conditions and have a healthy green color (Fig. 24–10). Plants either with or without nodules grow normally in soils containing an abundance of nitrogenous foods. The beneficial effect of nodulation is especially apparent in soils low in combined nitrogen. Not only is growth accelerated, particularly of the young plant, but the yield and protein content of the crop are also improved.

Symbiotic nitrogen fixation amounts to as

Figure 24–10. *Growth of a legume in nitrogen-poor soil;* left, *without root nodule bacteria;* right, *with root nodule bacteria. (From Fred, Baldwin and McCoy; Root Nodule Bacteria and Leguminous Plants. Madison, University of Wisconsin Press, 1932.)*

much as 200 pounds per acre per year under favorable conditions; 50 to 100 pounds is an average figure. The importance of the root nodule bacteria can be appreciated more easily when it is realized that this amount of nitrogen is contained in 500 to 1000 pounds of commercial fertilizer containing 10 per cent nitrogen.

The beneficial effects of leguminous plants upon soil fertility have been known for at least 2000 years. Vergil described crop rotation as practiced in ancient Rome; a non-legume was grown one year, alfalfa was grown the next year, and the same field was allowed to remain uncultivated the third year. Similar procedures in use today not only help to maintain the nitrogen content of the soil but also improve its texture and moisture-holding capacity.

Among the nonleguminous plants possessing nodules that contain nitrogen-fixing microorganisms are species of *Alnus,* the alder tree; *Myrica gale,* the bog myrtle; species of *Eleagnus;* and various other trees and shrubs in all parts of the world, from tropical to arctic regions. Nodules of these plants contain actinomycetes, fungi, or microorganisms as yet unidentified. That the nodules are essential for nitrogen fixation has been demonstrated by inoculating plants grown aseptically in nitrogen-poor soil with crushed nodules from infected plants: nodulation and normal growth then occurred.

The total amount of nitrogen added to soil by nitrogen-fixing microorganisms has been estimated to be nearly 100 million tons per year over the Earth's surface. Nitrogen fixation in the open oceans is largely a function of blue-green algae, and little is known about their productivity. Species of *Nostoc* are especially common in the Indian Ocean. *Nostoc* is also abundant in Antarctic waters and on rocks and soil surfaces. In Antarctica they fix nitrogen even at 0° C.

Nitrogen Transformation in Sewage. Most of the steps in the nitrogen cycle take place in water as well as in soil. Sewage contains plant and animal proteins, which decompose and form ammonia and other reduced nitrogenous compounds. Accumulation of these substances under anaerobic conditions creates unpleasant odors; moreover, some of the products are toxic and cannot be discharged into streams, lakes, or other bodies of water without danger to fish. In sewage treatment plants where sufficient aeration is provided, ammonia is oxidized to the inoffensive nitrate stage. The effectiveness of the treatment can be measured by quantitative determinations of nitrate.

The Sulfur Cycle. Sulfur is an essential component of proteins and hence is required by all organisms. It is found in three amino acids, cysteine, cystine, and methionine. Plants assimilate sulfates and reduce the sulfur to sulfhydryl (—SH) or disulfide (—S:S—). The sulfur cycle (Fig. 24–11) resembles the nitrogen cycle.

Dissimilation. Dissimilation of either plant or animal proteins through the usual series of intermediate breakdown products yields amino acids. The sulfur of the three amino acids mentioned above is completely reduced and removed as hydrogen sulfide. Many species of heterotrophic saprophytes bring about these dissimilation steps.

Oxidation of Sulfide and Sulfur. Sulfide oxidation by various facultative autotrophs produces globules of sulfur, which accumulate within the cells of some species, outside the cells of others. The photosynthetic sulfur bacteria, members of the Chromatiaceae and Chlorobiaceae, contain pigments similar to chlorophyll. Hydrogen sulfide (H_2S) serves as hydrogen donor in the photosynthetic process, and elemental sulfur is released. Nonphotosynthetic sulfur bacteria of the families of Achromatiaceae and Beggiatoaceae also may oxidize H_2S to free sulfur.

Sulfur must be oxidized to sulfate before it can be utilized by plants. This process is brought about by various strictly autotrophic bacteria, particularly species of *Thiobacillus, T. thioparus* and *T. thiooxidans.* Certain strains produce so much sulfuric acid that the reaction of the medium falls to less than pH 1.0.

Sulfate Reduction. A reversal in the sulfur cycle may also occur, just as it does in the nitrogen cycle. Various familiar bacteria, including species of *Clostridium, Proteus, Desulfovibrio,* and others, reduce sulfates to

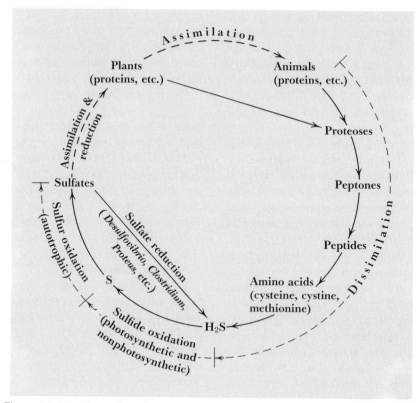

Figure 24–11. *The sulfur cycle. Microbial activities are indicated by solid arrows.*

H_2S. Sulfate reduction is not as critical a matter as reduction of nitrate, because the supply of sulfate in soils is more nearly adequate.

Interrelationships of the Carbon, Nitrogen, and Sulfur Cycles. The student must not think of the carbon, nitrogen, and sulfur cycles as separate and independent series of processes. Rather, each is a simplified representation of a sequence of interrelated processes, and no cycle can operate in the absence of the others. In the carbon cycle attention is focused on the transformations of carbon, but nitrogen and sulfur transformations also occur at the same time. Similarly, transformations of sulfur and carbon accompany transformations of nitrogen, and so forth. Protein dissimilation yields amino acids, and some of the same organisms set free ammonia (NH_3) and H_2S; moreover, part of the carbon of amino acids may be oxidized to CO_2. Soil conditions that favor oxidation of ammonia to nitrates also favor oxidation of H_2S to sulfates if appropriate bacteria are present, and conditions that favor nitrate reduction also favor sulfate reduction.

ROLES OF ENVIRONMENTAL MICROORGANISMS IN SANITATION

MICROORGANISMS AS SCAVENGERS

Microbial decomposition of plant and animal remains prevents the accumulation of vast amounts of dead organic matter and permits the biologic transformations by which food supplies are continuously made available.

Microbial Antagonism. The antagonistic activity of various microorganisms in soil and water is directly related to human and animal health. Few human pathogens can be isolated from soil, even though it is undoubtedly often contaminated by these organisms. Polluted streams and other bodies of water gradually lose their pathogenic flora. Many soil microorganisms are capable of producing antagonistic or even antibiotic substances, and any specimen of soil is likely to contain organisms antagonistic to one another. Bacterial viruses that kill and lyse various bacteria, including pathogenic species, are frequently present in soil and water. These and other factors limit the survival of many kinds of pathogenic bacteria.

Sewage Disposal. Disposal of sewage in sparsely settled areas is usually no problem if a suitable stream or body of water is available into which it may be dumped. In contrast, industrialization and urbanization and the consequent crowding of millions of people into relatively small areas create serious problems of disposal of waste materials. The amount of domestic sewage is approaching 150 gallons per person per day in most large cities of the United States.

Composition of Domestic Sewage. Domestic sewage consists of household wastes, often augmented by storm water, and is composed of 98.8 to 99.5 per cent water and 0.5 to 1.2 per cent solid matter in the form of large and small particles and colloidal and soluble substances. Colloidal and dissolved material is not removed by settling or by filtration. Various antiseptic detergents, often not biodegradable, pesticides, and other indigestible household chemicals are also present. The solids of sewage contain carbohydrates, fats, proteins, and various salts. The reaction of sewage varies from pH 6.0 to pH 8.5. Little if any dissolved oxygen is present.

Domestic sewage contains between 500,000 and 20,000,000 microorganisms per milliliter. These are largely intestinal bacteria, but soil microorganisms are also abundant. Most of the microorganisms of sewage are nonpathogenic. They include aerobic, facultative, and anaerobic types, both heterotrophic and autotrophic, and are principally mesophilic, but some psychrophiles and thermophiles are also present.

Industrial Wastes. Industrialization has created unusual problems of sewage disposal. Certain industries produce highly acid waste materials; others produce alkaline wastes; and some contribute troublesome chemicals, such as copper or silver salts, oils, or unusually large amounts of proteins, carbohydrates, or fats. A partial list of the types of industries will suggest the kinds of wastes that must be handled: meat packing, canning, brewing, dairy products manufacture, tanning of leather, manufacture of paper and textiles, oil refining, chemical manufacture.

Agricultural Wastes. Disposal of agricultural wastes is fast becoming a serious problem. Intensive livestock farming concentrates large numbers of animals—cattle, hogs, and chickens—in small areas, where they are fed, rather than being allowed to graze over a large territory. The wastes that accumulate must be treated or removed. Each pig, for example, produces about 3 gallons of semiliquid waste daily. Piggeries holding 20,000 pigs are not uncommon, and such an establishment necessitates a disposal operation of the same magnitude as a city of 40,000 people.

Objectives of Sewage Treatment. The objectives of sewage treatment are (1) to remove or decompose organic matter capable of supporting microbial growth, and (2) to remove or destroy pathogenic microorganisms. Pathogens of many types gain access to sewage, but those likely to survive long enough to cause human infection are principally intestinal in origin: typhoid and paratyphoid, dysentery, and Asiatic cholera bacteria.

The percentage of organic matter in sewage seems so small as to be insignificant. However, it should be recalled that nutrient laboratory media containing less than one per cent organic matter support bacterial populations of a billion or more per milliliter. Sewage contains all the nutrients necessary to nourish many kinds of bacteria. Lack of oxygen means that the chemical activities of these bacteria are largely anaerobic; hence foul smelling compounds are produced. Part of the aim of sewage treatment is to eliminate these offensive odors.

Small amounts of sewage discharged into a stream or large body of water are diluted enough that aerobic conditions are obtained. Soluble organic compounds are readily hydrolyzed and oxidized, and inoffensive products such as carbon dioxide, nitrates, and sulfates are formed. Complex compounds are hydrolyzed more slowly, but eventually the soluble products are also oxidized. Large amounts of sewage introduced into a body of water are not diluted sufficiently, and as a result conditions become highly anaerobic. Microbial respiration utilizes what little dissolved oxygen is present. Fish soon die, foul odors are produced and the water is useless for drinking or for recreational purposes.

Sewage Purification Processes. Sewage is treated by chemical or biologic means to remove organic matter. Chemical treatment is more expensive but can be carried out in a smaller area. Suspended and colloidal matter is flocculated by the addition of alum or iron salts; the precipitate entrains and sediments much of the organic material. The supernatant liquid or *effluent* can then be discharged into a large body of water or onto porous soil, where microbial oxidation of the remaining dissolved organic matter can take place.

Biologic treatment of sewage is most widely used. Processes range from that of the household septic tank, which handles at most a few hundred gallons per day, to municipal plants, through which many million gallons pass each day (Fig. 24–12). Details of the purification procedures vary greatly, but a few steps are common to most processes.

1. Gross objects and large particles are removed from the sewage by *screening* and *sedimentation*. Slow passage of the liquid through a large tank or other chamber permits particles to settle; the accumulated solid matter is known as *sludge*. Some systems also provide facilities for removing the scum of grease.

2. The liquid from which settleable solids have been removed contains colloidal and soluble carbohydrates, proteins, and fats, which are subject to *microbial digestion*. The products of fermentation and putrefaction include ammonium salts of organic acids, alcohols, amines, amino acids, glycerin and other chemicals, and the gases H_2S, CH_4, CO_2, and H_2. Most of the gases are combustible and are used as a source of power in large sewage treatment plants.

3. *Oxidation* of the products of fermentation and putrefaction is promoted by aeration, either by bubbling air through the digested sewage in a tank, by spraying the fluid intermittently onto a bed of coarse stones (a *trickling filter*; Fig. 24–13) over which it flows slowly in contact with air, or by distribution in drainage tiles through loose soil. The various organic compounds are oxidized by microorganisms, and the final products consist of CO_2 or carbonates, nitrates, and sulfates.

4. The *sludge* collected initially and also during subsequent stages is either removed periodically and dried for use as fertilizer, or it is allowed to digest, often as long as three months. The relatively indigestible residue or "stabilized" sludge that remains is also dried and used as fertilizer.

5. The *effluent* liquid from the treatment plant contains no organic matter; it does contain dissolved oxygen and nitrates, sulfates, and phosphates. There are also as many as a few hundred thousand bacteria per milliliter. The treated sewage can be discharged directly into a large body of water without seriously affecting animal life, but the salts will stimulate growth of aquatic plants such as algae. Few pathogens remain alive, but to be safe chlorine is usually added at the rate of two to five parts per million parts of liquid.

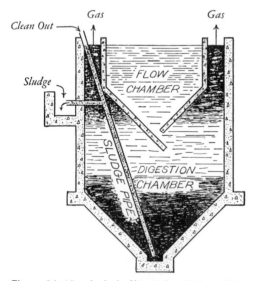

Figure 24–12. *An Imhoff tank, in which partial purification of sewage is effected. The sewage passes slowly through the flow chamber (in a direction at a right angle to the page). Settleable solid matter falls through the bottom opening and collects in the digestion chamber, where anaerobic processes gradually convert part of it to soluble form and gases. The gases may be collected and burned for heat and power. The undigested "sludge" is removed periodically. (From D. B. Swingle and W. G. Walter: General Bacteriology, 2d ed. Princeton, N.J., D. Van Nostrand Company, Inc., 1947.)*

A

Figure 24–13. *A trickling filter. A large bed of stones over which partially treated sewage is sprayed by a rotating "sparger" (shown stopped in this picture). The stones are covered by a gelatinous film containing microorganisms, which oxidize soluble compounds as the sewage trickles slowly down to collecting drains at the bottom of the bed. (A, Courtesy of Dr. Gordon M. Fair, in Rosenau: Preventive Medicine and Hygiene, D. Appleton-Century Co.; B, courtesy of Communicable Disease Center, U.S. Public Health Service, Atlanta, Ga.)*

B

Activated Sludge Process. One of the most efficient sewage treatment methods is the *activated sludge process.* Activated sludge is prepared by aerating sewage vigorously, whereupon floccules form that contain fine suspended and colloidal matter, including many bacteria and other microorganisms. When these floccules are allowed to settle and then used to inoculate another batch of aerated sewage, flocculation occurs more rapidly. After repetition of this process several times, a sedimented floc is obtained that is known as *activated sludge.* When added to fresh sewage and aerated vigorously, suspended and colloidal matter is adsorbed rapidly, forming large flocs that break up into small floccules, increase in size again, and break up continuously. The violent aeration promotes growth and activity of aerobic microorganisms and consequent oxidation of organic matter. Within 4 to 8 hours the mixture can be allowed to sediment, and the effluent is practically ready for discharge as completely treated sewage.

SANITARY QUALITY OF DRINKING WATER

Drinking water should be clear, cool, free from objectionable tastes and odors and

from harmful chemicals and microorganisms. Of these desired qualities, freedom from harmful microorganisms is most difficult to achieve. It is not impossible, but it demands constant vigilance and repeated testing.

The problem is made more acute because sewage disposal is usually completed by discharging effluent liquid into bodies of water, and necessity often dictates that the same bodies of water be used as sources of drinking water by other communities. Many cities empty sewage into Lake Michigan, but Chicago has to take its water supply from the lake. Rivers receive sewage, and a short distance downstream are used as sources of drinking water. Although sedimentation removes most particulate material including bacteria from both standing water and running water, it cannot be relied on to free water from harmful microorganisms. Purification is therefore necessary and must be controlled by constant testing.

Bacteria that Indicate Pollution. The ideal method of testing water for microbiologic safety would be to search for pathogens transmitted by water. Unfortunately, this is impractical. Water containing only a very few pathogens in each liter may be sufficiently polluted to cause many cases of disease. If the discharges from a single person with typhoid fever find their way into a reservoir used for drinking water, scores or hundreds of cases of typhoid fever may follow. The pathogens are relatively few and far between, and large samples must be examined in order to detect a single disease-producing organism. Moreover, pathogenic species in water contaminated with sewage are vastly outnumbered by harmless, normal intestinal bacteria and are quickly outgrown in ordinary cultural methods of examination. The detection of pathogens in water or sewage, then, is virtually impossible; it has been accomplished only a few times.

Various groups of bacteria that normally occur in the intestine of man or animals have been used to indicate the pollution of drinking water by sewage: the so-called fecal streptococci, certain spore-forming anaerobes, and the coliform bacteria. The coliform bacteria are the most widely accepted indicators of pollution, particularly in the United States.

The Coliform Bacteria. The coliform bacteria include *Escherichia coli* and certain other bacteria that resemble them morphologically and physiologically. These organisms frequently differ from one another in minor characteristics; at one time dozens of species were distinguished, but at present only four may be considered: *E. coli*, *K. pneumoniae*, *Enterobacter cloacae* and *E. aerogenes*. The first two will be discussed in greater detail.

Coliform bacteria usually occur in the intestinal tract of man and animals. *E. coli* is rarely, if ever, found outside the intestines, except where pollution by human or animal excreta has occurred. *K. pneumoniae* is widely distributed in nature and occurs in soil, water, grain, and in the intestinal tracts of man and animals.

Coliform bacteria are short, gram-negative rods that ferment lactose, forming acid and gas. They are facultative anaerobes and multiply most rapidly between 30° and 37° C. They grow luxuriantly upon ordinary media such as nutrient broth and nutrient agar. *E. coli* can be distinguished from *K. pneumoniae* by the appearance of its colonies on dif-

ferential plating media. Colonies of *E. coli* on E.M.B. (eosin methylene blue) agar are 2 to 4 mm. in diameter, possess a large, dark or even black center, and have a green, metallic sheen when observed by reflected light. Colonies of *K. pneumoniae* on the same medium are larger, very mucoid, and pinkish; they frequently have a small, brownish center (Plate IV).

Numerous other tests have been devised to distinguish the types of coliform bacteria. Four are so frequently used that their initials have been combined into the mnemonic *IMViC:* indole, methyl red (M.R.), Voges-Proskauer (V.P.), and citrate utilization. The indole test has already been described (page 76).

The M.R. and V.P. tests are performed on two to five day cultures in glucose-phosphate-peptone broth. A few drops of methyl red indicator added to some of the culture becomes red when the reaction is strongly acid; this result is called the positive test; a yellow color is the negative test.

The V.P. test is conducted by adding a small amount of 40 per cent KOH containing creatine to a portion of the glucose-phosphate-peptone culture, preferably with a catalyst such as alcoholic α-naphthol. The mixture is thoroughly shaken to aerate it and is then allowed to stand for 15 minutes. A positive test gives a pink color, which usually appears first at the top of the solution. This reaction is a test for acetylmethylcarbinol, which is oxidized to diacetyl under the conditions of test; diacetyl combines with the guanidine radical of creatine and produces a pink dye compound.

Utilization of citrate as a sole source of carbon is tested in a synthetic medium containing sodium citrate as the only carbon compound. Growth of the organism is the positive test. An acid-base indicator, which is sometimes added to the medium, assumes its alkaline color as the citrate radical is oxidized.

The IMViC reactions of a few coliform bacteria are listed in Table 24–3. The IMViC reaction of *E. coli*, $+ + - -$, means that the organism produces indole and is M.R. positive and V.P. and citrate negative. There are 16 possible combinations of positive and

**TABLE 24–3.   IMViC Reactions of
Coliform Bacteria**

Species	Indole	M.R.	V.P.	Citrate
Escherichia coli	+	+	−	−
Klebsiella pneumoniae	−	−	+	+
"Intermediates"	−	+	−	+
	+	−	+	+
		etc.		

negative tests of these four characteristics. Most of these combinations have been found, but the reactions of *E. coli* and *K. pneumoniae* are by far the most frequently encountered. The remaining 14 types are usually designated "intermediates." All coliform bacteria are considered significant in water from the sanitary viewpoint.

Sanitary Water Analysis. There are various methods of detecting coliform bacteria in water to be used for drinking. One standard procedure is based on the ability of these organisms to produce gas from lactose. Aliquots (five portions of 10 ml. each) of the sample are inoculated into lactose fermentation tubes and incubated at 37° C. for 48 hours (Fig. 24–14). The tubes are observed after 24 and 48 hours, and the appearance of gas at either time constitutes a positive *presumptive test.*

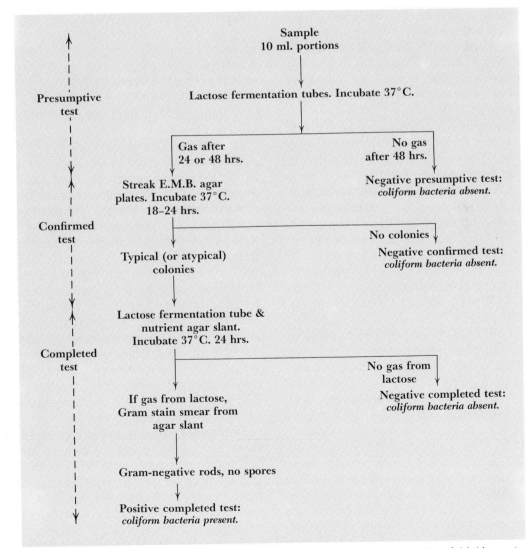

Figure 24–14.   *Abridged outline of the coliform tests used to detect probable pollution of drinking water with sewage. The three steps are called (1) the presumptive test, (2) the confirmed test, and (3) the completed test.*

Bacteria other than coliform organisms can produce gas in lactose; it is therefore necessary to isolate the lactose-fermenting species, and this is usually done by inoculating the surface of E.M.B. agar plates from one or more fermentation tubes containing gas. The appearance of typical coliform colonies is a positive *confirmed test*.

One or more colonies (typical, if present; otherwise any suspicious colonies) are then transferred to lactose fermentation tubes and nutrient agar slants and are incubated for 24 hours. If gas appears in the lactose tube and a stain from the slant reveals gram-negative rods with no spores present, the result is designated a positive *completed test* for coliform bacteria.

In some countries, positive presumptive tests are confirmed by subculture in tubes of MacConkey broth (lactose-peptone-bile) incubated at 44° C. Production of gas indicates the presence of fecal coliforms, because nonfecal coliforms are inhibited at this temperature.

Interpretation of the coliform test is arbitrary but is based upon experience. Each state or community in the United States may have its own requirements. The United States Public Health Service has established standards which are used as a guide by the various states or their subdivisions. The recommended standard requires essentially that acceptable water should contain, on the average, no more than one coliform organism per 100 ml. This means that when a single sample is examined by the above method none of the five 10 ml. portions should contain coliform bacteria. Experience has indicated that water conforming to this standard is almost certainly free from pathogenic bacteria.

Membrane Filter Coliform Count. A filtration technique for enumerating coliform bacteria in water, which was developed in Germany during World War II, has been accepted as a standard procedure for determining the sanitary quality of water. A portion of the sample is passed through a cellulose acetate filter membrane of such porosity as to retain bacteria, while permitting the water to pass freely. The filter membrane is then placed aseptically in a Petri dish on an absorbent pad saturated with a differential nutri-

ent solution such as M-Endo broth (buffered lactose-peptone-salts with bile salts and decolorized basic fuchsin) and incubated at 35° C. for 20 hours. The membrane is then examined by low-power microscopy, and purplish green colonies with a metallic sheen are counted. These are considered to be coliform bacteria (Fig. 24–15).

The amount of sample to be filtered varies according to the nature of the specimen: 100 to 500 ml. of finished, municipal water may be examined, whereas 0.1 to 10 ml. of well water may yield 20 to 80 coliform colonies (the recommended density for most accurate counting). Greater precision is possible by the membrane filter technique than by the multiple lactose tube method of estimating coliforms, because larger volumes of samples can be examined, and results are secured more quickly. The method is limited, however, by the clogging of the filters with algae, colloidal and other materials, and by the inhibition of coliforms in specimens containing excessively high, noncoliform populations.

The Plate Count. A plate count of aerobic and facultatively anaerobic bacteria is made on a standard plate count agar. Not more than 500 bacteria per ml. are permitted.

Purification of Water. Protection of the Supply. The first step in providing pure water, whether for a single household or for a large city, is to protect the source of supply against sewage pollution. Wells should be located at a considerable distance from septic tanks, barnyards, and other pollution; they should be carefully constructed to prevent seepage of surface water and should be capped with a concrete cover. Large watersheds from which water is collected into streams, ponds, or reservoirs should be carefully inspected; these are often fenced to exclude all sources of pollution.

Sedimentation. Purification by sedimentation as water flows slowly or stands in a reservoir is made more efficient by adding alum or iron salts; flocculent precipitates of the corresponding hydroxides entrain microorganisms and other suspended particles and settle rapidly. Sedimentation does not sterilize polluted water but markedly reduces its microbial population. It is often used as a first stage in purification.

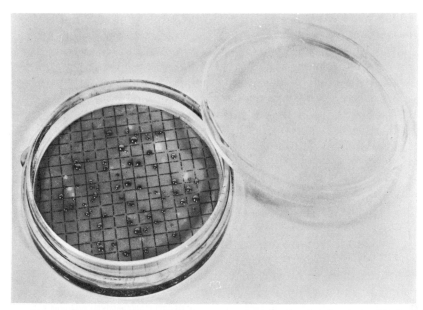

Figure 24–15. *Plastic Petri-type dish containing a broth selective for bacteria* (Escherichia coli) *that indicate sewage pollution of water. The membrane filter through which the water was filtered was removed and laid in the dish on a sterile pad soaked with broth, which the membrane absorbed. After incubation for 18 hours at 37°C the distinctively colored, glistening colonies of* E. coli *had developed. (Approximately actual size.) (Courtesy of Millipore Filter Corporation, Bedford, Mass.)*

Filtration. Filtration is an effective means of removing microorganisms and other suspended matter from water. Two types of sand filters are used in large scale filtration of water.

Slow sand filters are constructed of layers of fine sand, coarse sand, gravel, and rock. Water seeps through the filter slowly, and bacteria, algae, and protozoa are caught in the surface layers of fine sand. These microorganisms multiply and produce a gelatinous mass to which other microorganisms and suspended particles adsorb. The efficiency of the filter gradually decreases, and eventually the surface layer must be cleaned. Large filter beds are required because the rate of filtration is slow.

Rapid sand filters operate about 40 times faster than slow sand filters. They also consist of layers of sand, gravel, and rock, but a coagulant such as alum or ferrous sulfate is added to the water before filtration. The water passes through a settling tank in which most of the precipitate settles out and the remainder is removed by the filter. Rapid sand filters soon become clogged and must be cleaned by forcing water backward through the bed of gravel and sand. These filters are usually operated in bat-

teries so that some may be in operation while others are being cleaned.

Properly constructed and operated sand filters remove 90 to 99 per cent of the microorganisms and most of the suspended particles from water. Filtration does not sterilize water and cannot be relied upon to render it safe for human consumption; a final step is necessary.

Chlorination. Chlorination is the least expensive but most efficient means of rendering water safe for drinking. The amount of chlorine gas added depends on the degree of pollution of the supply and its organic matter content. Water is usually treated to contain 0.1 to 0.2 part per million of residual chlorine. Residual chlorine is the available chlorine remaining 20 minutes after its addition to the water. During this 20 minutes some of the chlorine combines with organic substances and with bacteria; the more heavily polluted the water or the greater its organic matter content, the more chlorine must be applied to ensure a safe residual. Preliminary sedimentation and filtration are therefore helpful.

Chlorine kills most nonspore-forming gram-negative bacteria, such as the intestinal pathogens, but in the concentrations employed it does not kill spores or many gram-positive bacteria. Chlorinated water is therefore not always sterile, but it is usually safe for human consumption.

Small amounts of water can be made safe by boiling for 10 minutes. This practice is often recommended for household use during floods or other disasters that disrupt the normal water purification system.

PATHOGENIC MICROORGANISMS IN THE AIR

The presence of mold spores in extramural air has already been mentioned. Updrafts over heated land masses carry clouds of spores many thousand feet into the air. The spores then travel great distances before they settle back to Earth. It will be recalled that air masses over the United States travel from west to east at a rate of 500 to 700 miles per day. The rate of settling of droplets of water of the same general dimensions as those of mold spores indicates that several hundred to several thousand hours are required for spores to fall to the ground from an elevation of 10,000 feet (Table 23–1).

A storm of spores of the mold *Alternaria*

covered the eastern third of the United States within a two day period in 1937. The spores originated in southern Minnesota and were carried aloft during the early hours of October 6. By the next day they were reported along a line from New York City to northern Texas and reached Florida and Georgia the following day (Fig. 24–16).

Outdoor Dissemination of Pathogens. There are two important aspects to the rapid and widespread distribution of spores. Some types of hay fever are caused by mold spores, and their general dissemination in this manner produces suffering among thousands of people. Even more important is the fact that plant pathogens can be distributed by the same method. Certain wheat rusts, for example, are carried northward through the central plains states by southerly winds from Texas to southern Canada in the spring and are returned by northerly winds during the winter. The destructive potential of this method of distributing plant pathogens in biologic warfare need not be elaborated.

Indoor Distribution of Pathogens. Most human pathogens that gain access to indoor air do not survive long unless covered by saliva or droplets of mucus. Organisms protected in this manner may, how-

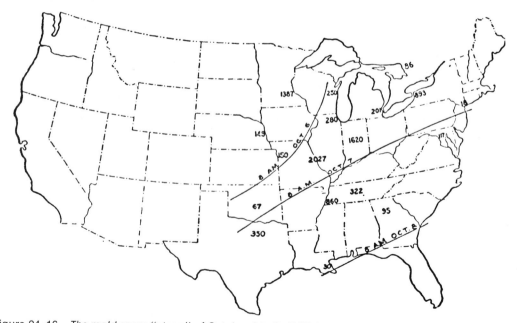

Figure 24–16.   *The mold spore "storm" of October 6 to 8, 1937. Numerals indicate the number of spores of* Alternaria *deposited on an area of 1.8 sq. cm. during the days of most intense activity. (From Aerobiology, Am. A. Adv. Sc., 1942.)*

ever, remain alive for several hours. Airborne diseases were discussed in Chapter 23.

Removal or Destruction of Airborne Microorganisms. Various means are employed to remove or destroy airborne microorganisms. Filtration through cotton or similar material is effective in the laboratory. Filtration through some air conditioning units removes many microorganisms.

Mercury vapor lamps that produce ultraviolet light are used in refrigerators, soda fountains, school rooms, and operating rooms in an attempt to kill microorganisms in the air. These installations apparently cause some reduction in airborne agents of spoilage or disease. Prolonged exposure to ultraviolet light is harmful to man; installations are therefore arranged in such a manner that part of the air in a room is treated without irradiating the occupants.

Many airborne microorganisms are killed by aerosols of disinfectant chemicals. The vapors of propylene glycol or triethylene glycol are strongly bactericidal and in the necessary concentrations are nontoxic to man. The vapor from as little as 0.5 mg. of propylene glycol can kill nearly all the microorganisms in a liter of heavily contaminated air within 15 seconds. Triethylene glycol is nearly 100 times as germicidal.

SUPPLEMENTARY READING

Alexander, M.: *Introduction to Soil Microbiology*. New York, John Wiley & Sons, Inc., 1961.

Alexander, M.: *Microbial Ecology*, New York, John Wiley & Sons, Inc., 1971.

American Association for the Advancement of Science: *Aerobiology* (Moulton, F. R. [ed.]). Washington, American Association for the Advancement of Science, 1942.

American Public Health Association: *Standard Methods for the Examination of Water and Wastewater,* 13th ed. Washington, D.C., American Public Health Association, 1971.

Brock, T. D.: *Microbial Ecology.* New York, John Wiley & Sons, Inc., 1971.

Gainey, P. L., and Lord, T. H.: *Microbiology of Water and Sewage.* Englewood Cliffs, N.J., Prentice-Hall, Inc., 1952.

Gray, T. R. G., and Williams, S. T.: *Soil Microorganisms*. Edinburgh, Oliver and Boyd, Ltd., 1971.

Gregory, P. H.: *The Microbiology of the Atmosphere*. London, Leonard Hill (Books) Limited, 1961.

Kuznetsov, S. I., Ivanov, M. V., and Lyalikova, N. N.: *Introduction to Geological Microbiology*. New York, McGraw-Hill Book Co., Inc., 1963.

Mitchell, R.: *Water Pollution Microbiology.* New York, John Wiley & Sons, Inc., 1972.

Prescott, S. C., Winslow, C.-E. A., and McCrady, M. H.: *Water Bacteriology,* 6th ed. New York, John Wiley & Sons, Inc., 1946.

Rosebury, T.: *Experimental Airborne Infection.* Baltimore, The Williams & Wilkins Co., 1947.

Sarles, W. B., Frazier, W. C., Wilson, J. B., and Knight, S. G.: *Microbiology, General and Applied,* 2nd ed. New York, Harper & Brothers, 1956.

Sieburth, J. M.: Bacterial substrates and productivity in marine ecosystems. Ann. Rev. of Ecology and Systematics, *7*:259–285, 1976.

Thimann, K. V.: *The Life of Bacteria,* 2nd ed. New York, The Macmillan Company, 1963.

Waksman, S. A.: *Principles of Soil Microbiology,* 2nd ed. Baltimore, The Williams & Wilkins Co., 1932.

Warren, C. E.: *Biology and Water Pollution Control*. Philadelphia, W. B. Saunders Co., 1971.

Wilber, C. G.: *The Biological Aspects of Water Pollution*. Springfield, Illinois, Charles C Thomas, Publisher, 1969.

25 MICROBIOLOGY OF FOODS AND DAIRY PRODUCTS

Microorganisms produce both desirable and undesirable changes in foods. Many products would not be possible without microbial assistance: sauerkraut, ripe olives, cocoa, cheese. Acids produced by microorganisms help to preserve certain foods, such as pickles and fermented milks, from unpleasant microbial activity. The undesirable changes known as spoilage are all too familiar.

MICROBIAL SPOILAGE OF FOODS

Spoilage is any change in the flavor, aroma, texture, or appearance of a food that renders it undesirable or unpalatable. The terms *undesirable* and *unpalatable* cannot be defined objectively; they depend on the customs and experiences of the individuals concerned. In general, however, each population group has certain standards or norms of palatability, and a food that fails to meet these standards is considered spoiled.

Microbial spoilage of foods is an ecologic problem. Many foods are produced or manufactured under conditions that ensure contamination with a variety of microorganisms, but which of these organisms survive and multiply depends on the composition of the food and the conditions of storage. Those organisms that can grow bring about changes characteristic of their metabolic patterns and alter the flavor, aroma, texture, or appearance of the product in a certain way.

Human and animal foods may be classified according to their source as (1) plant products, (2) animal products, (3) manufactured products.

INITIAL CONTAMINATION OF FOODS

Plant Products. Plant products are subject to microbial contamination from the soil in which they are grown, from the air, from insects, and from human handlers. The internal tissues of fresh plant products are usually free from microorganisms. Plant surfaces are relatively impermeable, and microorganisms do not readily penetrate to underlying tissues, where they can multiply rapidly, unless the surfaces are bruised or the organisms are "inoculated" by insect bites.

Root crops such as potatoes, beets, and carrots are liberally coated with soil microorganisms when harvested, but these vegetables possess an exceptionally impenetrable skin. Low-growing leafy vegetables like lettuce, spinach, and cabbage are also likely to be heavily contaminated; these products have a softer surface and are easily invaded by microorganisms. Vegetables and fruits that grow some distance above the ground are contaminated by insects and by microorganisms in the air. The latter are principally soil organisms. Insects tend to feed upon the same type of plant and hence distribute organisms from one to another so that all have a uniform flora. Yeasts, for example, are almost universally present upon grapes as a result of insect inoculation.

Microorganisms found upon plant products include molds, yeasts, spore-forming and some nonspore-forming rods, and various cocci.

Animal Products.   Animal products are subject to intrinsic as well as environmental and human contamination. The internal portions of a piece of meat are usually free from microorganisms if the animal has been properly slaughtered (i.e., killed quickly by a blow on the head or by cutting the jugular vein). The exposed surfaces are covered with bacteria derived from the animal's skin and intestines, the butchering equipment, and the air of the slaughter house. Fish fillets are even more likely to be covered by microorganisms, particularly when the fish are "cleaned" and cut on shipboard under poor handling conditions.

Microorganisms on meat include cocci, gram-negative rods such as *Achromobacter, Alcaligenes, Pseudomonas, Proteus,* and coliform species, anaerobic spore-forming bacteria, yeasts, and molds. Fish contain many of the same organisms, particularly nonspore-forming pigmented or nonpigmented rods.

Clean, fresh, uncracked eggs are usually free from microorganisms within the shell. Only about 8 per cent of fresh eggs contain microorganisms; the yolks frequently contain more bacteria than the whites. Dirty eggs are covered with microorganisms, which penetrate the shell under poor conditions of storage.

Milk.   Cow's milk as secreted by the glands of a healthy udder is sterile, but it frequently becomes contaminated by the micrococci and streptococci normally present in the milk ducts and cistern of the udder. The number of these bacteria is usually no more than a few score to a few hundred per milliliter of milk.

The udders of diseased cattle may be infected with pathogenic species of *Staphylococcus* or *Streptococcus* or with the tuberculosis or brucellosis organisms; these bacteria are also discharged in the milk.

Organisms from inside the udder constitute only a small fraction of those found in freshly drawn milk. Bacterial counts in fresh milk vary from a few hundred to several thousand per milliliter, and under poor conditions may be half a million or more. The sources of these organisms depend somewhat on whether milking is done by hand or by machine. In hand milking into open pails there is opportunity for contamination from the air of the stable, the animal's coat, and the hands and clothing of the worker. Machine milking reduces the significance of these sources of microorganisms but increases the chance of contamination by unsanitary equipment, which may add thousands or millions of bacteria per milliliter to milk if it is not properly cleaned and sterilized.

No process is better than the humans that conduct it, and careless dairy workers may contribute considerable numbers of pathogenic or nonpathogenic bacteria at any stage in milk production and handling.

Kinds of Microorganisms in Milk.   The bacteria ordinarily found in milk and other dairy products comprise four groups: (1) cocci, usually gram-positive, (2) gram-positive, nonspore-forming rods, (3) gram-negative, spore-forming rods, (4) gram-negative, nonspore-forming rods (Table 25–1).

The cocci of normal milk include various streptococci, notably *S. lactis,* which is almost always present in fresh milk. *Micrococcus* species are usually present, too. Both of these organisms may be derived from the healthy udder, and some are likely to withstand pasteurization. They may persist in utensils and other dairy equipment.

Lactobacilli are often found in milk and are important in the manufacture of fermented milks and many kinds of cheese. Their growth is favored by an acid medium, and for that reason they usually multiply better in milk that has already been partly soured by *S. lactis* or other organisms than in fresh milk.

Microbacterium lacticum is a thermoduric, nonspore-forming small rod that survives pasteurization and persists in milk equipment. It is frequently the cause of high bacterial counts in pasteurized milk.

Coliform bacteria are considered undesirable in milk and dairy products, because they indicate unsanitary conditions or practices. They are almost inevitably present in raw milk, because of their widespread occurrence in manure and on grains and other feeds, but they are easily killed by heat and should be absent from pasteurized milk. Their presence in pasteurized milk therefore

TABLE 25–1.   **Kinds and Sources of Microorganisms in Milk**

Organisms	Hay, feed	Manure	Equipment	Soil	Water	Udder	Remarks
Cocci							
Streptococcus	+	+	+			+	Early souring of milk, produces 0.8 to 1.0% lactic acid; used in butter and cheese starters.
Leuconostoc	+						Used in butter starters.
Micrococcus			+			+	May survive pasteurization.
Gram-positive non-sporeforming rods							
Lactobacillus	+	+					Produces 2 to 4% lactic acid; used in fermented milks and cheese.
Microbacterium		+	+				Resists pasteurization; survives 10 min. at 80° C.
Gram-positive sporeforming rods							
Bacillus	+	+		+			Survive pasteurization; late spoilage of dairy products.
Clostridium	+	+		+			
Gram-negative rods							
Coliforms	+	+		+			Usually in raw milk; not in properly handled pasteurized milk.
Pseudomonas	+	+	+	+	+		Low temperature spoilage.
Alcaligenes	+	+	+	+	+		Ropy milk, etc.
Achromobacter				+	+		Produces rancidity.
Flavobacterium				+	+		Produces rancidity.
Yeasts	+			+			Produce gassy fermentation.
Molds	+			+			Late utilization of acids.

indicates gross pollution of the raw milk, inadequate pasteurization, or recontamination after pasteurization.

Manufactured Products. The microbial flora of manufactured foods depends upon the nature of the food and the manufacturing process. Bakery products, for example, contain microorganisms derived from the various ingredients: flour, sugar, shortening, milk or milk powder, eggs or egg powder, and water. The equipment and the human handlers also contribute microorganisms. The baking process kills molds, yeasts, and nonspore-forming bacteria, but bacterial spores may survive. The outside of any product is subject to recontamination, and molds are particularly troublesome. Uncooked cream fillings or toppings are also likely to contain spoilage organisms.

Fermented foods, such as pickles, olives, and sauerkraut, contain microorganisms used in the manufacturing process. Undesirable organisms derived from the equipment, the air, and from man include molds, yeasts, and putrefactive bacteria.

Effects of Chemical Properties on Spoilage

The chemical properties of a food product influence the type of microorganisms that

TABLE 25–2. The Influence of the Chemical Properties of a Food on the Type of Microbial Spoilage to Which It Is Subject

Composition			Acidity		Osmotic Pressure	
Protein	Carbohydrate	Fat	Acid (<pH 4.5)	Nonacid (>pH 4.5)	Low	High
PREDOMINANT SPOILAGE ORGANISMS						
Bacteria	Yeasts	Molds	Molds	Bacteria	Molds	Molds
Molds	Molds	A few bacteria	Yeasts		Yeasts	
					Bacteria	

can grow and hence determine the nature of the spoilage process (Table 25–2).

Composition. Proteins are subject to bacterial spoilage. Many species can attack them, especially spore-formers, gram-negative rods such as *Pseudomonas* and *Proteus,* and a few cocci. Mold spoilage is also common.

Carbohydrate foods are spoiled particularly by yeasts and molds. Bacterial species of the genera *Streptococcus, Leuconostoc,* and *Micrococcus* are saccharolytic, and many other bacteria can also attack carbohydrates.

Fats undergo hydrolytic decomposition and become rancid as malodorous fatty acids are set free. Relatively few microorganisms are capable of digesting fats: some molds and a few gram-negative rod bacteria and cocci.

Acidity. The reaction of nearly all foods is below pH 7, and some may be as acid as pH 2 to 3. Foods are classified as *acid* or *nonacid*. The reaction of acid foods is below pH 4.5, and this group includes most fruits (Table 25–3). Nearly all vegetables, fish, meats, and milk products are nonacid.

The pH of acid foods is sufficiently low to prevent most bacterial spoilage, but yeasts and molds grow luxuriantly. Nonacid foods are particularly subject to bacterial spoilage, but will also support growth of molds under proper conditions.

Moisture and Osmotic Pressure. Foods that contain less than 10 to 13 per cent water do not support growth of microorganisms. Molds require the least free water, and bacteria require most. Many molds and some yeasts can tolerate salt concentra-

tions greater than 15 per cent, whereas bacteria are generally inhibited by 5 to 15 per cent salt. Sixty-five to 70 per cent sugar is required to inhibit molds; 50 per cent inhibits bacteria and most yeasts. Foods of high sugar or salt content are therefore most

TABLE 25–3. Approximate pH of Some Canned Food

Food	pH
Lemon juice	2.4
Cranberry juice	2.5
Rhubarb	3.1
Grapefruit	3.2
Apples	3.4
Cherries	3.4
Pineapple	3.5
Orange juice	3.7
Peaches	3.7
Apricots	3.8
Pears	4.3
Tomatoes	4.3
Carrots	5.1
Green beans	5.2
Pumpkin	5.3
Beets	5.4
Spinach	5.4
Asparagus	5.5
Broccoli	5.6
Tuna	5.8
Codfish	6.0
Peas	6.0
Lima beans	6.1
Duck	6.1
Mackerel	6.1
Chicken	6.2
Corn	6.3
Oysters	6.4
Clams	6.8
Crabmeat	6.8
Shrimp	7.0

likely to be spoiled by molds; foods of low salt or sugar content may be spoiled by any kind of organism.

Effects of Storage Conditions on Spoilage

Oxygen. The presence or absence of oxygen determines the types of organisms that can multiply and the kind of spoilage produced (Table 25–4). Molds and aerobic bacteria (species of *Bacillus* and *Pseudomonas*) grow only where there is plenty of air and cause chiefly surface spoilage; yeasts and facultative bacteria can grow in closed containers as well as when exposed to the air. Spoilage by the genus *Clostridium* is strictly anaerobic.

Temperature. Refrigerated foods are subject to spoilage by molds and by some yeasts and bacteria, including several species of gram-negative rods and a few micrococci. Low temperature retards spoilage, but even subfreezing temperatures do not prevent multiplication of all microorganisms until about $-7°$ C. is reached. Foods stored at $-18°$ C., the temperature of a home freezer, remain free from microbial growth, and a slow decrease in population may even occur.

Products stored in warm warehouses, ship holds, or other warm locations may be spoiled by thermophilic bacteria, most of which are spore-forming and hence resist heat sterilization in canning processes.

SPOILAGE OF PLANT PRODUCTS

Spoilage of plant products is determined principally by their acidity and chemical composition, because all are subject to similar initial microbial contamination.

Fruits. Fruits, including tomatoes, are highly acid, and about 90 per cent of their organic matter is carbohydrate, chiefly sugar (Table 25–5). Spoilage is therefore limited to molds and yeasts. Fruits usually become moldy after a few days at room temperature or even in the refrigerator, and crushed fruits or fruit juices not only become moldy but may develop gas and an alcoholic flavor as a result of yeast activity.

Vegetables. Most of the common vegetables contain less carbohydrate and more protein than do fruits. A high percentage of the carbohydrate of vegetables like corn and potatoes is starch, which can be digested by relatively few microorganisms. Protein amounts to as much as 40 per cent of the organic matter. Lack of acidity permits spoilage of vegetables by bacteria; yeast spoilage is relatively uncommon.

Aerobic bacteria and molds produce *decay* under conditions of sufficient aeration and humidity or free moisture; this type of spoilage is not accompanied by unpleasant odors. Anaerobic bacteria attack starch and proteins in the absence of air and produce the foul odors of *putrefaction*. Spoilage of home canned beans or corn illustrates the changes produced.

TABLE 25–4.   The Influence of Storage Conditions on the Type of Microbial Spoilage to Which a Food Is Subject

Air		Temperature		
Present	Absent	Low (<10° C.)	Moderate	High (>40° C.)
PREDOMINANT SPOILAGE ORGANISMS				
Molds	Bacteria	Molds	Molds	A few bacteria
Yeasts	Yeasts	A few yeasts and bacteria	Yeasts	
Bacteria			Bacteria	

TABLE 25–5. Approximate Composition of Various Types of Foods

Type of Food	Per Cent of Organic Matter		
	Protein	Carbohydrate	Fat
Fruits	2–8	85–97	0–3
Vegetables	15–30	50–85	0–5
Fish	70–95	0	5–30
Poultry	50–70	0	30–50
Eggs	51	3	46
Meats	35–50	0	50–65
Milk	30	40	30

SPOILAGE OF ANIMAL PRODUCTS

Animal products such as meat, fowl, fish, and eggs contain almost no carbohydrate. Proteins constitute 35 to 95 per cent of the organic matter; the remainder is fat. These products are therefore subject to spoilage by proteolytic bacteria and by molds. The nature and extent of spoilage depend on various environmental factors.

Meat. Spoilage is slow in meat that is properly refrigerated or stored under such conditions that the surfaces become dry, because the initial microbial contamination is confined to the surface; interior tissues are normally sterile. Microorganisms that penetrate slowly or that enter cut surfaces eventually produce anaerobic putrefaction. Putrefactive decomposition occurs rapidly in ground meat and flaked fish because the surface organisms are thoroughly distributed throughout the mass of food. Bacterial counts reach hundreds of millions per gram within a few days, even in the refrigerator, and obvious spoilage quickly results.

Storage at a temperature just above freezing greatly retards spoilage. Fresh beef and mutton are often "aged" for several weeks at low temperatures to permit autolysis, which improves their texture and flavor. The layer of fat covering these meats retards drying and protects the underlying muscle from microbial attack.

Fish. Microorganisms on the surface of fish fillets multiply rapidly, particularly if icing or refrigeration is delayed. Bacterial counts in fish stored for four days at 10° to 20° C. may be in the hundreds of millions per gram. The fish become slimy, and proteolysis occurs.

Shellfish such as oysters and clams present a peculiar problem, because they are frequently eaten raw or barely warmed, as in stews or chowders. These fish are taken from the mud at the bottom of shallow bays, and the shells are heavily loaded with bacteria. The bacterial content of the animal itself depends on that of the surrounding water, which is constantly taken into the animal, passed over its gills, and discharged. Shellfish that have been "fattened" in water containing sewage have a high bacterial count, including many sewage bacteria. Numerous outbreaks of typhoid fever have been traced to oysters from such sources. The bacterial content of oysters is decreased by removing them to clean water for a time, whereupon many bacteria are removed from the gills during normal passage of water through the animal's body. Shucked shellfish are likely to contain large numbers of bacteria and require prompt refrigeration to avoid putrefactive decomposition.

Milk. Milk presents a special problem because the temperature of freshly drawn milk is favorable for rapid multiplication of many bacterial species with which it is normally contaminated. Unless it is immediately cooled, the initial population of several hundred to a few hundred thousand bacteria per milliliter increases 20 times or more within two or three hours. Prompt refrigeration prevents bacterial multiplication for a day or two and inhibits it for several days thereafter (Fig. 25–1).

Normal Fermentation of Raw Milk. Raw milk stored at a moderate temperature supports a sequence of microorganisms (Fig. 25–2). S. lactis and related bacteria multiply promptly and rapidly and produce sufficient lactic acid to decrease the reaction to about pH 4.5, causing curdling. Lactobacilli start to multiply below pH 6 and continue the fermentation, producing 2 to 4 per cent lactic acid and bringing the reaction to pH 3.0 to 3.5. Few microorganisms can grow in a medium of this acidity. Under aerobic conditions, however, molds and film-forming yeasts can utilize the acid as a source of energy by oxidizing it to

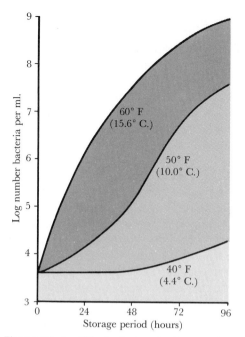

Figure 25–1. *Effect of storage temperature on bacterial multiplication in raw milk. (Data of Ayers et al.: U.S.D.A. Bull. 642, 1918.)*

carbon dioxide and water. The reaction therefore once more becomes nearly neutral. Most of the lactose has been utilized by this time, and further growth of lactic acid-producing bacteria does not occur. *Pseudomonas,* spore-formers, and other proteolytic and lipolytic bacteria, which were held in check by the rapid lactic fermentation, finally digest the casein and fat and reduce the milk to a dirty-looking, watery, putrid or rancid liquid. The initial stages of acid formation are completed in only a few

days; oxidation of the acid and decomposition of the protein and fat may require several weeks.

This sequence of microbial changes is fairly common and occurs with modifications in food fermentations, such as the manufacture of sauerkraut, and in the manufacture of cheese; some of the more highly ripened cheeses display obvious evidence of putrefaction.

SPOILAGE OF MANUFACTURED PRODUCTS

Baked Goods. Baking does not necessarily kill all bacterial spores within a loaf of bread despite the fact that a high temperature (190° C.) is used for 40 minutes or longer. The center of the loaf reaches only 97° to 100° C., and this temperature is maintained for about nine minutes in properly baked bread. Surviving spores of *Bacillus* species germinate if the bread is not cooled quickly or is stored at too high a temperature; these organisms produce a stringy decomposition known as *ropy bread*. Ropiness is prevented by acidifying the bread dough before baking to a reaction of about pH 5 with vinegar or lemon juice or with acetic, tartaric, lactic, or citric acid. Acidity inhibits the growth of spore-forming bacteria.

Mold spoilage is another problem in bakeries. Molds do not survive baking, but bread may easily be contaminated after it leaves the oven, particularly during slicing. Extreme care is therefore necessary to pre-

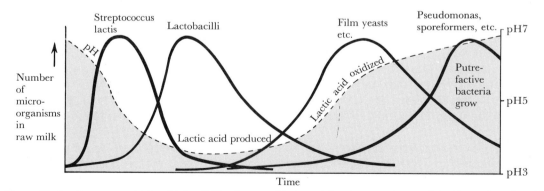

Figure 25–2. *Sequence of microbial activities in raw milk at a moderate temperature.* S. lactis *and* lactobacilli *multiply rapidly, ferment lactose, and produce sufficient lactic acid to bring the reaction to pH 3.5 or lower. Film yeasts, molds, etc., then oxidize the acid, and the pH rises, which permits putrefactive bacteria to grow.*

vent mold contamination before wrapping. Some bakers add sodium or calcium propionate to the bread dough to prevent mold growth on the loaf (propionates also prevent ropiness).

Cream or custard products such as cream puffs and eclairs are notorious, particularly during warm weather. The filling is an excellent culture medium and supports rapid multiplication of many kinds of bacteria, some of which are pathogenic. Many bakeries refuse to manufacture such products during the warm summer months.

Fermented Foods. Acid foods manufactured by fermentation or by the addition of vinegar (sauerkraut, pickles, olives, etc.) do not ordinarily support the growth of bacteria because their pH is too low. Molds and film-forming yeasts ("false yeasts") grow luxuriantly on such products, however, particularly when they are exposed to air in an opened jar or at the top of a fermentation or storage tank. As in the normal fermentation of milk, these organisms oxidize the acid and reduce the acidity to such an extent that putrefactive spore-forming and nonspore-forming bacteria can multiply. This type of spoilage is easily prevented by keeping the acid food in a tightly closed container so that molds, yeasts, or air cannot enter.

FOOD PRESERVATION

Food preservation practices range from the ancient use of drying to the ultramodern experimental application of gamma rays. There are five general methods: (1) control of moisture, (2) use of chemical preservatives, (3) storage at low temperature, (4) use of high temperatures, (5) treatment with radiation. Each method is suited to the preservation of certain products.

Asepsis is important in the successful application of any preservation procedure. This means that contamination of the food product must be reduced or prevented at all times, from its production, slaughter, or manufacture, through all the various processing and handling steps, to its distribution, sale, and final storage in home or restaurant.

CONTROL OF MOISTURE

Drying. The ancient practice of sun drying consisted of exposing fruits, vegetables, and small pieces of meat to the warm sun until they could no longer support bacterial, yeast, or even mold growth. The method is still used in favorable climates.

Commercial drying is usually carried out under controlled conditions of temperature, relative humidity, and air flow; the process is called *dehydration* or *desiccation*. Milk can be dried by being sprayed as a fine mist into a stream of warm air, which evaporates most of the water, or it may be sprayed upon hot rollers, from which the powder is scraped. Other products are heated at moderate temperatures, with or without forced air circulation.

Desiccation cannot be relied upon to sterilize a product, although some organisms are killed. Spores of bacteria, yeasts, and molds survive long periods in the dry state. However, products containing less than 10 per cent of free water generally keep indefinitely without spoilage, provided they are stored under dry conditions.

Addition of Salt or Sugar. Available moisture is reduced by the addition of salt or sugar. Salt is widely used to preserve fish and meat. In pickling, and in the manufacture of various fermented products, low concentrations of salt prevent the growth of spoilage organisms but permit the multiplication of desired fermentative types; this will be discussed later (see page 467).

Preservation of jellies, jams, maple syrup, and honey is attributed to their high sugar content (65 to 80 per cent), but poorly sealed or opened containers frequently allow the slow growth of molds. Osmophilic ("high osmotic pressure loving") yeasts occasionally grow in honey and produce sufficient carbon dioxide to burst the jar.

PRESERVATIVES

Various chemicals, including formaldehyde, boric acid, benzoic acid, and sulfur dioxide, have been used in the past to prevent spoilage of certain food products, including milk. Most of these chemicals are harmful, and their use is now prohibited or limited and strictly regulated (e.g., sodium benzoate).

Some fish and meats are preserved by smoking, often combined with salting. The smoke from sawdust or corncobs contains formaldehyde and pyroligneous acid (a mixture of creosote compounds), both of which inhibit bacteria; a small amount of these materials diffuses into the product. The smoking process also dries the surface of the meat or fish, and this helps to prevent microbial growth.

Organic acids are common preservatives that are particularly effective against putrefaction. They may be added directly (as is vinegar) or developed by fermentation of sugars in the food itself.

Approval has been given by the U.S. Food and Drug Administration for the use of certain antibiotics in the preservation of poultry. A condition of the granting of approval is that the antibiotic must be present in sufficiently low concentration to be destroyed by subsequent cooking. This is important because some individuals are or become sensitive (i.e., allergic) to certain antibiotics; moreover, continued use of an antibiotic may alter the microbial flora of the body and permit the establishment of antibiotic-resistant mutant bacteria. If the same resistant organisms subsequently cause an infection, it is impossible to treat the disease with the antibiotic in question. The deliberate or accidental introduction of antibiotics or other antibacterial chemicals into milk is considered adulteration and is forbidden by the federal government.

LOW TEMPERATURE

Ordinary Refrigeration. Low temperature retards food spoilage. The ordinary household refrigerator operating at 40° to 45° F. (4° to 7° C.) keeps most foods in a palatable condition for a few days. Temperatures only slightly above freezing are used for commercial storage of meats, fish, eggs, milk and other dairy products, and some vegetables and fruits that must be held several days or weeks before marketing.

Freezing and Cold Storage. Many foods can be kept several months in the frozen state. Quick freezing is preferred to slow freezing. Quick freezing implies a freezing time of 30 minutes or less and is accomplished in various ways, one of which consists of blowing cold air at 0° to −30° F. (−18° to −34° C.) across the materials being frozen. Ordinary or slow freezing requires three to 72 hours, and the temperature varies downward from 5°F. (−15°C.). Slow freezing is believed to produce large crystals of ice, which rupture cell walls and cause extensive "drip" or loss of fluid upon thawing, whereas quick freezing produces smaller ice crystals and less damage to the food tissues. Quick freezing reduces the loss of vitamins and stops tissue autolysis promptly. Frozen foods may be stored between 0° and 30° F. (−18° and −1° C.) with little further change.

Foods to be frozen should be prepared as carefully as if they were to be eaten directly. They should be sorted, trimmed, and washed, and in some cases "blanched" or scalded. Washing and blanching remove or destroy as many as 99 per cent of the microorganisms. Blanching consists of immersing the food in boiling water or exposing it to live steam for a very few minutes. The food is then immediately packaged and frozen as rapidly as possible. The interval between harvesting and freezing in many commercial operations is no more than two or three hours.

It should be emphasized that freezing does not improve the quality of any product. Only foods of high quality (that is, foods that would be acceptable if not frozen) should be preserved by this method. It is true that there may be some reduction in microbial count when the food is frozen and a slow reduction in count thereafter, but representatives of most species are likely to survive for months or years. This includes pathogenic as well as nonpathogenic types. All microbial activity ceases below −10° C., so that foods stored below this temperature remain free from spoilage indefinitely.

Quick freezing reduces but does not entirely prevent tissue damage. Frozen food is therefore highly susceptible to microbial invasion after thawing and should be used immediately, because the surviving bacteria begin to multiply as soon as they are warmed to their normal growth-temperature range. It is dangerous to refreeze frozen food that has thawed, because spoilage may have occurred during the interval of thawing.

HIGH TEMPERATURE

The rate at which high temperatures kill microorganisms varies with the species of organism, presence of spores, nature of the suspending medium (pH, consistency, etc.), temperature, and other factors. Heat treatments useful for preserving foods are subject to an upper limit determined by the characteristics of the food. Many foods acquire an overcooked taste or become soft and mushy when heated at too high a temperature or for too long an interval. The food processor must therefore select a heat process (temperature and time) that will yield a product that is safe from the health standpoint and capable of being stored without spoiling, yet retains its taste, texture, and nutritional properties insofar as is possible. Whether a product must be bacteriologically sterile depends upon the possibilities for growth of microorganisms during storage periods. For example, thermophilic bacteria do not necessarily have to be killed in a product that will be stored at a temperature below the growth range of these organisms; many bacteria cannot multiply in acid foods and hence do not necessarily have to be killed.

Pasteurization.   Heat treatment that kills some but not all microorganisms, usually at temperatures below 100° C., is known as *pasteurization*. It is employed in products whose quality would be adversely affected by higher temperatures and is often used in conjunction with other methods of preservation, such as drying, refrigeration, or cold storage.

Pasteurization was devised by Pasteur to prevent spoilage of wine and beer. It was later applied to the destruction of disease-producing bacteria in milk. It is used widely in the dairy industry, both in manufacturing processes in which certain specific microorganisms and none others are desired, and in the preservation of products such as processed cheese. Dried fruits, syrups, honey, apple juice, and other juices are also pasteurized. Spoilage of maple syrup by sugar-tolerant yeasts is prevented by heating the syrup to 93° C. at the time it is filled into cans and sealed. Similar spoilage of honey is prevented by heating at 71° C.

Pasteurization of Milk.   Pasteurization of milk is usually performed in special processing plants rather than on the dairy farm, because the equipment is expensive. Pasteurization has two important purposes: (1) to destroy all harmful microorganisms and (2) to improve the keeping quality of milk.

There are two procedures employed in pasteurizing milk: the low temperature–long time or "holding" method and the high temperature–short time or "flash" process. The holding method consists of heating the milk in covered tanks, where it is agitated constantly at not less than 145° F. (62.8° C.) for at least 30 minutes; it is then immediately cooled to approximately 40° F. (4.4° C.). Flash pasteurization is accomplished at 161° F. (71.7° C.) for at least 16 seconds. The milk flows continuously through a heated pipeline for the necessary time and then passes to cooling coils, where its temperature is brought down to 40° F. Either method of pasteurization, if properly performed, yields a safe product of good keeping quality. "Ultrapasteurization" is a more recently perfected process in which milk is heated for a very short time (e.g., a fraction of a second) at a still higher temperature.

Commercial pasteurization of milk began about 1890, and its acceptance was accelerated by knowledge that several diseases may be transmitted by raw milk. Investigators attempting to ascertain the best processing conditions for pasteurization soon found that *Mycobacterium tuberculosis* is the most resistant bacterial pathogen likely to be present in milk; subsequent studies were devoted principally to determining times and temperatures necessary to kill this organism in milk, cream, and other dairy products. A sample of the results is presented in Figure 25–3.

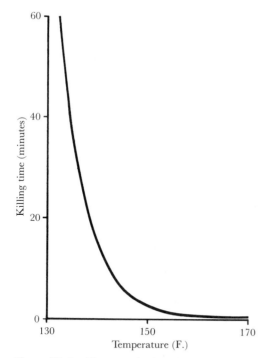

Figure 25–3. *Time required to kill* Mycobacterium tuberculosis *in milk at various temperatures. (Data of Park, Am. Rev. Tuberc., 15:399, 1927.)*

Economic factors dictated the upper limits placed on the heat treatment of milk. Excessive heating produces a "cooked" taste, which many consumers find undesirable. Heat also changes the "cream line," the height of the cream layer at the top of a bottle, by altering the physical condition of the butterfat so that the fat globules coalesce into a smaller volume. The vitamin content and other properties of milk change when heated. The cream line, however, is the factor that finally limited the time and temperature of pasteurization. This effect lost its significance when homogenization was introduced, because the finely dispersed fat globules do not separate and rise to the top of even unheated milk.

Pasteurization does not sterilize milk. It may reduce the bacterial count from a few hundred thousand to a few thousand per milliliter, but the surviving bacteria must be kept from multiplying by constant refrigeration. Pasteurized milk usually keeps better when refrigerated than raw milk of a comparable bacterial content, because pasteurization kills those bacteria that grow most readily at low temperatures. The surviving bacteria are

relatively thermoduric and often possess higher growth temperatures. Ultrapasteurization yields a product in which only resistant spores survive; its keeping properties are very good.

Canning.   Preservation by canning has been practiced for about 150 years. Early in the nineteenth century the French government offered a prize of 12,000 francs for a method of preserving food for use by the army. The prize was won in 1809 by a Paris confectioner, Nicholas Appert. His book on canning was published the following year, and the English translation, which appeared in 1811, bears the imposing title, *The Art of Preserving All Kinds of Animal and Vegetable Substances for Several Years.* Appert used wide-mouthed glass bottles, which were filled, corked, and heated in boiling water.

Tin-coated steel containers were introduced by Durand in England in 1810. Canning was apparently started in the United States in 1819 by Underwood in Boston, and by the next year Kensett in New York was engaged in the commercial production of canned foods. The Civil War caused a great expansion of the canning industry, as did the Spanish American War and the first World War.

It was recognized early that boiling water does not provide sufficient heat for sterilization of some kinds of food and that temperatures above 100° C. are necessary. In 1861, canneries increased the processing temperature by adding calcium chloride to water baths, and by 1874 steam pressure cookers or "retorts" were introduced. Methods of calculating heat processes from bacteriologic and physical data were perfected between 1923 and 1928, and since then the canning industry has been on a firm scientific basis.

The basic steps in canning are the same, whether at home or on a commercial scale (Fig. 25–4). The food should be fresh and of high quality. It is washed to remove gross dirt and as many microorganisms as possible; this reduces the burden on subsequent heat treatment. The second step is blanching or scalding in hot water or steam. The food is then filled into cans or jars while still hot. Commercial canners follow this step by ex-

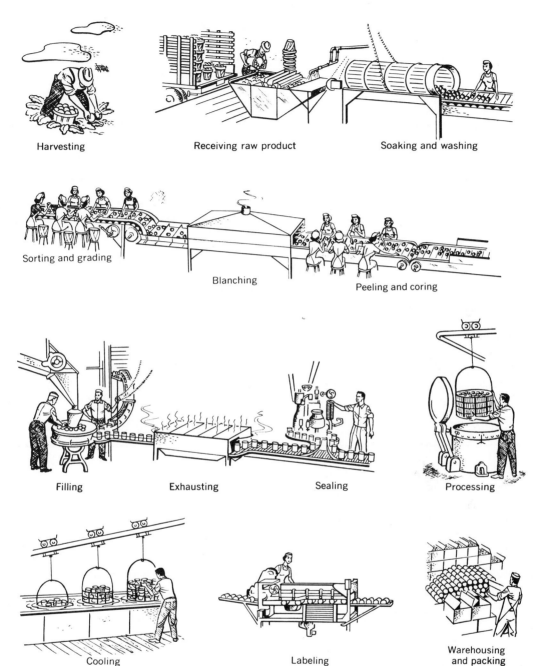

Harvesting

Receiving raw product

Soaking and washing

Sorting and grading

Blanching

Peeling and coring

Filling

Exhausting

Sealing

Processing

Cooling

Labeling

Warehousing and packing

Figure 25–4. *Commercial canning; a food passes through many steps between harvesting and final storage or shipment. (From Pelczar and Reid: Microbiology. New York, McGraw-Hill Book Co., Inc., 1958.)*

hausting or preheating the filled containers in a hot water or steam chest. Exhausting expands the food, drives off air or gas bubbles, and provides an atmosphere of steam in the "head space" at the top of the can. The can is then immediately sealed.

Heat processing follows at once so that the contents of the cans do not have an opportunity to cool. The treatment employed depends on the nature of the food and the size of the container. Acid foods require very little heat processing and are usually immersed in boiling water for only a few minutes (Table 25–6). Nonacid vegetables are processed in

TABLE 25–6. Heat Processes Recommended for Home Canning

Acid Foods (Pint or Quart Glass Jars, Packed Hot, Processed in Waterbath).

Food	Time at 212° F.
Applesauce	5 min.
Raspberries	5 min.
Cherries	5 min.
Rhubarb	5 min.
Tomatoes	5 min.
Peaches	15 min.

Nonacid Foods (Pint Glass Jars Processed in Pressure Cooker).

Food	Time at 240° F. (10 Lbs. Pressure)	Time at 250° F. (15 Lbs. Pressure)
Peas	45 min.	
Lima beans	50 min.	
Snap beans	30 min.	
Corn, whole kernel	60 min.	
cream style		75 min.
Greens		60 min.
Pumpkin		60 min.
Beef		85 min.
Chicken, boned		85 min.
Hamburger		90 min.
Lamb or mutton		85 min.

retorts heated by steam under pressure at 240° to 250° F. (116° to 121° C.) for periods as long as one or two hours. Meats, fish, and poultry are usually processed for one and one-half to two hours at 250° F. Viscous or solid foods, like pumpkin, into which heat penetrates only by conduction, require more thorough processing than fluid materials of comparable composition through which heat travels by convection. Large cans must be heated longer than small containers.

The processed cans are promptly cooled, either in the air or in cold water, to prevent overcooking and undesirable changes in texture and flavor. Rapid cooling also prevents germination and multiplication of the spores of highly resistant thermophiles, which might survive an inadequate heat treatment.

Canned foods are not always bacteriologically sterile but are considered "commercially sterile" if they contain no organisms capable of multiplication under usual condi-

tions of storage. Bacteriologic sterility is preferred if it can be achieved without sacrifice of the physical or chemical quality of the food material. Home canned products undoubtedly are often only commercially sterile.

The only highly resistant food poisoning organism of concern in canning is *Clostridium botulinum*. Spores may survive 10 minutes or longer at 240° F., but even the most resistant spores are killed in 15 minutes by steam at 250° F. A hot water bath process cannot be relied on to destroy this organism in nonacid foods, because boiling for six hours may be necessary.

Home use of the pressure cooker in canning has done much to reduce the incidence of disease as well as to enhance the keeping qualities of home canned foods. Careless use of the pressure canner, however, may create a false sense of security, and strict attention must be paid to the manufacturer's directions. It is particularly important to be sure that no air remains in the canner during the processing interval. A mixture of steam and air has a lower temperature than steam alone at the same pressure (see Figure 12–4, page 253). The pressure cooker should therefore be heated with the air vent wide open until *pure steam* issues for at least four minutes. Pressure may then be allowed to build up to the desired point. The best type of canner has a thermometer as well as a pressure gauge; the processing time should be determined by the thermometer.

IRRADIATION

The ancient method of preserving certain foods by drying in the sun owed part of its success to ultraviolet irradiation. Commercial attempts to utilize ultraviolet light have met with limited success. It will be recalled that this form of radiation is readily absorbed by particulate matter of all kinds and even by glass; therefore, only the surfaces of foods could be expected to be sterilized by this method.

Other radiations are being studied as sterilizing agents for foods. Gamma rays have been used successfully for the experimental preservation of hamburger patties in paper or plastic bags; the sterilized product could be stored at room temperature. Irradiation appears to offer advantages for preserving certain kinds of food, but at present it is expensive. The food industry is so highly competitive that continued research can be anticipated.

MANUFACTURED FOODS

Numerous foods are prepared by fermentation processes in which one or more kinds of microorganisms are responsible for the characteristic flavor or texture and sometimes for the keeping quality of the product. Fermented milks, alcoholic beverages, and other fermented foods have been used for thousands of years. Rule-of-thumb methods were handed down from one generation to the next, but why they worked—or on occasion failed to work—was not known. Only within the past hundred years have food fermentations been studied scientifically and the roles of the various microorganisms determined.

The manufacture of fermented vegetable products is a large industry, but it is also carried out on a small scale in homes in every country. Cabbages, cucumbers, lettuce, beets, turnips, and other vegetables can be used to make fermented products that are palatable and that possess greater freedom from spoilage than the natural vegetables. Most of the processes depend on the normal microflora of the vegetable to cause fermentation. They attack the natural sugars and yield organic acids, principally lactic, which serve as the preserving agents. The lactic fermentation of glucose by various species of bacteria is represented by the following equation:

$$C_6H_{12}O_6 \longrightarrow 2CH_3 \cdot CHOH \cdot COOH$$
Glucose   Lactic acid

This equation omits the numerous intermediate steps between glucose and lactic acid, but it indicates that the fermentation is an anaerobic intramolecular oxidation-reduction process. The homofermentative lactic bacteria that bring about this fermentation include certain cocci and lactobacilli. The heterofermentative lactic bacteria (*Leuconostoc* and some species of *Lactobacillus*) produce formic and acetic acids, ethyl alcohol, carbon dioxide, and other products, in addition to lactic acid. Both types of lactic bacteria participate in vegetable fermentations.

The original source of lactic bacteria is the soil, and these organisms are therefore universally present upon vegetable products, along with all other types of soil organisms. The initial problem is to limit microbial growth to the desired lactic bacteria. This is accomplished in part by the creation of anaerobic conditions and in part by the use of salt. Salt serves at least two functions: it helps to draw juices and sugars from the vegetable material and it increases the osmotic pressure of these juices to such an extent that most soil organisms cannot multiply. The sugars are fermented by the lactic bacteria.

SAUERKRAUT

Sauerkraut is fermented cabbage. The shredded cabbage is packed with about 2.5 per cent salt in containers, which may vary in size from a quart jar to a large tank. Weights are applied and the combined action of salt and pressure withdraws juice from the vegetable. Oxygen is soon exhausted, and a succession of bacteria ferment the plant sugar, producing lactic acid and small amounts of acetic acid, alcohol, and other products.

Leuconostoc mesenteroides and other cocci initiate fermentation and produce 0.7 to 1.0 per cent lactic acid. Lactobacilli then multiply and continue the fermentation, increasing the acidity to as much as 2.4 per cent lactic acid. The final reaction is approximately pH 3.5. Fermentation requires two to three weeks at 70° to 85° F. (21° to 29° C.).

Salt and anaerobic conditions prevent the growth of molds and aerobic bacteria throughout the fermenting mass, and the acidity that quickly develops inhibits most bacteria. Some halophilic yeasts are a cause of abnormal fermentation, and molds and film-forming yeasts can grow at the top of a fermentation tank; however, they spoil only the upper layers. These organisms oxidize the acids produced in the normal fermentation, and the decreased acidity then permits the growth of putrefactive bacteria.

PICKLES

Cucumber pickles have been made in the home for many years, but the commercial manufacture of pickles is a large industry. Homemade pickles are preserved by a combination of salt and vinegar. It is more economical for commercial manufacturers to allow cucumbers to undergo lactic fermentation and produce the acidity necessary for preservation.

Fresh cucumbers packed in tanks, whose capacity may be as great as 15,000 bushels, are covered with brine containing 10 to 20 per cent salt, which limits growth of microorganisms to *Lactobacillus plantarum* and other lactic bacteria. Fermentation proceeds under favorable conditions for 6 to 8 weeks, during which time the centers of the cucumbers change from an opaque white to a transparent green, and as much as 1 per cent lactic acid is produced.

Fermented cucumbers in brine are known as "salt stock" and can be kept for years without spoilage as long as the salt content is at least 10 per cent. Salt stock can be used to prepare sweet, sour, mixed, and other types of pickles according to the market demand.

Pickles are subject to loss of acidity by film-forming yeasts and subsequent spoilage by proteolytic bacteria.

DAIRY PRODUCTS

Manufactured dairy products include fermented milk, butter, and cheese. Those mentioned are produced with the aid of a lactic type of fermentation in which bacteria of the *S. lactis* group and the genus *Lactobacillus* participate.

The origin of these products is lost in antiquity, doubtless because lactic fermentation has long occurred naturally in milk. Later it was found that the acid flavor was produced more rapidly and consistently if a small amount of previously fermented product was added to fresh milk and the mixture kept at a suitable temperature. This was the origin of "starters."

A *starter* is a pure or mixed culture of microorganisms that is added to a substrate to initiate a desired fermentation. Starters are widely used in the dairy industry to produce characteristic changes in the manufacture of butter, cultured milks, and cheese. Many of the same products could be manufactured without the use of starters, but the processes would be wasteful because the proper mixture of microorganisms is not always present in a given batch of milk.

Butter starters are used in the manufacture of several products: they ripen cream to be used in making butter, they are used to manufacture cultured sour cream and buttermilk, and they improve the flavor and texture of cottage and cream cheese. Butter starters contain two types of bacteria: (1) vigorous lactic acid-producing species such as *S. lactis* and *S. cremoris* and (2) bacteria that produce flavor and aroma compounds—*Leuconostoc cremoris* or *L. dextranicum*. These two types of bacteria will grow indefinitely together if handled properly. The flavor and aroma of sour cream butter are attributed to diacetyl, which is produced by the *Leuconostoc* species from citrates normally present in small amounts in milk.

Cheese starters vary according to the cheese to be manufactured. Cheddar cheese, for example, may be manufactured by use of a single strain culture of *S. lactis* or *S. cremoris*. *Lactobacillus, Propionibacterium* or other bacterial species, or yeasts or molds assist in developing the flavor, aroma, and texture characteristic of other cheeses.

Fermented Milks. Fermented milks are prepared by cultivating lactic bacteria in milk. The lactic acid thickens or curdles the milk and produces the desired sour flavor. The nature of the product depends on the source of the milk (cows, goats, sheep, mares, buffaloes, etc.), the temperature to which it is heated before inoculation, the kinds of microorganisms in the starter, and the incubation temperature. Fermented milk products include cultured buttermilk, Bulgarian buttermilk, and acidophilus milk, all of which are used in the United States; yoghurt, also popular in this country but originally derived from the eastern Mediterranean area; the Armenian mazun, Egyptian leben, and Indian dadhi; kefir of the Balkan countries, and koumiss of southern Russia.

Yoghurt can be made from the milk of cows, goats, sheep, or buffalo. Originally the milk was concentrated by boiling, inoculated with part of a previous batch of yoghurt, and kept at 38° to 46° C. until a thick curd developed, usually within 10 to 12 hours. The acidity attained was 1 to 3 per cent, as lactic acid. The high incubation temperature limited fermentation to *Streptococcus thermophilus* and *Lactobacillus bulgaricus,* the latter producing the strong final acidity.

Yoghurt is made commercially in the United States from milk concentrated under vacuum or by adding milk powder or condensed milk. The concentrated milk is heated at 80° to 90° C. to kill nonspore-forming bacteria and is then inoculated with a starter containing equal numbers of *S. thermophilus* and *L. bulgaricus*. After thorough mixing it is dispensed into the final retail jars or cartons, incubated at 45° C. for two and one-half to three and one-half hours, and then refrigerated. The acidity attained is about 0.9 per cent as lactic acid. The product has a heavy, smooth, custardlike consistency, and a mildly sour, nutty flavor. It can be kept one to two weeks at refrigerator temperature. *S. thermophilus* initiates acid production and, at the end of the first hour of fermentation, outnumbers *L. bulgaricus,* which then grows rapidly and produces lactic acid and volatile products responsible for the characteristic flavor and aroma.

Butter. Butter contains approximately 80 per cent fat, small percentages of lactose and protein, and often 2.0 per cent salt. The remainder is water in the form of minute droplets dispersed throughout the butterfat. The salt is dissolved in this water.

Butter is made from either sweet or sour cream by churning, which separates the fat from most of the rest of the cream. The cream is usually pasteurized to destroy pathogenic bacteria and reduce the number of spoilage microorganisms. Butter culture is added if a more highly flavored and aromatic product is desired. After churning, the buttermilk is removed, and the butter is washed and finally "worked" to distribute the water droplets and salt, if added, uniformly.

Cheese. Cheese is the product made by separating the casein of milk from the liquid or whey. The butterfat often accompanies the casein, but most of the lactose and other soluble milk constituents remain in the whey. Approximately 400 kinds of cheese are known, and most of these can be prepared from any given batch of milk by properly regulating the conditions of manufacture.

Classification of Cheeses. Cheeses are classified according to their consistency and the use and nature of microbial ripening agents. The following outline lists the principal groups with examples of each:

A. *Unripened cheeses*
   1. Low fat (cottage cheese)
   2. High fat (cream cheese)
B. *Ripened cheeses*
   1. Hard cheeses (internal ripening)
      a. Ripened by bacteria (Cheddar cheese, Swiss cheese)
      b. Ripened by mold (Roquefort and other blue cheeses)

2. Soft cheeses (ripening proceeds from outside)
   a. Ripened by bacteria (Limburger cheese)
   b. Ripened by bacteria and molds (Camembert cheese)

Principal Steps in Cheese Manufacture.

A brief summary of the process of cheese making follows.

1. The milk is inoculated with a starter culture and warmed to a temperature favorable for acid production.

2. When a certain acidity has been reached, rennet extract is added. At the proper pH and temperature, curdling takes place within one-half to one hour.

3. The curd is cut into small cubes and the whey is drained off. Heat may be applied to hasten separation of the curd particles from the whey.

4. The curd is put into frames; it is then either pressed or allowed to stand to continue the removal of whey. The frames are removed as soon as the curd has set sufficiently to maintain its shape.

5. Salt is applied, either to the curd before it is placed in the frames or to the outside of the pressed cheese.

Cottage cheese is highly perishable. It contains insufficient acid or salt to prevent microbial spoilage and must be constantly refrigerated. It is subject to spoilage by all kinds of microorganisms, including molds, yeasts, and slime-producing bacteria.

The flavor of raw cheese curd is very bland, and it is rubbery in consistency. During ripening, cheese develops a distinctive flavor and aroma, and its texture changes; a hard cheese becomes crumbly, and a soft cheese may become smooth and semiliquid. Chemical changes that accompany ripening include a marked increase in soluble nitrogen compounds, such as amino acids and ammonia. The fatty acid content of some cheeses also increases as butterfat is hydrolyzed. The flavor and aroma of well-ripened cheese are attributed to these various compounds.

Lactic acid formation is important in the early stages of cheese manufacture and curing. Acidity hastens curdling; it suppresses the growth of undesirable gas-forming and putrefactive bacteria, activates the proteolytic enzyme pepsin, which is usually present in rennet extract, and helps the curd to fuse together and expel whey. Acidity therefore assists in the formation of the curd and in the texture and flavor changes of the young cheese.

Ripened cheeses are subdivided into hard varieties and soft varieties. Hard cheeses such as Cheddar, Swiss, and Roquefort contain no more than 39 per cent moisture. They are ripened by microorganisms growing throughout the cheese; bacteria are the principal ripening agents, but Roquefort and the other blue cheeses are ripened by molds. Ripening usually requires several months. Steps in the manufacture of Cheddar cheese are illustrated in Figure 25–5.

Soft cheeses contain more than 39 per cent water. They are ripened by molds, yeasts, and bacteria growing on the surface; these organisms produce hydrolytic enzymes, which diffuse inward, digesting the protein of the curd. Ripening is complete as soon as the enzymes reach the center of the cheese, which usually requires four to eight weeks. Cheeses of this kind are small because the greater water content of the curd prevents larger cheeses from holding their form and because the prolonged ripening period would result in over-ripening of the outer portions.

MAINTENANCE OF MILK QUALITY

High quality milk has a low bacterial count and contains no pathogenic bacteria; it is of good flavor and adequate keeping quality, normal in composition, and free from extraneous matter and toxic substances. The sanitary quality of milk is appropriately judged by its bacterial population. Federal regulations proposed by the United States Public Health Service apply to milk used in interstate commerce. The U.S.P.H.S. ordinance and code is recommended for adoption by states or smaller governmental agencies; many of these have adopted even more strict requirements than those recommended.

GRADES OF MILK

Several grades or classes of milk are distinguished on the basis of the number of bacteria they contain (Table 25–7). Many states or municipalities now permit the sale of only pasteurized milk. Grade A pasteurized milk must be prepared from raw milk that contains no more than 200,000 bacteria per milliliter; after pasteurization it must contain no more than 30,000 bacteria per milliliter, and the coliform count must not exceed 10 per milliliter. Certified milk is produced under conditions rigorously controlled by the American Association of Medical Milk Commissions, Inc.

METHODS OF TESTING MILK

Standard Plate Count. The plate count is the official method of counting bacteria in pasteurized milk and is often used in examining raw milk. Dilutions are prepared and plates poured with either yeast extract-tryptone-dextrose agar (Difco Laboratories) or milk-protein-hydrolysate-glucose agar (Baltimore Biological Laboratories). The plates are incubated at 32° C. or 35° C. and counted after 48 hours. Plates containing between 30 and 300 colonies are counted if available, and the results are multiplied by the proper dilution factor and expressed as "standard plate count" per milliliter (or gram). The expression standard plate count is used in preference to bacteria because milk contains chains or clumps of bacteria that yield single colonies.

Direct Microscopic Count. The number of bacteria in milk can be determined by direct microscopic examination. A special capillary pipette (Breed) or special loop is used to measure 0.01 ml. of milk, which is spread uniformly over an area of 1 sq. cm. on a clean microscope slide. After drying, the film is defatted by a solvent, such as xylene, and stained with methylene blue or another appropriate dye. The stained film is then examined under the oil immersion objective of the microscope. The area of the microscope field must first be standardized by measurement with a stage micrometer and a microscope factor is calculated, by means of which the number of bacteria seen per field can be translated into the number of bacteria per milliliter of the milk sample. A microscope whose oil immersion field has a diameter of 160 μm has a microscope factor of about 500,000; this means that each bacterium seen with the microscope represents 500,000 bacteria in the milk sample. Bacterial clumps are usually counted rather than individual cells, because the results are more nearly like those of the standard plate count.

The direct microscopic count has many advantages. It is more rapid than the standard plate count; results are obtained in a few minutes by a skilled operator, and the stained slide can be filed and kept for a permanent record. It is less expensive, requiring a smaller outlay for equipment, media and time. Moreover, the morphology of the organisms indicates improper practices or conditions, which should promptly be corrected. Large numbers of micrococci are often found when utensils are inadequately cleaned; improperly cooled milk usually contains many cocci in pairs or short chains; and the long-chained streptococci that may cause mastitis are easily recognized.

The microscopic count cannot be used satisfactorily with milk of low bacterial content, because many fields must be examined before a single bacterium is encountered. Moreover, recently heated milk contains bacteria still capable of retaining stains and hence appears to possess a higher bacterial count than is actually the case. The microscopic count is valuable principally as a method of rough grading, and is often used by dairies to classify incoming raw milk before pasteurization.

Dye Reduction Tests. The ability of bacteria to transfer hydrogen to dyes is utilized in the dye reduction test for grading raw milk. The amount of hydrogen transferred depends on the species and number of bacteria, the temperature, and other factors. The conditions for the test are kept constant, and the assumption is made that the microbial flora of raw milk samples is generally similar. The number of organisms is therefore the unknown factor, and the greater the number

Figure 25–5a. *The first step in making Cheddar cheese. The milk in the vat has just curdled and is ready to cut.*

⟵

Figure 25–5b. *The soft curd is cut into small cubes with "knives" such as this.* ⟶

Figure 25–5c. *The draining curd is cut into "mats."*
⟵

Figure 25–5d. *The "mats" are turned frequently to promote expulsion of whey, and the curd gradually fuses together until its consistency resembles that of chicken breast meat.* ⟶

Figure 25–5e. *The matted curd is milled and salted.*
$\longrightarrow$

Figure 25–5f. *The salted curd is packed in frames and pressed overnight. Part of the press is shown in the upper right corner.*
$\longleftarrow$

Figure 25–5g. *The pressed cheeses are removed from the frames and cured in a cool room for several weeks or months. The tall cheeses are a variety of Cheddar called longhorn. The flat cheeses with rounded edges at the right are Swiss.*
$\longrightarrow$

TABLE 25-7.   Bacterial Standards for Raw and Pasteurized Milk Recommended by the U. S. Public Health Service Milk Ordinance

Grade	Raw Milk for Pasteurization; Standard Plate Count Not to Exceed	Pasteurized Milk	
		Standard Plate Count Not to Exceed	Coliform Bacteria Not to Exceed
		PER MILLILITER	
A	100,000	20,000	10
Certified	10,000	500	1

of bacteria, the shorter is the time required to reduce the dye. The dyes used are those that decolorize (e.g., methylene blue) or change color characteristically (e.g., resazurin) as hydrogen is accepted. The time required for a given color change is noted.

The methylene blue reduction test is performed by mixing 10 ml. of milk with sufficient methylene blue solution to produce a final dye concentration of one part in 250,000 parts of milk. Test tubes containing the mixture are placed in a waterbath at 36° C. and observed periodically. The reduction endpoint is 80 per cent decolorization of the methylene blue; that is, the upper 20 per cent of the milk may retain the blue color. A milk sample that produces this endpoint within one-half hour contains many bacteria, probably millions per milliliter. The U.S. Department of Agriculture recommends that raw milk be classified into three groups based upon reduction times of 2½ and 4½ hours.

Resazurin undergoes a series of color changes during reduction from its slate-blue oxidized form through blue, purple, lavender, and pink; finally it becomes colorless. The resazurin test requires only about half as long as the methylene blue reduction test.

Dye reduction tests are used to grade raw milk to be pasteurized or evaporated. The tests can be performed by unskilled help following simple directions, and results are obtained within a few hours. Abnormal milks are often detected quickly enough to be diverted to other uses. The tests are not appropriate for final examination of pasteurized milk.

Phosphatase Test. The phosphatase test is used to check the adequacy of pasteurization and to detect any admixture of

raw milk with pasteurized milk. The enzyme phosphatase is secreted by the mammary gland of the cow and is always present in raw milk. Its normal action is hydrolysis of phosphoric acid esters.

Phosphatase is only slightly more resistant to heat than *M. tuberculosis* throughout the entire range of pasteurization conditions, both by the holding method and the flash method. Pasteurization practices can therefore be controlled by testing milk for its phosphatase content. Sensitive tests detect slight irregularities in the temperature or duration of heating or the addition of as little as 1 ml. of raw milk to 1000 ml. pasteurized milk.

The phosphatase test is performed by mixing milk and a buffer substrate containing disodium phenyl phosphate and incubating at 37° to 45° C. for a short time. Phosphatase hydrolyzes disodium phenyl phosphate, and the resulting phenol is detected by adding BQC indicator (2,6-dibromoquinone chloroimide), which produces a blue compound, indophenol.

Thermoduric Bacteria. Thermoduric bacteria as defined by dairy bacteriologists are those that survive pasteurization. Their significance for the milk processor derives from the fact that they do survive pasteurization and contribute to the bacterial count of pasteurized milk. They are not necessarily harmful, and they produce acid or digest proteins only slowly. Thermoduric bacteria are found in milk as the result of poor sanitation and carelessness. They are derived from dirty utensils, unclean cows, dirty milking barns, and also from unsanitary dairy plants.

The presence of thermoduric bacteria in milk is detected by laboratory pasteurization of samples of the raw milk. A 10 ml. specimen in a screw-capped vial is heated for 30 minutes at 61.7° C. and then plated. A standard of 5000 to 10,000 thermoduric bacteria per milliliter is often set; this limit can easily be met by proper attention to sanitary conditions.

INSPECTION

Inspection of milk-producing farms and processing plants is a necessary part of any program for maintaining the sanitary quality of milk. The U.S. Public Health Service Milk Ordinance contains numerous recommendations regarding conditions on farms and in milk plants: recommendations concerning removal of manure from barns, general sanitation, cooling of milk, and so forth. Farms and milk plants from which grade A milk is secured must be inspected at least once every six months. Both the raw milk from each farm and the finished product must be tested at least four times every six months.

TESTS OF CATTLE AND PERSONNEL

Detection of Infected Cattle. Tuberculosis, brucellosis, and mastitis in cattle are detectable by examination of the animals or by appropriate tests of their milk.

Tuberculosis. Past or present tuberculosis is detected by the tuberculin test.

Brucellosis. Brucellosis in cattle is indicated by the presence of *Brucella* antibodies in the blood or milk of infected animals. The disease is controlled by eliminating infected animals from herds.

Mastitis. Mastitis is more serious to the milk producer than to the consumer because milk production decreases. The presence of excessive leukocytes in milk and an alkaline reaction indicate probable mastitis; the causative organism is determined by microscopic examination and by isolation and identification.

Examination of Human Handlers. Human diseases spread by milk fall into two classes: intestinal and respiratory. Detection of undiagnosed or ambulatory cases and of carriers among the hundreds of thousands of farm and dairy personnel is an almost impossible task.

Carriers of intestinal infection can be detected by the isolation of pathogens from stool samples or rectal swabs; special enrichment media that suppress normal coliform bacteria are used, but even so, repeated examinations are necessary. Respiratory pathogens are detected in cultures from sputum samples and throat and nasal swabs. Blood agar and other enriched media are required to cultivate these organisms, many of which are nutritionally fastidious.

Employees of dairy plants from which the final pasteurized dairy product goes to the consumer should be examined periodically for intestinal and perhaps even respiratory pathogens. Regardless of whether this can be done, employee education will assist materially in controlling the spread of milkborne infection. Most individuals would be startled to know how much of the time their hands are contaminated by bacteria from their own oral, respiratory, and intestinal excretions.

SUPPLEMENTARY READING

American Public Health Association: *Standard Methods for the Examination of Dairy Products,* 12th ed. New York, American Public Health Association, 1967.

Elliker, P. R.: *Practical Dairy Bacteriology.* New York, McGraw-Hill Book Co., Inc., 1949.

Foster, E. M., Nelson, F. E., Speck, M. L., Doetsch, R. N., and Olson, J. C.: *Dairy Microbiology.* Englewood Cliffs, N.J., Prentice-Hall, Inc., 1957.

Frazier, W. C.: *Food Microbiology,* 2nd ed. New York, McGraw-Hill Book Co., Inc., 1967.

Hammer, B. W., and Babel, F. J.: *Dairy Bacteriology,* 4th ed. New York, John Wiley & Sons, Inc., 1957.

Jay, J. M.: *Modern Food Microbiology.* New York, Van Nostrand Reinhold Company, 1970.

Slanetz, L. W., Chichester, C. O., Gaufin, A. R., and Ordal, Z. J. (eds.): *Microbiological Quality of Foods.* New York, Academic Press, Inc., 1963.

Tanner, F. W.: *The Microbiology of Foods,* 2nd ed. Champaign, Ill., Garrard Press, 1944.

United States Department of Health, Education and Welfare, Public Health Service: *Grade "A" Pasteurized Milk Ordinance, 1965 Recommendations,* 1965.

United States Public Health Service: *Milk Ordinance and Code.* Public Health Bulletin 229, 1953.

26 INDUSTRIAL MICROBIOLOGY

The microbiologist plays a constant and important role in any microbial manufacturing process. He selects the microorganism to be used, devises the most favorable culture medium, and chooses proper cultural conditions (aeration, agitation, pH, temperature). He tries out the process in test tubes and flasks in the laboratory and then in the pilot plant. Methods that are satisfactory in the laboratory do not always give the best results on a larger scale, and readjustments may be required for commercial application. Constant control is necessary throughout the manufacturing process to ensure economy of materials and time and uniformly high quality of the product.

Industrial applications of microbiology include mass cultivation of microorganisms, manufacture of various chemicals, and textile manufacture.

MASS CULTIVATION OF MICROORGANISMS

Outside the laboratory, microorganisms are cultivated on a large scale principally for use in other industries. Bakers' yeast is used as a leavening agent by commercial bakers as well as in the home. Farmers inoculate legume seed with rhizobia, the root nodule bacteria, to ensure well nodulated plants. Butter and cheese starter cultures are necessary in dairy manufacture. In addition, yeasts and molds have been used as food or feed at one time or another, and mass cultures of pathogenic bacteria are required for the preparation of immunizing materials for the protection of man and domestic animals.

BAKERS' YEAST

Bakers' yeast is a strain of *Saccharomyces cerevisiae* carefully selected for its capacity to produce abundant gas quickly, its viability during ordinary storage, and its ability to produce a desirable flavor.

A pure culture of the chosen yeast must first be grown in the laboratory and gradually "built up" to larger and larger volume by transfer from test tube to small flask to large flask, and so forth, until eventually sufficient yeast is obtained to inoculate the main tank or fermenter. Great care is taken to avoid contamination at any stage of development of the culture.

The medium contains 0.5 to 1.5 per cent sugar, nitrogen in the form of peptones, peptides, amino acids, or ammonia, and mineral salts. The sugar is derived from molasses or from grains that have been cooked and treated with amylases to digest the starch. The reaction is adjusted to pH 4.4 to 4.6 by addition of sulfuric acid or by preliminary fermentation with *Lactobacillus delbrueckii*. Acidity favors growth of the yeast and discourages most bacteria that might cause spoilage. The pH of the medium is controlled during yeast multiplication by addition of ammonia or sulfuric acid as required.

The optimum temperature is 25° to 26° C.; the temperature frequently rises during fermentation, and cooling coils keep it from exceeding 30° C. Vigorous aeration provides the oxygen required for rapid growth. It will be recalled that a disaccharide such as sucrose or maltose is first hydrolyzed by yeast and then oxidized under aerobic conditions:

$$C_{12}H_{22}O_{11} + H_2O \rightarrow 2\ C_6H_{12}O_6$$

$$2\ C_6H_{12}O_6 + 12\ O_2 \rightarrow 12\ CO_2 + 12\ H_2O$$

This reaction yields the maximum available energy, a large part of which is utilized for synthesis of microbial protoplasm. The yeast cells multiply rapidly and exhaust the sugar supply within 10 or 11 hours.

The yeast is removed from the fermented medium by centrifugation, washed, and mixed with starch or corn meal before being pressed into cake form. The starch or corn meal helps to maintain the shape of the yeast cake. Yeast cakes must be kept cool to preserve the cells and to retard spoilage by other microorganisms. Yeast may also be dried to about 10 per cent moisture; dried yeast remains viable for several months without spoilage.

Yeast for animal feed can be manufactured from waste materials such as wood shavings and sawdust, straw, corn cobs, and other agricultural waste. The carbohydrates in these materials must first be converted into fermentable form, usually maltose or glucose, either by enzymic digestion or by acid hydrolysis.

MICROORGANISMS FOR MEDICAL USE

The theory and principles of artificial immunization of man and animals were discussed in Chapter 19. The immunizing agents, which are known as *vaccines,* are heavy suspensions of attenuated (i.e., weakened) or killed microorganisms. Microbial strains are carefully selected to possess the greatest possible immunizing power.

Vaccines are manufactured under strictly controlled conditions. Bacterial vaccines are prepared from cultures grown on agar or broth media. Cells on agar are suspended in saline (0.85 per cent sodium chloride); cells in broth are removed by centrifugation. The bacteria are washed by centrifugation with saline to eliminate extraneous material from the culture medium; they are finally suspended in saline at a standard concentration or density (e.g., one billion cells per milliliter).

Rickettsiae and viruses must be cultivated in living tissue: an animal body, chick embryo, or tissue culture. Vaccines prepared from these materials contain substances derived from the animal or tissue, which contribute nothing useful and may occasionally cause hypersensitivity or other undesirable side reactions. It is possible in some cases to reduce these effects by harvesting only portions of the infected tissue (e.g., certain organs of an inoculated animal, allantoic fluid from a chick embryo).

Many vaccines are killed or inactivated by heat (55° to 60° C. for 30 to 60 minutes), ultraviolet irradiation, or chemicals (formaldehyde, phenol, etc.). Other vaccines consist of organisms attenuated so that they are unable to produce disease. Pasteur attenuated *Bacillus anthracis* by cultivating it at 42° to 43° C. He also attenuated the virus of hydrophobia by drying infected rabbit spinal cords for several days. Other viruses may be attenuated by cultivation in unnatural hosts; for example, the yellow fever virus lost its normal pathogenicity for man when grown in mice.

The final product is tested for its content of the proper immunizing material, its immunizing potency, its freedom from contamination, and, in the case of a killed vaccine, its sterility. It is dispensed in bottles or vials and stored at low temperature until used. Chemical preservatives are added when possible to prevent growth of contaminants; otherwise, low temperature is relied on for preservation as well as maintenance of immunizing potency.

MANUFACTURE OF MICROBIAL PRODUCTS

In addition to their use in the manufacture of dairy products and other fermented foods, microorganisms are utilized in the manufacture of numerous chemicals that are useful domestically or industrially or for controlling disease: ethyl alcohol, acetic acid, solvents such as acetone and butyl alcohol, organic acids resulting from mold fermentations, antibiotics, enzymes, toxins, and toxoids.

ALCOHOLIC FERMENTATION

The equation that describes the net result of the alcoholic fermentation by yeast:

$$C_6H_{12}O_6 \rightarrow 2\ C_2H_5OH + 2\ CO_2$$

indicates that a sugar is the substrate and that the process is anaerobic. Ethyl alcohol and carbon dioxide accumulate in amounts as high as 90 per cent of the theoretical yield. Small amounts of other products usually are formed also.

The common yeasts can ferment the monosaccharides glucose and fructose and the disaccharides sucrose and maltose. Fruit juices, molasses, and other syrups can therefore be fermented with little preliminary treatment because they contain glucose, fructose, or sucrose. The polysaccharides starch and cellulose cannot be fermented directly by yeasts; they require preliminary hydrolysis by enzymes or acid to the disaccharide or monosaccharide stage.

Details of the process of alcoholic fermentation vary according to the desired product: beer, wine, distilled liquors, or industrial alcohol. Beer is manufactured from grains (i.e., a starchy source of carbohydrate), wines are made from fruit juices (i.e., sugar solutions), and distilled liquors and industrial alcohol may be made from either type of raw material.

Brewing. There are five major steps in the manufacture of beer or ale from grain: malting, mashing, fermenting, maturing, and finishing. Malting and mashing are concerned with the conversion of starch into fermentable form as maltose or glucose; fermentation is the actual production of alcohol and carbon dioxide; maturing is the aging process that improves the flavor of the beverage; finishing includes bottling and other steps necessary to market the product.

Malting. Malt is the chief raw material in the manufacture of beer and ale. It is germinated barley that has been dried and ground, and contains starch, proteins, and high concentrations of amylases and proteinases. Amylase converts the starch of barley and other grains such as wheat, corn, and rye into fermentable sugar. Most American beer is made from a mixture of grains in which barley malt represents 65 to 80 per cent of the raw material.

Barley grain is first *steeped* or soaked in water and then placed in a revolving drum to *germinate* at 15° to 21° C. for five to seven days. Germination is halted by drying when the sprout is about three-quarters of the length of the kernel, at which time the enzyme content is maximal. The dried malt can be stored without microbial spoilage, and the enzymes remain stable for a considerable period.

Barley malt is commonly used in Europe and America for converting starch into sugar for brewing. Mold amylase derived from *Aspergillus oryzae* is used for the same purpose in some countries; it is also used to produce sugars from grain and potatoes for manufacturing industrial alcohol.

Mashing. Mashing is the process by which the starch and proteins of malt and other grains are digested to produce *wort*. Wort contains dextrins, maltose and other sugars, protein breakdown products, minerals, and various growth factors.

Ground malt, with or without other cooked grain, is mixed with water at 65° to 75° C. and pH 5.0 to 5.8. After partial hydrolysis of the starch and protein, the solution is filtered, and hops are added if it is to be used for making beer or ale. Hops are the flowers of *Humulus lupulus* or *H. americana;* they contribute a characteristic flavor and mild antiseptic properties, which discourage the growth of certain spoilage bacteria.

Fermentation. Wort is inoculated heavily with a selected strain of *S. cerevisiae.* Yeasts used in brewing are classified as "top yeasts" or "bottom yeasts." Top yeasts float to the surface of a fermenting mixture; they are usually employed in making ale. Bottom yeasts settle in a fermentation tank; they are used in making beer. The beer fermentation continues for eight to 10 days at 6° to 12° C., whereas ale fermentation is complete in five to seven days at 14° to 23° C. The alcoholic content of beer is rarely more than six per cent; that of ale is often somewhat greater.

Maturing. Fresh beer or ale has a harsh flavor and other undesirable characteristics that are removed by maturing or aging. The fermented wort is refrigerated at approximately 0° C. for two weeks to several months. Unstable proteins, yeasts, resins, and other substances precipitate, the harshness disappears as esters are produced, and the beer becomes mellow. Some of the harshness is attributed to fusel oils, actually higher alcohols, which are oxidized or esterified during aging.

Finishing. Finishing consists of carbonation, cooling, filtering, and "racking" or dispensing into barrels, bottles, or cans. Bottled or canned beer is usually pasteurized at 60° to 61° C. for 20 minutes to kill yeasts and any undesirable microorganisms that may be present.

Wine Manufacture. Wine is, by definition, the product made by the "normal alcoholic fermentation of the juice of sound, ripe grapes and the usual cellar treatment." Beverages produced by the alcoholic fermentation of other fruits and berries are also often called wines; for example, orange wine, peach wine, blackberry wine.

Preparation of the "Must." Each wine is made from a particular type or variety of grape. In making red wines the grapes are crushed and stemmed, but the skins and seeds are left in the "must" or nutrient sugar solution of expressed juices. White wines are made from white grapes or from the juice of grapes from which the skins have been removed.

The pressed juice of grapes will undergo spontaneous alcoholic fermentation caused by yeasts normally present on grapes, and in fact the characteristic qualities of famous wines are attributed in part to strains of yeast found in certain localities. However, undesirable molds, wild yeasts, and bacteria are also likely to be present, and many wine makers now destroy these by adding sulfur dioxide to the must.

Fermentation. A starter of a selected strain of *Saccharomyces ellipsoideus* is added to the must; aeration promotes rapid early growth, but conditions are soon permitted to become anaerobic for the alcoholic fermentation. After three to five days at 21° to 32° C., the wine is drawn off from the pomace (skins, seeds, etc.), and further fermentation takes place for seven to 11 days. The yield of alcohol varies from 7 to 14 per cent according to the strain of yeast, the temperature of fermentation, and other factors.

Aging and Finishing. Fermentation is followed by *racking;* that is, the wine is drawn off from the sediment. It is then aged in wooden tanks for two to more than five years, during which time the wine gradually clears and develops bouquet and flavor as volatile esters are produced. Final *clarification* is accomplished by adding casein, gelatin, or Spanish clay and then filtering. The bottled wine may be pasteurized at 60° C. for 30 minutes.

Types of Wines. *Dry wines* are those in which "the fermentation of the sugars is practically complete." They contain too little sugar to be detected by the sense of taste. In *sweet wines*, "the alcoholic fermentation has been arrested," and the sugar content is great enough to be detected by taste. *Fortified wines* contain added alcohol in the form of brandy (i.e., the product resulting from distillation of wine). Fortified wines usually contain not less than 17 per cent alcohol, whereas the alcohol content of natural wines is less than 14 per cent. *Sparkling wines* contain carbon dioxide; the final stages of fermentation take place within the bottle.

Distilled Liquors. The characteristic flavor and aroma of a distilled liquor depend on the nature of the solution distilled. Whiskey is distilled fermented grain mash; it contains 40 to 55 per cent ethyl alcohol. Bourbon is whiskey prepared from a mash in which corn is the predominant grain; rye whiskey is manufactured from a mash in which rye grain predominates. Brandy usually contains 40 to 50 per cent alcohol and is prepared by distillation of fermented fruit juice, that is, wine. Rum is produced by distillation of fermented molasses or other sugar cane by-products and contains not less than 40 per cent alcohol. Gin is usually produced by extracting juniper berries with alcohol and distilling off the alcohol; it contains volatile extractives from the berries. Cordials and liqueurs are sweetened alcoholic distillates from fruits, flowers, leaves, etc.

ACETIC ACID PRODUCTION

The equation for the production of acetic acid (vinegar):

$$C_2H_5OH + O_2 \rightarrow CH_3COOH + H_2O$$

indicates that a source of ethyl alcohol is necessary and that the process is aerobic. The alcohol is usually derived from an alcoholic fermentation without distillation.

Vinegar is a solution containing at least 4 per cent acetic acid and small amounts of alcohol, glycerin, esters, sugars, and salts. Most vinegar is made from wine, apple cider, or fermented malt.

Acetobacter. The microorganisms that produce acetic acid from ethyl alcohol are species of *Acetobacter, A. aceti, A. orleanensis, A. schutzenbachii,* and others. They are widely distributed in the soil and hence in the air, and are almost universally present on grapes, apples, and other fruits; the juices therefore usually undergo acetic fermentation following spontaneous alcoholic fermentation unless precautions are taken to prevent it.

Home Method. Vinegar is commonly made at home from cider, grape juice, or miscellaneous pooled fruit juices in a barrel provided with two openings. Yeast naturally present or deliberately added ferments the sugar, and during this period the openings are closed except for a trap to release gas pressure. Cessation of gas evolution is a sign that the alcoholic fermentation has ceased. The barrel should be laid on its side and the openings unstoppered to permit air circulation. Several weeks or months are required for spontaneous vinegar fermentation, and periodic sampling is necessary to ascertain when the product is of suitable strength.

More satisfactory results may be secured if the yeast is allowed to settle after the alcoholic fermentation. The solution is carefully drawn off and transferred to a vinegar barrel and inoculated and acidified by adding 10 to 25 per cent of pure vinegar. When

fermentation is complete the vinegar should be bottled and stoppered tightly to prevent continued growth of *Acetobacter,* because these organisms can oxidize acetic acid after the alcohol concentration drops below 1 or 2 per cent, thus reducing the strength of the vinegar. Aging for a year or more greatly improves the flavor and aroma of the product. The vinegar may finally be pasteurized.

During acetic fermentation the bacteria grow as a slimy zoogleal mat or membrane over the surface of the solution, where they have access to both ethyl alcohol and oxygen. If the mat is disturbed and sinks to the bottom ("mother of vinegar"), acetification stops until another mat forms.

Orleans Method. The French or Orleans process of manufacturing vinegar employs casks of about 200 liters (50 gallons) capacity. These are one-third filled with good vinegar, and 10 to 15 liters of wine are added at weekly intervals. After five weeks, 10 to 15 liters of vinegar are drawn off each week and the same amount of wine is added. Several holes are drilled in the barrel for air circulation, and often a grating or lattice of wood is provided to support the film of *Acetobacter.* This process is more or less continuous but requires constant attention and maintenance.

Generator Method. Vinegar or commercial acetic acid is made by the German or generator method in a tank that may be as large as 15 feet in diameter and 20 feet high (Fig. 26–1). A perforated false bottom supports beechwood shavings, and a perforated false top provides an opportunity for air to pass upward through the shavings.

The shavings are thoroughly soaked in good vinegar to inoculate them with *Acetobacter* organisms. An alcohol solution distributed over the false top trickles slowly over the shavings. The bacterial oxidation of alcohol to acetic acid on the shavings evolves heat, and the generator behaves like a chimney, drawing air in at the bottom. Cooling coils are sometimes necessary to keep the temperature within the favorable range of 25° to 30° C.

Several passages through a generator are required to produce vinegar of legal (4 per cent) strength. This is accomplished by re-

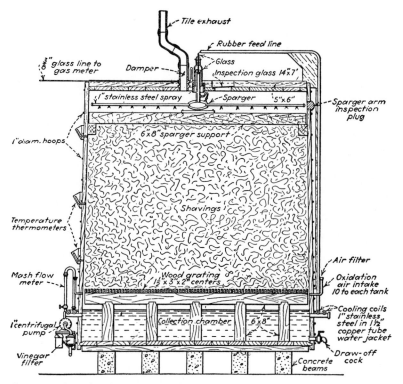

Figure 26–1. *Cross section of a vinegar generator. The alcohol solution is sprayed over the shavings by the rotating sparger of stainless steel. It trickles slowly down and accumulates in the collection chamber at the bottom, from which it may be recirculated or flow to another generator. (From Frobisher: Fundamentals of Microbiology, 9th ed., Philadelphia, W. B. Saunders Co., 1974.)*

circulation or by use of several generators in series. Eight to 10 days are required for complete acetification in a recirculating generator. The yield under favorable conditions is 50 to 55 gm. of acetic acid for every 100 gm. of sugar. This is about 80 per cent of the theoretical yield.

INDUSTRIAL SOLVENTS

Acetone-Butyl Alcohol Fermentation. The acetone-butyl alcohol fermentation is one of several important microbiologic processes employed in the manufacture of useful solvents. In addition to butyl alcohol and acetone, ethyl alcohol, carbon dioxide, hydrogen, and small amounts of acetic and butyric acids are produced. Acetone is used in making explosives, cellulose acetate, and adhesives and is a common chemical solvent. Butyl alcohol is used in lacquers.

The principal raw materials used in the United States are molasses and corn. Sterile diluted molasses or cooked corn mash in 50,000 gallon fermentation tanks is inoculated with *Clostridium acetobutylicum* (Fig. 26–2). Fermentation under anaerobic conditions is complete after 48 to 72 hours at 37° C. (Fig. 26–3). The carbon dioxide and hydrogen gases that evolve during the fermentation account for approximately 60 per cent of the fermentable carbohydrate; they are recovered for industrial use. The neutral solvents, butyl alcohol, acetone, and ethyl alcohol, are recovered by fractional distillation.

MOLD FERMENTATIONS

Molds manufacture numerous organic acids by partial oxidation of glucose or other substrate: gluconic, oxalic, citric, etc. Gluconic acid is produced by certain strains of *Aspergillus niger;* calcium gluconate is sometimes prescribed as a source of cal-

Figure 26–2. *The lower portion of 50,000 gallon fermentation tanks. (Courtesy of Commercial Solvents Corp.)*

cium for children and pregnant women. Gallic acid, produced by *A. niger* from tannin or tannic acid, is used in manufacturing inks and dyes.

Citric Acid Manufacture. More than

10,000 tons of citric acid are produced each year in the United States. It is used in soft drinks and other foods and in medicinal preparations.

Many molds are able to produce citric

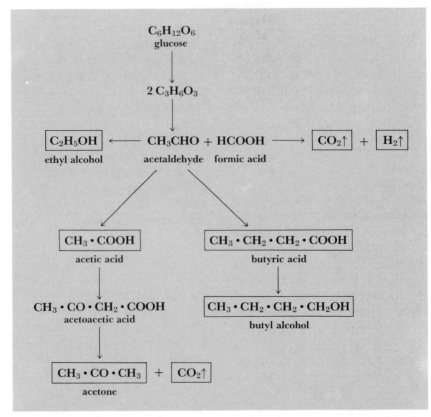

Figure 26–3. *Flow sheet of the acetone-butyl alcohol fermentation (abridged).*

acid; strains of *A. niger* are most satisfactory commercially. The medium contains 14 to 20 per cent glucose or sucrose, ammonium nitrate as the source of nitrogen, and other salts; the pH is adjusted to 1.6 to 2.2. Fermentation in shallow pans at 25° to 30° C. requires seven to 10 days. A shallow pan provides a large mat of mold, which is necessary for converting sugar into citric acid. An aerated submerged tank process is also used; fermentation is more rapid by this method. The citric acid is precipitated as calcium citrate and recovered by treating with sulfuric acid.

ANTIBIOTICS

The manufacture of antibiotics is an industry that did not exist in 1941, but whose products sold for more than $300,000,000 ten years later. Fleming had discovered penicillin in 1929, and toward the end of the next decade Florey, Chain, Heatley, and Abraham devised methods of producing it in small amounts and found it amazingly effective in treating staphylococcal and streptococcal bloodstream infections. Since England was at this time devoting all her energies to the second World War, Florey and Heatley came to the United States in 1941 and enlisted the assistance of governmental, industrial, and educational research laboratories. The rapidity with which all agencies tackled the problems of producing and testing penicillin is shown by the fact that in September, 1943, there was sufficient drug for the armed forces of the western allies.

In 1945, total production of penicillin was 12,000 pounds, and the price was $3870 per pound. By 1963, production had increased over 100-fold, and penicillin sold for $56 per pound. The rise in streptomycin production was comparable: 44,000 pounds were sold in 1950 and 870,000 in 1963.

During this time thousands of antibiotics were isolated, and dozens proved to be more or less useful. Waksman listed over 70 newly reported antibiotics from *Streptomyces* species in a two year period (1961–1962). Many new antibiotics are found to be identical with others previously announced, and the majority are too toxic or of too limited effectiveness for practical application.

Production of Penicillin. The mold from which Fleming isolated penicillin was subsequently identified as *Penicillium notatum*. Other strains were later found to yield greater amounts of the antibiotic, and a strain of a different species, *P. chrysogenum,* is now used for commercial production.

Early in World War II penicillin was produced by a surface culture method. Flasks or bottles containing a shallow layer of medium were inoculated with spores and incubated at 24° C. for five to eight days. The penicillin was then harvested from the medium. Strict asepsis was necessary, because contamination by other microorganisms reduced the yield; this may have been caused by the widespread occurrence of penicillinase-producing bacteria, which inactivate the antibiotic. The surface culture method of manufacturing penicillin was expensive; hundreds of thousands of flasks or bottles were required, and each was inoculated, incubated, and harvested individually.

Submerged culture methods were introduced by 1943 and are now employed almost exclusively. The medium is constantly aerated and agitated, and the mold grows throughout as pellets. Deep tanks with a capacity of several thousand gallons are filled with a culture medium consisting of corn-steep liquor, lactose, glucose, nutrient salts, phenylacetic acid or a derivative, and calcium carbonate as buffer. Corn-steep liquor is an extract obtained during the manufacture of starch and other corn products; it supplies organic nitrogen, minerals, reducing sugar, and lactic acid. Phenylacetic acid and its derivatives are precursors of penicillin and increase the yield of the antibiotic. After fermentation the penicillin is extracted, concentrated, crystallized, dried, and assayed to determine its potency before being bottled and sold.

Assay of Penicillin. The potency of a lot or batch of penicillin is determined by a biologic assay in which the unknown is compared with a standard preparation of crystalline sodium penicillin G. One international unit of penicillin activity is contained in 0.6 μg. of the standard; that is, one mg. of the international standard contains 1667 units of crystalline sodium penicillin G.

The official U.S. Food and Drug Adminis-

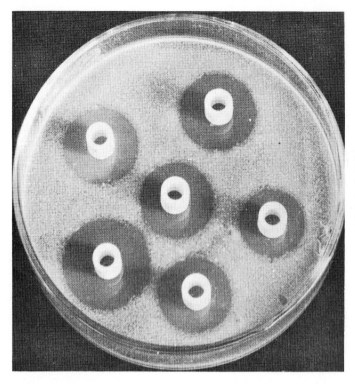

Figure 26–4. *Penicillin assay by the agar cup plate method. The agar is inoculated heavily with* Staphylococcus aureus *and poured. Open cups are pressed into place and filled with penicillin solutions. Zones of inhibition appear after a few hours' incubation. (From Grant: Microbiology and Human Progress. New York, Rinehart & Co., Inc., 1953.)*

tration cylinder-plate method of assaying penicillin is performed in Petri dishes containing a nutrient agar previously inoculated with a specified strain of *S. aureus* (Fig. 26–4). Stainless steel cylinders open at both ends are placed on the agar and filled with dilutions of standard penicillin and of the unknown sample. The plates are incubated at 37° C. for 16 to 18 hours, and the zones of inhibition of bacterial growth are measured. The antibiotic activity of the unknown sample is determined by comparing its zones of inhibition with those of the standard penicillin.

Penicillin is also assayed by a serial dilution method. Dilutions of penicillin are prepared in a liquid medium inoculated with the test organism; the tubes are incubated, and the inhibition of growth is noted. Results with an unknown solution are compared with those produced by a preparation of known strength.

Streptomycin. Streptomycin was discovered in 1943 in Dr. Waksman's laboratory at Rutgers University. It is produced by *Streptomyces griseus,* one of more than 400 species in a bacterial genus characterized by moldlike mycelial growth and the formation of conidia. These are primarily sap-

rophytic soil bacteria, particularly active in the decomposition of organic matter.

Streptomycin is produced commercially by a submerged culture method in 10,000 to 15,000 gallon tanks. The culture medium contains hydrolyzed protein and sugar. Growth continues at 25° to 30° C. with vigorous aeration until the maximum possible yield of streptomycin has been attained. The mycelium is removed by filtration, and the antibiotic is adsorbed onto activated carbon, eluted, purified, sterilized, dried, and packaged.

ENZYMES

Four principal types of microbial enzyme are manufactured for industrial use: amylases, invertase, proteinases, and pectinase. In general, the proper microorganism is cultivated under conditions favorable to enzyme formation, and the enzyme is then extracted and purified by precipitation or other means. Amylase, for example, is often derived from species of *Aspergillus* by the mold-bran process. Wheat bran, moistened with a suitable nutrient solution and sterilized in shallow trays, is inoculated with the mold spores and incubated under optimum temperature and moisture conditions until satisfactory growth has been obtained. The enzyme is then extracted from the bran with a suitable solvent such as alcohol, and is filtered, concentrated or precipitated, and dried. For some purposes mold-bran containing amylase can be dried without extraction, a process that effects a considerable reduction in expense. Amylases are secured from various genera of molds and from a variety of *Bacillus* species.

Amylases are used to hydrolyze starch to dextrins or sugars or both in making adhesives, in preparing materials for sizing, in desizing textiles, in clarifying fruit juices, and in saccharifying starchy solutions for fermentation.

Invertase from the yeast, *Saccharomyces cerevisiae,* is used to hydrolyze sucrose to glucose and fructose in the manufacture of noncrystallizable syrups, as in the production of liquid-centered candies.

Proteinases from *Aspergillus* and *Bacillus* species digest proteins and are used in meat tenderizers, in leather manufacture, in whiskey making, and in clarifying beer.

Pectinases, derived from various species of *Aspergillus,* are used to clarify fruit juices and to ret flax by digesting the cement that holds the fibers together in the plant stem.

DEXTRANS

Dextrans, which are polymers of glucose that are useful as a blood plasma substitute in combating shock, are formed in considerable quantity by certain capsulated bacteria such as *Leuconostoc mesenteroides*. They are produced when the bacteria grow in a sucrose medium, according to the equation:

$$n \text{ sucrose} \rightarrow (\text{glucose})_n + n \text{ fructose}$$
$$\text{dextran}$$

Dextrans are not produced in a glucose medium. Certain strains of *L. mesenteroides* convert over 35 per cent of the supplied sucrose into dextran that is recoverable by appropriate precipitation. The organism is grown in a sucrose-tryptone-yeast extract broth at pH 6.7 and 25° C. until the reaction falls to pH 4.5. Dextran is then precipitated by the addition of an equal volume of methyl alcohol. Partial hydrolysis with hydrochloric acid yields compounds with a molecular weight of 50,000 to 100,000, which are spray-dried and packaged.

GIBBERELLINS

Gibberellins are plant growth regulators formed by the mold *Fusarium moniliforme*. Diseased rice seedlings infected with this organism grow unusually tall and spindly and are light in color. The responsible mold product has been isolated and is an optically active crystallizable acid with the empirical formula $C_{19}H_{22}O_6$. In small amounts, it is a powerful plant growth stimulant and can be used to hasten maturation and to extend the geographic boundaries within which certain crops can be grown.

Gibberellic acid, the active agent, is produced in aerated cultures in a buffered glucose synthetic medium with NH_4Cl as the source of nitrogen. Cultures are incubated at 25° C. for approximately 65 hours. About 7.5 gm. of crude crystalline product can be obtained from 100 gallons of culture.

STEROID TRANSFORMATIONS

Steroids are complex organic compounds with the following basic structure:

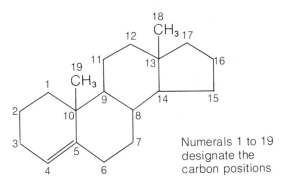

Numerals 1 to 19 designate the carbon positions

They are physiologically active substances, normally produced by animals or plants.

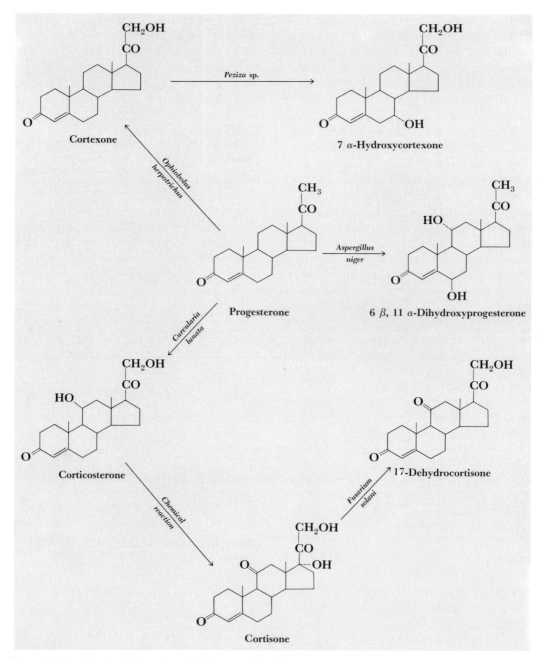

Figure 26–5. *Structural formulae of a few steroids. Progesterone occurs naturally; it is transformed into a number of other steroids by enzymes of various fungi.*

The various steroids differ from one another according to the presence and location of small radicals such as $=O$, $—OH$, $—CO \cdot CH_3$, $—CO \cdot CH_2OH$ (Fig. 26–5). About 1950 it was discovered that compounds of a steroid nature from natural sources could be transformed by various fungi into substances of different activity, and an important new branch of research

developed: the production and evaluation of steroid drugs.

Steroid transformations differ from the activity usually associated with the manufacture of chemicals by microbial metabolism. Acids, neutral solvents, and most other fermentative or synthetic products are formed in the course of normal metabolic activity, perhaps modified by some change in cultural conditions. Steroids, however, are transformed by the action of one or a very few enzymes—often oxidases or dehydrogenases, which are frequently produced by only a single species of organism—and the reaction is completely separate from the normal metabolic pattern of the organism.

The proper species is cultivated on a sugar medium containing organic or inorganic nitrogen and mineral salts, either pure or in the form of corn-steep liquor, for 17 to 48 hours with aeration and at the optimum temperature for the production of the desired enzyme. The steroid to be transformed is then added to the medium, and further incubation under controlled conditions of pH, temperature, aeration, and agitation, usually for 24 to 48 hours, is allowed. The steroid is extracted with a suitable solvent such as chloroform, methylene chloride, or ethylene chloride, and further purified for use.

Progesterone is normally produced by the corpus luteum of the ovary at ovulation and causes characteristic changes during the latter half of the menstrual cycle. Cortisone and related corticosteroids are active in carbohydrate and protein metabolism: they increase the deposition of glycogen in the liver, cause a marked decrease in circulating lymphocytes and eosinophilic leukocytes and degeneration of the thymus gland, and inhibit the inflammatory response.

Steroid transformations have provided compounds of new or enhanced pharmacologic activity, as well as chemicals from which other useful steroids can be made by further transformations.

BACTERIAL TOXINS (EXOTOXINS) AND TOXOIDS

Bacterial exotoxins are poisonous proteins secreted by the living cells of certain species. Their toxicity is very great; 1 mg. of tetanus toxin, for example, contains sufficient poison to kill about four million guinea pigs. Fortunately only a few species produce exotoxins; most of these are gram-positive bacteria.

Antigenicity of Toxins and Toxoids. Exotoxins are highly antigenic; that is, they vigorously stimulate the human or animal body to produce antibodies known as antitoxins, which neutralize and destroy the toxic property. Exotoxins are also unstable. They lose toxicity on aging and are gradually converted into toxoids. The transformation into toxoid is accelerated by heat, formaldehyde, and other chemicals. Toxoids retain the antigenic power of the original toxins and, because they lack toxicity, can be used to produce immunity against the corresponding toxins.

Diphtheria and tetanus toxoids are commonly administered to infants as part of their routine immunization during the first year of life; they are often combined in a triple immunizing agent with pertussis (whooping cough) vaccine.

Production of Toxin. Conditions for maximal laboratory production of toxin are frequently critical, and are often not those that favor best growth. Temperature of incubation and the pH and composition of the culture medium are important. The diphtheria organism must be grown aerobically in order to produce toxin, whereas the tetanus, gas gangrene, and botulism bacilli require highly anaerobic conditions.

Manufacture of Toxoid. The proper bacterium is cultivated under optimal conditions until the greatest possible yield of toxin is obtained. The cells are removed and the toxin in the broth is converted into toxoid by treatment with 0.4 to 0.5 per cent formalin until animal tests show that no toxicity remains. This may require a month or more at 37° C. The immunizing power of the toxoid is then determined by inoculating animals and challenging them with potent toxin after an appropriate interval (e.g., two weeks.)

Toxoid is partially purified by precipitating the protein from other broth constituents with ammonium sulfate. The protein, redissolved in buffered saline, is known as "fluid toxoid" or "plain toxoid."

Toxoid adsorbed to a precipitate of aluminum hydroxide or aluminum phosphate is preferred in some situations. It is made by adding an aluminum salt to the toxoid solution and precipitating by appropriate chemical treatment. The precipitate contains the toxoid and when suspended in saline is known as "alum precipitated toxoid."

HYDROCARBON FERMENTATION

Hydrocarbons, both aliphatic and aromatic, with the exception of certain compounds of low molecular weight like methane, have generally been considered resistant to microbial attack. Since about 1950 there has been increasing evidence that microorganisms can utilize, or at least degrade, larger hydrocarbons such as those found in petroleum and its products. The central problem for the organism appears to be the transport of a hydrophobic substance to an intracellular site of enzyme activity. Petroleum and jet fuel are relatively crude mixtures in which emulsifiers may form and assist this process, particularly if traces of water are also present. In consequence, extensive microbial growth produces a slime that clogs fuel lines and causes other trouble. Over 100 species of microorganism have been isolated from such slime and demonstrated to be capable of attacking hydrocarbons. Species of *Pseudomonas* are particularly common among the bacteria.

Considerable research is under way to develop methods of decomposing hydrocarbons, which are often present in industrial sewage and challenge the capabilities of treatment plants. Among the bacteria, members of the genus *Pseudomonas* can utilize the widest range of substrates and offer promise for practical use.

TEXTILE MICROBIOLOGY

There are two principal aspects of the microbiology of textiles. One is the use of microorganisms in preparing fibers such as flax and hemp. The other is the deterioration of textiles, including cordage and ropes, and the preservation of such materials.

RETTING

Fibers of flax and hemp are loosened from the plant stems by retting. The fiber bundles of flax are held just within the outer layers of cells and outside the central pithy and woody layers by an intercellular cement of pectin

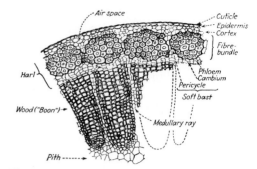

Figure 26–6. *Cross section of part of a flax stem showing fiber bundles just inside the outer few layers of cells (diagrammatic). (From Prescott and Dunn: Industrial Microbiology. New York, McGraw-Hill Book Co., 1959.)*

(Fig. 26–6). Numerous bacteria and molds can digest pectin and permit the fiber bundles to be separated mechanically from the stems and from each other.

Anaerobic retting is accomplished by immersing the plant stalks in natural or artificial ponds or in tanks of water, where a variety of bacteria including *Clostridium felsineum* digest the pectin.

An aerobic process known as *dew retting* relies mainly on molds. The plant material is spread in thin layers on the ground and exposed to the elements. Microorganisms from the air, the soil, and the plant itself slowly hydrolyze the pectin, subject to continuous temperature and moisture changes. The fibers obtained by this method are frequently of poor quality and the yield is small.

Tank methods, either aerobic or anaerobic, are more predictable but require close control because overretting may damage the fibers. One tank process requires only 50 hours, whereas dew retting requires several weeks.

DETERIORATION OF TEXTILES

Textile fibers in common use may be classified as follows:

A. Natural fibers
  1. Plant (principally carbohydrates)
     Examples: cotton, flax, hemp, jute
  2. Animal (principally protein)
     Examples: wool, silk

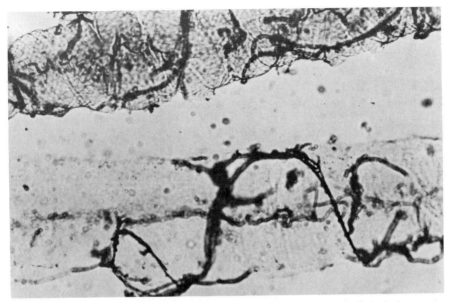

Figure 26–7. *Mildewed fiber of cotton. (Photograph by B. Prindle.)*

B. Artificial fibers
   1. Semisynthetic
      Examples: viscose rayon, cellulose acetate
   2. Synthetic
      Examples: Nylon, Orlon, Dacron

Microorganisms weaken and discolor textiles and alter their affinity for dyes. Mildew is the growth of fungi, often noted as a discoloration but actually accompanied by slow loss of strength (Fig. 26–7).

Cotton is composed of cellulose. The microbial flora of raw cotton is that of the soil and may amount to 10,000,000 bacteria and 100,000 molds per gram. Nearly 200 mold species have been found on cotton fabrics. Bacteria are not significant causes of deterioration, because only a few aerobic cellulose-digesting species are known, and they multiply slowly. Many molds, however, can digest cellulose. They need only humid atmosphere and nutrients usually present in a cotton fabric.

Wool is protein, principally keratin. Microorganisms found on wool include molds, actinomycetes, and true bacteria. Several mold species weaken and discolor wool fabrics; some aerobic spore-forming bacilli cause deterioration, others produce discoloration.

The treatment that a fabric receives during manufacture often affects its susceptibility to deterioration. Starch sizing provides added nutrients for some microorganisms. Bleaching and dyeing, on the other hand, may kill many organisms.

Semisynthetic fibers are partially resistant to deterioration; the resistance of cellulose acetate, for example, is roughly proportional to the degree of acetylation. Synthetic fibers are almost completely unaffected by microorganisms. The natural fibers in mixed fabrics such as cotton-Dacron are, of course, subject to deterioration.

PRESERVATION OF TEXTILES

The only sure method of preventing microbial deterioration of textiles is to maintain the moisture content at less than 8 per cent and the relative humidity less than 75 per cent.

Some materials, such as tents, tarpaulins, fish nets, cordage, and ropes, cannot be kept dry and must therefore be protected by antiseptic chemicals. For centuries sailors have treated ropes with tar to reduce deterioration. Copper compounds are widely used, and a phenolic compound, 2,2'-methylenebis (4-chlorophenol), has been increasingly employed since World War II, when it was used to protect textiles and leather goods for the armed services.

SUPPLEMENTARY READING

Casida, L. E., Jr.: *Industrial Microbiology*. New York, John Wiley & Sons, Inc., 1968.

Miller, B. M., and Litsky, W. (eds.): *Industrial Microbiology*. New York, McGraw-Hill Book Co., Inc., 1976.

Prescott, S. C., and Dunn, C. G.: *Industrial Microbiology,* 3rd ed. New York, McGraw-Hill Book Co., Inc., 1959.

Rainbow, D., and Rose, A. H. (eds.): *Biochemistry of Industrial Microorganisms*. New York, Academic Press, 1963.

Rhodes, A., and Fletcher, D. L.: *Principles of Industrial Microbiology*. Oxford, Pergamon Press, 1966.

Umbreit, W. W. (ed.): *Advances in Applied Microbiology.* New York, Academic Press, Inc., 1959–1964.

GLOSSARY

Abbé condenser: A lens system beneath the stage of the microscope; it focuses and concentrates light on the object and produces a cone of light of sufficient angle to fill the aperture of the objective lens.

Acidfast: Capable of retaining a stain despite washing with dilute acid; e.g., *Mycobacterium* species.

Active acquired immunity: Immunity acquired by having a subclinical or frank case of disease or by injection of the infectious agent, either dead or attenuated, or of a product or component of the infectious agent.

Active transport: Passage of a substance across the cell membrane against a concentration gradient; energy must be supplied.

Adenosine triphosphate (ATP): An organic compound containing the purine, adenine, the five-carbon sugar, ribose, and three phosphate groups; a major carrier of phosphate and energy in biologic systems.

Aerobic (Gr. *aer* air + *bios* life): Growing or metabolizing in the presence of free oxygen.

Agar (agar-agar): A polysaccharide (galactan) extracted from *Gelidium* and other seaweeds, used as a solidifying agent in culture media.

Agglutination (L. *agglutinare* to glue to): Aggregation of particles into clumps; specifically, the aggregation of a cellular antigen by antibody.

Algae (L. *algor* cold): A group of chlorophyll-containing plants that are either unicellular or consist of a plant body, called a thallus, which is not differentiated into roots, stems, or leaves.

Allergy (Gr. *allos* other + *ergon* work): Exaggerated or unusual sensitivity to a substance that is harmless to most individuals of the same species; e.g., hay fever.

Allograft (Gr. *allos* other): Homograft.

Allosteric enzyme (Gr. *allos* other + *stereos* solid, three-dimensional): An enzyme whose activity is altered when its structure is distorted by an organic compound or metal ion at a nonsubstrate site.

Amensalism (L. *a* away + *mensa* table): A relationship between species in which one is suppressed by toxic products of the other; the latter is not benefited directly, but may benefit indirectly by elimination of a competing population.

Anaerobic: Having the ability to live or metabolize without free oxygen. **Obligate anaerobe,** one which can live only in the absence of free oxygen. **Facultative anaerobe,** one which usually lives in its presence but can live without it.

Anaphylaxis (Gr. *ana* back + *phylaxis* protection): The untoward and possibly fatal response of a hypersensitive animal to contact with the antigen to which it is sensitive; attributed to reaction of the antigen with antibody present in small amounts in association with certain body cells.

Antibiotic (Gr. *anti* against + *bios* life): A product of cellular (usually microbial) activity that inhibits or kills certain specific microorganisms.

Antibody: A protein (immunoglobulin), usually found in serum, whose presence can be demonstrated by its specific reactivity with an antigen or hapten.

Antigen (Gr. *anti* against + *genan* to produce): A substance (usually a protein or protein-polysaccharide complex) that incites a specific immunologic response in an animal, such as the formation or liberation of antibodies.

Antimetabolite: An inactive substance that is similar to an essential metabolite and that tends to replace the metabolite.

Antiseptic (Gr. *anti* against + *sepsis* putrefaction): An agent used to prevent sepsis or putrefaction; an agent that inhibits multiplication of microorganisms; a bacteriostatic or fungistatic chemical.

Antiserum: Serum containing antibodies; serum from an animal that has been immunized against a specific antigen.

Antitoxin (Gr. *anti* against + *toxikon* poison): Antibody that neutralizes (i.e., destroys the toxicity of) toxin, particularly exotoxin.

Apoenzyme (Gr. *apo* from + *en* in + *zyme* yeast): The protein portion of an enzyme to which a coenzyme can attach.

Archetista (Gr. *arche* first): Proposed kingdom for the viruses.

Ascospore (Gr. *askos* bag + *sporos* seed): Sexual spore of ascomycetes; spore contained within a sac (ascus).

Asepsis: Prevention of the presence of microorganisms (e.g., by use of sanitary technique).

Assimilation (L. *ad* to + *similare* to make like): The process of manufacturing protoplasmic constituents.

Autograft (Gr. *autos* self): Tissue transplant from one site to another of the same individual.

Autolysis (Gr. *autos* self + *lysis* solution): Digestion of a cell or cell material by some of the cellular enzymes.

Autotroph (Gr. *autos* self + *trophe* nourishment): 1. An organism that requires only carbon dioxide and other inorganic nutrients and derives its energy from light or from oxidation of inorganic compounds. 2. An organism that performs the carboxydismutase reaction.

Auxotroph (Gr. *auxe* increase + *trophe* nourishment): A mutant that requires one or more growth factors that the wild-type organism can synthesize.

Bacteremia (Gr. *bakterion* little staff + *haima* blood): The presence of bacteria in the blood (the blood is not damaged).

Bactericide (Gr. *bakterion* little staff + L. *caedere* to kill): An agent that kills bacteria (usually within a specified time, such as 1 hour).

Bacterin: A suspension of killed or attenuated bacteria used for artificial immunization of man or animals.

Bacteriophage (Gr. *bakterion* little staff + *phagein* to eat): Bacterial virus; a virus parasitic upon bacteria.

Bacteriostasis (Gr. *bakterion* little staff + *stasis* stoppage): Inhibition of bacterial multiplication without immediate killing.

Bacterium (Gr. *bakterion* little staff): 1. A unicellular procaryote. 2. A member of the Schizomycetes or fission fungi.

Bacteroid: Bizarre, irregularly shaped bacterial form, often found under special conditions (e.g., *Acetobacter* species in vinegar or *Rhizobium* species in legume root nodules).

Benthic (Gr. *benthos* bottom of the sea): On or attached to the bottom beneath a body of water.

Binomial system (L. *bi* two + *nomen* name): The system of naming organisms using two names: genus and species.

Bioassay: Estimation of the amount or activity of a biologically active substance by measuring its effect on a living organism.

Biosphere: Regions of the Earth that support or contain life: the upper layers of soil, bodies of water, and the lower atmosphere.

Blepharoplast (Gr. *blepharis* eyelash + *plastos* formed): Cytoplasmic granule from which a bacterial flagellum arises.

Brownian movement (Robert Brown, British botanist, 1773–1858): Motion of small particles in suspension as a result of being bumped by water molecules.

Budding: An asexual reproductive process in which a small bit of protoplasm containing nuclear material pinches off from a cell.

Buffer: A chemical that prevents marked change in the pH of a solution when moderate amounts of acid or alkali are added (e.g., phosphates, proteins).

Calorie (L. *calor* heat): The amount of heat necessary to warm 1 gm. of water 1 degree C., between 15° and 16° C. (1000 cal. = 1 kcal.).

Capsid (L. *capsula* little box): The protein coat surrounding the nucleic acid core of a virus particle.

Capsomere: A protein subunit of a virus capsid.

Capsule (L. *capsula* little box): Thick, viscous, jellylike structure surrounding cells of certain bacterial species; polysaccharide or (less frequently) polypeptide.

Carrier: An individual who harbors and may disseminate an infectious agent but who does not display signs of disease.

Catalyst (Gr. *katalysis* dissolution): A substance that changes the velocity of a reaction without being part of the final product.

Chemosynthesis: Use of chemical bonds as a source of energy for synthetic reactions (i.e., assimilation).

Chemotaxis (Gr. *chemo* chemistry + *taxis* arrangement): Movement toward (positive) or away from (negative) a chemical substance.

Chemotherapy (Gr. *chemo* chemistry + *therapeia* treatment): Treatment of infectious disease by chemicals that inhibit or kill the etiologic agent but do not harm the host in the concentrations employed.

Chlamydospore (Gr. *chlamys* cloak): A thick-walled, resistant, asexual spore produced by certain fungi.

Chloroplast (Gr. *chloros* green + *plastos* formed): A chlorophyll-bearing intracellular organelle.

Chromatin (Gr. *chroma* color): The readily stainable substance (DNA and protein) in the nuclei of cells; in contrast to nonstainable constituents such as the nuclear membrane (of eucaryotes) and nuclear sap.

Chromatophore (Gr. *chroma* color + *phoros* bearing): A pigment cell or color-producing plastid; a bacterial chlorophyll-containing granule.

Cilia (L. *cilium* eyelash): Small, hairlike cytoplasmic appendages on the outer surfaces of cells; their coordinated beating moves the cell or its environment.

Cistron: The sequence of codons that specifies the order of amino acids in a polypeptide chain.

Classification: Systematic grouping of organisms according to a predetermined scheme. **Phylogenetic (natural) c.,** grouping based on origin and evolutionary development. **Phenetic (artificial) c.,** grouping based on easily recognized characteristics.

Clone: A population derived from a single cell or stock by vegetative propagation.

Clostridium (Gr. *kloster* spindle): A spindle-shaped cell with a greatly enlarged central endospore.

Coccus (Gr. *kokkos* berry): A spherical bacterium.

Codon: Nucleotide triplet containing the information necessary to specify a particular amino acid to be used in assembly of a polypeptide chain.

Coenocytic (Gr. *koinos* common + *kytos* vessel, cell): Nonseptate; mold or algal filaments consisting essentially of tubes containing protoplasm with numerous nuclei scattered throughout.

Coenzyme: A low molecular weight organic compound that binds to an enzyme and permits its reaction with the substrate (e.g., NAD).

Cofactor: A metallic ion necessary for enzyme activity (e.g., magnesium, zinc, copper, iron).

Coliform bacteria: Bacteria that resemble *Escherichia coli* in morphology and certain other characters (e.g., lactose fermentation).

Commensalism (L. *cum* with + *mensa* table): An association between species in which one organism benefits but the other is neither benefited nor harmed.

Communicable: Capable of being transmitted from one individual to another.

Competition: An association between two species, both of which require some limiting environmental factor, such as a nutrient.

Complement: A thermolabile, nonspecific complex of at least nine components in the sera of most normal warm-blooded animals; it participates in many immunologic reactions, such as cytolysis and phagocytosis.

Conidiophore (conidium + Gr. *phoros* bearing): A mold hypha bearing conidia on its tip.

Conidium (Gr. *konidion* particle of dust): Exposed or unprotected asexual spore, borne on the tip of the fertile hypha of certain fungi.

Conjugation (L. *cum* with + *jugum* yoke): The act of joining together; the temporary union of one organism with another while nuclear material is exchanged or transferred from one to another; occurs in many ciliate protozoa and in some bacteria.

Constitutive enzyme: Enzyme produced by a cell whether or not its substrate is present.

Contagious (L. *contagio* contact): Readily transmitted by direct or indirect contact or through the air.

Culture: Any growth or cultivation of microorganisms. **Pure c.,** a culture containing only a single kind of organism. **Culture medium,** see *Medium*.

Cytochrome (Gr. *kytos* cell + *chroma* color): An iron-containing protein that transfers electrons along an oxidative pathway.

Cytoplasm (Gr. *kytos* cell + *plasma* plasm): The protoplasm of a cell exclusive of the nucleus.

Cytopyge (Gr. *kytos* cell + *pyge* buttock): Anal opening of protozoa.

Cytostome (Gr. *kytos* cell + *stoma* mouth): Oral opening of protozoa.

Darkfield microscope: A microscope with a special condenser lens that passes only a thin cone of light, some of which reflects from very small or slender objects, such as spirochetes, and can be seen against a black field.

Decay: Microbial decomposition of organic matter under conditions sufficiently aerobic to prevent formation of products with unpleasant odors.

Definition: Sharpness of the image produced by a microscope or other optical system.

Dehydrogenase: An enzyme that catalyzes the transfer of electrons and hydrogen from a substrate, usually to a hydrogen carrier.

Dehydrogenation: A form of oxidation in which hydrogen atoms, including electrons, are removed from the substrate or hydrogen and electron donor.

Deoxyribonucleic acid (DNA): A macromolecule consisting of a double helix of paired nucleotides, the sequence of which constitutes the genetic code.

Dermatophyte (Gr. *derma* skin + *phytos* plant): A fungus parasitic upon the skin, hair, or nails.

Diploid (Gr. *diploos* double): Having two sets of chromosomes.

Disinfection (L. *dis* not + *inficere* to corrupt): Destruction or removal of agents capable of causing infectious disease.

Dissimilation (L. *dis* not + *similis* alike): Intracellular breakdown of food materials.

DNA: See *Deoxyribonucleic acid*.

Dysentery (Gr. *dys* difficult + *enteron* intestine): Acute infection with inflammation of the lower ileum and colon, accompanied by abdominal pain, tenesmus (straining), and frequent stools containing blood and mucus.

Ecology (Gr. *oikos* house + *logos* word, treatise): Interrelationships between organisms and their environment.

Ecosystem: The complex of living and nonliving components in a specified location that comprise a stable system in which materials follow a circular path between the living and nonliving parts.

Electrophoresis (electricity + Gr. *phoresis* being borne or carried): Migration of a molecule (e.g., a protein) or a particle (e.g., a cell) in an electric field.

Endemic (Gr. *en* in + *demos* people, district): The normal, relatively constant incidence of a disease in a population or area.

Endergonic (Gr. *endon* within + *ergon* work): A reaction in which energy is absorbed; requires energy.

Endoplasmic reticulum (Gr. *endon* within + *plasma* anything formed; L. *reticulum* little net): A network of membranes within the cytoplasm of eucaryotic cells; the site of much enzyme activity, including protein synthesis.

Endospore (Gr. *endon* within + *sporos* seed): Highly resistant body formed within the cells of certain bacteria, especially species of *Bacillus* and *Clostridium*.

Endotoxin (Gr. *endon* within + *toxikon* poison): Polysaccharide-protein-lipid complex that comprises part of the cell wall of gram-negative bacteria; released upon autolysis of the dead cells.

Enteritis (Gr. *enteron* intestine + *itis* inflammation): Inflammation of the intestine.

Enzyme (Gr. *en* in + *zyme* yeast): An organic catalyst that alters the rate of a chemical reaction but remains unchanged at the end of the reaction.

Epidemic (Gr. *epi* on + *demos* people, district): Sudden, widespread occurrence of large numbers of cases of an infectious disease in a population or area.

Episome (Gr. *epi* on + *soma* body): A genetic structure containing DNA, found in the cytoplasm of a cell.

Eucaryotic (Gr. *eu* good, true + *karyon* nucleus): Highly evolved cell type; nucleus consists of DNA molecules that comprise the chromosomes and is bounded by a membrane; cell walls, if present, are composed of simple polysaccharides or inorganic substances; membranes contain sterols; many distinct organelles.

Exergonic (Gr. *ex* out, away from + *ergon* work): A reaction in which energy is released.

Exotoxin (Gr. *exo* outside + *toxikon* poison): Soluble protein poison secreted by the living cells of a few bacteria, plants, and animals.

Feedback inhibition: Inhibition of the first enzyme in a pathway by the product of that pathway.

Fermentation (L. *fermentum* leaven): Anaerobic decomposition of an organic substance by an enzyme system in which the final hydrogen acceptor is an organic compound.

Fission (L. *fissio* cleave): Process of asexual reproduction in which an organism divides into two approximately equal parts.

Flagellar antigen: Antigenic substance (protein) of bacterial flagella.

Flagellum (L. *flagellum* whip): A long, whip-like process containing a contractile protein; the organ of locomotion of many microorganisms.

Fluorescence microscope: A microscope in which the object stained with a fluorescent dye, is illuminated with ultraviolet light and emits visible light.

Focal infection: A local infection from which microorganisms continuously or intermittently enter the bloodstream.

Focal length: The distance at which parallel rays entering a lens are brought to a focus.

Fomite: An object contaminated by a diseased individual.

Fungi: Proposed kingdom of multinucleate and sometimes multicellular eucaryotes with rigid cell walls and an absorptive mode of nutrition (e.g., slime molds, molds, yeasts, mushrooms, toadstools).

GC ratio: The percentage of guanine + cytosine in DNA.

Genome (Gr. *gennan* to produce + *oma* mass): The self-replicating portion of a cell; a complete set of hereditary factors.

Genotype: The fundamental hereditary make-up of an individual.

Genus: A group of related species.

Germ theory: Theory that fermentation, putrefaction, and some diseases are caused by microorganisms.

Germicide (L. *germen* germ + *caedere* to kill): Disinfectant.

Glycolysis (Gr. *glykys* sweet + *lysis* solution): Dissimilation of glucose to smaller compounds by the Meyerhof-Embden pathway.

Gnotobiotic animals (Gr. *gnotos* known + *bios* life): Germ-free animals that have been deliberately inoculated with one or more known organisms.

Growth factor: A substance that must be supplied in small amounts for growth of an organism (e.g., vitamins or amino acids that the organism cannot synthesize).

Halophile (Gr. *hals* salt + *philein* to love): An organism that cannot grow in a salt-free medium.

Haploid (Gr. *haploos* simple): Having a single set of chromosomes.

Hapten (Gr. *haptein* to touch, seize): A partial or incomplete antigen; a substance that can react specifically with an antibody, but cannot by itself incite its appearance.

Heterograft (Gr. *heteros* other): Tissue transplant from one species to another.

Heterotroph (Gr. *heteros* other + *trophe* nourishment): An organism that requires complex organic compounds for its carbon source.

Homograft (Gr. *homos* same): Tissue transplant from one individual to a genetically nonidentical individual of the same species; allograft.

Humus: A dark mass of decayed plant and animal matter that gives soil a loose texture and brown or black color.

Hydrogen bond: A weak bond between two parts of a molecule in which a hydrogen atom is shared between two other atoms, one of which is often oxygen (e.g., $\equiv$N—H$\cdots$O$=$C$\equiv$).

Hypersensitivity (Gr. *hyper* excessive): Exaggerated reactivity of an animal that has become unusually sensitized to an antigen.

Hypertonic (Gr. *hyper* excessive + *tonos* tension): Having a higher concentration of solute and hence a higher osmotic pressure than another solution.

Hypha (Gr. *hyphe* web): A segment of mycelium; one of the filaments composing a fungal mycelium.

Hypotonic (Gr. *hypo* under + *tonos* tension): Having a lower concentration of solute and hence lower osmotic pressure than another solution.

Immunity (L. *immunis* safe): Defense against infection that depends upon specifically reactive cells (immunocytes) or blood proteins (antibodies, immunoglobulins).

Immunoglobulin: A protein of animal origin with known antibody activity, or a chemically related protein.

Index of refraction: The number obtained by dividing the angle of a light ray in air by the angle of the ray after passing into another medium (e.g., light passing from air into a glass microscope slide).

Inducible enzyme: Enzyme produced by a cell only when its specific substrate or a chemically related compound is present.

Infectious disease: Disease caused by pathogenic organisms such as bacteria, viruses, protozoa, or fungi; it may or may not be contagious.

Inflammation (L. *inflammare* to set on fire): The response of tissues to injury, characterized by pain, heat, redness, swelling, and loss of function.

Interferon: A nonspecific antiviral agent produced in the course of virus infection or in response to injection of foreign nucleic acid.

Invasiveness (L. *in* into + *vadere* to go): The ability of a microorganism to establish itself in a host.

Involution form: Cell of bizarre shape formed under abnormal cultural conditions.

Isograft (Gr. *isos* same): Tissue transplant from one individual to another genetically identical individual.

Isotonic (Gr. *isos* same + *tonos* tension): Having the same osmotic pressure as another solution.

Lamella: A thin leaf or plate.

Leukocyte (Gr. *leukos* white + *kytos* cell): One of the colorless, nucleated, more or less ameboid cells of the blood.

Leukocytosis (leukocyte + Gr. *osis* increase): Abnormally large number of leukocytes in the blood.

Leukopenia (leukocyte + Gr. *penes* poor): Abnormally small number of leukocytes in the blood, fewer than 5000 per cu. mm.

Lophotrichous (Gr. *lophos* tuft + *thrix* hair). Type of bacterial flagellation; more than one flagellum located at one or both ends of the cell; flagella have one or two curves.

Lyophilization (Gr. *lyein* to dissolve + *philein* to love): The process of quick-freezing a substance (serum, bacterial suspension, etc.) and then rapidly dehydrating the frozen material in a high vacuum.

Lysogeny (Gr. *lysis* dissolution + *gennan* to produce): A genetic property that confers on a cell the ability to produce virus at some future time.

Medium (culture medium): A solution or other substrate used for the cultivation of microorganisms.

Meiosis (Gr. *meiosis* diminution): Nuclear division that results in daughter cells with the haploid number of chromosomes (i.e., one-half the number in the original cell).

Meningitis (Gr. *meninx* membrane): Inflammation of the meninges (membranes that cover the brain and spinal cord).

Mesophile (Gr. *mesos* middle + *philein* to love): An organism that grows best at moderate temperatures (20° to 45° C.).

Mesosome: An organelle in procaryotes bounded by a membrane that is continuous with the plasma membrane; a probable site of respiratory activity.

Metabolism (Gr. *metabole* change): The chemical activities by which an organism synthesizes its constituents and converts energy from outside sources into energy-rich chemical bonds.

Metabolite: A substance produced by metabolism. **Essential metabolite,** a necessary constituent of a metabolic process.

Metachromatic granules (Gr. *meta* change + *chroma* color): Granules in the cells of certain bacteria that stain intensely with basic dyes.

Microaerophile (Gr. *mikros* small + *aero* air + *philein* to love): An organism that is apparently inhibited by oxygen at normal atmospheric pressure but not at reduced pressure.

Micrometer (Gr. *mikros* small + *metron* measure): Metric unit of length used in describing microorganisms; one-millionth of a meter; μm.

Micron: Micrometer, 0.001 mm. = 10^{-6} m.

Microorganism: An organism so small that an independent unit cannot be seen by the unaided eye; a microscopic organism; microorganisms include bacteria, viruses, protozoa, and unicellular fungi and algae.

Microscope (Gr. *mikros* small + *skopein* to view): An instrument for producing a magnified image of a small object. **Simple m.,** a microscope consisting of a single lens or lens system (e.g., a reading glass). **Compound m.,** a microscope containing two or more lenses or lens systems.

Mitochondrion (Gr. *mitos* thread + *chondros* cartilage): Granular or rod-shaped intracellular organelles that contain the electron transport system and certain other enzymes; site of oxidative phosphorylation.

Mitosis (Gr. *mitos* thread): Cell or nuclear division that yields two daughter nuclei with exactly the same chromosome make-up as the parent nucleus.

Monera (Gr. *moneres* singular, unicellular): Proposed kingdom that includes procaryotes (unicellular and sometimes colonial organisms); predominantly absorptive, but some are photosynthetic; blue-green algae, bacteria.

Monotrichous (Gr. *monos* single + *thrix* hair): Type of bacterial flagellation; a single flagellum situated at one or both ends of the cell; flagella have more than two curves.

Mucopeptide: A sugar-amino acid complex.

Multitrichous (L. *multus* many + Gr. *thrix* hair): Type of bacterial flagellation; more than one flagellum at or near one or both ends of the cell; flagella have more than two curves.

Mutagenic agent (L. *mutare* to change + Gr. *gennan* to produce): An agent that increases the rate of spontaneous mutation (e.g., radiations, nitrogen mustard).

Mutation (L. *mutare* to change): Sudden, permanent, random genetic change.

Mutualism: An association in which two different species are obligately dependent upon each other for life in an ecosystem or for performance of some reaction.

Mycelium (Gr. *myke* fungus + *helos* nail): The vegetative body of a fungus consisting of a mass of filaments (hyphae).

Natural resistance: Nonspecific defense against infection, associated with physical or physiologic characteristics of an individual.

Neutralism: Absence of interaction between two populations residing together; neither benefits nor harms the other.

Nonsusceptibility: Complete protection against a particular disease attributed to genetic factors characteristic of the species.

Nucleotide: One of the subunits into which nucleic acids are split by the action of nucleases; composed of a nitrogenous base (a purine or a pyrimidine), a pentose (ribose or deoxyribose), and phosphate.

Numerical aperture (N.A.): Mathematical expression of resolving power of a lens; N.A. = $\eta \sin \theta$, where η (eta) is the refractive index of the medium between the object and the lens and θ (theta) is the angle between the optical axis of the lens and the most divergent rays passing through the lens.

Oidium: A hyphal fragment of certain molds, capable of reproducing the plant.

Oligodynamic action (Gr. *oligos* little + *dynamis* power): Inhibition or killing of microorganisms by very small amounts of a chemical (e.g., heavy metals such as silver and mercury).

Operon: A group of cistrons controlling the enzymes in a single biosynthetic pathway.

Opportunistic microorganism: A normally harmless organism of the endogenous flora that produces disease as a result of fortuitous events affecting the host.

Opsonin (Gr. *opsonein* to prepare food for): Antibody that increases the rate or extent of phagocytosis of bacteria and other cells.

Organelle: Cell substructure that performs a specific function.

Osmophile (Gr. *osmos* impulse + *philein* to love): An organism that is adapted to media of high osmotic pressure.

Osmosis (Gr. *osmos* impulse): Diffusion of solvent molecules through a semipermeable membrane from a dilute solution into a more concentrated solution.

Osmotic pressure: The pressure with which a solvent tends to pass through a semipermeable membrane; it varies directly with the difference in solute concentrations on opposite sides of the membrane.

Oxidative phosphorylation: The conversion of inorganic phosphate to the energy-rich phosphate of ATP by reactions associated with the electron transfer system.

Pandemic (Gr. *pan* all + *demos* people): A widespread epidemic.

Parasite (Gr. *para* beside + *sitos* food): An organism that lives in or on and at the expense of another, the host.

Parasitism: Direct attack by a smaller organism upon a larger; the parasite benefits and the host is harmed.

Passive immunity: Immunity acquired either naturally from the mother (transplacentally or in the colostrum) or artificially by injection of blood, serum, or an antibody-containing serum fraction.

Pasteurization: Heat treatment that kills undesired microorganisms (e.g., pathogens, spoilage organisms) but not necessarily all microorganisms.

Pathogenesis (Gr. *pathos* disease + *genesis* origin): The development of disease.

Pathogenic: Disease-producing.

Pathogenicity: The ability of a group of organisms (species, genus, etc.) to produce disease.

Pellicle (L. *pellicula* thin skin): A thin membrane of microbial growth on the surface of a liquid culture.

Peptonization: Enzymatic digestion (hydrolysis) of protein and formation of peptones.

Peritrichous (Gr. *peri* around + *thrix* hair): Type of bacterial flagellation; numerous flagella extend from all sides of the cell.

Permease (L. *per* through + *meare* to pass): An enzyme that transports a nutrient from the medium through the cell membrane; a translocase.

pH: Negative logarithm of hydrogen ion concentration.

Phagocyte (Gr. *phagein* to eat + *kytos* cell): A cell that ingests other cells (e.g., effete blood cells), microorganisms, and foreign particles.

Phagocytosis: Ingestion of particulate agents (e.g., bacterial cells) by various wandering cells of the blood (leukocytes) or by fixed cells lining the capillaries.

Phenol coefficient: A figure comparing the dilution of a disinfectant chemical that kills a test organism under specified conditions with the dilution of phenol that kills the organism under the same conditions.

Phosphorylation: Introduction of phosphate into an organic compound.

Photophosphorylation (Gr. *photos* light + phosphorylation): Conversion of light energy to that of high-energy phosphate.

Photoreactivation (Gr. *photos* light): "Revival" of organisms that have apparently been killed (e.g., by ultraviolet light) after exposure to visible light.

Photosynthesis (Gr. *photos* light + *synthesis* putting together): The synthesis of carbohydrate from carbon dioxide and water using the energy of light secured with the aid of chlorophyll.

Phototaxis (Gr. *photos* light + *taxis* influence): Movement influenced by light.

Phytoplankton (Gr. *phyton* plant + *planktos* wandering): Freely floating algae.

Pilus (L. *pilus* hair): Fimbria; a straight, nonflagellar, protein appendage, found particularly on some gram-negative bacteria.

Plankton (Gr. *planktos* wandering): Minute, freely floating plant or animal organisms.

Plasma (Gr. *plassein* to mold): The fluid portion of blood after removal of formed elements, the cells and platelets.

Plasma membrane: Discrete, differentiated outer layer of the cytoplasm immediately beneath the cell wall; a triple-layered "unit" membrane that regulates the passage of material into and out of the cell.

Plasmid (Gr. *plasma* formed): An extranuclear body containing DNA; episome.

Plasmodesm (Gr. *plassein* to mold + *desmos* band): The protoplasmic link between two adjacent cells that do not quite separate after transverse fission.

Plasmolysis (Gr. *plassein* to mold + *lysis* solution): Shrinkage of the cytoplasmic mass within a cell owing to loss of water by osmotic action in a hypertonic solution.

Plasmoptysis (Gr. *plassein* to mold + *ptyein* to spit): Bursting of a cell in a hypotonic solution owing to inward movement of water.

Plastid (Gr. *plastis* formed): Any specialized organelle of the cell other than the nucleus and centrosome (e.g., a chloroplast).

Pleomorphic form (Gr. *pleon* more + *morphe* form): Irregular or variant morphologic form.

Pleomorphism: Variation in shape of an organism.

Pneumonia (Gr. *pneumon* lung): Inflammation of the lungs.

Polysome (Gr. *polys* many + *soma* body): A chain of ribosomes held together temporarily by mRNA.

Precipitation (serologic) (L. *praecipitare* to cast down): Aggregation of a soluble antigen (e.g., protein) by antibody.

Preservative: A chemical that prevents growth of spoilage microorganisms.

Procaryotic (Gr. *pro* before, primordial + *karyon* nucleus): Comparatively undifferentiated cell type; nucleus consists of a single, circular molecule of DNA, not bounded by a membrane; cell walls are rigid, composed of mucopeptide and often lipid; membranes lack sterols; few distinct organelles.

Properdin: A normal serum euglobulin that has antimicrobial activity in the presence of Mg^{++} ions and complement components.

Prophage (Gr. *pro* before + *phagein* to eat): A bacteriophage or some part of a bacteriophage present in a dormant condition within a bacterial cell; a temperate bacteriophage.

Proteinase: An enzyme that catalyzes hydrolysis of protein.

Protist (Gr. *protista* the very first): Relatively simple organism. **Lower p.,** procaryotic organisms (bacteria and blue-green algae). **Higher p.,** eucaryotic organisms (protozoa, fungi, higher algae).

Protista: Haeckel's third kingdom, including the relatively simple organisms—bacteria, protozoa, fungi, and algae—that are unicellular and coenocytic or, if multicellular, not differentiated into distinct tissue regions.

Protocooperation (Gr. *protos* first + L. *cooperatio* work): A mutually advantageous association between two species; may be essential for their existence or activity in a particular ecosystem, but the species in the partnership may change.

Protoplast (Gr. *protos* first + *plastos* formed): A cell. With reference to bacteria, a cell that lacks a cell wall.

Prototroph (Gr. *protos* first + *trophe* nourishment): Wild-type; the "normal" or usually encountered form of an organism; does not require added growth factors (cf. auxotroph).

Protozoa (Gr. *protos* first + *zoon* animal): Single-celled animals.

Pseudopod (Gr. *pseudos* false + *poys* foot): A temporary protrusion of the cytoplasm of an ameba or an ameboid cell that serves for locomotion and feeding.

Psychrophile (Gr. *psychros* cold + *philein* to love): An organism that can grow at a low temperature; its generation time at 0°C. is less than 48 hours.

Putrefaction: Microbial decomposition of organic compounds, especially proteins, under anaerobic conditions with the production of foul smelling compounds.

Pyogenic (Gr. *pyon* pus + *gennan* to produce): Pus-producing.

Resolution: The ability of a lens to produce separate images of parts of an object that are only a small distance apart; the ability to distinguish fine detail; measured by numerical aperture (N.A.).

Respiration (L. *respirare* to breathe): 1. Cellular utilization of oxygen with production of carbon dioxide and conservation of energy in a biologically useful form such as ATP. 2. Biologic oxidation in which molecular oxygen or oxygen from inorganic compounds serves as terminal hydrogen and electron acceptor.

Rhizoid (Gr. *rhiza* root + *eidos* form): Rootlike absorptive "holdfast" of certain fungi and other plants.

Rhizosphere (Gr. *rhiza* root): The region immediately surrounding plant roots.

Ribonucleic acid (RNA): Nucleic acid containing the pentose ribose; present in both cytoplasm and nucleus; intimately concerned in synthesis of protein.

Ribosome: Cytoplasmic granule containing protein and RNA, either free or attached to the membranes of the endoplasmic reticulum; site of protein synthesis.

Sanitization (L. *sanitas* health): The process of making an object sanitary or safe to use.

Saprophyte (Gr. *sapros* rotten + *phyton* plant): An organism that lives upon dead and decaying organic matter; it absorbs nutrients through the cell membrane following extracellular digestion of the nonliving material.

Schizomycetes (Gr. *schizein* to divide + *mykes* fungus): Formerly, the plant class containing the bacteria, or "fission fungi."

Septicemia (Gr. *septikos* putrefaction + *haima* blood): Bacterial invasion of the bloodstream and destruction of blood components.

Septum: Crosswall dividing a filamentous plant or fungus into cells.

Serology: Study of the nature and behavior of serum antibodies.

Serum (L. *serum* whey): The fluid portion of blood after removal of cells, platelets, and fibrin.

Somatic antigen (Gr. *soma* body): Bacterial cell wall antigen (protein-lipopolysaccharide complex).

Species (L. *species* sort, kind): 1. A group of actually or potentially interbreeding organisms that do not crossbreed with other organisms. 2. A group (of bacteria) possessing the same genetic constitution.

Spheroplast (Gr. *sphaira* sphere + *plastos* formed): A cell possessing only a partial or modified cell wall.

Sporangiophore (sporangium + Gr. *phoros* bearing): A mold hypha with a sporangium at its tip.

Sporangium (Gr. *sporos* seed + *angeion* vessel): 1. A bacterial cell containing an endospore. 2. The sac containing spores of certain fungi (e.g., *Rhizopus*).

Spore (Gr. *sporos* seed): 1. A reproductive body of one of the lower organisms, such as a protozoon or a fungus, that can develop directly into an adult. 2. An inactive resting or resistant form produced at the tip of a hypha (conidiospore) or within a sac (sporangiospore), or within a bacterial cell (endospore).

Starter: A pure or mixed culture of microorganisms used to initiate a desired process, as in cheese manufacture.

Sterile: Free from all living organisms.

Sterilization: Destruction or removal of all living organisms.

Strain: A microbial isolate or culture.

Supersonic vibrations (L. *super* above + *sonus* sound): Sound of 9000 to 200,000 vibrations per second.

Suppuration (L. *sub* under + *puris* pus): The formation of pus.

Symbiosis (Gr. *syn* together + *bios* life): The living together of two dissimilar organisms; the association may be mutualistic, commensal, amensal, or parasitic.

Synergism (Gr. *syn* together + *ergon* work): An association in which two species together effect a change that neither can accomplish alone.

Taxonomy (Gr. *taxis* arrangement + *nomos* law): The description, classification, and naming of plants, animals, and microorganisms.

Temperate bacteriophage: A bacterial virus whose genetic material can associate intimately with and replicate at the same rate as that of the host bacterium.

Thallophyte (Gr. *thallos* green shoot + *phyton* plant): One of a group of plants that do not form embryos during development nor have vascular tissues (i.e., algae, fungi, lichens, bacteria).

Thallus (Gr. *thallos* green shoot): A simple plant body, not differentiated into root, stem, and leaf; the body of a higher, multicellular fungus.

Thermoduric (Gr. *thermos* heat + L. *durus* enduring): An organism that is unusually resistant to heat; the term is ordinarily limited to nonspore-formers.

Thermophile (Gr. *thermos* heat + *philein* to love): An organism that grows at high temperatures (e.g., 55° C.); its optimum growth temperature is greater than 45° C.

Thylakoid: Organelle consisting of leaflike lamellar plates in the cytoplasm of blue-green algae; the site of photosynthesis.

Toxemia (Gr. *toxikon* poison + *haima* blood): A condition resulting from absorption of a toxin formed at a local site of infection (e.g., tetanus) or ingested (e.g., botulism).

Toxigenicity: The ability of a microorganism to produce toxic substances (exotoxins, endotoxins, etc.).

Toxoid: Toxin that has been made harmless by aging or by treatment with formaldehyde, heat, or other agents, but which retains antigenicity.

Transcription (L. *trans* across + *scribere* to write): Formation of messenger RNA, whose nucleotide sequence is dictated by that of DNA.

Transduction (L. *trans* across + *ducere* to lead): Transfer of a genetic fragment from one cell to another by a virus (e.g., transfer from one bacterium to another by a bacteriophage).

Transformation: 1. Genetic change in bacteria resulting from absorption and incorporation of exogenous DNA into the bacterial genome. 2. Alteration in growth characteristics of mammalian cells *in vitro* by oncogenic viruses; transformed cells are altered in shape, cease to display contact inhibition (and hence pile up over one another), divide indefinitely in serial culture, produce increased amounts of organic acids and acid mucopolysaccharides, possess new antigens, and produce tumors when inoculated into host animals.

Translation (L. *trans* across + *latus* carried): Formation of a polypeptide chain containing amino acids in the sequence dictated by messenger RNA.

Trichome (Gr. *trichos* hair): Chain or filament of cells so closely associated with one another that they rarely live separately.

Trophozoite (Gr. *trophe* nourishment + *zoon* animal): Vegetative cell of a protozoon; active, motile, feeding stage.

Tyndallization (John Tyndall, Irish physicist, 1820–1893): Intermittent sterilization accomplished by heating at 100° C. for one-half hour on each of three successive days.

Vaccination (L. *vacca* cow): Artificial immunization against smallpox by inoculation with cowpox virus; loosely, any artificial, active immunization.

Vaccine (L. *vacca* cow):   1. Lymph containing the virus of cowpox (vaccinia) used in artificial immunization against smallpox. 2. A suspension of killed or attenuated bacteria used for artificial immunization of man or animals.

Virion: A complete, infectious virus particle, consisting of a nucleic acid core with its protein coat, or capsid.

Virulence (L. *virus* poison):   The capacity of a given strain or pure culture of a species to produce disease; it is a function of invasiveness and toxigenicity, and is measured with reference to a given host.

Virus (L. *virus* poison):   Submicroscopic infectious agent composed of nucleic acid and protein; obligately parasitic and hence replicates only within a living host cell.

Xenograft (Gr. *xenos* foreign):   Heterograft.

Zooplankton (Gr. *zoon* animal + *planktos*, wandering):   Freely floating small animals.

Zoospore (Gr. *zoon* + *sporos* seed):   An asexual, flagellated, motile spore.

Zygospore (Gr. *zygon* yoke + *sporos* seed):   Spore formed by the union of two apparently identical cells.

INDEX

SOME PREFIXES AND SUFFIXES USED IN MICROBIOLOGY (Continued)

Prefix or Suffix	Meaning	Example
myo-	muscle	*myo*cardium (heart muscle)
myx-	mucus	*myx*omycete (slime mold)
neo-	new	*neo*natal (new born)
-nom-	law	tax*onom*y (arrangement and classification of living organisms)
oligo-	few, small	*oligo*saccharide (a compound containing a few sugar residues—less than a polysaccharide)
-ose	denotes a sugar	lact*ose*
-osis	disease of, abnormal increase	lymphocyt*osis* (an increased number of lymphocytes)
osmo-	impulse	*osmo*sis (passage of solvent through a semipermeable membrane separating solutions of different concentration)
osteo-	bone	*osteo*myelitis (inflammation of the bone marrow)
pan-	all	*pan*demic (widespread epidemic)
para-	beside	*para*site (an organism that feeds in and at the expense of a host; literally, beside food)
patho-	disease	*patho*genic (disease-producing)
-penia	need, lack	leuko*penia* (lack of leukocytes)
per-	through	*per*meable (able to be passed through)
peri-	around	*peri*trichous (having flagella on all sides)
-phag-	eat	*phag*ocyte (a cell that ingests other cells and substances)
-phil-	like, have affinity for	eosino*phil*ic (staining intensely with eosin)
-pher-, -phor-	bear, support	conidio*phore* (conidium-bearing)
-phot-	light	*phot*osynthesis
-phyt-	plant	*phy*otxin (poison derived from a plant)
-phyll	leaf	chloro*phyll* (green leaf pigment)
-pleo-	more	*pleo*morphic (occurring in more than one form)
-pod-	foot	pseudo*pod*
poly-	many, much	*poly*morphonuclear (having a many-shaped nucleus)
post-	after, behind	*post*natal
pro-	before	*pro*dromal (indicating the approach of a disease)
proto-	first, primitive	*proto*zoa
pyo-	pus	*pyo*genic (pus-producing)
re-	back, again	*re*infect
rhiz-	root	*rhiz*oid (rootlike)
sapro-	rotten	*sapro*phytic (living upon decaying organic matter)
schizo-	divide	*Schizo*mycetes (fission fungi)
-scope	look at, observe	micro*scope*
-som-	body	*som*atic antigen (antigen of bacterial cell body, as opposed to flagellar antigen)
-spor-	seed	endo*spore*
-sta-	make stand, stop	bacterio*sta*tic (inhibiting bacterial multiplication)
sym-, syn-	together	*sym*biosis (life together)
-taxis	order, arrangement, influence	chemo*taxis* (movement influenced by a chemical)